BIODETERIORATION RESEARCH 4

Mycotoxins, Wood Decay, Plant Stress, Biocorrosion, and General Biodeterioration

BIODETERIORATION RESEARCH 4

Mycotoxins, Wood Decay, Plant Stress, Biocorrosion, and General Biodeterioration

Edited by

Gerald C. Llewellyn
The George Washington University
Washington, D. C.

William V. Dashek
Clark Atlanta University
Atlanta, Georgia

and

Charles E. O'Rear
The George Washington University
Washington, D. C.

PLENUM PRESS • NEW YORK AND LONDON

Library of Congress Cataloging-in-Publication Data

Pan American Biodeterioration Society. Meeting (4th : 1991)
Biodeterioration research 4 : mycotoxins, wood decay, plant stress, biocorrosion, and general biodeterioration / edited by Gerald C. Llewellyn, William V. Dashek, and Charles E. O'Rear.
p. cm.
"Proceedings of the Fourth Meeting of the Pan American Biodeterioration Society, held August 20-25, 1991, as an electronic symposium"--T.p. verso.
Includes bibliographical references and indexes.
ISBN 0-306-44638-3
1. Materials--Biodeterioration--Congresses. 2. Mycotoxins--Congresses. 3. Biodegration--Congresses. 4. Wood--Biodegradation--Congresses. I. Llewellyn, Gerald C. II. Dashek, William V. III. O'Rear, Charles E. IV. Title. V. Title: Biodeterioration research four.
TA418.74.P36 1991
620.1'1223--dc20 94-5031
CIP

Proceedings of the Fourth Meeting of the Pan American Biodeterioration Society, held August 20–25, 1991, as an electronic symposium

ISBN 0-306-44638-3

A Division of Plenum Publishing Corporation
233 Spring Street, New York, N.Y. 10013-1578

Printed in the United States of America

SPONSORS

THE GEORGE WASHINGTON UNIVERSITY

THE SMITHSONIAN INSTITUTION

CLARK ATLANTA UNIVERSITY, DEPARTMENT OF BIOLOGY

UNITED STATES DEPARTMENT OF AGRICULTURE, RESEARCH SERVICE

THE UNIVERSITY OF GEORGIA

MORRIS BROWN COLLEGE

SPELLMAN COLLEGE

MOREHOUSE COLLEGE

NORTH CAROLINA STATE UNIVERSITY AT RALEIGH

FOOD AND DRUG ADMINISTRATION, TOXICOLOGICAL STUDIES

UNITED STATES DEPARTMENT OF AGRICULTURE, FOREST SERVICE

PENNSYLVANIA STATE UNIVERSITY

UNIVERSITY OF LA PLATA - ARGENTINA

PROGRAM PLANNING AND ORGANIZING CO-CHAIRMEN

WILLIAM V. DASHEK
Department of Biology
Clark Atlanta University
J.P. Brawley Drive at Fair Street, S.W.
Atlanta, GA 30314, USA

GERALD C. LLEWELLYN
Department of Forensic Sciences
The George Washington University
Washington, DC 20052, USA

CHARLES E. O'REAR
Department of Forensic Sciences
The George Washington University
Washington, DC 20052, USA

PROCEEDINGS COORDINATOR

ASHLEY W. HANCOCK, JR.
Department of Forensic Sciences
The George Washington University
Washington, DC 20052, USA

SECTION CHAIRPERSONS

WILLIAM V. DASHEK
Department of Biology
Clark Atlanta University
J.P. Brawley Drive at Fair Street, S.W.
Atlanta, GA 30314, USA

LEONARD FRIEDMAN
Division of Toxicological Studies
Center for Food Safety and Applied Nutrition
U.S. Food and Drug Administration
Beltsville, MD 20708, USA

WINSTON M. HAGLER
Mycotoxin Laboratory
Department of Poultry Science
North Carolina State University
Raleigh, NC 27695-7600, USA

TERRY L. HIGHLEY
United States Department of Agriculture
Forest Service
Forest Products Laboratory
Madison, WI 53705-2398, USA

BETTY R. JONES
Laboratory of Parasitology
Department of Biology
Morris Brown College
Atlanta, GA 30314, USA

GERALD C. LLEWELLYN
Department of Forensic Sciences
The George Washington University
Washington, DC 20052, USA

CHARLES E. O'REAR
Department of Forensic Sciences
The George Washington University
Washington, DC 20052, USA

IN RECOGNITION OF SCHOLARLY CONTRIBUTIONS

TO BIODETERIORATION-RELATED RESEARCH

The Editors acknowledge the contributions to biodeterioration and biodegradation research by the late Rufus K. Guthrie. He was an outstanding scientist, a dedicated teacher, and an amiable colleague who will be missed by all who knew him.

RUFUS K. GUTHRIE

Died November 5, 1992

Department of Human Ecology
School of Public Health
University of Texas
Houston, Texas

PREFACE

This volume, unlike the three preceding it, represents the collected papers from an experiment with an "electronic symposium". Co-participators in this symposium included The George Washington University, The Smithsonian Institution, Clark Atlanta University, the Agriculture Research Service of the United States Department of Agriculture, The University of Georgia, Morris Brown College, Spellman College, Morehouse College, North Carolina State University at Raleigh, The United States Food and Drug Administration, and the Forest Service of the United States Department of Agriculture among others.

This unusual "electronic symposium" concept was developed by members of the Program, Planning and Organizing Committee as an alternative to the more costly convention-type symposium. As before, leading scientists in specific topic areas were invited to participate. Topic Session chairpersons were encouraged to arrange their own method of communication by telephone, electronic mail, or conference call, and report their findings back to the symposium center at The George Washington University. Additional papers were accepted from individuals and laboratories who are actively involved in relevant areas of research and study. Participation was also arranged for internationally established scientists. International authors are represented herein from Nigeria, Italy, Spain, Brazil and Argentina. Our goal was to present a research composite volume that reflected current developments, informed reviews, new and recently developing areas of the present state of knowledge as it relates to these proceeding topics.

All of the reports included in this volume have undergone scientific, technical and editorial peer review. The efforts of these outside reviewers were invaluable, and their contributions are acknowledged and appreciated. It was the topic chairperson in conjunction with the authors, who spearheaded the reviews, alterations and revisions.

This publication is intended to be a general contribution to the broad science of Biotoxins, Biodegradation and Biodeterioration, but in no way encompasses or represents all of the aspects of these topics.

We hope that this experiment with an "electronic symposium" and especially the resulting proceedings, will be of substantial value to a broad range of scientists.

The Editors

ACKNOWLEDGMENTS AND APPRECIATIONS

WILLIAM V. DASHEK

JAMES J. DEPLOEY

LEONARD FRIEDMAN

WINSTON HAGLER, JR.

ASHLEY W. HANCOCK, JR.

TERRY L. HIGHLEY

ROBERT E. HODSON

BETTY R. JONES

NANCY KERR

GERALD C. LLEWELLYN

GLADYS LLEWELLYN

CHARLES E. O'REAR

ALBERT POHLAND

ALAN POSTLETHWAITE

MARION E. SIMPSON

RICHARD SMUCKER

MARY TRUCKSESS

HECTOR A. VIDELA

The cover illustrations were modified from Biodeterioration Research 1, 2 and 3.

CONTENTS

SECTION I

MYCOTOXINS AND MOLDS AS BIODETERIOGENS

SECTION II

PROCESSES OF WOOD DECAY AND DETERIORATION

SECTION III

BIODETERIOGENS AND PLANT STRESS INJURY

SECTION IV

PARASITES AS BIODETERIOGENS AND HEALTH RELATED ISSUES

SECTION V

GENERAL BIOTOXINS, BIODEGRADATION AND BIODETERIORATION

SECTION I

MYCOTOXINS AND MOLDS AS BIODETERIOGENS

PART A

CITRININ: A MYCOTOXIN OF INTEREST

PART B

MYCOTOXICOSES: FUNGAL PATHOLOGY, TOXICOLOGY AND ECOLOGY

MYCOTOXINS AND MOLDS AS BIODETERIOGENS

INTRODUCTION

This section, probably due to the interests of the co-editors, became one of the several dominant features of the proceedings. There are two distinct parts for the section concerned with fungal toxins and their detrimental effects. Part A functions as a review of a specific mycotoxin, citrinin. Part B is much more encompassing having reports ranging from mycotoxin-related deterioration of feed and food products to fungal taxonomy to disposal of mycotoxin-contaminated materials.

Mycotoxins as biodeteriogens have a historical basis. It was mycotoxins, specifically the aflatoxins produced by *Aspergillus flavus*, which were found in animal feeds in the early 1960s that served as the basis for the organization and subsequent publications of proceedings from several biodeterioration/biodegradation meetings. The role as a biodeteriogen played by these toxins needs to be recognized, if not emphasized. Over the years, numerous fungi were added to the list of mycotoxin producers. These fungi and/or their toxins were significant, directly or indirectly, for decreasing the quality of feeds and foods. Mycotoxins were also important considerations in the short and long-term health of the animals or humans who consumed the products. It was during the decades of the 1970s and 1980s that scientific reports became more specialized and represented the findings beyond those of prime interest to the mycologist. New specialists and experts became more involved. Mycotoxins began to illustrated the true need and application of an interdisciplinary approach. Some such interdisciplinary scientists who became leaders in this topical area were the chemist, the veterinarian, the biochemist, the toxicologist, the nutritionist, the agronomist, the epidemiologist, and so on. But, along the way, all of these specialists, whether they knew it or not, continued to contribute and provide information relevant to the original theme of biodeterioration. Often their studies related to the food/feed-substrate on which the mold grew and how the toxin was produced. Or, their efforts noted how the specific mycotoxin affected an exposed organism, be it man, animal, plant, or microbe.

Some thirty years later, the pieces of the puzzle continue to fall into place, the base of mycotoxin knowledge has been expanding, and aspects of the original problem(s) are routinely revisited via new technology and ideas. Albeit, the importance of this topic in general has fluctuated during the last three decades, mycotoxins as an area of scientific interest has not disappeared. Rather, one renaissance of the area of study after another has occurred. It could be said that fungal toxinology/toxicology has expanded even as a tool for scientific study. These topical rebounds of scientific interest have surfaced and resurfaced on a continual and creative basis. At times it has been the notoriety of a new or revisited mold or its toxin(s) or the finding of the toxin in a commodity which has served to stimulate interest. These rebirths have included, for example, the present interest in the *Fusarium*-toxins as well as the *Rhizoctonia*-toxins. Several of the reports herein include recent work related to these specific fungi.

For Part A, it was the need to consolidate the information relevant to citrinin that served as a basis for the development of this section. The organizer wished to gather information and then attempt to evaluate future needs relevant to research concerning this mycotoxin. Oftentimes citrinin occurs in the presence of other and seemingly more important and probably more notorious fungal toxins. Certain aspects relevant to citrinin could have been passed over as unimportant, but an opportunity was provided for evaluation to occur. The reports herein were able to indicate not only the present status of this mycotoxin but specific needs for future studies. The conclusion was that studies regarding citrinin should be directed toward evaluating its potential carcinogenicity as well as the development of better detection methods.

Part B for this section considered mycotoxicoses and specifically fungal pathology, toxicology, and ecology. Several of the papers report toxic responses in animals caused by mycotoxins. Toxins of interest include slaframine, swainsonine, aflatoxin, *Fusarium*-toxins, and deoxynivalenol among others. Mycotoxin-contaminated feeds were studied in beef, dairy heifers, dairy cows, rats, poultry, horses, and dogs. Agricultural substrates reported related primarily to corn, grains, and cottonseeds. Allergenicity of spores was reported in one study while another report presented results of mycotoxic effects upon other fungal spores.

Overall, the topics presented were diverse, yet typical of the wide impact and study interests of the mycotoxicologist or mycotoxinologist. The common thread of biodeteriogenicity is evident as the molds grow and produce toxins on a variety of substrates; subsequently as the toxins affect plant material used as feed or food; which in turn affects the exposed organism. Even the disposal of mycotoxins found in contaminated products presents problems relevant to biodeterioration.

In conclusion, whether it be citrinin, slaframine, aflatoxin, or zearalenone, there continues to be a very diverse and significant interest in the occurrence, control, effects/mode-of-action, chemical characterization, and detection of mycotoxins. The numerous mycotoxic fungi, their mycotoxin metabolites, the fungi themselves, the wide economic importance, and the relevant health aspects pertinent to this topic are well illustrated by the reports herein. It is anticipated that this diversity of topics will stimulate discussion, encourage the sharing of information, and result in cross-fertilization of ideas for solutions to problems at hand. The importance of this ever-evident interdisciplinary subject of interest, mycotoxins, cannot be underestimated.

Gerald C. Llewellyn, Ph.D.
Department of Forensic Science
The George Washington University
Washington, DC 20052

PART A

CITRININ: A MYCOTOXIN OF INTEREST

CITRININ: A MYCOTOXIN OF INTEREST

INTRODUCTION

Citrinin (3, 4-dihydro-8-hydroxy-3, 4, 5-trimethyl-6-oxo-3H-2-benzopyran-7-carboxylic acid), a secondary fungal metabolite produced by several species of Penicillium and Aspergillus, was first isolated in 1931. It has been detected in a number of agriculture products demonstrated to be nephrotoxic in animals and implicated as a possible causative agent in the outbreak of yellow rice toxicosis in Japan (1951). However, the significance of citrinin as a potential threat to human health remains obscure and open to speculation.. Much of this obscurity is due to the lack of precise quantitative methods for analysis of the toxin in food and feed commodities. An additional explanation may be the potential confounding fact that citrinin is frequently found in the presence of other mycotoxins, e.g., ochratoxins. In spite of the large gaps in our knowledge about a mold product that is present to varying degrees in our food supply, it appears that the amount of research on citrinin has diminished over the last decade.

To determine whether this reduced research represents shortsightedness or whether it is justified, it was decided that perspective could be gained by compiling much of the information available regarding citrinin in a series of reviews. A panel of scientists prominent in the field was selected on the basis of their research publications and other writings related to citrinin. The panel was asked to review and evaluate the research carried out in their own and other laboratories, and to indicate the kinds of future work warranted.

Four areas of work on citrinin were covered. The first paper, by Dr. David Wilson, concerns the analysis and occurrence of citrinin. It highlights specific analytical problems regarding citrinin in foods and approaches that have been attempted to address these problems.

The review by Dr. Christine Hanika and William W. Carlton considers evidence that citrinin produces mycotoxicosis in farm animals, emphasizing experimental studies to toxicology and pathology, and demonstrating species and organ specificity of the compound. A discussion of the biochemical effects of citrinin by Robert C. Braunberg illustrates the varied effects of this compound on the metabolic machinery of cells acting both in vitro and in vivo.

Finally, the distribution and metabolism of citrinin and discussed by. Drs. Bonnie B. Dunn and Leonard Friedman. Correlations are developed concerning the mode of handling of the toxin by the animal, particularly the rat, and its reported tissue-specific toxicity.

The format of the symposium did not readily permit discussions among the authors interacting as a panel. However, the exchange of ideas between some of the individuals did take place and proved to be beneficial. The consensus was that work on citrinin should continue toward improving the methodology for its analysis in foods and feeds and investigating its potential carcinogenic or co-carcinogenic properties.

Leonard Friedman, Ph.D.
Beltsville Research Facility
U. S. Food and Drug Administration
Laurel, MD 20709 USA

THE BIOCHEMICAL EFFECTS OF CITRININ: A Review

ROBERT C. BRAUNBERG, Food and Drug Administration, Beltsville Research Laboratories, Beltsville, MD 20708, USA

INTRODUCTION

The implication of citrinin by Krogh et al. (1973) as one of the causative agents of Balkan porcine necropathy stimulated a number of workers to initiate studies of the biochemical effects of this mycotoxin. Much of the work involved in vitro measurements carried out with tissue slices or with cell cultures. In many of the earlier studies, various doses of citrinin were administered to the test animals and measurements were made in vivo or on tissues removed from the animal. In subsequent studies, the exposure to the tissue has been carried out in vitro together with the measurement of the effect.

BIOCHEMICAL EFFECTS OBSERVED AFTER DOSAGE IN VIVO

Respiration

Succinic dehydrogenase activity of tissue homogenates from rats dosed subcutaneously with citrinin diminished in a dose-related manner after treatment (Hashimoto and Morita, 1957). The effective dose for depression of respiration in kidney slices and for the activity of succinic dehydrogenase in kidney homogenates was 10 mg/kg. This effect was observed in renal slices for only 2 or 3 days after the animals were dosed.

In a 6-weeks study of the effect upon the enzymes and substrates of the Embdon-Myerhoff pathway in mice after intraperitoneal administration of citrinin at 15, 25, or 35 mg/kg, Gupta et al. (1986) observed that adenosine triphosphatase, hexokinase, lactic dehydrogenase, lactic acid, and pyruvic acid decreased in the blood, liver, kidney, and brain. Glucose levels in blood, kidney,

Biodeterioration Research 4, Edited by G.C. Llewellyn *et al.*, Plenum Press, New York, 1994

and brain increased while the glycogen level of the liver decreased. Ramadoss and Shanmugasundaram (1973) found that isolated rabbit liver mitochondria from citrinin-treated animals showed a decrease in the rate of substrate oxidation compared to control mitochondria.

Siraj and co-workers (1981) administered citrinin at 25 mg/Kg to rat neonates by gastric intubation at day 1 after birth; the kidneys and liver were removed and homogenized on day 13. Hepatic cytochrome P_{450} content increased 15% in liver homogenates in citrinin-treated rats. In the rats treated simultaneously with citrinin and ochratoxin A at 1 mg/Kg, however, the cytochrome P_{450} level was reduced to 88% of the control value. Citrinin had no effect upon levels of NADPH-dehydrogenase or ADPH cytochrome C reductase. There was a significant interaction with those enzymes, however, when the rats were treated with citrinin and ochratoxin A together. Citrinin had no effect upon pentobarbital hydroxylase and aminopyrine N-demethylase activities, but inhibited aniline hydroxylase by 33%.

Active Ion Transport

Berndt and Hayes (1978), after dosing rats with citrinin, prepared renal cortex slices and measured the transport of various ions. They found that p-aminohippurate (PAH) transport was markedly inhibited by day 4. The effect was less dramatic with tetraethylammonium (TEA) transport, and there was no effect upon α-aminoisobutyrate (AIB) transport (Phillips et al., 1978). In a study by Baggett and Berndt (1984b), citrinin was given intraperitoneally to rats at doses (35-55 mg/kg) which had no effect upon ion transport by the cortical slices but markedly potentiated the depression of both TEA and PAH transport produced by potassium dichromate (10 mg/kg) or mercuric chloride (4 mg/kg).

In further studies (Berndt, 1983; Berndt and Hayes, 1982), ^{14}C-citrinin was used to monitor transport of the mycotoxin. Male Sprague-Dawley rats were administered citrinin intraperitoneally before the preparation of renal slices. It was found that 60 min after administration of the mycotoxin at 50 mg/kg body weight, about 2% of the ^{14}C-labeled material was taken up by the renal cortex. If, on the other hand, the rats were pre-treated with probenecid, a specific, competitive inhibitor of transport, only 0.48% of the administered dose of citrinin was recovered in the renal cortex. This inhibition of renal transport of citrinin was similar to the effect of dinitrophenol, a metabolic inhibitor as well as a competitive inhibitor of anion transport (Berndt, 1983). Citrinin has also been observed to preferentially inhibit Na-K-ATPase in rats (Patil et al., 1980). Siraj et al. (1981) found that citrinin at 25 mg/kg inhibited renal Mg^{+2}-ATPase, but not hepatic ATPase, in rat neonates.

After pre-treatment of rats with citrinin at 60 mg/kg intraperitoneally, membrane function by both isolated basolateral membrane vesicles and isolated brush border vesicles were markedly reduced at 3 hr (Ansari et al., 1991).

Effects on the Synthesis of Macromolecules

In studies carried out by Sansing et al. (1976), the incorporation of ^{14}C orotic acid into liver and kidney RNA was elevated after the administration of citrinin at 22 mg/kg, in contrast to ochratoxin A (6 mg/kg), which inhibited this process. When citrinin (3 mg/kg) and ochratoxin-A (3 mg/kg) were administered together, RNA synthesis was inhibited. Citrinin (10 mg/kg) and penicillic acid (6 mg/kg), when administered together, stimulated RNA synthesis in the kidney while inhibiting it in the liver. The authors identified this as a regenerative process elicited after initial mycotoxin-induced damage.

Clinical Effects Related to Biochemical Function

Clinical biochemical changes evident of renal damage were observed in young beagle dogs fed citrinin at doses of 20 and 40 mg/kg (Carlton et al., 1974). These changes included increases in BUN, urinary lactic dehydrogenase, glutamic oxalacetic transaminase, isocitric dehydrogenase, glucosuria, and proteinuria.

Endo and Kuroda (1976) administered citrinin to rats for 7 days and noted a significant reduction in both plasma and liver cholesterol levels. They compared the hypocholestrolemic activity of citrinin and clofibrate (p-chlorophenoxyisobutyrate). The activity of orally administered citrinin at 5.0 mg/kg (plasma cholesterol level, 61.8 mg/dl) was comparable to the effect of clofibrate at 100 mg/dl (plasma cholesterol level, 62.1 mg/100 ml; control cholesterol value, 82.0 mg/dl).

Kitchen et al. (1977a,b,c) observed renal damage in beagle dogs in which nephrosis had been induced with doses of citrinin up to 10 mg/kg body weight. This damage was reflected in an increase of glutamic oxalacetic transaminase and lactic dehydrogenase in the urine and in elevated serum lactic dehydrogenase. Serum electrolyte levels were also elevated.

Serum enzyme levels of rabbits treated for seven days with citrinin at 2 mg/kg showed no change; kidney alkaline phosphatase, however, decreased significantly. Immunological studies showed that the decrease in enzyme activity was due to the presence of inactive enzyme protein. Cyclic AMP and hormones such as prostaglandin E_1 and parathyroid hormone reversed the decreased enzyme activity due to citrinin poisoning. These findings have led Ramadoss and Shanmugasundaram (1977) to postulate the existence of active and inactive forms of alkaline phosphatase.

Haberman et al. (1987) observed an increase in glucose levels in the urine of rats treated with citrinin at 35 mg/Kg together with chromate at 10 mg/Kg; the increase was greater than the sum of the effects obtained when the animals were exposed to the same levels of the nephrotoxins separately. There was no increase in the blood glucose level.

Significant change has been reported in blood urea content and both glutamic oxalacetic transaminase and alkaline phosphatase activities after the administration of citrinin at a feeding level of 100 ppb in chickens (Abdelhamid and Dorra, 1990).

EFFECTS OF CITRININ UPON BIOCHEMICAL FUNCTION FOLLOWING IN VITRO TREATMENT WITH THE MYCOTOXIN

Respiration

The effect of citrinin upon the endogenous respiration of slices of kidney, liver, and brain was reduced in proportion to the dose (Hashimoto and Morita, 1957). When respiration of brain tissue was uncoupled by the addition of 10^{-5} M 2,4-dinitrophenol and then 1 mM citrinin was added, the increased Q_{O2} disappeared completely 1 hr after the addition. Noncompetitive inhibition of succinic dehydrogenase activity was observed with kidney homogenate by either the manometric or dye-reduction techniques. When kidney slices and kidney homogenates prepared from the tissue of normal (untreated) animals were treated with citrinin at 60 μg/ml in vitro, both respiration and activity of succinic dehydrogenase were inhibited. Similar measurements of the respiration of slices of brain disclosed no effect on rats poisoned with citrinin (10 mg/kg). When tissues of control animals were treated with 1 mM citrinin, however, respiration and succinic dehydrogenase activity were again diminished (Hashimoto and Morita, 1957).

Table 1. Studies of <u>in vivo</u> Exposure to Citrinin.

Species	System Measured	Reference
Beagle dogs	glucosuria, elevated	Carlton et al., 1974
Beagle dogs	proteinuria, elevated	Carlton et al., 1974
Beagle dogs	urinary lactic dehydrogenase levels, elevated	Carlton et al., 1974
Beagle dogs	urinary glutamic oxalacetic transaminase levels, elevated	Carlton et al., 1974
Beagle dogs	urinary isocitric dehydrogenase levels, elevated	Carlton et al., 1974
Beagle dogs	glutamic oxalacetic transaminase levels, elevated	Kitchen et al., 1977b
Beagle dogs	lactic dehydrogenase levels, elevated	Kitchen et al., 1977b
Beagle dogs	serum electrolyte levels, elevated	Kitchen et al., 1977b
Laying hens	BUN, elevated	Abdelhamid & Dorra, 1990
Laying hens	glutamic oxalacetic transaminase levels, elevated	Abdelhamid & Dorra, 1990
Laying hens	alkaline phosphatase levels, elevated	Abdelhamid & Dorra, 1990
Mice	adenosine triphosphatase hexokinase, decreased	Gupta et al., 1986
Mice	lactic dehydrogenase, decreased	Gupta et al., 1986
Mice	lactic acid levels, decreased	Gupta et al., 1986
Mice	pyruvic acid levels, decreased	Gupta et al., 1986
Mice	glucose levels, increased	Gupta et al., 1986
Mice	LD50	Sansing et al., 1976
Mice	macromolecule synthesis	Sansing et al., 1976
Mice	RNA synthesis	Sansing et al., 1976
Rabbits	liver glutamate dehydrogenase, inhibited	Ramadoss & Mukherjee, 1977
Rabbits 1977	kidney alkaline phosphatase, decreased	Ramadoss & Shanmuga-sundaram,
Rats	cholesterol levels,	Endo & Kuroda, 1976
Rats	reduced glucose levels	Haberman et al., 1987
Rats, miniature	Na-K-ATPase, inhibited	Patil et al., 1980
Rats, neonatal	LD50 (25 mg/kg)	Siraj et al., 1981
Rats, neonatal	ATPase, unaffected	Siraj et al., 1981
Rats, neonatal	pentobarbital hydroxylase, unaffected	Siraj et al., 1981
Rats, neonatal	aminopyrene <u>N</u>-demethylase, unaffected	Siraj et al., 1981
Rats, neonatal	Na-K-ATPase, unaffected, synergistic effect with ochratoxin A	Siraj et al., 1981
Rats, neonatal	hepatic cytochrome P450, elevated	Siraj et al., 1981
Rats, neonatal	inhibition	Siraj et al., 1981
Tissue homogenates	succinic dehydrogenase, reduce	Hashimoto & Morita, 1957

Slices of rat renal cortex treated with citrinin showed glutathione (GSH) depletion (Baggett and Berndt, 1984a). A similar effect was seen in liver slices. In rats treated with the glutathione depleting agent, dimethylmalate (DEM), and/or citrinin 72 hr before the preparation of liver and renal cortical slices, renal and hepatic GSH were lower when citrinin was given after DEM than when DEM alone was given. A synergistic effect was observed. When the nephrotoxin, hexachloro-1,3-butadiene, was given in combination with citrinin, the effect upon GSH was additive rather than synergistic.

In studies carried out in vitro, liver glutamate dehydrogenase was inhibited by citrinin (Ramadoss and Mukherjee, 1977). Preparations of bovine liver glutamate dehydrogenase, citrate synthetase, and phosphofructokinase were incubated with 1 mM citrinin at 35°C for 15 min. Only glutamate dehydrogenase was irreversibly inhibited by the mycotoxin.

Tanzawa et al. (1977) showed that 3-hydroxy-3-methylglutaryl coenzyme A reductase is inhibited by citrinin in vitro. Ramadoss and Mukherjee (1977) have demonstrated the uncoupling of oxidative phosphorylation in dog heart mitochondria by the mycotoxin.

Aleo et al. (1991a,b) have reported mitochondrial dysfunction in rat renal proximal tubule suspensions after exposure to citrinin at concentrations ranging from 63 to 500 μM. Tubule viability decreased in a concentration- and time-dependent manner, although evidence on the basis of lipid peroxidation (which only increased after cell death) indicated that iron-mediated oxidative damage was not the cause of citrinin-induced cell death. Basal oxygen consumption showed a transitory rise at all concentrations, and ATP content declined in proportion to the mycotoxin concentration. Nystatin-stimulated respiration, an indirect measure of mitochondrial state 3 respiration, was decreased slightly by citrinin. Although respiration of renal proximal tubules with site I respiration substrates (glutamate/malate) was stimulated by citrinin, succinate-supported respiration was decreased. Similarly, with isolated renal cortical mitochondria, state 4 respiration was increased, but state 3 and uncoupled respiration were reduced by 25 μM citrinin. The authors concluded, therefore, that citrinin has multiple effects on mitochondrial function in renal proximal tubule cells which may have contributed to the lethality of citrinin toward those cells.

Braunberg et al. (1988) reported that the respiration of rat renal cortical explants was inhibited nearly 50% by citrinin in the culture medium at a concentration of 1×10^{-4} M.

Lipid Metabolism and Growth

Using cell-free systems from rat liver and also from yeast, Kuroda et al. (1977) observed a strong inhibition of the incorporation of labeled acetate into nonsaponifiable lipids by citrinin; labeled mevalonate incorporation was not inhibited, however. Two enzymes involved in cholesterol synthesis were specifically inhibited: acetoacetyl-COA-thiolase (ED_{50}, 0.2 mM) and 3-hydroxy-3-methylglutaryl-COA-reductase (ED_{50}, 0.5 mM).

An inhibitory effect upon growth of Tetrahymena pyriformis accompanied by a decrease in cell size was observed in the presence of citrinin at a concentration of 50-100 μg/ml (Hayes et al., 1976). The respiration rate was marginally diminished, but RNA, DNA, and protein synthesis were substantially reduced at 25 μg/ml. The greatest decrease was in RNA. A bioassay for citrinin using T. pyriformis in which the lower limit for detection was 1-5 μg/ml was developed by Hayes and co-workers (1976).

When 3 strains of mouse leukemia cells were treated with citrinin (Yamato et al., 1985) and inhibition of growth was assessed for each strain, the following IC_{50} values (mg/ml medium) were revealed: L1210, 8.4; L5178, 4.3; p388, 3.3. In the in vitro assay, K_i for ornithine decarboxylase inhibition was found to be 0.15 mM.

Yoneyama and Sharma (1987) have reported that the growth of Mandin-Darby bovine kidney cells was inhibited in a concentration-dependent manner when treated with citrinin. The number of surviving cells at the end of 24 hr decreased significantly upon exposure to citrinin at 3×10^{-5} M, and almost complete detachment of cells occurred at 1×10^{-4} M.

Active Ion Transport and Related Enzymes

In observations of the effects of citrinin upon Mandin-Darby bovine kidney cells, Yoneyma and colleagues (1986, 1987) reported that K^+ dependent phosphatase was inhibited to 77% of control values at 1×10^{-4} M and to 45% at 5×10^{-4} M citrinin concentrations. A significant reduction was shown by succinic dehydrogenase at 5×10^{-4} M and acid phosphatase at 1×10^{-4} M. When treated in the same manner, primary isolated fetal bovine cells were generally less sensitive to citrinin than the Mandin-Darby line. Viability as indicated by trypan blue exclusion was significantly reduced at 3×10^{-5} M citrinin for the kidney cell line and at 5×10^{-4} M for the fetal bovine kidney cells. The lower sensitivity of the fetal cells may be a reflection of Tune's finding (1974) that the nephrotoxity of cephaloridine was low in newborn animals because of incomplete development of the renal transport mechanism.

When Berndt and Hayes (1978) challenged renal cortex slices from untreated rats with citrinin _in vitro_, AIB transport was increased significantly. This effect was strain-dependent. It occurred in renal cortical slices from Sprague-Dawley rats but not those from the Fischer 344 strain, which were generally less sensitive to the mycotoxin.

Macromolecules

DNA, RNA, and Protein

When cells of a recombination-deficient mutant of _Bacillus subtilis_ M45(rec-) were tested in the presence of citrinin for DNA-attacking ability, the result was positive (Ueno and Kubota, 1976). The investigators related this finding to the reported carcinogenicity of the mycotoxin. Wurgler et al. (1991) observed a lack of mutagenicity when _Salmonella typhimurium_ TA102 was challenged with citrinin. Neither citrinin nor a combination of citrinin and ochratoxin A produced reverse mutations.

Single strand breaks were observed in chromosomal DNA of _Escherichia coli_ as an effect of citrinin at 100 μg/ml. _In vitro_ studies with permeabilized cells have shown that replication of DNA is inhibited at 200 μg/ml and synthesis of stable RNA is slightly diminished at 300 μg/ml (Martin et al., 1986). These studies suggest the involvement of free radicals in the mechanism of action of citrinin upon DNA.

Lorkowski and Roschenthaler (1979) compared the actions of citrinin and ochratoxin A upon several strains of bacteria. They found that citrinin inhibited RNA and DNA synthesis more strongly than did ochratoxin A. Replicative DNA synthesis was inhibited at 200 μM in permeabilized _E. coli_ cells by citrinin, and the synthesis of stable RNA was slightly diminished at 300 μg/ml; however, protein synthesis was not affected at concentrations up to 450 μg/ml (Martin et al., 1986).

Citrinin present in the culture medium of hepatoma tissue culture cells at 100 μg/ml inhibited RNA synthesis after 10 min exposure, protein synthesis after

Table 2. Studies of in vitro Exposure to Citrinin.

System	Measured	Reference
Brain slices	respiration, decreased	Hashimoto & Morita, 1957
Drosophila	induction of somatic mosaicism, no effect	Belitski et al., 1983
Enzyme 1977 in vitro	bovine liver glutamate dehydrogenase	Ramadoss & Mukherjee,
Enzyme 1977 in vitro	3-hydroxy-3-methylglutaryl coenzyme A reductase	Tanzawa et al.,
Hepatocyte cultures	DNA repair, no effect	Belitski et al., 1983
Hepatocyte cultures	viability, no effect	Belitsky et al., 1985
Hepatocyte cultures	viability	Robbana-Barnat, 1989
Leukemia cell lines, mouse	inhibition of growth	Yamato et al., 1985
Leukemia cell lines, mouse	ornithine decarboxylase	Yamato et al., 1985
Liver homogenate rat	inhibition of acetate incorporation in lipid	Kuroda et al., 1977
Liver homogenate rat	inhibition of mevalonate incorporation in lipids	Kuroda et al., 1977
Liver homogenate rat	acetoacetyl-COA-thiolase	Kuroda et al., 1977
Liver homogenate rat	3-hydroxy-3-methylglutaryl-COA-reductase	Kuroda et al., 1977
Liver slices	respiration, reduced	Hashimoto & Morita, 1957
Renal cell line Mandin-Darby bovine	succinic dehydrogenase	Yoneyama & Sharma, 1987; Yoneyama et al., 1986
Renal cell line Mandin-Darby bovine	viability	Yoneyama & Sharma, 1987; Yoneyama et al., 1986
Renal cells, primary bovine fetal	succinic dehydrogenase	Yoneyama & Sharma,1987; Yoneyama et al., 1986
Renal cells, primary bovine fetal	viability	Yoneyama & Sharma, 1987; Yoneyama et al, 1986
Renal cortical explants,rat	respiration	Braunberg et al., 1988
Renal cortical explants,rat	macromolecule synthesis	Braunberg et al., 1988
Renal cortical explants,rat	active ion transport	Braunberg et al., 1988
Renal proximal tubule suspension	respiration	Aleo et al., 1991b
Renal proximal tubul suspension	ATP content	Aleo et al., 1991b
Renal slices (in vivo dosing)	GSH levels	Baggett & Berndt, 1984a
Renal slices (in vivo dosing)	active ion transport	Baggett & Berndt, 1984a,b
Renal slices	active ion transport	Berndt & Hayes, 1978
Renal slices	respiration, decreased	Hashimoto & Morita, 1957
Renal slices (in vivo dosing)	citrinin transport	Berndt & Hayes, 1978
Renal vesicles, isolated	active ion transport	Ansari et al., 1991
T. pyriformis cultures	viability	Hayes et al., 1976
T. pyriformis cultures	respiration	Hayes et al., 1976
T. pyriformis cultures	macromolecule synthesis	Hayes et al., 1976
Tissue homogenates	succinic dehydrogenase, reduced	Hashimoto & Morita, 1957

20 min, and DNA synthesis after 120 min (Creppy et al., 1980). Compared to the time of exposure required in the same system for the inhibition of macromolecule synthesis by ochratoxin A, citrinin inhibited protein synthesis preferentially, leading these authors to the conclusion that the two mycotoxins have different modes of action. However, when both compounds were added simultaneously to cultures of hepatoma cells, both RNA and protein synthesis ceased immediately, suggesting a cooperative effect for both toxins.

Lorkowski et al. (1980) reported that citrinin inhibited DNA synthesis in rat hepatoma cells. Robbana-Barnat et al. (1989), on the other hand, reported that citrinin was inactive ($IC_{50} > 10$ μg/ml) upon DNA synthesis in LF hepatoma cells. When the effect of citrinin upon AWRF rat fibroblast cells and the CHO line of Chinese hamster ovary cells was compared (Stetina and Votava, 1986), citrinin caused 50% inhibition of colony formation in the ovary cells at 31 μg/ml and in cultures of the fibroblasts at 6.7 μg/ml. Citrinin concentrations above 5 μg/ml caused a 20% inhibition in DNA synthesis. At 200 μg/ml, however, there were a significant number of DNA breaks in the ovary cells, whereas none were observed in cultures of the AWRF cells.

Microsomal fractions prepared from tissues obtained from both rats and humans were found to metabolize citrinin, yielding a potent inducer of chromosomal aberrations in the clastogenicity assay (Thust and Kneist, 1979). Complex translocations were produced by citrinin at 5×10^{-4} M. Yoneyama and Sharma (1987) found that DNA synthesis in cultured kidney epithelial cells was significantly inhibited by citrinin at 3×10^{-5} M. RNA and protein synthesis were significantly inhibited at 3×10^{-5} M and 5×10^{-5} M, respectively. The synthesis of each class of macromolecule was completely inhibited at 1×10^{-4} M. However, the authors speculate that this was the result of inhibition of the transport of precursor compounds into the cell.

Macromolecule biosynthesis by explants of rat renal cortex was diminished by citrinin in a dose-related manner (Braunberg et al., 1988). The lowest concentrations of the mycotoxin at which significant inhibition could be noted were 1×10^{-4} M for RNA (44.7%), 1×10^{-6} M for DNA (20.3%), and 1×10^{-5} M for protein (22.7%).

Belitski et al. (1983) studied the effects of citrinin upon DNA repair in mammalian liver cell cultures and also upon the induction of somatic mosaicism in Drosophila. Citrinin elicited no effect upon either system.

RNA, DNA, and protein synthesis were found to be substantially reduced by citrinin at 25 μg/ml (Hayes et al., 1976). The greatest decrease was in RNA.

Table 3. Studies Involving Measurement of Synthesis of Macromolecules.

System	Measurement	Reference
Bacteria 1979	macromolecule synthesis	Lorkowski & Roschenthaler,
Bacteria, B. subtilis mutant	DNA attacking ability	Ueno & Kubota, 1976
Bacteria, E. coli DNA	strand breaks	Martin et al., 1986
Clastogenicity assay	chromosomal aberrations	Thust & Kneist, 1979
Fibroblast cells, rat	colony formation	Stetina & Votava, 1986
Fibroblast cells, rat	DNA synthesis	Stetina & Votava, 1986
Hepatoma cells	RNA synthesis	Creppy et al., 1980
Hepatoma cells	protein synthesis	Creppy et al., 1980
Hepatoma cells, rat	DNA synthesis	Lorkowski et al., 1980
Ovary cells, Chinese hamster	colony formation	Stetina & Votava, 1986
Ovary cells, Chinese hamster	DNA synthesis	Stetina & Votava, 1986
Renal cell line, Mandin-Darby bovine	growth	Yoneyama & Sharma, 1987
Renal cell line, Mandin-Darby bovine	DNA synthesis	Yoneyama & Sharma, 1987
Renal cortical explants	macromolecule synthesis	Braunberg et al., 1988

SUMMARY

The biochemical effects of citrinin after both in vivo dosing of the animal and in vitro exposure of the test system to the mycotoxin have been reviewed. Toxic effects were measured in cell lines, primary cell isolates, tissue explants and slices, tissue homogenates, and mitochondria. Citrinin inhibits active ion transport. The effect is enhanced in the presence of additional mycotoxins and certain xenobiotic compounds such as the GSH depleter, diethylmaleate. Differences in degree of toxicity between cell lines have been attributed to differences in transport rate for the mycotoxin and to differing mechanisms of detoxification. Both in vitro and in vivo exposure to the mycotoxin leads to depressed respiration activity of kidney slices, and citrinin in vitro depresses the activity of succinic dehydrogenase in kidney homogenates. Perturbation in renal tissue respiration caused by citrinin has been related to alterations in the function of mitochondria. Among the enzyme systems upon which citrinin produces profound effects are kidney alkaline phosphatase, glutamate dehydrogenase, and the succinic oxidase system. DNA synthesis in mammalian cells was reported to be inhibited by citrinin at levels as low as 10^{-6} M. Higher levels of the toxin also exert inhibitory activity on RNA and protein synthesis. Damage to DNA by citrinin has also been found in bacterial cells and in mammalian cells in culture.

REFERENCES

Abdelhamid, A.M. and Dorra, T.M. (1990). Study of the effects of feeding laying hens on separate mycotoxin (aflatoxins, patulin or citrinin)-contaminated diets on the egg quality and tissue constituents. Arch. Tierernaehr., 40(4), 305-316.

Aleo, M.D., Wyatt, R.D., and Schnellmann, R.G. (1991a). Mitochondrial dysfunction is an early event in ochratoxin A but not oosporein toxicity to rat renal proximal tubules. Toxicol. Appl. Pharmacol., 107, 73-80.

Aleo, M.D., Wyatt, R.D., and Schnellmann, R.G. (1991b). The role of altered mitochondrial function in citrinin-induced toxicity to rat renal proximal tubule suspensions. Toxicol. Appl. Pharmacol., 109(3), 455-463.

Ansari, R.A., Thakran, R.S., and Berndt, W.O. (1991). The effects of potassium chromate and citrinin on rat renal membrane transport. Fundam. Appl. Toxicol., 16(4), 701-709.

Baggett, J.M. and Berndt, W.O. (1984a). Renal and hepatic glutathione concentrations in rats after treatment with hexachloro-1,3-butadiene and citrinin. Arch. Toxicol., 56, 46-49.

Baggett, J.M. and Berndt, W.O. (1984b). Interaction of potasium dichromate with the nephrotoxins mercuric chloride and citrinin. Toxicology, 33(2), 157-169.

Belitski, G.A., Khovanova, E.M., Budunova, I.V., and Sharunich, E.G. (1983). Induction by mycotoxins of somatic mosaicism in Drosophila and DNA repair in mammalian liver cell cultures. Byull. Eksp. Biol. Med., 96(7), 83-86.

Belitsky, J.R., Khovanova, E.M., Budonova, I.V., and Sharupts, H.G. (1985). Mycotoxin induction of somatic mosaicism in Drosophila and DNA repair in mammalian liver cell cultures. Cell Biol. Toxicol., 1(3), 133-143.

Berndt, W.O. and Hayes, A.W. (1978). Effects of citrinin on renal tubular transport functions in the rat. J. Environ. Pathol. Toxicol., 1, 93-103.

Berndt, W.O. (1983). Transport of citrinin by rat renal cortex. Arch. Toxicol., 54(1), 35-40.

Berndt, W.O., and Hayes, A.W. (1982). The effect of probenecid oncitrinin-induced nephrotoxicity. Toxicol. Appl. Pharmacol., 64, 118-124.

Braunberg, R.C., Gantt, O., Gaines, D., Shinn, J., and Friedman, L. (1988). Biochemical effect of two nephrotoxins, ochratoxin A and citrinin, on minipig renal cortical explants. FASEB J., 2, A963.

Carlton, W.W., Sansing, G., Szczech, G.M., and Tuite, J. (1974). Citrinin mycotoxicosis in beagle dogs. Food Cosmet. Toxicol., 12, 479-490.

Creppy, E.E., Lorkowski, G., Beck, G., Roschenthaler, R., and Dirheimer, G. (1980). Combined action of citrinin and ochratoxin A on hepatoma tissue culture cells. Toxicol. Lett., 5, 375-380.

Endo, A. and Kuroda, M. (1976). Citrinin, an inhibitor of cholesterol synthesis. J. Antibiot. (Tokyo), 29(8), 841-843.

Gupta, M., Dolui, A.K., Dey, S.N., Mukherjee, S., Majumder, U.K., and Batabyal, S.K. (1986). Effect of the mycotoxin citrinin on some hormones and on enzymes and substrates of the Embden-Myerhof pathway in mice. Toxicology, 24, 519-523.

Haberman, P.J., Baggett, J.M., and Berndt, W.J. (1987). The effect of chromate on citrinin-induced renal dysfunction in the rat. Toxicol. Lett., 38(1-2), 83-91.

Hashimoto, K. and Morita, Y. (1957). Inhibitory effect of citrinin on the dehydrogenase system of rat's kidney, liver and brain tissue, specially concerning the mechanism of polyuria observed in the poisoned animal. Jpn. J. Pharmacol., 7, 48-54.

Hayes, G.A., Birkhead, H.E., and Hall, E.A. (1976). Antiprotozoal activity in citrinin. Bull. Environ. Contam. Toxicol., 15(4), 429-436.

Kitchen, D.D., Carlton, W.W., and Hinsman, E.J. (1977a). Ochratoxin A and citrinin induced nephrosis in beagle dogs. III. Terminal renal ultrastructure alterations. Vet. Pathol., 14(4), 392-406.

Kitchen, D.D., Carlton, W.W., and Tuite, J. (1977b). Ochratoxin A and citrinin induced nephrosis in beagle dogs. I. Clinical and clinicopathological features. Vet. Pathol., 14(2), 154-172.

Kitchen, D.D., Carlton, W.W., and Tuite, J. (1977c). Ochratoxin A and citrinin-induced nephrosis in beagle dogs. II. Pathology. Vet. Pathol., 14, 154-172.

Krogh, P., Hald, B., and Pederson, E.J. (1973). Occurrence of ochratoxin A and citrinin in cereals associated with mycotoxic porcine nephropathy. Acta Pathol. Microbiol. Scand., 81, 689-695.

Kuroda, M., Hazma-Shimada, Y., and Endo, A. (1977). Inhibition of sterol synthesis by citrinin in a cell-free system from rat liver and yeast. Biochim. Biophys. Acta, 486(2), 254-259.

Lorkowski, G., Creppy, E.E., Beck, G., Dirheimer, G., and Roschenthaler, R. (1980). Inhibitory action of citrinin on cultured hepatoma cells. Food Cosmet. Toxicol., 18(5), 489-491.

Lorkowski, G. and Roschenthaler, R. (1979). Comparison of some actions of ochratoxin A and citrinin on bacteria. In: 4th International IUPAC Symposium, 32 pp. (abstract 604). Chem. Rundsch., Lausanne.

Martin, W., Lorkowowski, G., Creppy, E.E., Dirheimer, G., and Roschenthaler, R. (1986). Action of citrinin on bacterial chromosomal and plasmid DNA in vivo and in vitro. Appl. Environ. Microbiol., 52(6), 1273-1279.

Patil, T.N., Hayes, A.W., and Berndt, W.O. (1980). Effects of citrinin on ATPase activities. Fed. Proc., 39, 1101.

Phillips, R.D., Berndt, W.O., and Hayes, A.W. (1978). Distribution and excretion of [^{14}C]-citrinin in rats. Abstract. Toxicol. Appl. Pharmacol., 45, 276.

Ramadoss, C.S. (1973). Mycotoxicosis--Studies on Secondary Metabolites of Aspergillus candidus. Cited by Ramadoss and Mukerjee (1977), University of Madras, India.

Ramadoss, C.S. and Shanmugasundaram, E.R.B. (1973). Effect of citrinin on liver metabolism in rabbits. Indian J. Biochem. Biophys., 10, 296.

Ramadoss, C.S. and Mukherjee, A. (1977). In vitro inhibition of bovine liver glutamic dehydrogenase by citrinin, a mycotoxin. J. Antibiot., 30(2), 172-174.

Ramadoss, C.S. and Shanmugasundaram, E.R. (1977). The mechanism of action of citrinin on rabbit kidney alkaline phosphatase activity in vivo. J. Biochem., 81(6), 1825-1831.

Robbana-Barnat, S., LaFarge-Frayssinet, C., and Frayssinet, C. (1989). Use of cell cultures for predicting the biological effects of mycotoxins. Cell Biol. Toxicol., 5, 217-226.

Sansing, G.A., Lillehoj, E.B., Detroy, R.W., and Miller, M.A. (1976). Synergistic toxic effects of citrinin, ochratoxin A and penicillic acid in mice. Toxicon, 14, 213-220.

Siraj, M., Phillipa, T.D., and Hayes, A.W. (1981). Effects of the mycotoxins citrinin and ochratoxin A on hepatic mixed function oxidase and adenosinetriphosphatase in neonatal rats. J. Toxicol. Environ. Health, 8, 131-140.

Stetina, R. and Votava, M. (1986). Induction of DNA single-strand breaks and DNA synthesis inhibition by patulin, ochratoxin A, citrinin and aflatoxin B_1 in cell lines CHO and AWRF. Folia Biol. (Prague), 32, 128-144.

Tanzawa, K., Kuroda, M., and Endo, A. (1977). Time-dependent, irreversible inhibition of 3-hydroxy-3-methylglutaryl- coenzyme A reductase by the antibiotic citrinin. Biochim. Biophys. Acta, 488(1), 97-101.

Thust, R. and Kneist, S. (1979). Activity of citrinin metabolized by rat and human microsome fractions in clastogenicity and SCE assays on Chinese hamster V79-E cells. Mutat. Res., 67(4), 321-330.

Tune, B.M. (1974). Relationship between the transport and toxicity of cephalosporins in the kidney. J. Infect. Dis., 132, 189-194.

Ueno, Y. and Kubota, K. (1976). DNA-attacking ability of carcinogenic mycotoxins in recombination-deficient mutant cells of Bacillis subtilis.Cancer Res., 36, 445-451.

Wurgler, F.E., Friederich, U., and Schlatter, J. (1991). Lack of mutagenicity of ochratoxin A and B, citrinin, patulin and cnestine in Salmonella typhimurium TA102. Mutat. Res., 261, 209-216.

Yamato, M., Hamada, M., Naganawa, H., Maeda, K., Takeuchi, T., and Umezawa, H. (1985). Screening of microbial inhibitors of mammalian ornithine decarboxylase. J. Antibiot., 38, 442-443.

Yoneyama, M. and Sharma, R.P. (1987). Biochemical alterations induced by citrinin in cultured kidney epithelial cell systems. Arch. Environ. Contam. Toxicol., 16, 765-770.

Yoneyama, M., Sharma, R.P., and Kleinschuster, S.J. (1986). Cytotoxicity of citrinin in cultured kidney epithelial cell systems. Ecotoxicol. Environ. Safety, 11, 100-110.

DISTRIBUTION AND METABOLISM OF CITRININ: A Review

BONNIE B. DUNN*, Positron Emission Tomography Department, National Institutes of Health, Bethesda, MD 20892, USA.

LEONARD FRIEDMAN, Division of Toxicological Studies, Center for Food Safety and Applied Nutrition, Food and Drug Administration, Beltsville, MD 20708, USA

INTRODUCTION

Citrinin was first isolated by Hetherington and Raistrick in 1931 from a culture filtrate of Penicillium Thom. Other species of Penicillium as well as several species of Aspergillius also produce citrinin (Timonin and Rouatt, 1944; Krogh et al., 1970; Mislivec and Tuite, 1970; Ito et al., 1973; Pohland and Mislivec, 1976; Stoloff, 1976; Nelson et al., 1980; Mislivec, 1981). It is a potentially important mycotoxin; it had been found as a natural contaminant of peanuts (Diener, 1960), corn (Mislivec and Tuite, 1970), rice (Saito et al., 1971); wheat, rye, barley, and oats in Canada (Scott et al., 1972); barley and oats in Denmark (Krogh et al., 1973); naturally rotted apples containing Penicillium expansum (Harwig et al., 1973); and Indian groundnuts infected with A. flavus, P. citrinum, and A. terreus (Subrahmanyam and Rao, 1974).

Citrinin was thought to be the cause of a yellow color in rice imported from Thailand to Japan in 1951 (Tatsuno et al., 1955). Saito et al. (1971) reported that the rice was contaminated with P. citrinum; although the presence of citrinin was never definitely established, colored portions were reported to fluoresce under ultraviolet light. Its reported antibiotic activity against Gram-positive bacteria (Timonin, 1942) was not developed because of its nephrotoxic properties. It has been suggested that the occurrence of porcine nephropathy, termed "mold nephrosis," in Denmark was caused by ingestion of

Biodeterioration Research 4, Edited by G.C. Llewellyn *et al.*, Plenum Press, New York, 1994

moldy feed contaminated with citrinin- and ochratoxin-producing strains of P. viridicatum. The kidney degeneration observed in Danish pigs was similar to the experimentally induced damage found in rats and rabbits, suggesting that ochratoxin A or citrinin or both may cause this naturally occurring mycotoxicosis.

The major reported pathological change in animals exposed to citrinin is kidney damage (Ambrose and DeEds, 1946; Carlton and Tuite, 1977), although hepatotoxic action has also been reported (Carlton and Tuite, 1977; Ramadoss and Shanmugasundaram, 1973; Phillips and Hayes, 1978). Ueno and Kubota (1976) found that citrinin gave a positive response in a presumptive screening test for carcinogens and mutagens measured by DNA-complexing ability in bacteria. It has not been shown to be carcinogenic in any animal feeding studies. Scott et al. (1972) have reported that citrinin, together with ochratoxin A, occurs in various stored grains; the citrinin concentrations were between 0.07 and 80 $\mu g/g$. Feeds associated with porcine nephropathy in Denmark were reported to contain citrinin and ochratoxin A at 0.16-2.00 $\mu g/g$ (Hald and Krogh, 1973; Krogh et al., 1973). Similar foodstuff contamination led Krogh (1972) to conclude that ochratoxin A and citrinin were responsible for an endemic human nephropathy in the Balkan countries. Although citrinin has never been documented as the sole contaminant in a mycotoxicosis of public health significance, its potential for harm is greater, since it elicited synergistic lethal responses when coupled with ochratoxin A or penicillic acid (Sansing et al., 1976). It may play its most toxic role as a synergist.

Citrinin is slightly soluble in water, and readily soluble in dilute solutions of NaOH, Na_2CO_3, or sodium acetate and in many organic solvents. Photodecomposition occurs under fluorescent light both in solution and in the solid state (Neely et al., 1972). Citrinin is thermolabile in acid or alkaline solutions. Aqueous solutions change color with changes in pH, ranging from lemon-yellow at pH 4.6 to cherry-red at pH 9.9. Citrinin has been detected and quantitated by colorimetry and fluorescence. Its color reactions include formation of an orange color in alkaline solution, brown with $FeCl_3$ (Hetherington and Raistrick, 1931), green with TiCl (Haese, 1963), and deep wine-red with water followed by NaOH (Wang and Ting, 1949).

Citrinin fluoresces yellow in longwave ultraviolet light on silica gel TLC and is detectable in amounts as low as 0.01 $\mu g/mg$ (Hald and Krogh, 1973; Neely et al., 1972). The basis for measurement at low concentrations in grain extracts is visual or densitometric comparison with a standard on a TLC plate. In chloroform solution the lower limit of fluorometric determination is 0.5 μg/ml (Kawashiro et al., 1955; Neely et al., 1972). Quantitation by TLC is usually performed as a densitometric comparison with a standard (Kawashiro et al., 1955; Scott et al., 1972; Durackova et al., 1976). Scott et al. (1972) confirmed citrinin by spraying the TLC with ammonia, which removes the yellow fluorescence, or by spraying with BF_4-methanol, which changes the fluorescence from yellow to green. Development of the chromatogram with ethyl acetate-acetone-water (5 + 5 + 2) gives a compact fluorescent spot at R_f 0.5 (Hald and Krogh, 1973).

Various thin layer chromatographic procedures have been used for the determination of citrinin (Stubblefield, 1979; Wilson et al., 1976). Stubblefield reported that silica gel plates prepared as a slurry with aqueous 0.05M Na_2EDTA (ethylenediaminetetraacetic acid) and developed in benzene-acetic acid (95 + 5) allowed fluorodensitometric measurement of the spot (365 nm excitation, 505 nm emission) and reduced streaking. In the TLC procedure of Wilson et al. (1976), silica gel and ethyl acetate-acetone-0.5% HCl (50 + 50 + 20) were used as adsorbent and developing solvent, respectively. Ultraviolet absorption in 95% methanol occurs at 222 (ϵ = 22,280 M-1 X cm-1), 253 (ϵ = 8279), and 319 (ϵ = 4719) nm. The infrared spectrum (KBr) has maxima at 3484, 2985, 1675, and 1639 cm.

High performance liquid chromatographic (HPLC) techniques suitable for determination of citrinin and metabolites in tissues and body fluids have been described by Phillips et al. (1980), Dunn (1982), and Dunn et al. (1983). Other HPLC techniques, using reverse-phase or ion-pair chromatography and UV or fluorescence methods of detection, have also been applied to analysis of grains (Marti et al., 1978; Lepom, 1986; Zimmerli et al., 1989).

A limited number of early studies (Table 1) have investigated the fate of citrinin administered to animals. Ambrose and DeEds (1946), on the basis of toxicity data, reported that citrinin is rapidly absorbed. Chu (1946), using a bacteriological method, demonstrated the apparent presence of citrinin in the blood of cats following intramuscular injection or incorporation of citrinin into the stomach or small intestine. Wang et al. (1950) and Wang and Ting (1951), using a colorimetric method of detection involving reaction with $FeCl_3$, found that peak citrinin concentrations in the blood of rabbits after parenteral injection were reached within 5 min and citrinin was still detectable after 24 hr. After oral administration, the peak was reached in approximately 3 hr. When dogs were given single intravenous injections of 70 and 83 mg of citrinin, 22% of the citrinin was recovered in the urine within 48 hr. These same investigators found that only 80% of citrinin added to plasma or whole blood could be recovered and that free citrinin was equally distributed between the red cells and plasma (Wang et al., 1950). Other workers, including Phillips et al. (1978) and Damodaran and Shanmugasundaran (1978), reported studies on the distribution of radioactivity from administered [^{14}C]-citrinin in rat and mouse tissues, respectively.

Table 1. Studies on Fate of Citrinin Administered to Animals.

Researcher, Year	Animal	Mode of Dosing	Method of Measurement
Ambrose & DeEds (1945, 1946)	mice, guinea pigs, rabbits	sc, ip, iv	toxicity data
Chu (1946)	rats	im, po	bacteriological
Wang & Ting (1951)	rabbits, dogs	sc, iv, im, po	colorimetric (reaction with $FeCl_3$
Damodaran & Shanmugasundaram (1978)	mice	ip	radioactivity (^{14}C-citrinin)
Phillips et al. (1979a,b)	rats	iv	radioactivity (^{14}C-citrinin) & TLC
Phillips et al. (1980)	rats	iv	radioactivity (^{14}C-citrinin)& HPLC
Dunn (1982)	rats	po	radioactivity (^{14}C-citrinin) TLC & HPLC
Dunn et al. (1983)	rats	po	radioactivity (^{14}C-citrinin) & HPLC

ABSORPTION, DISTRIBUTION, AND EXCRETION

^{14}C-Citrinin is rapidly absorbed from the gastrointestinal tract after it has been administered. The tissue distribution and excretion of an oral dose of ^{14}C-radiolabeled citrinin was determined; organs with the highest concentration of radioactivity per unit weight were the kidney, followed in order of decreasing concentration by the liver, urinary bladder, whole blood, lungs, sex organs, and spleen (Dunn, 1982; Damodaran and Shanmugasundaram, 1978; Phillips et al.,

1978). The liver and kidney have a high capacity to bind chemicals and they probably concentrate more toxicants than any other organs. This ability to concentrate chemicals might be related to their importance in the elimination of toxicants from the body. Although the precise mechanism by which the liver and kidney remove toxicants from the blood has not yet been established, active transport or binding to tissue components is likely to be involved. The tissue distribution results suggest that a major storage site for ^{14}C-citrinin is the plasma (Dunn, 1982). It can therefore be speculated further that the blood-rich organs such as the kidneys, liver, lung, and spleen are possible storage sites as a result of exposure to the citrinin and/or degradation products carried by the plasma. Citrinin appears to be absorbed intact from the gastrointestinal tract, metabolized by the liver, excreted by the bile, returned to the gastrointestinal tract, reabsorbed from the gut into the liver, transferred to the blood, and eventually passed to the kidney for excretion (Dunn, 1982).

All tissues except the liver showed a bimodular absorption curve, with the maxima a half hour and 2-4 hr after reabsorption of the radiolabeled compound from the blood back into the tissues and gastrointestinal tract (Dunn, 1982). Phillips et al. (1979a) reported two plasma elimination rates of radioactivity after intravenous administration of ^{14}C-citrinin, with half lives of 2.6 and 14.9 hr, respectively. The extremely high levels of radioactivity in the bile indicate the importance of the bile in the excretion pathway, and the bimodal pattern offers further evidence in support of the hypothesis that citrinin is reabsorbed in the gastrointestinal tract. Phillips et al. (1979b) also observed considerable radioactivity (32% at 3 hr) in the bile of ^{14}C-citrinin-dosed rats with only modest amounts in the feces, also pointing to considerable enterohepatic circulation of citrinin or its metabolites. This is a much faster rate of absorption than the rate for patulin (48 hr) (Dailey et al., 1977); it compares favorably with the fast rate of absorption of penicillic acid (peaks at 2-4 hr) (Park et al., 1977). These two mycotoxins are of similar structure, origin, and toxicity parameters. The reported data for citrinin suggest a relationship between tissue distribution and kidney and liver toxicity. Animals of both sexes were found to have decreased radioactivity in the tissues, gastrointestinal tract, and blood and increased ^{14}C-radioactivity in the urine 4 hr after administration of ^{14}C-citrinin. The liver appeared to absorb increasing amounts of the radiolabeled compound until it reached a maximum at

2 hr and then steadily declined until the termination of the experiment (Dunn, 1982).

A significant difference in the rate of absorption between the male and female animals was found for the kidney ($P<0.02$) and the sex organs ($P<0.001$) (Dunn, 1982). Male rats reached a maximum absorption in the kidneys at 2 hr after administration (3 hr for the female rats) and retained less radioactivity than did female rats 6 hr after administration. A half hour after administration of ^{14}C-citrinin, female rats absorbed approximately ten times more radioactivity in the ovaries and fallopian tubes than did the testes of the male rats. The amount of radioactivity remained higher for the female than for the male rat sex organs at all time intervals after administration of the mycotoxin. As in the kidneys, the testes of male rats reached a maximum level of radioactivity 2 hr after oral administration, whereas the peak level of radioactivity in the female sex organs was reached at 3 hr (Dunn, 1982).

Damodaran and Shanmugasundaram (1978) found that citrinin was rapidly absorbed after intraperitoneal injection and taken up by the liver and kidneys. This observation is similar to that of Ambrose and DeEds (1945, 1946), who concluded from toxicity data that citrinin was rapidly absorbed after intraperitoneal injection and taken up by the liver and kidneys. It can be concluded that citrinin is rapidly absorbed irrespective of the route of administration. In 1951 Wang and Ting found that oral administration of citrinin at 20-45 mg/kg body weight elicited a peak level of citrinin in blood (of rabbits and dogs) at 3 hr. Phillips et al. (1979a) found that in rats dosed with ^{14}C-citrinin (3 mg/kg), 14.7-7.5% of the total activity was detected in the liver, 5.6-4.7% in the kidneys, and 9.2-4.7% in the serum at 30 min and 6 hr, respectively, after intravenous administration.

The high concentrations of radioactivity in the liver and kidney were also evident in patulin (Dailey et al., 1977) and penicillic acid metabolism studies (Ciegler et al., 1971; Park et al., 1977) although these two mycotoxins also showed high concentrations of radioactivity in the red blood cells, whereas ^{14}C-citrinin-dosed animals showed high concentrations of radioactivity in the plasma portion of the blood (Dunn, 1982; Damodaran and Shanmugasundaram, 1978). Damodaran and Shanmugasundaram (1978) reported that ^{14}C-citrinin binds to human serum albumin in vitro, which confirms the ability of citrinin to bind to serum protein, possibly as a means of transport. Phillips et al. (1979a) confirmed that this binding occurs in vivo; they found only 1.65% of total plasma citrinin in the plasma ultrafiltrate after intravenous administration of ^{14}C-citrinin (6 mg/kg). Galtier (1978) found that ochratoxin A was also present in higher concentrations in the plasma portion of the blood, and further equilibrium dialysis tests showed that 95% of the ochratoxin A in rat plasma was bound to protein for up to 8 or 48 hr after dosing. Dunn (1982) found that none of the orally administered dose of labeled citrinin was recovered as ^{14}C-CO_2 during a 6-hr period. This information suggests that little or no catabolism occurred during this time.

The main excretory route of the compound was the urine. Phillips et al. (1979a) reported that 73.9% of the radioactivity was detected in the urine of rats within 24 hr after intravenous treatment with ^{14}C-citrinin, but only 4% appeared in the feces during this time. In 1980 these same researchers reported that 36% of the ^{14}C-radioactivity was detected in rat urine within 2 hr. Additionally Phillips et al. (1979a) found that renal elimination was by glomerular filtration and tubular excretion. Dunn (1982) reported that little radioactivity was eliminated during the initial time periods of 0.5 and 1.0 hr after administration of the labeled compound. At 24 hr after oral administration, however, 41% of the administered dose was excreted in the urine. Negligible amounts of radioactivity were detected in the feces. The slow rate of elimination initially observed may be due to the early toxic insult to the kidney.

Dunn (1982) observed that rats given citrinin orally at 30 mg/kg frequently showed signs of lethargy. Autopsy observations revealed that the organs most affected by citrinin treatment were the kidneys, liver, lungs, and spleen. The kidneys of dosed rats, 4-6 hr after administration, were diffusely dark red-brown and sometimes swollen in rats (Krogh et al., 1970; Dunn, 1982). These autopsy observations are similar to those found by Thacker et al. (1977) in guinea pigs, swine (Friis et al., 1969), and dogs (Carlton et al., 1974). In a second study, Phillips et al. (1978) by pretreating rats with citrinin at 50 mg/kg provided evidence for the impact of citrinin-induced toxicity on elimination rates of radioactivity from ^{14}C-citrinin. In those animals that recovered from nephrotoxicity 4 days after dosing, 66% of the citrinin appeared in the urine within 24 hr; in the animals that were still nephrotoxic, the amount was only 13%. By 72 hr, nearly 80% of total radioactivity was eliminated in the urine of the recovered or normal rats, whereas only 40% was found in the urine of the rats with evidence of nephrotoxicity. In addition, at 74 hr 8.6% of total radioactivity was retained in the tissues of the nephrotoxic rats, whereas only 1.3% remained in the tissues of the recovered animals. It is not surprising that the excretion of citrinin via the urinary tract is impaired as a result of its nephrotoxic action in view of the more recent finding by Berndt (1983) that citrinin is transported by the classical renal organic anion system.

METABOLITES

HPLC and TLC analysis of both urine and urine chloroform extracts revealed that about 58% of the radioactivity excreted in the urine within 24 hr was unchanged citrinin in addition to the presence of a major citrinin-related metabolite (Dunn, 1982). The metabolite is more polar than citrinin (R_t 6.2 min); it has a retention time of 4.4 min on HPLC (10 μm μBondapak/C-18, Waters Associates; 0.25N MeOH-H_3PO_4-EtOAc (55 + 40 + 5); flow 1 ml/min; 332 nm) and R_f 0.4-0.5 on Adsorbosil-1-EDTA TLC plates (benzene-acetic acid, 5 + 99) and fluoresces blue under UV radiation (excitation 365 nm, emission 505 nm). The HPLC elution fraction containing the metabolite was further studied for characterization by Dunn et al. (1983). Structural identification was performed by mass spectrometry, which revealed M+ at m/e 266 (20%) and major peaks at 248 (100), 230 (45), 215 (34), 202 (25), 187 (11), and 176 (17). The loss of three waters would indicate that an oxygen (MW 16) had been added to the citrinin molecule during metabolism in an effort to excrete the molecule. The UV spectrum of the urinary metabolite was identical to that of citrinin, an indication that the urinary metabolite absorbs at the same wavelengths as authentic citrinin. The urinary metabolite, however, fluoresced blue, rather than yellow.

Based on its mass spectrum, the urinary metabolite produced by *in vivo* metabolism by the rat appears to be dihydrocitrinone, 3,4-dihydro-6,8-dihydroxy-3,4,5-trimethylisocoumarin-7-carboxylic acid. This phenolic compound had previously been isolated from a mutant strain of *Aspergillus terreus* Thom (Hassall and Jones, 1962), which produced the dihydrocitrinone but not citrinin in a fluid culture. These researchers isolated a phenol with a molecular formula differing from that of citrinin only in the oxygen content ($C_{13}H_{14}O_6$). Dunn (1982) found that the major urinary metabolites of ^{14}C-citrinin metabolism are unchanged in citrinin and dihydrocitrinone.

These data in conjunction with no detectable levels of ^{14}C-CO_2 support the hypothesis that no or only limited fragmentation of the parent compound had occurred. This finding was supported by the earlier study of Phillips et al. (1979a) in which the urine from ^{14}C-citrinin administered intraperitoneally to rats was analyzed by TLC. The only major spot visible under UV light showed a bright yellow fluorescence of citrinin corresponding to the R_f of authentic citrinin, and contained all the radioactivity. These workers also found that the radioactivity extractable into chloroform from urine was not altered by the presence of β-glucuronidase or sulfatase, indicating the absence of glucuronide or sulfate conjugates. Also, on the basis of extractability of acidified urine, they calculated that approximately 14% of the radioactivity from ^{14}C-citrinin was a metabolite(s) of citrinin. In a preliminary communication, Phillips et al. (1980) reported detecting two unidentified metabolites of ^{14}C-citrinin in the urine of rats administered the labeled compound. Finally, using probenecid, an inhibitor of organic ion secretion, Berndt and Hayes (1982) showed a correlation between reduced renal accumulation of citrinin and reduced mortality and adverse renal function, supporting the hypothesis that citrinin and not a metabolite is the nephrotoxic moiety. Therefore, the reported low carcinogenic potential for citrinin might be explained by the nearly complete (and rapid) excretion of the intact molecule. Further studies are recommended to determine if this metabolite retains the toxicity of citrinin, or if it had been detoxified.

SUMMARY

The absorption, distribution, metabolism, and excretion of citrinin have been studied in several species including rats, mice, rabbits, and dogs. This work has been reviewed. Regardless of the mode of administration, the toxin is rapidly absorbed into the blood and generally rapidly excreted. Aside from the blood the major sites of uptake are the kidney and liver. The predominant mode of excretion is via the kidney, and although a significant amount of radioactivity from administered ^{14}C-citrinin is found in the bile, only small amounts of citrinin or its metabolites are excreted in the feces, indicating considerable enterohepatic circulation. A major metabolite has been isolated from the urine of rats and identified as dihydrocitrinone.

REFERENCES

Ambrose, A.M. and DeEds, F. (1945). Some toxicological and pharmacological properties of citrinin. Abstract. Proc. Soc. Exp. Biol. Med. 59, 289.

Ambrose, A.M. and DeEds, F. (1946). Some toxicological and pharmacological properties of citrinin. J. Pharmacol. Exp. Ther., 88, 173-186.

Berndt, W.O. (1983). Transport of citrinin by rat renal cortex. Arch. Toxicol., 54(1), 35-40.

Berndt, W.O. and Hayes, A.W. (1982). The effect of probenecid on citrinin-induced nephrotoxicity. Toxicol. Appl. Pharmacol., 64, 118-124.

Carlton, W.W., Sansing, G., Szczech, G.M., and Tuite, J. (1974). Citrinin mycotoxicosis in beagle dogs. Food Cosmet. Toxicol., 12, 479-490.

Carlton, W.W. and Tuite, J. (1977). Metabolites of P. viridicatum toxicology. In: Mycotoxins in Human and Animal Health (J. Rodricks, C. Hesseltine, & M. Mehlman, eds.), Pathotox Publishers, Inc., Park Forest South, IL.

Chu, W.C. (1946). Miscellaneous pharmacologic actions of citrinin. J. Lab. Clin. Med., 31, 72.

Ciegler, M.M., Detroy, R.W., and Lillihoj, E.B. (1971). Patulin, Penicillic Acid and Other Carcinogenic Lactones. Academic Press, New York.

Dailey, R.E., Blaschka, A.M., and Brouwer, E.A. (1977). Adsorption, distribution and excretion of ^{14}C-patulin by rats. J. Toxicol. Environ. Health, 3, 479-489.

Damodaran, C. and Shanmugasundaram, E.R.B. (1978). Distribution of radioactive citrinin in tissues and serum proteins. J. Radioanal. Chem., 46, 373-377.

Diener, U.L. (1960). The mycoflora of peanuts in storage. Phytopathology, 50, 220-223.

Dunn, B.B. (1982). Adsorption, Distribution and Excretion of ^{14}C-Citrinin. Doctoral Dissertation, University of Maryland, College Park, MD.

Dunn, B.B., Stack, M.E., Park, D.L., Joshi, A., Friedman, L., and King, R.L. (1983). Isolation and identification of dihydrocitrinone, a urinary metabolite of citrinin in rats. J. Toxicol. Environ. Health, 12, 283-289.

Durackova, Z., Betina, V., and Nemec, P. (1976). Systematic analysis of mycotoxins by thin layer chromatography. J. Chromatogr., 116, 141.

Friis, P., Hasselager, E., and Krogh, P. (1969). Isolation of citrinin and oxalic acid from Penicillium viridicatum Westling and their nephrotoxicity in rats and pigs. Acta Pathol. Microbiol. Scand., 77, 559-560.

Galtier, P. (1978). Contribution of pharmacokinetic studies to mycology--Ochratoxin A. Vet. Sci. Commun., 1, 349-358.

Haese, G. (1963). Uber Antimycin. Arch. Pharmacol. (Weinheim), 296, 227-232.

Hald, B. and Krogh, P. (1973). Analysis and chemical confirmation of citrinin in barley. J. Assoc. Off. Anal. Chem., 56, 1440-1443.

Harwig, J., Chen, Y.K., Kennedy, B.P.C., and Scott, P.M. (1973). Occurrence of patulin and patulin-producing strains of Penicillium expansum in natural rots of apple in Canada. Can. Inst. Food Sci. Technol. J., 6, 22-25.

Hassall, C.H. and Jones, D.W. (1962). The biosynthesis of phenols. Part 5. A new metabolic product of Aspergillus terreus Thom. J. Chem. Soc., 4189-4191.

Hetherington, A.C. and Raistrick, H. (1931). Studies in the biochemistry of microorganisms. Part XIV. On the production and chemical constitution of a new yellow coloring matter, citrinin, produced from glucose by Penicillium citrinum Thom. Philos. Trans. R. Soc. (London) Ser. B., 220, 269-295.

Ito, Y., Kawai, K., and Nozawa, Y. (1973). Biochemical studies of pigments from the pathogenic fungus Microsporum cookei. J. Biochem., 74, 805.

Kawashiro, I., Yanabe, H., Takenchi, H., and Nishimura, C. (1955). Determination of citrinin contained in rice by fluorometry. Bull. Natl. Hyg. Lab. (Tokyo), 73, 191-196.

Krogh, P. (1972). Mycotoxic porcine nephropathy: A possible model for Balkan nephropathy in endemic nephropathy. In: 2nd International Symposium on Endemic Nephropathy, pp. 266-270. Bulgarian Academy of Science, Sofia.

Krogh, P., Hald, B., and Pederson, E.J. (1973). Occurrence of ochratoxin A and citrinin in cereals associated with mycotoxic porcine nephropathy. Acta Pathol. Microbiol. Scand., 81, 689-695.

Krogh, P., Hasselager, E., and Friis, P. (1970). Studies on fungal nephrotoxicity isolation of two nephrotoxic compounds from Penicillium viridicatum

Westling: Citrinin and oxalic acid. Acta Pathol. Microbiol. Scand., 78, 401-413.

Lepom, P. (1986). Simultaneous determination of the mycotoxins citrinin and ochratoxin A in wheat and barley by high-performance liquid chromatography. J. Chromatogr., 352, 335.

Marti, L.R., Wilson, D.M., and Evans, B.D. (1978). Determination of citrinin in corn and barley. J. Assoc. Off. Anal. Chem., 61, 1353.

Mislivec, P.B. (1981). Toxic species of Penicillium common in food. J. Food Prot., 44, 723-726.

Mislivec, P.B. and Tuite, J. (1970). Species of Penicillium occurring in freshly harvested and stored Dent corn kernals. Mycologia, 62, 67-74.

Neely, W.C., Ellis, S.P., Davis, N.D., and Diener, V.L. (1972). Spectroanalytical parameters of fungal metabolites. I. Citrinin. J. Assoc. Off. Anal. Chem., 55, 1122-1127.

Nelson, T.S., Blasley, J.N., Kirby, L.K., Johnson, Z.B., and Ballam, G.C. (1980). Isolation and identification of citrinin produced by Penicillium lanosum. Poultry Sci., 59, 2055-2059.

Park, D., Dailey, R., Friedman, L., and Heath, J. (1977). The adsorption, distribution and excretion of ^{14}C-penicillic acid by rats. Ann. Nutr., 31, 919-934.

Phillips, R.D., Berndt, W.O., and Hayes, A.W. (1978). Distribution and excretion of [^{14}C]-citrinin in Rats. Abstract. Toxicol. Appl. Pharmacol., 45, 276.

Phillips, R.D., Berndt, W.O., and Hayes, A.W. (1979a). Distribution and excretion Of [^{14}C]-citrinin in rats. Toxicology, 12(3), 285-298.

Phillips, R.D., Berndt, W.O., and Hayes, A.W. (1979b). Renal and hepatic clearance studies of citrinin in rats. Abstract. Toxicol. Appl. Pharmacol., 48(II), A14.

Phillips, R.D., Berndt, W.O., and Hayes, A.W. (1980). Urinary excretion of two metabolites of the mycotoxin ^{14}C-citrinin in rats. Abstract. Toxicol. Appl. Pharmacol., 45, A114.

Phillips, R.D. and Hayes, A.W. (1978). Effect of the mycotoxin citrinin on composition of mouse liver and kidney. Toxicon, 16, 351-359.

Phillips, R.D. Hayes, A.W., and Berndt, W.O. (1980). High performance liquid chromatographic analysis of the mycotoxin citrinin and its application to biological fluids. J. Chromatogr., 190, 419-427.

Pohland, A.E. and Mislivec, P.B. (1976). Metabolites of various Penicillium species encountered on foods. In: Mycotoxins and Other Fungal Related Food Problems. American Chemical Society, Washington, D.C.

Ramadoss, C.S. and Shanmugasundaram, E.R.B. (1973). Effect of citrinin on liver metabolism in rabbits. Indian J. Biochem. Biophys., 10, 296.

Saito, M., Enomota, M., and Tatsuno, T. (1971). Yellowed Rice Toxins. Citrinin. Academic Press, New York.

Sansing, G.A., Lillehöj, E.B., Detroy, R.W., and Miller, M.A. (1976). Synergistic toxic effects of citrinin, ochratoxin A and penicillic acid in mice. Toxicon, 14, 213-220.

Scott, P.M., van Walbeck, W., Kennedy, B., and Anyeti, D. (1972). Mycotoxins (ochratoxin A, citrinin, and sterigmatocystin) and toxigenic fungi in grains and other agricultural products. J. Agric. Food Chem., 20, 1103-1109.

Stoloff, L. (1976). Occurrence of Mycotoxins in Foods and Feeds. American Chemical Society, Washington, D.C.

Stubblefield, R.D. (1979). Thin layer chromatographic determination of citrinin. J. Assoc. Off. Anal. Chem., 62, 201-202.

Subrahmanyam, P. and Rao, A.S. (1974). Occurrence of aflatoxins and citrinin in groundnut (Arachis hypogaea L.) at harvest in relation to pod condition and kernel moisture content. Curr. Sci., 43, 707-710.

Tatsuno, T., Tsukioka, M., Sakai, Y., and Asami, Y. (1955). The toxic substance in yellowed rice. Chem. Pharm. Bull., 3, 476.

Thacker, H.L., Carlton, W.W., and Sansing, G.A. (1977). Citrinin mycotoxicosis in the guinea pig. Food Cosmet. Toxicol., 15, 553-561.

Timonin, M.I. (1942). Another mould with anti-bacterial ability. Science, 96, 494.

Timonin, M.I. and Rouatt, J.W. (1944). Production of citrinin by Aspergillus species of the Candidus group. Can. J. Publ. Health, 35, 80.

Ueno, Y. and Kubota, K. (1976). DNA-attacking ability of carcinogenic mycotoxins in recombination-deficient mutant cells of Bacillis subtilis. Cancer Res., 36, 445-451.

Wang, Y. and Ting, H.S. (1949). Colorimetric determination of citrinin. Chem. Abstr., 43, 4723f.

Wang, Y. and Ting, H.S. (1951). Metabolism of citrinin. II. Adsorption, retention and excretion of citrinin. Chem. Abstr., 45, 3009d.

Wang, Y., Ting, H.S., and Mann, W. (1950). Metabolism of citrinin. I. Determination of free citrinin in blood. Clin. J. Physiol., 17, 259-270.

Wilson, D.M., Tabor, W.H., and Trucksess, M.W. (1976). Screening method for the detection of aflatoxin, ochratoxin A, zearalenone, penicillic acid, and citrinin. J. Assoc. Off. Anal. Chem., 59, 125-127.

Zimmerli, B., Dick, R., and Banmann, V. (1989). High performance liquid chromatographic determination of citrinin in cereals using an acid buffered silica gel column. J. Chromatogr., 462, 406-410.

TOXICOLOGY AND PATHOLOGY OF CITRININ

CHRISTINE HANIKA*, Department of Veterinary Pathobiology, School of Veterinary Medicine, Purdue University, West Lafayette, IN 47907, USA
WILLIAM W. CARLTON, Department of Veterinary Pathobiology, School of Veterinary Medicine, Purdue University, West Lafayette, IN 47907, USA

INTRODUCTION

Mycotoxins are secondary metabolites of filamentous fungi or molds that may colonize crops in the field or feeds or foodstuffs during storage. Production of mycotoxins is influenced by a variety of environmental factors, and different strains of fungi are capable of producing different, and often multiple toxins (Pier, 1981). Some toxic effects of fungi have been recognized in man and animals for centuries, while investigations continue on other suspected or potential effects (Butler, 1984; Pier, 1981). Factors influencing the occurrence of mycotoxicoses are complex, including interactions among multiple contaminating fungi, multiple opportunities at which contamination of foods or feeds may occur, environmental factors, plant genetic factors, influence of agricultural biocides on toxin formation, and synergistic action of multiple toxins (Moss, 1984).

Citrinin is a secondary metabolite of various species of the genera Aspergillus and Penicillium (Pier, 1981). It has been found as a contaminant of wheat, oats, barley, rye and corn (Pier, 1981) and has been identified in feedstuffs associated with outbreaks of porcine mycotoxic nephropathy in Denmark (Krogh et al., 1973).

The toxicology of citrinin, including pharmacologic, carcinogenic, teratogenic, immunotoxic and pathologic effects, has been studied in a variety of animal species and will be reviewed here.

PHARMACOLOGIC EFFECTS

Citrinin given parenterally has produced effects similar to parasympathetic stimulation. Rabbits given citrinin intravenously had clinical signs of miosis, salivation, increased bronchial secretion and lacrimation (Ambrose and DeEds, 1946). These effects were partially ameliorated by atropine. In the dog, intravenous administration of citrinin resulted in decreased blood pressure, decreased heart rate and peripheral vasodilation which were reversed by atropine (Ambrose and DeEds, 1946; Chu, 1946). Gastrointestinal motility was increased and respiration was stimulated.

In isolated organ studies, citrinin produced bronchial constriction, increased tonus of cardiac muscle, frog gastrocnemius and intestine, and decreased uterine tonus (Ambrose and DeEds, 1946). Solutions of 0.8-2.0% citrinin caused some sensory nerve depression, reflected in increased reflex times (Chu, 1946).

Dry citrinin and concentrations of citrinin in solution as low as 1% produced conjunctival irritation in the rabbit and cat (Ambrose and DeEds, 1946; Chu, 1946). Citrinin dust and dilute solutions irritated nasal mucosa. Citrinin in various forms was not irritating to the skin, but was mildly irritating when injected into the rabbit (Chu, 1946).

CARCINOGENIC EFFECTS

Male F344 rats fed a diet containing 0.1% citrinin developed renal clear cell adenomas after 40 weeks (Arai and Hibino, 1983). Adenomas developed in 35 of 48 (72.9%) rats fed citrinin, while no control rats developed similar neoplasms.

Other studies have suggested a synergistic effect of citrinin on renal tumorigenesis. Shinohara et al. (1976) studied the effects of citrinin in combination with N-(3,5-dichlorophenyl) succinimide (NDPS) and N-nitrosodimethylamine (DMN). DMN alone induced renal neoplasms in rats, while NDPS was not carcinogenic but caused severe interstitial nephritis. Citrinin was fed to rats at 0.02% or 0.05% for 20 weeks after exposure to one of these compounds. Neoplasms developed in 94.7% and 86.7% of the rats fed 0.02% and 0.05% citrinin, respectively, compared to 57.1% in rats treated with DMN alone. The increased incidence of neoplasms was primarily due to an increased incidence of renal cell adenoma/carcinoma, with a slight increase in embryonal cell tumors. Neither citrinin nor NDPS alone induced renal neoplasms, but when NDPS treatment was followed by citrinin treatment, 4 of 18 rats developed renal cell adenomas.

Kanisawa (1984) reported a synergistic effect of citrinin on renal tumorigenesis in mice simultaneously fed ochratoxin A. Male ddY mice were fed diets containing 25 ppm ochratoxin A alone and in combination with 100-200 ppm citrinin. The combination diet resulted in increased numbers of renal neoplasms; however, this trend toward greater incidence of neoplasms was not statistically significant. Sequential administration of ochratoxin A and citrinin did not affect tumorigenesis.

REPRODUCTIVE AND TERATOGENIC EFFECTS

Teratogenic effects of citrinin have been investigated in the rat, mouse and chick embryo. Reddy et al. (1982) studied embryocidal, fetotoxic and teratogenic effects of a single dose of 35 mg/kg given to rats on days 3 or 6-15 of pregnancy. Although numbers of implants were not affected, resorption rates were greater in treated animals than in controls, and fetal growth was decreased. Since maternal mortality was also increased, it was difficult to completely separate fetal and maternal toxicity. Soft tissue anomalies attributed to citrinin exposure included cleft palate, enlarged kidneys and internal hydrocephalus. The effect of combined subthreshold doses of citrinin and ochratoxin A was studied in rats injected on gestational days 5-14 (Mayura et al., 1984). Significant incidence of skeletal deformities and soft tissue anomalies such as hydronephrosis was reported with the combination treatment, suggesting an interaction between the two toxins.

In the mouse, the effect of citrinin given at various doses (up to 40 mg/kg) on one of days 7-10 of pregnancy was investigated by Hood et al. (1976). Doses of 30 and 40 mg/kg decreased fetal survival to gestation day 18 and decreased fetal growth rate. However, these doses also caused maternal toxicity, suggesting a lack of differential effect on mother and offspring.

Ciegler et al. (1977) studied the effects of citrinin on 40-day-old chick embryos. A dose-response relationship was implied in the incidence of malformations of the extremities, exencephaly, exophthalmia, crossed beaks and twisting of the head and neck to the left instead of the right. The LD50 for citrinin in the 4-day-old chick embryo was 80.5 μm/egg (54.3-131 μm/egg).

When given to chick embryos in combination with ochratoxin A, citrinin appeared to have a strictly additive effect in producing malformations (Vesela et al., 1983).

NATURALLY OCCURRING TOXICITY

A possible occurrence of citrinin mycotoxicosis in feeder cattle was reported by Lloyd et al. (1978). Over several months 200 of 9600 cattle died. Clinical signs included anorexia, depression, mucous and hemorrhagic diarrhea, coughing, and fever. Lesions consistent with bovine viral diarrhea were accompanied by fibrosing nephritis and tubular nephrosis, with tubular hyaline casts and birefringent crystals. Some cattle were uremic, with elevated BUN and serum phosphorus values. High moisture corn fed to the cattle was moldy and some samples contained 2-3 ppm citrinin. Penicillium citrinum and Aspergillus niger were the predominant contaminating fungi. BVD virus was also isolated from tissues of some affected cattle. After the feed was changed, affected animals recovered and there was no recurrence of the problem.

There are no well-documented reports of naturally occurring citrinin mycotoxicosis, but citrinin may be a contributing factor in other mycotoxicoses such as porcine mycotoxic nephropathy which is primarily a result of ochratoxin A toxicity (Krogh, 1976, 1978).

EXPERIMENTAL STUDIES OF TOXICOLOGY AND PATHOLOGY

The toxicopathology of citrinin has been studied in several laboratory animals using various dosages and routes of administration. Results and data from several studies are summarized in Table 1.

Rat

The acute toxicity of citrinin administered intraperitoneally and subcutaneously was investigated in the rat (Ambrose and DeEds, 1945, 1946). Hyperemia of the ears and feet occurred within 30 minutes after administration and persisted about 2 hours. At doses of 50 mg or more subcutaneously, respiratory stimulation, dyspnea or prostration occurred. Nasal discharge also occurred in rats given 50 mg or more intraperitoneally.

Table 1. Effects of Citrinin in Various Laboratory Amimals

Animal	Dose and route of citrinin administration	Effects	Reference
Rat	25-200 mg/kg, intravenous intraperitoneal	14 day LD_{50} = 67 mg/kg	Ambrose and DeEds 1945, 1946
	Intraperitoneal	LD_{50} = 17 mg/kg Thoracic, pulmonary and hepatic hemorrage. Repeated doses resulted in increased tolerance.	Wang et al., 1947
	5-48 mg/kg, repeated, per os	Acute tubular nephrosis, fibrosis	Krogh et al., 1970
	50-80 mg/kg, intraperitoneal	72 hr LD_{50} = 64 mg.kg Renal tubular necrosis	Jordan et al., 1978a
	32 or 48 mg/kg, daily for 7 days, intraperitoneal	Decreased susceptibility to challenge by LD_{50} dose	
Rabbit	5-150 mg/kg, intravenous	14 day LD_{50} = 19 mg/kg	Ambrose and DeEds, 1945, 1946
	20 mg/kg, intravenus	Lacrimation, nasal discharge, salivation, "drowsiness", hypothermia	Wang et al., 1947
	Intraperitoneal	LD_{50} = mg/kg Thoracic, pulmonary and hepatic hemorrage. Repeated doses resulted in increased tolerance.	
	40-55 mg/kg, intraperitoneal	72 hr LD_{50} = 50 mg/kg	Hanika et al., 1983
	125-150 mg/kg, per os	72 hr LD_{50} = 134 mg/kg	
	80-100 mg/kg, per os	Acute renal failure, tubular dysfunction	Hanika et al., 1984
Ginea Pig	50-75 mg/kg, subcutaneous	14 day LD_{50} = 37 mg/kg Nephrosis, mycarditis	Amrrose aand DeEds 1946
	25-50 mg/kg/day, oral	14 day LD_{50} = 43 mg/kg/day	Thacker et al., 1977
Animal	Dose and route of citrinin administration	Effects	Reference
Syrian Hamster	40-80 mg/kg, intraperitoneal, DMSO-50% ethanol solvent	72 hr LD_{50} = 66 mg/kg	Jordan et al., 1978
	60-90 mg/kg, intraperitoneal, 0.5 N NaOH solvent	72 hr LD_{50} = 75 mg/kg	
	190-240 mg/kg, oral, o.5 N NaOH solvent	72 hr LD_{50} = 220 mg/kg	
	35-65 mg/kg/day, intraperitoneal, 0.5 N NaOH solvent	7 day LD_{50} = 43 mg/kg/day	
	50-120 mg/kg/day, oral 0.5 N NaOH solvent	7 day LD_{50} = 67 mg/kg/day Nephrosis	

(continued)

Table 1. (Continued)

Swine	20-100 mg/kg/day, dietary	Renal degeneration, lesions consistent with porcine mycotoxic nephropathy	Friis et al., 1969
Dog	Penicillium citrinin culture, dietary, 12,5-50%	Renal degeneration	Carlton et al., 1974
	2.5-5 mg/kg/day, oral 6 weeks	No effect	
	20-40 mg/kg/day, intraperitoneal	Emesis, intussusception, renal degeneration, death by day 5	
	5 mg/kg/day, intraperitoneal, 14 days	No effect	Kitchen et al. 1977 a,b,c
	10 mg/kg/day, intraperitoneal	Death in 7-11 days	
Animal	Dose and route of citrinin administration	Effects	Reference
Mouse	Intraperitoneal, subcutaneous	14 day LD_{50} = 35 mg/kg	Ambrose and DeEds 1946
	25-200 mg/kg, intraperitoneal	72 hr LD_{50} = 89 mg/kg	Sansing et al., 1976
	Intraperitoneal, DMSO-50% ethanol solvent	Males-72 hr LD_{50} 58 mg/kg Females-72 hr LD_{50} = 62 mg/kg	Jordan et al. 1977
	Intraperitoneal, 0.5 N NaOH solvent	Males-72 hr LD_{50} = 80 mg/kg Females-72 hr LD_{50} = 87 mg/kg	
	Subcutaneous, DMSO-50% ethanol solvent	Males-72 hr LD_{50} = 73 mg/kg	
	Oral, DMSO-50% ethanol Solvent	Males-72 hr LD_{50} = 112 mg/kg	
	Oral, 0.5 N NaOH solvent	Males-72 hr LD_{50} = 105 mg/kg	
	Various	Tubular nephrosis	
	35 mg/kg, intraperitoneal	Hepatic fatty degeneration	Phillips and Hayes 1978
	30 mg/kg/day, intrapertioneal, for 4 days	50% mortality	
Laying Hen	50-250 ppm, dietary, 3 weeks	Diarrhea	Ames et al., 1976
Broiler Chick	62.5-500 ppm, dietary, 3 weeks	Enlarged kidneys, diarrhea	Ames et al., 1976
	33-260 ppm, dietary, 6 weeks	Enteritis, renal and hepatic degeneration	Roberts and Mora, 1978
	200-600 ppm, dietary, intermittently	Polydipsia, polyuria	Gustavson ey al., 1981
	47.5-110 mg.kg, oral	72 hr LD_{50} = 95 mg/kg	Mehdi et al., 1981
	100-500 ppm, dietary, 3 weeks	Diarrhea, polydipsia, renal enlargement	
	50-100 ug/ml in drinking water	Increased water intake	Burditt et al., 1983

Table 1. (Continued)

Animal	Dose and route of citrinin administration	Effects	Reference
Broiler Chick	300 ug/g, dietary	Increased water intake, decreased body weight	Manning et al., 1985
	50-250 ppm, dietary	Increased water intake and urine output with total dose of 3,68 mg	Kirby et al., 1987
Turkey poult	50-80 mg.kg, oral	72 hr LD_{50} = 56 mg/kg	Mehdi et al., 1983
Duckling	30-110 mg/kg, oral	72 hr LD_{50} = 57 mg/kg	Mehdi et al., 1983 1984a
	110-500 ppm, dietary, 15 days	Nephropathy	Mehdi et al., 1984b

Repeated doses of citrinin for 2-3 weeks in rats caused renal enlargement and proximal tubular degeneration (Krogh et al., 1970). Degenerate tubules had hydropic degeneration, loss of brush border and nuclear pyknosis. Dilated cystic tubules were lined by attenuated epithelial cells with basophilic cytoplasm and hyperchromic nuclei with prominent nucleoli. Basement membrane thickening and interstitial cell proliferation with collagen deposition were features.

Jordan et al. (1978a) investigated the effects in the rat of intraperitoneal administration of citrinin dissolved in dimethylsulphoxide (DMSO) and 50% ethanol. Principal lesions involved the kidneys. Grossly, kidneys were pale and swollen with pinpoint pale foci on the cortical surface corresponding to pale radiating cortical streaks on cross section. These alterations were apparent as early as 24 hours after treatment. Histopathologic alterations in tubular epithelial cells evident at 3 hr. included cytoplasmic vacuolation or condensation, nuclear pyknosis, karyorrhexis, karolysis and cytoplasmic dissolution. Extensive tubular necrosis was apparent 36-72 hours after treatment. After six days, regeneration was the predominant feature. Phillips et al. (1980) found similar gross and histopathologic alterations in rats given intraperitoneal injections of 50 mg/kg citrinin dissolved in 5% sodium bicarbonate. The initial portions of proximal convoluted tubules were identified as the primary site of damage. At 24 hours after treatment, alterations included increased cytoplasmic PAS-staining affinity, cytoplasmic vacuolation, hyaline droplet formation, and nuclear pyknosis.

Desquamation of tubular epithelium was present 24-96 hr after treatment, and tubulorrhexis at 72-96 hr. Tubular regeneration was seen by 72 hr.

Ultrastructural alterations in the kidneys of rats given 50 mg/kg citrinin intraperitoneally have been described (Lockard et al., 1980). Alterations involved clusters of proximal convolutions, with affected areas separated by normal areas. After 24 hours, there was cytoplasmic vacuolation, with most vacuoles appearing empty and a few containing granular material. Less marked alterations included mitochondrial swelling with loss of cristae, dilatation of rough endoplasmic reticulum and Golgi apparatus, and presence of myelin figures. After 48 hours, tubular necrosis was extensive, with severe mitochondrial degeneration, dilatation of endoplasmic reticulum and Golgi apparatus, focal areas of loss of microvilli, rupture of cell membranes and nuclear pyknosis and karyorrhexis. Cellular and mineralized debris were present in tubular lumens. Tubular basement membranes appeared intact. Regeneration was evident 72 hr after treatment, with some kidneys appearing normal by 96 hr.

Citrinin's effect on renal function has been studied in vivo and in vitro using tissue slice techniques. Clinicopathologic alterations were studied over 48 hours by Jordan et al. (1978b) in rats given 50 mg/kg citrinin intraperitoneally. Alterations consistent with nephrosis included blood urea nitrogen concentrations of 5 times normal, elevated serum potassium values, and decreased serum sodium and chloride values. Glucosuria and hematuria were most severe 3 hours after treatment and proteinuria was evident. Urinary sodium and chloride values and glutamic-oxalacetic transaminase and lactic dehydrogenase activities were increased. Urinary sediment contained epithelial cells and erythrocytes as early as 3 hr after treatment, and granular and hyaline casts were present between 6 and 48 hours.

Berndt and Hayes (1977) found that rats which survived a dose of 70 mg/kg citrinin developed increased urine output and decreased urine osmolality, a "high output" type of renal failure. Proteinuria and glucosuria were other indications of renal dysfunction, and effects were most severe on day 4 after administration. Rats that survived regained normal renal function in 5-8 days. In rats that failed to survive, death was preceded by a period of oliguria or anuria. Results of renal slice studies were inhibition of p-amino-hippurate (PAH) and tetraethylammonium (TEA) transport (decreased tissue slice accumulation), also with maximum effect on day 4, suggesting that citrinin affected proximal tubular function. In studies by Suzuki et al. (1975), citrinin decreased PAH transport by rat kidney in vitro. As reviewed by Berndt et al. (1980), other effects included elevation of BUN and decrease in glomerular filtration rate.

Renal function and clearance studies were done by Phillips et al. (1980) on normal rats treated with citrinin and on "nephrotoxic" (pretreated 4-5 days earlier with 50 mg/kg citrinin) rats. In the normal rats, citrinin clearance was greater than inulin clearance, indicating active tubular secretion of citrinin. Fractional K+ reabsorption was increased in non-pretreated rats, but net tubular secretion of K+ occurred in pretreated, "nephrotoxic" rats. PAH/insulin clearance ratios were decreased in the test groups, suggesting possible competition by citrinin with PAH for the organic anion transport system.

Using renal cortical slices, Berndt (1983) reported that probenecid, a specific inhibitor of anion transport, and dinitrophenol, a metabolic inhibitor and competitive inhibitor of anion transport, inhibited 14C-citrinin uptake. Lactate, a promoter of organic anion uptake, increased 14C-citrinin accumulation. These findings were consistent with secretion of citrinin by the renal organic anion secretory system. Probenecid pretreatment reduced renal citrinin accumulation in rats given 55 mg/kg citrinin, resulting in decreased mortality (Berndt and Hayes, 1982). The probenecid pretreatment also had a protective effect against citrinin-produced PAH and TEA transport alterations as determined in renal cortical slices.

Rabbit

The toxicity of citrinin in the rabbit was investigated by Ambrose and /4 rabbits given 100 mg/kg died within 24 hours. Three of 5 rabbits given 20 mg/kg and 5/5 given 50 mg/kg died within 14 days of citrinin administration. A single intravenous dose of 10 or 20 mg/kg caused miosis and hyperemia of the ears, with salivation, nasal discharge lacrimation resulting from the higher dose. Defecation, micturition and respiratory stimulation were less consistent features. Administration of atropine failed to obviate the miosis. Rabbits given repeated daily doses had similar clinical signs following administration of 10, 15 and 20 mg/kg, but no gross abnormalities were found at necropsy. Urinalysis was negative for protein and glucose. Renal lesions in a rabbit receiving a single dose of 20 mg/kg intravenously primarily involved proximal convoluted tubules. Swelling, necrosis and desquamation of tubular epithelium with protein casts and mineralization of necrotic debris were features of the nephrosis. Eosinophilic masses within glomerular capillaries suggested fibrin thrombi. In rabbits receiving the repeated doses of citrinin, renal lesions were localized to clusters of tubules separated by normal parenchyma. Necrosis was not evident, but affected tubules were dilated and lined by attenuated or vacuolated epithelial cells. A dose-response relationship in the severity of renal lesions was evident, except the lesions were not more severe in rabbits given repeated doses. Wang et al. (1947) reported that repeated daily injections in the rabbit resulted in increased tolerance to citrinin.

Citrinin's toxicity in the rabbit was further investigated by Hanika et al. (1983). Rabbits developed diarrhea after oral administration of citrinin (125-150 mg/kg). Cytoplasmic vacuolation of renal tubular epithelial cells, seen 8 hr after dosing, progressed to necrosis of proximal convoluted tubules and straight segments. Slight tubular cell necrosis and tubular regeneration were evident 7 days after a single dose of citrinin and were the characteristic features following multiple sublethal doses.

Ultrastructually, citrinin caused alterations in the S2 and S3 segments of the proximal tubules, straight segments of distal tubules and collecting ducts (Hanika et al., 1986). Brush border disruption, cytoplasmic rarefaction, swelling and disorganization of interdigitating processes, and distention of intercellular spaces were suggestive of damage to membrane structure and/or transport functions. Mitochondrial alterations implied a possible interference with cellular bioenergetics.

Rabbits receiving a single oral dose of 120 mg citrinin/kg had acute clinicopathologic alterations of azotemia, metabolic acidosis, hemoconcentration and hyperkalemia, suggesting nephrotoxicity as well as other less specific metabolic derangements (Hanika et al., 1984). Rabbits given a single dose of 80-100 mg/kg developed acute renal failure (azotemia, decreased creatinine clearance) with tubular dysfunction as indicated by glucosuria, isosthenuria and cylindruria (Hanika et al., 1984). Rabbits that became oliguric died, while those surviving recovered normal or near-normal renal function within 7 days, correlating with histopathologic evidence of tubular regeneration.

<u>Guinea Pig</u>

The primary alterations found by Ambrose and DeEds (1946) in guinea pigs given 50 or 75 mg/kg citrinin were in the kidneys. Cytoplasmic vacuolation and cellular swelling, necrosis and desquamation involved the proximal convoluted tubules. Tubular protein and mineralized casts were present as well. Focal areas of acute myocarditis were found in guinea pigs receiving the higher dose.

Thacker et al. (1977) studied the effects of daily repeated oral doses of 25-50 mg/kg citrinin. Gross lesions involved the kidneys, liver, intestines and skeletal muscles. The kidneys were pale and swollen with radial streaking of the cortex. Histopathologically, there was a dose-related incidence of epithelial cell degeneration and necrosis in the proximal tubules. Mineralization, tubular casts and tubular dilation were present. The distribution of affected tubules was patchy. Livers were pale and mottled with pale discrete foci; histopathologic alterations included focal necrosis, mild biliary hyperplasia and focal mineralization. Mucosal inflammation, necrosis and ulceration were present in various regions of the intestine, particularly the cecum. A dose-related incidence of gastric mucosal ulceration occurred in female guinea pigs. Skeletal muscle degeneration and necrosis, often with mineralization, were in the diaphragm and other muscles and were most prominent in animals treated with larger doses.

Hamster

The effects of single and multiple doses of citrinin in the Syrian hamster were studied by Jordan et al. (1978c). Gross lesions of pale swollen kidneys with pale foci on the cortical surface corresponding to pale radiating cortical streaks were present. Renal tubular necrosis was patchy in distribution and primarily involved straight segments and distal convoluted tubules, with necrosis of proximal tubules in the most severely affected kidneys. Nuclear pyknosis, cytoplasmic condensation or lysis and epithelial cellular desquamation with mineralized and protein casts and tubular dilatation were features. Lesions in hamsters treated with multiple doses of citrinin were no more severe than in those receiving a single dose. Mild evidence of regeneration was seen in some hamsters treated with multiple doses. Extrarenal lesions attributed to citrinin included mild focal myocardial necrosis and mineralization, mild lymphoid necrosis, and, in repeatedly dosed hamsters, mild gastric ulceration. Hamsters dying within 10 hours of citrinin treatment had no significant histopathologic lesions. The LD50 value for daily repeated doses was less than that for a single dose of citrinin (Table 1), suggesting an additive effect of repeated doses of the toxin.

Swine

Dietary citrinin was given to 30 kg swine at various doses for up to 70 days (Friis et al., 1969). Pigs fed 100 mg/kg daily died in renal insufficiency after only 1 or 2 days, with elevated serum creatinine values and glucosuria. The lower dose pigs had growth depression and loss of weight. Azotemia and serum creatinine elevation were greatest after 1 week of treatment. Glucosuria, proteinuria and celluria were present. At necropsy, the kidneys were markedly enlarged and grey-yellow in color, with small subcapsular cysts. Histopathologically, affected proximal convoluted tubules were lined by epithelial cells with basophilic cytoplasm and polygonal hyperchromic nuclei. Tubular basement membranes were thickened and accompanied by interstitial cell activation and fibrosis. Cysts were formed by dilated tubules. These lesions were consistent with those of porcine mycotoxic nephropathy, a naturally occurring disease of Danish swine fed moldy grain.

Dog

Carlton et al. (1974) studied the effects on young Beagle dogs of feeding citrinin-containing fungal cultures and of pure citrinin administered orally and intraperitoneally. Dogs given daily oral doses of 2.5-5.0 mg/kg citrinin were unaffected. Larger daily doses (20-40 mg/kg/day) resulted in emesis whether given orally or intraperitoneally. Clinical signs following intraperitoneal dosing included defecation, urination, tenesmus, lacrimation, and excitement followed by depression. Treated dogs were moribund by day 5, and 5 of 6 dogs had intussusception of the terminal ileum. Renal disease developed in dogs treated intraperitoneally and in dogs fed the fungal culture. Dogs became polydypsic and polyuric, dehydrated and thin. Grossly, the kidneys were swollen and pale. Degeneration and necrosis spared the proximal tubules and involved the distal convoluted tubules, the ascending thick limb of Henle, and collecting ducts. Alterations ranged from cytoplasmic swelling and vacuolation to necrosis, cytolysis and desquamation of epithelial cells.

Clinicopathologic alterations were similar in dogs given 20 or 40 mg/kg citrinin daily and in those fed the fungal culture. Decreased urinary specific gravity, glucosuria, celluria and cylindruria indicated renal damage. Blood urea nitrogen was slightly increased. The earliest clinicopathological indication of renal damage, in dogs given citrinin injections, was increased urinary activities of lactic dehydrogenase and glutamic-oxalacetic transaminase, with less marked elevations of glutamic-pyruvic transaminase and isocitric dehydrogenase activities.

Kitchen et al.(1977a,b,c) studied the effects of 5-10 mg/kg citrinin given daily by intraperitoneal injection in young Beagle dogs. Citrinin was given alone or in combination with ochratoxin A. After 14 days, no apparent effect was produced by the 5 mg/kg dose of citrinin, but dogs given the larger dose died within 7-11 days. Pathologic and clinicopathologic alterations were similar to those described by Carlton et al.(1974). Intestinal intussusception and renal degeneration were observed. Tubular degeneration and necrosis were limited to the distal tubules and collecting ducts. Clinical alterations and mortality were increased in dogs given the combined mycotoxins, suggesting synergism.

Terminal renal ultrastructural alterations in Beagle dogs given 10 mg/kg/day citrinin intraperitoneally were described by Kitchen et al.(1977a). Alterations in distal tubules and collecting ducts were primarily characterized by the accumulation of lipid-filled perinuclear vacuoles. Mild mitochondrial swelling and distortion, cytoplasmic disarray, and accumulation of floccular material within dilated basilar membranes also occurred. Ultrastructural alterations were also seen in proximal tubules and included cytoplasmic disarray, disorganization and dilation of the endoplasmic reticulum and the presence of myelin figures. Mitochondria were occasionally swollen and had fragmented cristae.

Mouse

The toxicity of citrinin in the mouse was investigated by Jordan et al. (1977). Differences in toxicity were compared when citrinin was dissolved in dimethylsulphoxide and 50% ethanol or in 0.5 NaOH (Table 1). Principal lesions were confined to the kidneys and were characterized by mild to moderate necrosis of tubular epithelium, tubular dilatation, the presence of protein casts in tubular lumens and increased mitotic activity. Epithelial cell necrosis was most prominent in the straight segments of cortical tubules but also occurred in convoluted and medullary tubules. Renal lesions following multiple injections of citrinin were not more severe than those following a single injection. Dietary administration of citrinin to mice failed to produce renal lesions. The mouse was considered relatively resistant to citrinin toxicosis.

No renal lesions were observed in mice given intraperitoneal injections of 35 mg/kg citrinin dissolved in propylene glycol (Phillips and Hayes, 1978). Kidney weight was increased 24 and 48 hours after treatment. Histopathologic lesions seen 2-14 days after treatment were limited to the liver and included fatty degeneration and glycogen depletion. Repeated daily doses of citrinin resulted in decreased body weight and liver:body weight ratios. Bilgrami et al. (1988) reported degeneration of glomeruli and renal tubules as well as hepatic fibrosis and proliferation of central veins in mice given daily doses of citrinin for 120 days.

In renal slice studies, citrinin increased PAH uptake and did not affect TEA uptake at moderate doses, indicating that the mouse responded to citrinin in a different manner than the rat (Berndt and Hayes, 1980). The non-inhibitory effects were associated with lack of covalent binding of citrinin in the mouse kidney.

Hematological changes caused by repeated doses of citrinin in mice were studied by Gupta et al. (1983). Anemia and thrombocytopenia associated with decreased numbers of bone marrow precursors were reported, but the mechanism of the anemia was not fully defined since peripheral blood morphology and reticulocyte numbers were not evaluated. Repeated exposure to small doses of citrinin did not depress the immune response (Reddy et al., 1988).

Chicken

The effect of citrinin on laying and broiler chickens has been studied by several investigators. Dietary citrinin fed to laying hens at 250 ppm for three weeks caused "diarrhea" (Ames et al., 1976). Occasionally, fluid would flow from the mouths of affected birds. These signs resolved within a few days after discontinuation of the treatment. Nelson et al. (1981) determined that less than 330 ppm of dietary citrinin caused decreased growth rate in chicks. Ames et al. (1976) also investigated the effects of dietary citrinin in newly hatched broiler chicks. Citrinin treatment resulted in improved feed utilization, possibly due to its antibiotic activity, but birds fed 500 ppm had decreased body weight compared to controls. Polydipsia and diarrhea occurred and relative weights of kidneys and liver were increased at most doses. Serum sodium values were elevated in chicks fed 125 and 250 ppm citrinin. In other studies, water intake and/or urine excretion were increased in chicks fed citrinin (Burditt et al., 1983; Gustavson et al., 1981; Kirby et al., 1987; Manning et al., 1985). Dietary citrinin was unrelated to the incidence of leg deformities in broiler chickens grown on wire floors (Washburn et al., 1976) and had little effect on egg quality in laying hens (Abdelhamid and Dorra, 1990), although it was detected in both egg yolk and white.

Roberts and Mora (1978) studied the effects of feeding citrinin to broiler chicks for 6 weeks. Growth depression, polydipsia, diarrhea and dehydration were reported in birds fed 130 or 260 ppm. Hydropic degeneration of renal epithelial cells, glomerular alterations and increased numbers of mitotic figures were present in the kidney. Hepatic alterations included vacuolar degeneration, periportal and centrilobular necrosis and biliary hyperplasia with multifocal leukocytic infiltration. Hemorrhagic jejunitis was an additional lesion.

Diarrhea and polydipsia, as well as renal, hepatic and lymphoid alterations were found in broiler chicks given single or multiple oral doses of citrinin by Mehdi et al.(1981). Degeneration and necrosis of renal tubular epithelial cells occurred in both proximal and distal convoluted tubules and involved both the renal cortical units and medullary cones. Protein and cellular cast formation and mineralization of necrotic debris were present. The hepatic lesions were foci of hemorrhagic necrosis, and the lymphoid changes were necrosis and depletion. The effects of multiple sublethal doses of citrinin were similar to those produced by the LD50 dose. Dietary citrinin administered over 3 weeks caused renal enlargement but no histopathological alterations.

Renal ultrastructural alterations were studied after feeding 300 ppm citrinin to chicks from 0-21 days of age (Brown et al., 1986). Principal alterations were in the proximal tubular epithelial cells and were characterized by ring-shaped mitochondria, cytoplasmic aggregates of smooth endoplasmic reticulum, membrane-bound intranuclear inclusions, and an increase in size and number of secondary lysosomes and peroxisomes. Lesions in birds fed citrinin plus ochratoxin A were similar to those in birds receiving ochratoxin A alone, but sometimes included smooth endoplasmic reticulum aggregates similar to those seen in birds treated with citrinin.

The effect on renal function of citrinin alone and in combination with ochratoxin A has been studied in pullets. Unilateral renal portal infusion of 200 ppm citrinin caused increases urine flow rate, free water clearance and fractional sodium, potassium and phosphate excretion, with decreased urine osmolality, in the infused kidney (Hnatow and Wideman, 1985). Renal function apparently returned to normal 30 minutes after cessation of the infusion, and no histopathologic alterations were observed. In a similar study, Glahn and Wideman (1987) were unable to confirm the phosphaturic or kaliuric effects of citrinin; however, the renal portal infusion method demonstrated that citrinin acts directly on the kidney, altering some tubular transport processes. Pullets given 10 repeated daily injections of 6 mg citrinin/kg body weight had only increased manure moisture, increased renal PAH clearance rates and decreased plasma Ca and P concentrations (Glahn et al., 1989). The seeming discrepancy between the latter and former (Hnatow and Wideman, 1985) studies may be explained on the basis of effective dose rate of citrinin, time between citrinin administration and testing, or a difference between acute and multiple exposures to the toxin. In pullets pre-treated with a single dose of ochratoxin A, the effects of renal portal infusion with citrinin were diminished (Glahn et al., 1988).

Witlock et al. (1977) found alterations in the morphology of the intestinal villi in broiler chicks fed 250-500 ppm citrinin for 3 weeks. In the mid-portion of the small intestine, villi were fused at the apical surface, forming ridges and ribbon-like formations. Interdigitation of fused villi sometimes formed herringbone patterns.

Dietary exposure of broiler chickens to citrinin (125-500 ppm) had no effect on cell-mediated and humoral immune responses (Campbell et al., 1981).

Turkey

Mehdi et al. (1983, 1984a,b) studied the toxicity of citrinin in turkeys. Turkey poults had polydipsia and watery droppings, as early as 12 hours after treatment, and necrosis of proximal and distal renal tubules. Focal areas of hemorrhagic necrosis and biliary hyperplasia were in the liver. Lymphoid depletion and necrosis were most severe in turkeys given the largest dose (80 mg/kg). In poults given the oral LD50 dose of citrinin, histopathologic renal alterations were evident 3 hours after treatment, and regeneration had appeared by 48 hours (Mehdi et al., 1984a). Repeated sublethal doses of citrinin did not have an additive effect; in fact, repeatedly-treated turkeys appeared less susceptible to the toxic effects of citrinin (Mehdi et al., 1984b).

Duckling

Citrinin was also nephrotoxic in the duckling, causing renal tubular necrosis primarily in medullary cone regions (Mehdi et al., 1983). The duckling was less susceptible to nephrotoxicity than the turkey. Most ducklings dying following a single oral dose of citrinin did so within 12 hours of treatment; hyperkalemia and metabolic acidosis were considered contributing factors (Mehdi et al., 1984a). Mild to moderate splenic lymphoid necrosis was a prominent lesion in the duckling. Citrinin, fed at concentrations up to 500 ppm for 15 days, caused decreased weight gain and produced renal tubular necrosis, most severe in the medullary regions (Mehdi et al., 1984b). Other renal alterations were medullary interstitial fibrosis, mineralization of necrotic tubules and regeneration. Renal lesions in ducklings fed the largest dose were more severe than in ducklings receiving a single dose of citrinin. Repeated sublethal doses of citrinin did not have an additive effect.

SUMMARY

In mammalian species, citrinin had parasympathomimetic pharmacologic activity and was primarily a nephrotoxin causing degeneration and necrosis of tubular epithelial cells. In the rat, rabbit, guinea pig and pig, the proximal tubules were most severely damaged. Clinicopathologic studies produced data that citrinin caused acute renal failure with tubular dysfunction. Specific proximal tubule damage may be related to interactions of citrinin and the proximal tubular organic anion transport system. Ultrastructural alterations in rat and rabbit kidneys implied effects on membrane structure/function, with mitochondrial changes indicating possible interferences with cellular bioenergetics. The mouse was relatively resistant to citrinin toxicosis, but developed renal and hepatic lesions in some studies. In poultry, polydipsia and diarrhea were consistent findings. Citrinin was nephrotoxic and hepatotoxic in poultry, also causing morphologic abnormalities in intestinal villi of chickens which may have been related to the diarrhea in this species.

Animals surviving the initial exposure to citrinin generally had renal tubular regeneration within several days and recovered to regain renal function. Repeated exposures to citrinin caused evidence of ongoing necrosis and regeneration in some studies, while in other studies animals had no renal lesions.

Results of feeding trials in chickens implied that the toxic effects were reversible after withdrawal of the citrinin. However, morphologic alterations of tubular cysts and interstitial fibrosis seen in swine fed citrinin suggested that irreversible renal damage might occur. At least an additive effect on nephrotoxicity has been produced when citrinin was combined with other toxins.

There appeared to be species variations in susceptibility to citrinin. In some species, repeated doses had an additive effect, which in others, repeated doses appeared to result in decreased susceptibility. Mechanisms of toxicity may vary with species as well since results of renal slice studies indicated different handling of citrinin by the rat and mouse kidney. Citrinin caused maternal toxicity in rodents as well as teratogenic effects. Teratogenicity was also demonstrated in chick embryos.

Citrinin has produced increased incidence of renal tumors in rats, and appeared to have a co-carcinogenic effect with some other chemicals. Hematologic alterations were described in the mouse, but were not well defined. No decrease in cell-mediated or humoral immune functions has been produced by citrinin in vivo. An additive to synergistic effect was seen in various types of studies when citrinin was combined with ochratoxin A.

Acknowledgements

The authors thank Ms. Colleen Mitchell for clerical preparation of the manuscript.

REFERENCES

Abdelhamid, A.M. and Dorra, T.M. (1990). Study of effects of feeding laying hens on separate mycotoxins (aflatoxins, patulin or citrinin) - contaminated diets on the egg quality and tissue constituents. Arch. Anim. Nutr., 40, 305-316.

Ambrose, A.M. and DeEds, F. (1945). Acute and subacute toxicity of pure citrinin. Proc. Soc. Exp. Biol. Med., 59, 289-291.

Ambrose, A.M. and DeEds, F. (1946). Some toxicological and pharmacological properties of citrinin. J. Pharmacol. Exp. Ther., 88, 173-186.

Ames, D.D., Wyatt, R.D., Marks, H.L. and Washburn, K.W. (1976). Effect of citrinin, a mycotoxin produced by Penicillium citrinum, on laying hens and young broiler chicks. Poult. Sci., 55, 1294-1301.

Arai, M. and Hibino, T. (1983). Tumorigenicity of citrinin in male F344 rats. Cancer Lett. 17, 281-287.

Berndt, W.O. (1983). Transport of citrinin by rat renal cortex. Arch. Toxicol., 54, 35-40.

Berndt, W.O. and Hayes, A.W. (1977). Effects of citrinin on renal tubular transport functions in the rat. J. Environ. Pathol. Toxicol., 1, 93-103.

Berndt, W.O. and Hayes, A.W. (1980). Effects of citrinin on renal transport by mouse kidney. Pharmacologist, 22, 157.

Berndt, W.O. and Hayes, A.W. (1982). Effect of probenecid on citrinin-induced nephrotoxicity. Toxicol. Appl. Pharmacol., 64, 118-124.

Berndt, W.O., Hayes, A.W., and Phillips, R.D. (1980). Effects of mycotoxins on renal function: mycotoxic nephropathy. Kidney Int., 18, 656-664.

Bilgrami, K.S., Sinha, S.P., and Jeswal, P. (1988). Nephrotoxic and hepatotoxic effects of citrinin in mice (Mus musculus). Proc. Indian Natn. Sci. Acad., B54, 35-37.

Brown, T.P., Manning, R.O., Fletcher, O.J., and Wyatt, R.D. (1986). The individual and combined effects of citrinin and ochratoxin A on renal ultrastructure in layer chicks. Avian Dis., 30, 191-198.

Burditt, S.J., Hagler, W.M. Jr., and Hamilton, P.B. (1983). Survey of molds and mycotoxins for their ability to cause feed refusal in chickens. Poult. Sci., 62, 2187-2191.

Butler, W.H. (1984). Pathological aspects of mycotoxicoses. Chem. Ind., No. 15, 530-532.

Carlton, W.W., Sansing, G.A., Szczech, G.M. and Tuite, J. (1974). Citrinin mycotoxicosis in Beagle dogs. Fd. Cosmet. Toxicol., 12, 479-490.

Campbell, M.L., Doerr, J.A., and Wyatt, R.D. (1981). Immune status in broiler chickens during citrinin toxicosis. Poult. Sci., 60, 1634. (Abstr.)

Chu, W.C. (1946) Miscellaneous pharmacologic actions of citrinin. J. Lab. Clin. Med., 31, 72-78.

Ciegler, A., Vesonder, R.F., and Jackson, L.K. (1977). Production and biological activity of patulin and citrinin from Penicillium expansum. Appl. Environ. Microbiol., 33, 1004-1006.

Friis, P., Hasselager, E., and Krogh, P. (1969). Isolation of citrinin and oxalic acid from Penicillium viridicatum Westling and their nephrotoxicity in rats and pigs. Acta Pathol. Microbiol. Scand., 77, 559-560.

Glahn, R.P., Shapiro, R.S., Vena, V.E., and Wideman, R.F. Jr. (1989). Effects of chronic ochratoxin A and citrinin toxicosis on kidney function of Single Comb White Leghorn pullets. Poult. Sci., 68, 1205-1212.

Glahn, R.P. and Wideman, R.F. Jr. (1987). Avian diuretic response to renal portal infusions of the mycotoxin citrinin. Poult. Sci., 66, 1316-1325.

Glahn, R.P., Wideman, R.F. Jr., and Evangelisti, J.W. (1988). Effects of ochratoxin A alone and in combination with citrinin on kidney function of Single Comb White Leghorn pullets. Poult. Sci., 67, 1034-1042.

Gupta, M., Sasmal, D., Bandyopadhyay, S., Bagchi, G., Chatterjee, T., and Dey, S. (1983). Hematological changes produced in mice by ochratoxin A and citrinin. Toxicology, 26, 55-62.

Gustavson, S.A., Cockrill, J.M., Beasley, J.N., and Nelson, T.S. (1981). Effect of dietary citrinin on urine excretion in broiler chickens. Avian. Dis., 25, 827-830.

Hanika, C., Carlton, W.W., Boon, G.D., and Tuite, J. (1984). Citrinin mycotoxicosis in the rabbit: clinicopathologic alterations. Food Chem. Toxicol., 22, 999-1008.

Hanika, C., Carlton, W.W., Hinsman, E.J., and Tuite, J. (1986). Citrinin mycotoxicosis in the rabbit: ultrastructural alterations. Vet. Pathol., 23, 245-253.

Hanika, C., Carlton, W.W., and Tuite, J. (1983). Citrinin mycotoxicosis in the rabbit. Food Chem. Toxicol., 21, 487-493.

Hnatow, L.L. and Wideman, R.F. Jr. (1985). Kidney function of Single Comb White Leghorn pullets following acute renal portal infusion of the mycotoxin citrinin. Poult. Sci., 64, 1553-1561.

Hood, R.D., Hayes, A.W., and Scammell, J.G. (1976). Effects of prenatal administration of citrinin and viriditoxin to mice. Food Cosmet. Toxicol., 14, 175-178.

Jordan, W.H., Carlton, W.W., and Sansing, G.A. (1977). Citrinin mycotoxicosis in the mouse. Food Cosmet. Toxicol. 15, 29-34.

Jordan, W.H., Carlton, W.W., and Sansing, G.A. (1978a). Citrinin mycotoxicosis in the rat. I. Toxicology and pathology. Food Cosmet. Toxicol., 16, 431-439.

Jordan, W.H., Carlton, W.W. and Sansing, G.A. (1978b). Citrinin mycotoxicosis in the rat. II. Clinicopathological observations. Food Cosmet. Toxicol., 16, 441-447.

Jordan, W.H., Carlton, W.W., and Sansing, G.A. (1978c). Citrinin mycotoxicosis in the Syrian hamster. Food Cosmet. Toxicol., 16, 355-363.

Kanisawa, M. (1984). Synergistic effect of citrinin on hepatorenal carcinogenesis of ochratoxin A in mice, In: Toxigenic Fungi - Their Toxins and Health Hazard, pp. 245-254 (H. Hurata and Y. Ueno, ed.), Elsevier, Amsterdam, Netherlands.

Kirby, L.K., Nelson, T.S., Halley, J.T., and Beasley, J.N. (1987). Citrinin toxicity in young chicks. Poult. Sci., 66, 966-968.

Kitchen, D.N., Carlton, W.W., and Hinsman, E.J. (1977a). Ochratoxin A and citrinin induced nephrosis in Beagle dogs. III. Terminal renal ultrastructural alterations. Vet. Pathol. 14, 392-406.

Kitchen, D.N., Carlton, W.W., and Tuite, J. (1977b). Ochratoxin A and citrinin induced nephrosis in Beagle dogs. I. Clinical and clinicopathological features. Vet. Pathol., 14, 154-172.

Kitchen, D.N., Carlton, W.W., and Tuite, J. (1977c). Ochratoxin A and citrinin induced nephrosis in Beagle dogs. II. Pathology. Vet. Pathol., 14, 261-272.

Krogh, P. (1976). Mycotoxic nephropathy. Adv. Vet. Sci. Comp. Med., 20,147-170.

Krogh, P. (1978). Causal associations of mycotoxic nephropathy. Acta Pathol. Microbiol. Scand. Sect. A, Suppl. No. 269, 1-28.

Krogh, P., Hald, B., and Pedersen, E.J. (1973). Occurrence of ochratoxin A and citrinin in cereals associated with mycotoxic porcine nephropathy. Acta Pathol. Microbiol. Scand. Sect. B, 81, 689-695.

Krogh, P., Hasselager, E., and Friis, P. (1970). Studies on fungal nephrotoxicity. 2. Isolation of two nephrotoxic compounds from Penicillium viridicatum Westling: citrinin and oxalic acid. Acta Pathol. Microbiol. Scand. Sect. B, 78, 401-413.

Lloyd, W.E., Holter, J.A, Hill, H.T. Wohlgemuth, K. and Classick, L. (1978). A case of probable citrinin toxicosis in a herd of Iowa feeder cattle. Am. Assn. Vet. Lab. Diag., 21st Ann. Proc., 147-160.

Lockard, V.G., Phillips, R.D., Hayes, A.W., Berndt, W.O., and O'Neal, R.M. (1980). Citrinin nephrotoxicity in rats: a light and electron microscopic study. Exp. Mol. Pathol. 32, 226-240.

Manning, R.O., Brown, T.P., Wyatt, R.D., and Fletcher, O.J. (1985). The individual and combined effects of citrinin and ochratoxin A in broiler chicks. Avian Dis., 29, 986-997.

Mayura, K., Parker, R., Berndt, W.O., and Phillips, R.D. (1984). Effect of simultaneous prenatal exposure to ochratoxin A and citrinin in the rat. J. Toxicol. Environ. Health, 13, 553-561.

Mehdi, N.A.Q., Carlton, W.W., Boon, G.D., and Tuite, J. (1984a). Studies on the sequential development and pathogenesis of citrinin mycotoxicosis in turkeys and ducklings. Vet. Pathol., 21, 216-223.

Mehdi, N.A.Q., Carlton, W.W., and Tuite, J. (1981). Citrinin mycotoxicosis in broiler chickens. Food Cosmet. Toxicol., 19, 723-733.

Mehdi, N.A.Q., Carlton, W.W., and Tuite, J. (1983). Acute toxicity of citrinin in turkeys and ducklings. Avian Pathol. 12, 221-223.

Mehdi, N.A.Q., Carlton, W.W., and Tuite, J. (1984b). Mycotoxicoses produced in ducklings and turkeys by dietary and multiple doses of citrinin. Avian Pathol., 13, 37-50.

Moss, M.O. (1984). Conditions and factors influencing mycotoxin formation in the field and during the storage of food. Chem. Ind., No. 15, 533-535.

Nelson, T.S., Beasley, J.N., Kirby, L.K., Johnson, Z.B., Ballam, G.C., and Campbell, M.M. (1981). Citrinin toxicity in growing chicks. Poult. Sci., 60, 2165-2166.

Phillips, R.D. and Hayes, A.W. (1978). Effect of the mycotoxin citrinin on composition of mouse liver and kidney. Toxicon, 16, 351-359.

Phillips, R.D., Hayes, A.W., Berndt, W.O., and Williams, W.L. (1980). Effects of citrinin on renal function and structure. Toxicology, 16, 123-137.

Pier, A. (1981). Mycotoxins and animal health. Adv. Vet. Sci. Comp. Med., 25, 185-243.

Reddy, R.V., Mayura, K., Hayes, A.W., and Berndt, W.O. (1982). Embryocidal, teratogenic and fetotoxic effects of citrinin in rats. Toxicology, 25, 151-160.

Reddy, R.V., Taylor, M.J., and Sharma, R.P. (1988). Evaluation of citrinin toxicity on the immune functions of mice. J. Food Prot., 51, 32-36.

Roberts, W.T. and Mora, E.C. (1978). Toxicity of Penicillium citrinum AUA-532 contaminated corn and citrinin in broiler chicks. Poult. Sci., 57, 1221-1226.

Sansing, G.A., Lillejoh, E.B., Detroy, R.W., and Miller, M.A. (1976). Synergistic toxic effects of citrinin, ochratoxin A and penicillic acid in mice. Toxicon, 14, 213-220.

Shinohara, Y., Arai, M., Hirao, K., Sugihara, S., Nakanishi, K., Tsunoda, H., and Ito, N. (1976). Combination effect of citrinin and other chemicals on rat kidney tumorigenesis. Gann, 67, 147-155.

Suzuki, S., Kozuka, Y., Satoh, T., and Yamazaki, M. (1975). Studies on the nephrotoxicity of ochratoxin A in rats. Toxicol. Appl. Pharmacol., 34, 479-490.

Thacker, H.L., Carlton, W.W., and Sansing, G.A. (1977). Citrinin mycotoxicosis in the Guinea pig. Food Cosmet. Toxicol., 15, 553-561.

Vesela, D., Vesely, D., and Jelinek, R. (1983). Toxic effects of ochratoxin A and citrinin, alone and in combination, on chick embryos. Appl. Environ. Microbiol., 45, 91-93.

Wang, Y., Hong, F.K., Hwang, F.T., and Fan, C.S. (1947). Citrinin as an antibiotic. Science, 106, 291-292.

Washburn, K.W., Ames, D., and Wyatt, R.D. (1976). Effects of dietary citrinin and aflatoxin on leg problems of broilers grown on wire floors. Poult. Sci., 55, 1977-1979.

Witlock, D.R., Wyatt, R.D., and Ruff, M.D. (1977). Morphological changes in the avian intestine induced by citrinin and lack of effect of aflatoxin and T-2 toxin as seen with scanning electron microscopy. Toxicon, 15, 41-44.

CITRININ: ANALYSIS AND OCCURRENCE

DAVID M. WILSON, Department of Plant Pathology, Coastal Plain Experiment Station, University of Georgia, Tifton, GA 31793, USA

INTRODUCTION

Citrinin was first isolated and identified from cultures of the fungus Penicillium citrinum. Citrinin [(3R-trans)-4,6-dihydro-8-hydroxy-3,4,5-trimethyl-6-oxo-3H-2-benzopyran-7-carboxylic acid] has a molecular weight of 250.24 and an empirical formula of $C_{13}H_{14}O_5$. The chemistry and occurrence of citrinin and the fungi and plants reported to produce citrinin have been reviewed by Betina (1984) and Wilson and Abramson (1992). As a mycotoxin, citrinin is of interest because it acts as a nephrotoxin and often occurs together with ochratoxin A. Citrinin is not extremely stable, is an acid, and can bind metal ions. Therefore, it is not easy to develop analytical methods for its determination in foods and feeds.

ANALYTICAL METHODS

Citrinin methods that have been developed include several thin-layer chromatography (TLC) and high-pressure liquid chromatography (HPLC) procedures and a fluorometric screening technique. The fluorometric screening method is based on an acidified acetonitrile extraction of corn, barley and peanuts followed by acid-base partitions with quantitation using a fluorometer (Trantham and Wilson, 1984). The primary problems with citrinin analysis have been variable recoveries, instability and tailing during chromatography.

Thin layer methods

A thin layer method for citrinin analysis in barley was developed by Hald and Krogh (1973). The method used a chloroform extraction of acidified grain. Thin Layer Chromatography was accomplished using silica gel plates developed with ethyl acetate-acetone-water (5+5+2). Citrinin was detected as a bright yellow fluorescent spot with longwave UV light, excitation at 365 nm. The recovery was above 80% with a detection limit of 200 ng/g. Hald and Krogh (1973) and Golinski and Grabarkiewicz-Szczesna (1984) developed TLC confirmatory tests using the acetate derivative.

Chalam and Stahr (1979) used an aqueous methanol extraction which was rendered alkaline after filtration. The alkaline solution was extracted with chloroform to remove interferences and then acidified. Chloroform was used to partition the citrinin and the chloroform phase was used for TLC. The TLC separation was accomplished using $CHCl_3$-acetone-ethanol-H_2O (60+40+10+1). Wilson et al. (1976) developed a TLC screening method for detection of aflatoxin, ochratoxin, zearalenone, penicillic acid and citrinin using a 0.5 N H_3PO_4-$CHCl_3$ (1+10) extraction. The method gave qualitative results for citrinin in corn and beans but not peanuts. Reversed phase TLC plates were employed by Abramson et al. (1989) to detect a variety of mycotoxins including citrinin.

Gimeno (1979, 1984) and Gimeno and Martins (1983) developed multi-mycotoxin TLC methods which included citrinin determination in grains, mixed feeds, apples and pears. These procedures have been useful for qualitative detection of citrinin.

The major problem with TLC of citrinin on silica gel plates has been tailing of the peak. Stubblefield (1979) prepared TLC plates using a slurry containing 0.05 M aqueous Na_2EDTA to control tailing. The dried TLC plates were developed in acetic acid-benzene (5+95). This method had a limit of 10 ng citrinin per spot with a linear range of 10 to 100 ng per spot. Marti et al. (1978) impregnated TLC plates with oxalic acid before development in order to minimize tailing. Silica gel TLC plates impregnated with glycolic acid were used by Gimeno (1984). The plates were impregnated by dipping pre-coated plates in 10% glycolic acid in ethanol, drying and activating for 20 minutes at 65 C. The glycolic acid plates were developed with ethyl ether-hexane-ethyl acetate-90% formic acid (70+90+40+2), tolulene-ethyl acetate-$CHCl_3$-90% formic acid (70+50+50+20) or ethyl ether-hexane-ethyl acetate (50+100+50) (Gimeno, 1984). All of these TLC methods eliminated tailing of citrinin on TLC plates.

HPLC methods

The HPLC separation and detection methods generally used acidified organic extraction solvents followed by an acid-base partition cleanup. Marti et al. (1978) extracted corn and barley samples with an aqueous acidified acetonitrile solvent with preliminary purification by several acid-base partitions. The reversed phase HPLC separation was accomplished using either a 0.25 N H_3PO_4-methanol (45+55) or 0.25 N H_3PO_4-acetonitrile (50+50) mobile phase with fluorescence detection (excitation 325-385 nm, emission above 451 nm). This method resulted in a 65-80% recovery of citrinin ranging from 50-4000 ng/g for corn or barley. Phillips et al. (1980) developed a reversed phase HPLC method for plasma, bile and urine using a 0.25 N H_3PO_4-acetonitrile-2-propanol (55+35+10) mobile phase with UV detection at 340 or 365 nm. This procedure had a detection limit less than 2 ng and peak areas were linear between 2 and 50 ng at 340 nm and 5 to 200 ng at 365 nm.

Both citrinin and ochratoxin A are nephrotoxic secondary metabolites that occur together in grains at times. Lepom (1986) developed an HPLC method for citrinin and ochratoxin A in wheat and barley. The samples were extracted with an aqueous acetonitrile solution (90% acetonitrile). The aqueous portion contained 4% KCl and 0.4 ml H_2SO_4 per sample. The extract was filtered and 25 ml H_2O was added to the filtrate. The citrinin and ochratoxin A were extracted with $CHCl_3$ and from there into 5% sodium hydrogen carbonate. The basic phase was acidified to pH 1.5 with HCl and partitioned into $CHCl_3$. The $CHCl_3$ extract was evaporated to dryness and the residue taken up with mobile phase of acetonitrile-isopropanol-$8x10^{-2}$M H_3PO_4. The reversed phase separation was accomplished on a C_{18} column with fluorescence detection. The recovery was 70% for citrinin and 84% for ochratoxin A from wheat. The recovery was 62% for citrinin and 86% for ochratoxin A from barley. Zimmerli et al. (1989) described an acid-buffered silica gel HPLC column suitable for citrinin quantitation. The citrinin was extracted using $CHCl_3$-0.1 M H_3PO_4 (15+2). The extract was layered onto on a Extrelut$^{(R)}$ column impregnated with 1% sodium bicarbonate and washed with $CHCl_3$. The citrinin was eluted with $CHCl_3$ + 1% formic acid and evaporated before being taken up with $CHCl_3$ for HPLC analysis. The HPLC silica column was prepared using 0.2 M citric acid adjusted to pH 2.5, dried and activated at 80° C for 30 hr. The mobile phase was n-hexane-$CHCl_3$ (60+40) with fluorescence detection (excitation-360 nm, emission-500 nm).

However, the retention time of citrinin was not stable. The retention time became shorter over time with repeated use. The method was successfully applied to wheat, wheat bran, rice, barley, corn, oats, oat groats and pastas with a detection limit of 0.1 ng/g.

Frisvad and Thrane (1987) and Kuronen (1989) developed HPLC screening methods for mycotoxins, including citrinin, based on retention indices and UV-visible spectra using diode array detection. Fristvad and Thrane (1987) tested 182 metabolites. Kuronen (1989) based the retention index scale on a homologous series of alkanones as internal standards.

Stability

Repeatability and low recoveries are the two problems most often encountered with citrinin analysis. The chemistry of citrinin makes the analysis sensitive to pH, metal ions, surface factors, light and heat. All of these on occasion have been reported to interfere with analytical results (Wilson and Abramson, 1992; Betina, 1984). Citrinin, produced by Penicillium lanosum, could not be extracted with various organic solvents from inoculated corn dried at 65 C but was readily extracted from air-dried corn (Nelson et al., 1980). Nelson et al. (1984) later reported that in other experiments citrinin could be extracted with 0.1 M H_3PO_4 and $CHCl_3$ from citrinin-contaminated corn that had been oven-dried at 65 C. Thus, matrix effects greatly influenced the extraction of citrinin, and this may explain many of the reported difficulties with citrinin analyses.

Dunn et al. (1983) isolated dihydrocitrinone in urine and free citrinin in blood of rats fed citrinin. Patterson and Damoglou (1987) found that P. viridicatum converted citrinin into dihydrocitrinone and ochratoxin A when the fungus was actively growing. El-Banna (1987) reported that 94% of citrinin was destroyed during the germination of barley.

The stability, bioconversion and complexes of citrinin all need further study because each may influence citrinin analyses. Further analytical studies are required to develop reliable inexpensive citrinin methods. Those currently available are useful for screening citrinin incidence in nature. But, at the present time, there is no validated quantitative method.

OCCURRENCE OF CITRININ

Citrinin is one of only seven mycotoxins that have been found to occur significantly in naturally contaminated foods and feeds. Other mycotoxins that are of concern are the aflatoxins, ochratoxin A, patulin, zearalenone, trichothecenes, and penicillic acid (Jelinek et al., 1989). Jelinek et al. (1989) reported that citrinin occurs widely in foods and feeds but that the incidence has not been quantified to date, and more data are needed on the incidence of citrinin.

Scott et al. (1972) reported that citrinin occurred with ochratoxin A in 13 of 27 ochratoxin A contaminated grain samples collected in Saskatchewan, Canada. The citrinin concentration ranged from 0.7 to 80 ppm. Citrinin and ochratoxin A were reported to occur together in cereals associated with porcine nephropathy in Denmark. Nine percent of the samples contained citrinin and ochratoxin A while 58 percent contained ochratoxin A (Krogh et al., 1973). Szebiotko et al. (1981) reported that citrinin occurred along with ochratoxin A, sterigmatocystin and penicillic acid in corn, barley and wheat in Poland. Pande et al. (1990) reported 50-90 ppb citrinin in rice in India. Citrinin has also been reported to occur in Indian coconut products, dry copra and copra meal and peanuts (Kumari and Nusrath, 1987; El-Maghraby and El-Maraghy, 1987). Abdelhamid (1990) found that 15% of 52 Egyptian food and feed samples contained citrinin. The products included rice bran, rice germ, corn, wheat bran, cottonseed cake and fish meal. Jonsyn (1988) indicated that sesame seed may be contaminated in Sierra Leone upon occasion.

Ochratoxin A and citrinin developed in wheat stored for 60 weeks at 15 and 19% initial moisture in Manitoba, Canada (Abramson et al., 1990). The maximal concentration of citrinin, 80 ppm, was seen at week 48 in the 19% moisture wheat and 0.65 ppm was seen in the 15% moisture wheat. The effects of time and initial moisture content and their interactions were significant factors in citrinin accumulation in storage. Madhyastha et al. (1990) evaluated corn, wheat, peanut, rapeseed and soybean for their ability to support ochratoxin and citrinin accumulation by *Penicillium verrucosum*. Corn and wheat supported citrinin accumulation while peanut, rapeseed and soybean did not. Takahashi et al. (1989) found that citrinin accumulated in rice stored at 90% relative humidity for 16-60 days but not in rice stored at 85% relative humidity for 60 days.

There is sufficient worldwide data that indicate citrinin contamination of foods and feeds is widespread. However, much more work on improved methods and systematic surveys are needed before the importance of citrinin contamination can be properly assessed.

SUMMARY

The analysis and occurrence of citrinin are discussed. This isolate of Penicillium citrinum is a mycotoxin which acts as a nephrotoxin. Citrinin is not an extremely stable compound , making determinations in foods and feeds difficult. It also tends to be acidic and binds with metal ions. Both citrinin and ochratoxin A occur in grains. Two chromatography detection procedures are discussed, thin layer chromatography and high pressure liquid chromatography. Detailed procedures are presented for these tow procedures. In addition, citrinin and ochratoxin A occurrence are presented and discussed relevant to feeds and foods including national and international examples.

Acknowledgments

I thank Darlene Morrison for her help in typing the manuscript and Susan Wilson for her help in proofreading the manuscript. Also, I thank Bonnie Evans for her help in the literature retrieval. This work was supported by State and Hatch funds allocated to the Georgia Agriculture Experiment Station.

REFERENCES

Abramson, D., Mills, J. T., and Sinha, R. N. (1990). Mycotoxin production in amber durum wheat stored at 15 and 19% moisture content. Food Add. Contamin. 7,617-627.

Abramson, D., Thorsteinson, and Forest, D. (1989). Chromatography of mycotoxins on pre-coated reverse-phase thin-layer plates. Arch. Environ. Contam. Toxicol. 18, 327-330.

Abdelhamid, A. M. (1990). Occurrence of some mycotoxins (aflatoxin, ochratoxin A, citrinin, zearalenone and vomitoxin) in various Egyptian feeds. Arch. Tierernahr. 40, 647-664.

Betina, V. (1984). Citrinin and related substances. In: "Mycotoxins - Production, Isolation, Separation and Purification" (V. Betina, ed.), pp. 217-236. Elsevier Science Publishers, Amsterdam.

Chalam, R. V., and Stahr, H. M. (1979). Thin layer chromatographic determination of citrinin. J. Assoc. Office Anal. Chem. 62, 570-572.

Dunn, B. D., Stack, M. E., Park, D. L., Joshi, A., Friedman, L., and King, R. L. (1983). Isolation and identification of dihydrocitrinone, a urinary metabolite of citrinin in rats. J. Toxicol. Environ. Health 12, 283-289.

El-Banna, A. A. (1987). Stability of citrinin and deoxynivalenol during germination process of barley. Mycotoxin Res. 3, 37-41.

El-Maghraby, O. M. O., and El-Maraghy, S. S. M. (1987). Mycoflora and mycotoxins of peanut (Arachis hypogaea L.) seeds in Egypt. 1. Sugar fungi and natural occurrence of mycotoxins. Mycopathologia 98, 165-170.

Frisvad, J. C., and Thrane, U. (1987). Standardized high-performance liquid chromatography of 182 mycotoxins and other fungal metabolites based on alkylphenone retention indices and UV-vis spectra (Diode array detection). J. Chromatography 404, 195-214.

Gimeno, A. (1979). Thin layer chromatographic determination of aflatoxins, ochratoxins, sterigmatocystin, zearalenone, citrinin, T-2 toxin, diacetoxyscirpenol, penicillic acid, patulin and penitrem A. J. Assoc. Office Anal. Chem. 62, 579-585.

Gimeno, A. (1984). Determination of citrinin in corn and barley on thin layer chromatographic plates impregnated with glycolic acid. J. Assoc. Office Anal. Chem. 67, 194-196.

Gimeno, A., and Martins, M. L. (1983). Rapid thin layer chromatographic determination of patulin, citrinin and aflatoxin in apples and pears, and their juices and jams. J. Assoc. Office Anal. Chem. 66, 85-91.

Golinski, P., and Grabarkiewicz-Szczesna, J. (1984). Chemical confirmatory tests for ochratoxin A, citrinin, penicillic acid, sterigmatocystin and zearalenone performed directly on thin layer chromatographic plates. J. Assoc. Office Anal. Chem. 67, 1108-1110.

Hald, B., and Krogh, P. (1973). Analysis and chemical confirmation of citrinin in barley. J. Assoc. Office Anal. Chem. 56, 1440-1443.

Jelinek, C. F., Pohland, A. E., and Wood, G. E. (1989). Worldwide occurrence of mycotoxins in foods and feeds - an update. J. Assoc. Office Anal. Chem. 72, 223-228.

Jonsyn, F. E. (1988). Seedborne fungi of sesame (Sesamum indicum L.) in Sierra Leone and their potential aflatoxin/mycotoxin production. Mycopathologia 104, 123-127.

Krogh, P., Hald, B., and Pederson, E. J. (1973). Occurrence of ochratoxin A and citrinin in cereals associated with porcine nephropathy. Acta. Path. Microbiol. Scand. Sec. B. 81, 689-695.

Kumari, C. K., and Nusrath, M. (1987). Natural occurrence of citrinin and ochratoxin A in coconut products. Nat. Acad. Sci. Letters 10, 303-305.

Kuronen, P. (1989). High-performance liquid chromatographic screening method for mycotoxins using new retention indexes and diode array detection. Arch. Environ. Contam. Toxicol. 18, 336-348.

Lepom, P. (1986). Simultaneous determination of the mycotoxins citrinin and ochratoxin A in wheat and barley by high-performance liquid chromatography. J. Chromatogr. 355, 335-339.

Madhyastha, S. M., Marquardt, R. R., Frohlich, A. A., Platford, G., and Abramson, D. (1990). Effects of different cereal and oilseed substrates on the growth and production of toxins by Aspergillus alutaceus and Penicillium verrucosum. J. Agric. Food Chem. 38, 1506-1510.

Marti, L. R., Wilson, D. M., and Evans, B. D. (1978). Determination of citrinin in corn and barley. J. Assoc. Office Anal. Chem. 61, 1353-1358.

Nelson, T. S., Beasley, J. N., Kirby, L. K., and Johnson, Z. B. (1985). Effect of heat drying of corn on the extraction of citrinin. Poultry Sci. 64, 866-870.

Nelson, T. S., Beasley, J. N., Kirby, L. K., Johnson, Z. B., and Ballam, G. C. (1980). Isolation and identification of citrinin produced by Penicillium lanosum. Poultry Sci. 59, 2055-2059.

Pande, N., Saxena, J., and Pandey, H. (1990). Natural occurrence of mycotoxins in some cereals. Mycoses 33, 126-128.

Patterson, M. F., and Damoglou, A. P. (1987). Conversion of the mycotoxin citrinin into dihydrocitrinone and ochratoxin A by Penicillium viridicatum. Appl. Microbiol. Biotechnol. 27, 574-578.

Phillips, R. D., Hayes, A. W., and Berndt, W. O. (1980). High-performance liquid chromatographic analysis of the mycotoxin citrinin. J. Chromatogr. 190, 419-427.

Scott, P. M., Van Walbeek, W., Kennedy, B., and Anyeti, D. (1972). Mycotoxins (ochratoxin A, citrinin, and sterigmatocystin) and toxigenic fungi in grains and other agricultural products. J. Agr. Food Chem. 20, 1103-1109.

Stubblefield, R. D. (1979). Thin layer chromatographic determination of citrinin. J. Assoc. Office Anal. Chem. 62, 201-202.

Szebiotko, K., Chelkowski, J., Dopierala, G., Godlewska, B., and Radomyska, W. (1981). Mycotoxins in cereal grain. Part I. Ochratoxin, citrinin, sterigmatocystin, penicillic acid and toxigenic fungi in cereal grain. Nahrung 25, 415-421.

Takahashi, H., Yazaki, H., Manabe, M., Matsuura, S., and Kimura, S. (1989). Distribution of aflatoxin, citrinin, and invading fungal mycelium in rice kernels inoculated with Aspergillus flavus and Penicillium citrinum. Cereal Chem. 66, 337-341.

Trantham, A. L., and Wilson, D. M. (1984). Fluorometric screening method for citrinin in corn, barley and peanuts. J. Assoc. Office Anal. Chem. 67, 37-38.

Wilson, D. M., and Abramson, D. (1992). Mycotoxins. In: "Storage of Cereal Grains and Their Products, Fourth Edition" (D. B. Sauer, ed.), pp. 341-391. American Association of Cereal Chemists, St. Paul, MN.

Wilson, D. M., Tabor, W. H., and Trucksess, M. W. (1976). Screening method for the detection of aflatoxin, ochratoxin, zearalenone, penicillic acid and citrinin. J. Assoc. Office Anal. Chem. 59, 125-127.

Zimmerli, B., Dick, R., and Baumann, U. (1989). High-performance liquid chromatographic determination of citrinin in cereals using an acid-buffered silica gel column. J. Chromatog. 462, 406-410.

PART B

MYCOTOXICOSES: FUNGAL PATHOLOGY, TOXICOLOGY AND ECOLOGY

MYCOTOXICOSES: FUNGAL PATHOLOGY, TOXICOLOGY AND ECOLOGY

INTRODUCTION

It is ironic that in mycotoxicology, a field begun because of moldy feed toxicity, much of the research currently conducted on toxigenic molds and mycotoxins has little relationship to animal health and productivity. The divergence between epidemiology and controlled experiments seems to be the basis for this disparity. As a result, the cumulative evidence of the effects of feed-borne mycotoxins on animal productivity and health under field conditions has received less attention than it deserves.

This section is comprised of fifteen research papers. No less than eight of these reports deal directly with mycotoxicoses or diagnostic concepts surrounding mycotoxin-induced problems in animal health and productivity. The first in this category, a grocery store survey for aflatoxin and zearalenone in commercial dog foods, suggests that mycotoxin contamination of pet foods may be minimal. Second is an investigation of "slobbers" outbreaks in several midwestern states which yielded new outbreaks of Rhizoctonia leguminicols that provide both slaframine and swainsonine in copious amounts. Third is a particularly productive paper which examines the utility of feed Fusarium counts in confirming or predicting production problems in poultry under field conditions. Fourth is an important paper which reports identification of at least two unknown toxins in both contaminated feed samples and in cultures of Fusarium moniliforme. Fifth in this series, a shipment of moldy cottonseed was toxic to both dairy and beef heifers; even after extensive analysis, no mycotoxins were identified. Sixth, a significant, dose-related correlation of deoxynivalenol in feeds with decreased milk production in dairy cattle, is reported. This information is very exciting, in light of the general belief that deoxynivalenol is nontoxic in all agricultural species save swine. Seventh, the report on aflatoxin production in high moisture corn silos in Minnesota may be the only documented case of aflatoxin production in grain during storage. This investigation was initiated because of morbidity and loss of productivity in a dairy herd. Finally, a paper on the interaction of aflatoxin with dietary manganese represents nutritional mycotoxicology.

The remaining seven articles cover topics from crop production to fungal allergens. An interesting report describing effects of mycotoxins on spore germination in thermophilic fungi, which are important in recycling of organic materials through decay in nature, is a significant ecological contribution. The practical impossibility of preventing or managing pre-harvest aflatoxin contamination of corn by application of fungicides is the topic of one report. Given the difficulties and importance of identifying Fusarium isolates, the molecular approach to Fusarium taxonomy proposed in one of these papers offers an objective, reproducible identification system with potential application in other fungal genera. Another paper describes an apparently safe and effective method of disposal of aflatoxin-contaminated corn by burial in the soil. Fungal pathogenicity is represented in a report of a method using antibodies from allergic people to identify specific allergens in fungal spores. This approach has application in better understanding hypertensive reactions to fungi and their prevention. Fungal physiology is represented be a paper describing potential methods for optimizing zearalenone fermentations. The last paper in this category reports differential virulence of Aspergillus flavus/ parasiticus isolated from soybeans and corn. This has not been previously reported.

The quality of the papers contributed to this section is impressive and greatly appreciated. Some of the reports may be pivotal in mycotoxicology. The willingness of the scientists represented here, to publish their work via an electronic symposium reflects their pioneering spirit and is also greatly appreciated.

Winston M. Hagler, Jr., Ph.D.
Professor and Director
Mycotoxin Laboratory
Department of Poultry Science
North Carolina State University
Raleigh, NC 27695-3228 USA

SURVEY OF COMMERCIAL DOG FOODS FOR AFLATOXIN B_1 AND ZEARALENONE

JOHN M. CULLEN*, Department of Microbiology, Pathology, and Parasitology, College of Veterinary Medicine, North Carolina State University, Raleigh, NC 27695-8401, USA

WINSTON M. HAGLER, JR., Mycotoxin Laboratory, Department of Poultry Science, College of Agriculture and Life Sciences, North Carolina State University, Raleigh, NC 27695-7608, USA

INTRODUCTION

Mycotoxin contamination of grain, peanut, and soybean crops, while a recurrent problem, varies in severity from year to year due to environmental factors which modulate susceptibility of host plants to fungal pathogens (Council for Agricultural Science and Technology, 1989). Improper storage of crops, including allowing insect outbreaks in storage facilities, also plays an important role in mycotoxin contamination of grain (Council for Agricultural Science and Technology, 1979). Because plant products are used as protein and carbohydrate sources in commercially prepared animal feeds, mycotoxin contamination of prepared pet foods may occur.

Aflatoxin B_1, produced by Aspergillus flavus Link ex. Fries and A. parasiticus Speare, is a hepatotoxic carcinogen commonly found in corn, peanuts, and tree nuts at harvest or after storage (Council for Agricultural Science and Technology, 1979, 1989). Zearalenone, a fungal estrogen produced by certain species of Fusarium, is also a common contaminant of feedstuffs causing reproductive problems (hyperestrogenism) in swine and possibly other livestock and poultry (Christensen, 1979). Sorghum, wheat, soybeans, and soybean meal, all potential ingredients in pet foods, have been reported to be susceptible to contamination with zearalenone and other mycotoxins (Hagler et al., 1984; Hagler et al., 1987; Hagler et al., 1989).

In the 1950's, outbreaks of acute hepatic necrosis and death in dogs were attributed to moldy corn, now believed to have been contaminated with aflatoxin, that was used in the production of commercial dog food (Newberne et al., 1955). More recently, fatal acute aflatoxicosis has been reported in two groups of dogs fed rations contaminated with aflatoxin B_1 (Ligget and Colvin, 1986). While it is well recognized that dogs are sensitive to the acute hepatotoxic effects of aflatoxin (Newberne et al., 1960; Ligget and Colvin, 1986), acute intoxication most likely represents only a small proportion of the health risk posed by aflatoxin ingestion. Regular exposure to low levels of dietary aflatoxin is more often encountered in domestic animals and may be responsible for chronic liver injury in dogs (Newberne and Butler, 1969).

Reproductive losses have been reported in a variety of large animals, but swine are particularly affected when zearalenone concentration in feed is excessive. The effects of zearalenone are not known in dogs, but its estrogenic effects have the potential to cause reproductive problems and hepatic injury.

Acute and chronic canine hepatic disease of unknown etiology continues to be a diagnostic dilemma for veterinary pathologists and clinicians. Although aflatoxicosis is often suggested as a differential diagnosis, for these diagnostic challenges there are little or no data that link aflatoxin exposure to commercial dog food. Consequently, we investigated the aflatoxin content of 100 samples of both brand-name and inexpensive dry dog foods to determine the likelihood that contamination of dog food with aflatoxin and/or zearalenone was a source of liver injury in dogs. The objectives of this work, then, were to examine expensive and inexpensive dry dog food purchased over a period of several months for the presence of aflatoxin and zearalenone and to determine if there was a relationship between purchase price and mycotoxin contamination.

MATERIALS AND METHODS

One hundred samples of commercial dry dog food were purchased from local grocery stores over a period of about six months in 1989. Bags were opened and samples collected in paper bags; samples were refrigerated until analysis. Both aflatoxin and zearalenone were quantitatively determined using AGRI-SCREEN® kits, commercially available enzyme-linked immunosorbent assays

(ELISA) specific for the respective mycotoxins (NEOGEN, Inc., Lansing, MI). Briefly, samples were milled to pass a 20-mesh screen, then extracted in a blender with methanol-water in accordance with recommendation of the test kit manufacturer. The limit of detection for aflatoxin was 5 ng/g and for zearalenone, 100 ng/g. A standard curve was generated using standards provided by NEOGEN, Inc. Samples which contained >20 ng/g of aflatoxin or >100 ng/g of zearalenone were to be confirmed by chromatographic and/or mass spectral techniques.

RESULTS

Analysis of the 100 dog food samples by ELISA revealed that none of the samples contained >20 ng/g of aflatoxin. One of the samples was shown by ELISA to contain 250 ng/g of zearalenone. Zearalenone was confirmed in this sample by the thin layer chromatography technique of Swanson et al. (1984).

DISCUSSION

In spite of the relatively small number of samples and the fact that we sampled during only one year, the data obtained suggested that aflatoxin B_1 and zearalenone are not likely to be encountered in commercial dry dog foods. Consequently, it appears that commercial dry dog foods are not likely to present a risk of hepatic injury or reproductive abnormalities associated with contamination with these mycotoxins. The lack of aflatoxin contamination following 1988, a year marked by drought and aflatoxin contamination of corn in the Midwestern United States, the area that produces much of the corn and wheat used in the formulation of dog food was a welcome result.

SUMMARY

One hundred samples of commercial dry dog foods were purchased from local stores and analyzed by ELISA techniques utilizing antibodies specific for aflatoxin B_1 and zearalenone. No sample contained in excess of 20 ng/g of aflatoxin. One sample was confirmed to contain zearalenone, but the concentration was only 250 ng/g. Despite the limited number of samples examined or the inclusion of only one year in the survey, commercial dry dog foods would appear not to be a source of aflatoxin or zearalenone which could cause idiopathic hepatic injury or reproductive problems in dogs.

Acknowledgments

The authors wish to thank Dr. Richard Dillman for helpful suggestions in the initiation of this study. The technical assistance of Faye Suggs, Jim Brewer, and Kathy Albano is appreciated. This project was supported in part by a grant from SmithKline Beecham. The use of trade names in this publication does not imply approval by North Carolina State University nor criticism of similar ones not mentioned.

REFERENCES

Christensen, C.M. (1979). Zearalenone. In: Conference on Mycotoxins in Animal Feeds and Grains Related to Animal Health. PB-300 300, FDA 031, pp. 1-75 (W. Shimoda, ed.), Bur. Vet. Med., Food and Drug Admin. U.S. Dept. of Commerce, Nat. Tech. Info. Ser., Washington, DC.

Council for Agricultural Science and Technology. (1989). Mycotoxins: Economic and Health Risks. Task Force Report No. 116. CAST, Des Moine, IA.

Hagler, W.M., Jr., Swanson, S.P., Babadoost, M., Haney, C.A., and D.T. Bowman. (1987). Aflatoxin, zearalenone, and deoxynivalenol in North Carolina grain sorghum, 1981-1985. Crop Sci., 27, 1273-1278.

Council for Agricultural Science and Technology. (1979). Aflatoxin and Other Mycotoxins: An Agricultural Perspective. Task Force Report No. 80. CAST, Des Moines, IA.

Hagler, W.M., Jr., Jones, F.T., and Bowman, D.T. (1989). Mycotoxin contamination in North Carolina 1985 crop soybeans. I. Zearelenone and deoxynivalenol in soybeans and soybean meal. In: Biodeterioration Research 2, pp. 337-349 (G.C. Llewellyn and C.E. O'Rear, eds.), Plenum Publishing Co., New York.

Hagler, W.M., Jr., Tyczkowska, K., and Hamilton, P.B. (1984). Simultaneous occurrence of deoxynivalenol, zearalenone, and aflatoxin in 1982 scabby wheat from the midwestern United States. Appl. Environ. Microbiol., 47, 151-154.

Ligget, A.D. and Colvin, B.M. (1986). Canine aflatoxicosis: A continuing problem. Vet. Human Toxicol., 28, 428-430.

Newberne, P.M., Butler, W.H. (1969). Acute and chrocnic effects of aflatoxin on the liver of domestic and laboratory animals: A Review. Cancer Res., 29:236-250.

Newberne, P.M., Bailey, W.S., and Siebold, H.R. (1955). Notes on a recent outbreak and experimental reproduction of hepatitis X in dogs. J. Am. Vet. Med. Assoc., 127, 59-62.

Newberne, P.M., Russo, R., Wogan, G.N. (1960). Acute toxicity of aflatoxin B in the dog. Path. Vet., 3, 331-340.

Swanson, S.P., Corley, R.A., White, D.G., and Buck, W.B. (1984). Rapid thin layer chromatographic method for determination of zearalenone and zearalenol in grains and animal feeds. J. Assoc. Offic. Anal. Chem., 67, 580-582.

SLAFRAMINE AND SWAINSONINE PRODUCTION BY *RHIZOCTONIA LEGUMINICOLA* ISOLATED FROM SIX OUTBREAKS OF SLOBBERS IN CATTLE AND HORSES

LINDA R. DANIEL*, Department of Animal Science, North Carolina State University, Raleigh, NC 27695-7621, USA
WINSTON M. HAGLER, JR., Mycotoxin Laboratory, Department of Poultry Science, North Carolina State University, Raleigh, NC 27695-7636, USA
WARREN J. CROOM, JR., Department of Animal Science, North Carolina State University, Raleigh, NC 27695-7621, USA

INTRODUCTION

O'Dell et al. (1959) in Missouri reported their investigations into "slobbers," a syndrome characterized by slobbering and other toxic symptoms in ruminants and horses consuming red clover (Trifolium pratense L.) and, rarely, other legume forages, particularly alfalfa (Medicago sativa L.). Death of an animal was unusual, but was reported in a few cases. Slobbers (salivary syndrome) in cattle and horses was shown during the late 1960's to be due to the consumption of legume forages infected with a non-spore producing fungus Rhizoctonia leguminicola (Gough and Elliot, 1956). Smalley et al. (1962) first isolated the causative fungus and Gardiner et al. (1968) published the strcuture of an indolizidine alkaloid (1-acetoxy-6-aminooctahydro- indolizine) given the trivial name, slaframine, which reproduced the syndrome in livestock and laboratory animals (Aust et al., 1968) when administered in purified form. Hagler and Behlow (1981) were the first to directly identify the presence of slaframine in samples of red clover hay that caused salivation in horses. Broquist et al. (1985) identified swainsonine (1,2,8-trihydroxyoctahydroindolizine), a structurally related alkaloid, in the same hay sample.

Earlier, R. leguminicola had been shown by Schneider et al. (1983) to produce swainsonine. Swainsonine is also found in certain members of the legume genera Astragalus and Oxytropis and is believed responsible for the "locoism" syndrome in cattle and horses consuming these plants (Broquist, 1985). However, the exact involvement of swainsonine, a potent inhibitor of certain alpha-mannosidases, in either slobbers or locoism remains unclear (Croom et al., 1986). In a previous study (Bartlett et al., 1987), our laboratories compared the ability of five new isolates of R. leguminicola to produce slaframine and swainsonine. The significance of the isolation and screening of new R. leguminicola strains is obvious. Only a relatively few isolates are known to exist and both slaframine and swainsonine have been tentatively identified as having therapeutic value in animal production and in human disease. Slaframine has growth- promoting activity in food animals through up-regulation of digestion and growth hormone release [see reviews by Hagler and Croom (1989) and Croom et al. (1990)]. Additionally, swainsonine has been tentatively identified as an antimetastatic and tumor shrinking agent (Newton et al., 1989). The characterization of isolates producing copious quantities of either of these metabolites is important to their development as drugs.

The objective of this study was to characterize six new fungal isolates of R. leguminicola from forage samples associated with cases of excessive salivation in cattle and horses. Of special interest were microscopic identification of the fungal strains and the testing of their ability to produce slaframine and swainsonine in pure culture.

MATERIALS AND METHODS

Cultures of R. leguminicola were isolated in 1988 and 1989 from six outbreaks of slobbers in cattle: Wisconsin (2 cases), Illinois (2 cases), Idaho (1 case), or in 1 case, horses, in Pennsylvania. Red clover hay was involved in 4 cases and alfalfa hay was involved in cases reported from Illinois and Idaho. Forage samples were examined microscopically for the presence of the characteristic mycelium of R. leguminicola. Rhizoctonia leguminicola strains were isolated by plating forage pieces on 2% water agar in Petri dishes and incubating at 25C for up to 5 days. The isolates of R. leguminicola obtained from the toxic forages and the control strain, R. leguminicola (ATCC 26280), were maintained on 1.5% red clover infusion slants (10%, w/v, second cutting red clover hay in tap water plus 1.5% agar). Isolates were cultured as described by Bartlett et al. (1987) in Roux bottles at ambient temperature (25 ± 3C) on 0.3% yeast extract-supplemented Czapek-Dox broth. Three replicate inoculations were made for each strain and cultured for 14 days. The mats were blotted dry, weighed for wet weight, air-dried at room temperature to a constant weight and dry weights recorded. Mats were ground and subsequently Soxhlet-extracted with ethanol for 24 hr. After extraction the ethanol was removed in vacuo and residue redissolved in deionized water. The aqueous extract was defatted with CH_2Cl_2, adjusted to pH 10 with Na_2CO_3 and slaframine isolated by partitioning into CH_2Cl_2. Swainsonine remained in the aqueous phase. Slaframine was quantitated using gas-liquid chromatography and gravimetric techniques (Bartlett et al., 1987). Retention times of slaframine standards versus samples were recorded and purity was determined by area percent under the SF peak. All slaframine values tabulated were corrected for purity. Swainsonine was quantitated by measuring jack bean alpha-mannosidase inhibition (Tulsiani et al., 1985). Statistical analysis was conducted using an analysis of variance with strain and inoculation batches as main effects (GB-Stat 2.1, 1990, Dynamic Microsystems, Inc., Silver Spring, MD). When overall significance ($P < .05$) was indicated, differences between the new isolates and ATCC 26280 were determined using a protected t test.

RESULTS AND DISCUSSION

Rhizoctonia leguminicola was microscopically identified in all of the forage samples examined and subsequently isolated onto 1.5% water agar. Mat weights and in vitro production of slaframine and swainsonine by these new R. leguminicola isolates and ATCC 26280 are shown in Tables 1 and 2. Wet mat weights varied between 17.0 and 26.9 g with ATCC 26280 yielding significantly heavier ($P < .01$) wet mats than all other strains. Similarly, the range of dry mat weights was 3.26 g to 4.21 g with ATCC 26280 yielding more dry mycelium than other strains. These results are similar to those observed in earlier studies by Bartlett et al. (1987) with other R. leguminicola strains where mat dry weights varied between 1.79 g and 3.27 g.

Table 1. Wet and Dry Mat Weights of Isolates of Rhizoctonia leguminicola.

Strain	Paramater		
	Weight (Mean ± SE[a])		Number
	wet (g)	dry (g)	
ATCC 26280	26.90 ± 0.74	4.21 ± 0.04	50
7852	19.15 ± 0.47[c]	3.67 ± 0.04[c]	49
7543	18.79 ± 0.74[c]	3.26 ± 0.04[c]	60
8040	18.73 ± 0.71[c]	3.27 ± 0.05[c]	52
7691	17.57 ± 0.70[c]	3.32 ± 0.06[c]	53
Wis 88	22.01 ± 0.71[c]	3.20 ± 0.06[c]	63
ILL	19.59 ± 0.86[c]	3.35 ± 0.04[c]	57

[a]Mean ± standard error of the mean.
[b]Replicate inoculation batch effect ($P<.01$).
[c]Significantly different from ATCC ($P<.01$).

There was some variation in slaframine and swainsonine production between inoculation batches for all strains. The average within strain coefficients of variation for slaframine and swainsonine production per gram of mat dry weight were 86% and 36%, respectively. Both slaframine and swainsonine production exhibited significant differences between inoculation batches, whether expressed on whole mat or per gram of mat dry matter basis for ILL and ATCC 2628 (Table 2). Similarly, slaframine purity varied by inoculation batch, although there were no significant differences in purity between strains (Table 2). In the present study, slaframine purity ranged from 39.3% to 68.4% compared to 16.1% to 80.2% in the study of Bartlett et al. (1987). The variation in data from both studies suggests that unidentified environmental and nutritional factors greatly affect the *in vitro* culture of *R. leguminicola* and its production of slaframine and swainsonine.

Slaframine production did not differ significantly between strains with ranges of 0.51 g to 1.4 g and 0.13 g to 0.39 g for mg/mat and mg/g mat, respectively. Swainsonine production ranged from 3.73 to 14.46 mg/mat and from 1.01 to 4.47 mg/g mat (Table 2). Strain ILL produced 154% ($P<.01$) more swainsonine per mat and 222% ($P<.01$) more swainsonine per g mycelium than strain ATCC 26280 (Table 2). There were no statistically significant differences in the slaframine/swainsonine ratio among strains, although they varied between 4.8 and 23.2. The strains isolated and cultured in the present study generally produced less slaframine and more swainsonine than those strains described by Bartlett et al. (1987). It is interesting that the reference strain, ATCC 26280, produced significantly more slaframine and swainsonine in the earlier study (Bartlett et al., 1987; Table 2). The reasons for this are not clear since the two studies were conducted within the same laboratory under virtually identical conditions. To date, ten of the eleven R. leguminicola strains that we have tested (Bartlett et al., 1987; present study) produce both slaframine and swainsonine. One strain, CNC, reported by Bartlett et al. (1987) produced little of either alkaloid and may not be R. leguminicola. In the present study, six new, geographically diverse isolates of R. leguminicola, including two isolated from alfalfa hay, were identified. This work has greatly increased the number of known isolates of this fungal plant pathogen, expanding the known genetic reserves of R. leguminicola as their potential use in animal production and human health continues to be explored.

Table 2. Slaframine and Swainsonine Production by Isolates of *Rhizoctonia leguminicola*.

Strain	Paramater					
	-----Mean ± SE[a]-----					
	Slaframine[b]			Swainsonine[b]		Slaframine/ Swainsonine
	purity (%)[c]	mg/mat[c]	mg/g mat	mg/mat[d]	mg/g mat[d]	
ATCC 26280	39.3± 12.9	0.51± 0.33	0.13± 0.08	5.19± 0.73	1.39± 0.11	23.2± 10.6
7852	68.4± 9.8	1.40± 0.66	0.39± 0.19	3.73± 0.72	1.01± 0.17	4.8± 3.0
7543	50.8± 12.7	1.12± 0.44	0.37± 0.13	6.11± 1.33	2.09± 0.36	9.0± 4.8
8040	41.7± 0.8	0.53± 0.37	0.17± 0.12	8.14± 1.18	2.57± 0.62	46.9± 34.2
7691	52.4± 7.8	0.83± 0.32	0.25± 0.10	10.21± 2.86	3.05± 0.87	18.8± 8.3
Wis 88	37.1± 11.1	0.52± 0.34	0.16± 0.11	7.10± 0.83	2.23± 0.32	47.1± 34.8
ILL	56.5± 23.5	0.80± 0.17	0.25± 0.06	14.46± 3.76[*]	4.47± 1.21[*]	22.3± 11.1

[a]Mean standard error of the mean.
[b]Dry matter basis
[c]Replicate inoculation batch effect (P<.10).
[d]Replicate batch effect (P<.05).
[*]Significantly different from ATCC (P<.01).

SUMMARY

Six new isolates of R. leguminicola were cultured from samples collected during the investigations of the occurrence of slobbers syndrome. The new isolates of R. leguminicola were cultured, with ATCC 26280 as a reference strain, on yeast extract-supplemented Czapek-Dox medium in Roux bottles. Mycelial mats were weighed, then extracted for slaframine and swainsonine determination. Variation was found in slaframine and swainsonine production between batches within all strains cultured One isolate, ILL, produced large amounts of swainsonine compared to all other strains. There were no significant differences in the swainsonine/slaframine ratios between isolate cultures. These data suggest that unidentified environmental factors influence slaframine and swainsonine production by R. leguminicola.

Acknowledgments

The use of trade names in this publication does not imply endorsement by the North Carolina Agr. Res. Serv. of the products named, nor criticism of similar ones not mentioned. The contributions of Faye Suggs, Jim Brewer, and David Crowther are sincerely appreciated.

REFERENCES

Aust, S.D., Broquist, H.P., and Rinehart, K.L., Jr. (1968). Slaframine: A parasympathomimetic from Rhizoctonia leguminicola. Biotech. Bioeng., 10, 403-412.

Bartlett, H.S., Wilson, M.E., Croom, W.J., Jr., Broquist, H.P., and Hagler, W.M., Jr. (1987). Slaframine and swainsonine production by Rhizoctonia leguminicola: Strain comparison, In: Biodeterioration Research 1, pp. 135-148 (G.C. Llewellyn and C.E. O'Rear, eds.), Plenum Publishing Co., New York, NY.

Broquist H.P. (1985). The indolizidine alkaloids, slaframine and swainsonine contaminants in animal forages. Ann. Rev. Nutr., 5, 391-409.

Broquist, H.P., Mason, P.S., Hagler, W.M., Jr., and Harris, T.M. (1985). Identification of swainsonine as a probable contributory mycotoxin in moldy forage mycotoxicosis. Appl. Environ. Microbiol., 48, 386-388.

Croom, W.J., Jr., Hagler, W.M., Jr., Broquist, H.P., Grindem, C., Cullen, J.M., Bartlett, H.S., Rakes, A., Touster, D., and Tulsiani, D. (1986). Effect of chronic administration of swainsonine in lactating dairy cows. Toxin Rev., 5(2), 260.

Croom, W.J., Jr., Froetchel, M.A., and Hagler, W.M., Jr. (1990). Cholinergic regulation of digestive function in ruminants and other domestic livestock: A review. J. An. Sci., 68, 3023-3032.

Gardiner, R.A., Rinehart, K.L., Jr., Synder, J., Broquist, H.P. (1968). Slaframine. Absolute stereochemistry and a revised structure. J. Am. Chem. Soc., 90, 5639-5640.

GBStat 2.1 (1990). Dynamic Microsystems, Inc., Silver Spring, MD.

Gough, F.J., and Elliot, E.S. (1956). Blackpatch of red clover and other legumes caused by Rhizoctonia leguminicola. West Virginia Agr. Exp. Sta. Bull., 387T, 1-23.

Hagler, W.M., Jr. and Behlow, R.F. (1981). Salivary syndrome: Identification of slaframine in red clover hay. Appl. Environ. Microbiol., 42, 1067-1073.

Hagler, W.M., Jr., and Croom, W.J., Jr. (1989). Slaframine: Occurrence, chemistry, and physiological activity, In: Toxicants of Plant Origin, Volume I, Alkaloids, pp. 257-279 (P.R. Cheeke, ed.), CRC Press, Boca Raton, FL.

Newton, S.A., White, S.L., Humphries, M.J., and Olden, K. (1989). Swainsonine inhibition of spontaneous metastasis. J. Nat. Cancer Inst., 81, 1024-1028.

O'Dell, B.L., Regan, W.O., and Beach, T.J. (1959). A study of the toxic principle in red clover. Missouri Agr. Exp. Sta. Res. Bull., 702, 3-12.

Schneider, M.J., Ungemach, F.S., Broquist, H.P., and Harris, T.M. (1983). (1S,2R, 8R, 8AR)-1,2,8-trihydroxyoctahydroindolizine (swainsonine), an alpha-mannosidase inhibitor from Rhizoctonia leguminicola. Tetrahedron, 39, 29-32.

Smalley, E.B., Nichols, R.E., Crump, M.H., and Henning, J.N. (1962). A physiological disturbance in animals resulting from ingestion of Rhizoctonia leguminicola-infested red clover forage. Phytopathology, 52, 753.

EFFICACY OF COLONY FORMING UNIT DATA IN DETECTION OF A FUSAROTOXICOSIS PROBLEM ASSOCIATED WITH POST-MANUFACTURING GROWTH OF *FUSARIUM SPP.* IN POULTRY FEEDS

F. T. JONES* and **M.J. WINELAND**, Department of Poultry Science, North Carolina State University, Box 7608, Raleigh, NC 27695, USA

INTRODUCTION

Although species of Fusarium were pervasive in commercial corn (Kommedahl and Windels, 1981) and many species of Fusarium were known mycotoxin producers (Joffe, 1965), the adverse effects of T-2 toxin (a Fusarium mycotoxin) in poultry were not described under laboratory conditions until the early 1970s (Wyatt et al. 1973, 1975). Since these early descriptions, other laboratory studies have documented the effects of a number of other Fusarium mycotoxins on poultry (Hoerr et al., 1981; Mirocha, 1983; Chu et al., 1988; Walser et al., 1988; Ademoyero and Hamilton, 1989; Marijanovic et al., 1991; Wu et al., 1991). While valuable for their basic physiological and toxicological information, these laboratory studies have been of limited value in the diagnosis of field outbreaks of fusarotoxicosis since the simultaneous occurrence of Fusarium mycotoxins (Ciegler, 1978) and "mycotoxin hot spots" (Shotwell et al., 1975) often produced vague, confusing, or conflicting symptoms. These confusing symptoms along with the need for prompt answers in commercial animal production situations and a lack of suitable field analytical methods for the detection of Fusarium mycotoxins have hampered efforts to study fusarotoxicosis under field conditions. Furthermore, it is clear that the study of field fusarotoxicosis incidents is beset with a set of obstacles entirely different from those encountered in laboratory studies. Thus, techniques or procedures

Biodeterioration Research 4, Edited by G.C. Llewellyn *et al.*, Plenum Press, New York, 1994

which work well in laboratory situations may be entirely inappropriate in field situations.

While the use of precise chemical methods capable of simultaneously detecting numerous Fusarium mycotoxins (Kamimura et al., 1981) would provide direct evidence of fusarotoxicosis field cases, methods such as these are often not available on a timely basis. Moreover, there may be Fusarium metabolites involved in field situations which have not yet been described in the scientific literature. Therefore, less direct methods such as Fusarium colony-forming units (CFU) data may be used in the diagnosis of field fusarotoxicosis incidents. Christensen and Meronuck (1974) report that PCNB medium was originally developed to isolate Fusarium from soil, but is also effective at isolation of Fusarium from feed and grains. The present report documents the use of PCNB medium in the diagnosis of a field fusarotoxicosis situation.

MATERIALS AND METHODS

Situation Background

The company which cooperated with this study was an typical broiler production complex which consisted of breeder flocks, broiler flocks, a feed mill, a hatchery and a processing plant. The company was in the process of evaluating efforts to control molds and mycotoxins when it was noticed that compared to the previous year, egg production rates were similar but, hatchability was slightly depressed company wide. It was observed that Fusarium CFU data obtained from the previous year's corn crop were over 10-fold lower than those obtained from the present year's crop (Table 1). Thus, it was suggested that fungi of the genus Fusarium and/or fusarotoxicosis were involved in the situation. It was also noted that the hatchability was

Table 1. Fusarium CFU obtained from Corn of Previous and Present Harvest Years.

Harvest Year	No. of Assays	Ave. FUSARIUM CFU/g
Previous	15	25,803.3
Present	16	392,643.7

severely more depressed in some flocks than in other flocks. Continued observation of symptoms permitted the classification of flocks as either "normal" or "affected". While symptoms were varied in affected flocks, dark green to black colored droppings and mouth lesions were similar to those observed in other fusarotoxicosis cases (Hamilton, 1989). Feed consumption was also severely depressed. In view of these observations, 4 normal and 4 affected flocks of approximately the same age and size were chosen for study. Composite feed samples were collected weekly for 9 weeks from at least 10 feeder pans in each of the 8 flocks involved.

Laboratory Procedures

Fusarium counts (CFU) were done in duplicate by aseptically weighing one 10-g portion sample into each of two sterile blender jars. The blender jars were then placed on the blender base and 90 ml of sterile 0.85% saline with 0.005% Triton X-100 (Fisher Scientific, Atlanta, GA 30091) was poured over the sample just prior to blending at 8000 rpm for 1 min. Decimal dilutions were done in sterile saline before plating on PCNB medium which was made according to the directions which follow. Plates were incubated at room temperature (ca. 24-26 C) for 120±4 hrs. The following medium ingredients were autoclaved at 121 C for 15 min:

	g/l
Bacti-Peptone (Gibco Laboratories, Madison, WI)	15.0
KH_2PO_4	1.0
$MgSO_4$	0.24
Agar	16.0
Bacti-Bile (Gibco Laboratories, Madison, WI)	0.5
Chloramphenicol	0.04

Autoclaved ingredients were cooled to ca. 50-55 C and 0.5 g/l pentachloronitrobenzene (PCNB, Terrachlor, Uniroyal Chemical Co. Inc., Middlebury, CT 06749) as added to the medium prior to pouring plates.

Feed moistures (oven dry weights) were done according to the procedure described by Jones et al. (1982). Egg production and hatchability records were obtained from the poultry company involved.

Detection of Deoxynivalenol (DON)

Determination of DON was done by the N.C. State University Mycotoxin Laboratory using the extraction and purification method of Trenholm et al.

(1985) coupled with packed-column gas chromatography separation and electron-capture detection of trimethylsilyl derivatives produced by the method of Rizzo et al. (1986).

RESULTS

Data collected from samples collected from the feed mill suggest that Fusarium growth and production of DON were occurring in dust within the corn, soybean meal, and pellet supply storage tanks (Table 2). Due to the design of the feed manufacturing equipment, it is almost certain that Fusarium CFU and DON were transmitted to feeds. Data collected from the feed mill along with data obtained from present year corn samples (Table 1) suggest that Fusarium counts obtained from present year feeds were higher than those of feeds manufactured the previous year. Yet when viewed on a company wide basis, these enhanced counts had no apparent effect on egg production and little effect on hatchability (data not shown). Clearly, these data suggest that additional fungal growth or mycotoxin production was required to produce measurable effects on hatchability.

Table 2. Fusarium CFU, Moisture and Deoxynivalenol (DON) Analyses Obtained Samples Collected at the Feed Mill.

Sample Description	Moisture %	FUSARIUM CFU/g	DON ng/g
Dust - Soy Tank 1	13.46	270,000	77
Dust - Soy Tank 2	15.19	103,000	271
Dust - Corn Tank 1	31.44	310,000,000	242
Dust - Corn Tank 2	32.20	258,000,000	130
Dust - Pellet Supply Tank	19.14	10,000,000	77
Unpelleted (Mash) Feed	12.35	71,000	150
Pelleted Feed	13.49	000	96

During the period observed, egg production was not reduced in affected flocks (Figure 1), but the hatchability of eggs from affected flocks was 1 to 5% lower than that of eggs from normal flocks (Figure 2). Fusarium CFU obtained from normal flocks averaged over two-fold lower than those obtained from affected flocks, while the few DON samples analyzed averaged over three-fold higher in affected flocks (Table 3). These observations suggest that decreased hatchability of eggs from affected flocks was related to Fusarium growth in feeds. Interestingly, the elevated DON levels found in feed from affected flocks were found the week following the detection of elevated Fusarium CFU. While the data are incomplete, the fact that one of the four affected flocks refused consumption of affected feeds strongly suggests the involvement of DON and possibly other Fusarium mycotoxins in hatchability losses.

It was observed that in comparison to normal farms the affected farms had: 1) more caked or clumped feed in the feed handling system, 2) damper poultry house litter conditions and 3) larger feed storage tanks. The following scenario may have occurred: Fusarium CFU increased through the feed handling system, additional moisture was obtained from damp environmental conditions, and feeds remained on the affected farms longer. Since feed system sanitation, feed moisture, and feed residence time are all linked to enhance fungal activity (Hamilton, 1975; Jones et al. 1982; Good and Hamilton, 1981), these data suggest that Fusarium growth and mycotoxin production occurred on affected farms.

Table 3. Fusarium CFU and Deoxynivalenol (DON) Analyses Obtained from Samples Collected from Normal and Affected Poultry Flocks.

	Normal Flocks			Affected Flocks		
Week of Yr	No.[b]	FUSARIUM CFU/g	DON[a] ng/g	No.	FUSARIUM CFU/g	DON ng/g
9	4	39,925		4	48,000	
10	4	40,000		4	287,000	
11	4	42,000	173	4	70,000	750[c]
12	4	27,733		4	33,500	
13	4	33,067		4	260,000	
14	4	72,000	185	4	132,500	464[c]
15	4	115,075		4	136,750	
16	4	31,000		4	123,000	
17	4	36,500		4	88,000	
Ave.	36	48,589	179	36	130,972	607

[a] One composite sample of feed was analyzed to obtain the deoxynivalenol (DON) values listed.
[b] Number of composite samples (one per farm).
[c] One flock refused to eat these feeds.

DISCUSSION

The data presented here are clearly similar to those of Smith and Hamilton (1970) and Jones et al. (1982) who found that the percentage of feed samples contaminated with aflatoxin increased following manufacture. Apparently there are niches within the average farm feed delivery system which contain the conditions required by Fusarium and other fungi for growth. Therefore, the following steps outlined by Hamilton (1975), Good and Hamilton (1981) and Jones and Hamilton (1986) should be taken to prevent the growth of fungi in farm feeder systems.

While CFU have been called circumstantial evidence for the existence of a Fusarium problem (Ciegler, 1978), CFU were in this case vital evidence that the problem was related to Fusarium mycotoxins. Furthermore, since the toxicology, metabolism, and detection of Fusarium mycotoxins are not yet clearly understood (Chu et al., 1988; Walser et al., 1988; Ademoyero and Hamilton, 1989; Hamilton, 1989; Marijanovic et al., 1991; Wu et al., 1991), in field cases of suspected fusarotoxicosis, CFU may be a more effective diagnostic tool than mycotoxin analyses. However, CFU must be correctly employed to obtain accurate information. Based upon the experience obtained from this and other field applications, the following suggestions are advanced as to the correct use of CFU in the diagnosis of field fusarotoxicosis cases:

The correct application of the CFU procedure is heavily dependant on analyst technique. Therefore, CFU data obtained by persons without proper training are of questionable value.

Even when analysts are correctly trained, scores of minute details can effect the results obtained. Thus, decisions should never be based upon a single CFU determination. Counts should always be done at least in duplicate.

Since mycotoxins are known to occur in "hot spots" (Shotwell et al., 1975), results from single feed samples can be deceptive. If possible, multiple feed samples should be run. If only single feed samples are possible, prepare a composite sample by collection of sub-samples from at least five locations within the lot. Thoroughly mix composite samples prior to any determination.

Since CFU determinations enumerate both toxigenic and non-toxigenic Fusarium isolates, counts in the absence of detectable clinical symptoms are of doubtful value.

CFU data can be a link in the "chain of evidence" leading from the affected

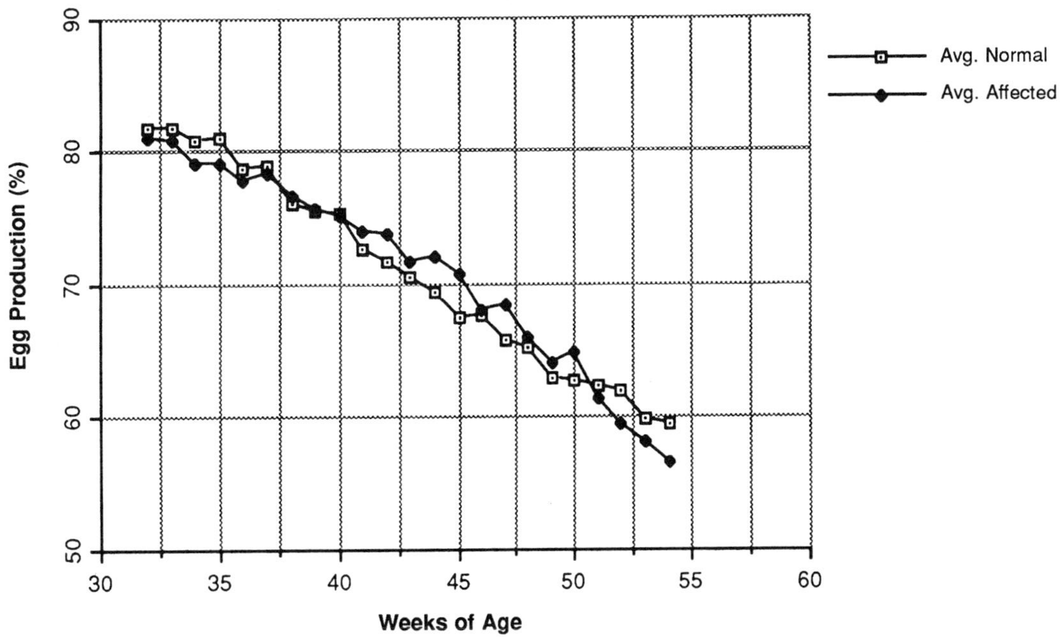

Figure 1. Egg Production in Normal and Affected Broiler Flocks.

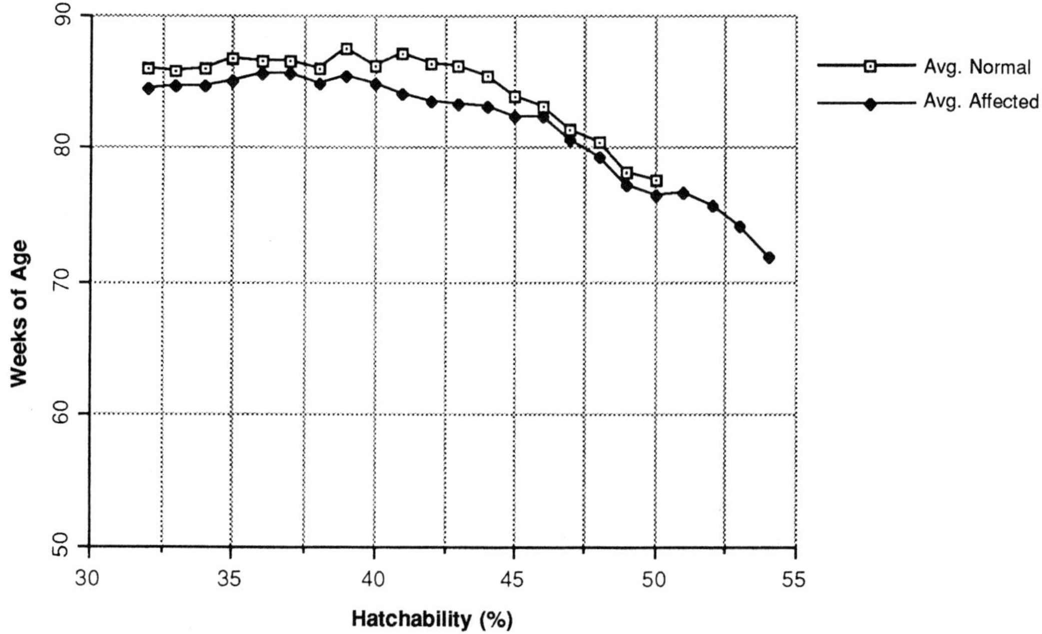

Figure 2. Hatchability of Eggs from Normal and Affected Broiler Breeder Flocks.

animal to the mycotoxin involved. However, since Fusarium spp. are destroyed by heat (Table 2) the "chain of evidence" may not apply if the feed involved has been heat-treated (e.g. pelleted or extruded).

While threshold levels can be difficult and dangerous recommendations, the following guidelines have been used with some success in suspected field fusarotoxicosis situations. Feed samples not subject to heat treatment should be <75,000 CFU/g, while corn or feed ingredient samples should be <100,000 CFU/g.

SUMMARY

Fusarium counts (colony forming units, CFU) were used to identify a field fusarotoxicosis outbreak in the broiler breeder flocks of one broiler company. Although CFU data suggested that Fusarium growth was occurring in dust within the corn, soybean meal, and pellet supply storage tanks, enhanced counts had no apparent effect on company average egg production and little effect on company average hatchability. Yet certain flocks were severely affected while other flocks were not influenced. Thus, feed samples, egg production and hatchability data were collected weekly for 9 weeks from four "normal" flocks and four "affected" flocks. While egg production was not reduced in affected flocks, the hatchability of eggs from affected flocks was 1 to 5% lower than that of eggs from normal flocks. Fusarium CFU obtained from normal flocks averaged over two-fold lower than those obtained from affected flocks. These observations suggest that decreased hatchability of eggs from affected flocks was related to Fusarium growth in feeds. While CFU have been called circumstantial evidence for the existence of a Fusarium problem, CFU were in this case vital evidence that the problem was related to Fusarium mycotoxins.

REFERENCES

Ademoyero, A.A. and Hamilton, P.B. (1989). Research note: Influence of degree of acetylation of scirpenol mycotoxins on feed refusal by chickens. Poultry Sci., 68, 854-856.

Christensen, C.M. and Meronuck, R.A. (1974). Manual of fungi in feeds, foods and cereal grains. Agric. Ext. Serv. Pub., Univ. Minn., St. Paul, MN.

Chu, Q., Cook, M.E., Wu, W., and Smalley, E.B. (1988). Immune and bone properties of chicks consuming corn contaminated with Fusarium that induces dyschondroplasia. Avian Dis., 32, 132-136.

Ciegler, A. (1978). Trichothecenes: Occurrence and toxicoses. J. Food Prot., 41, 399-403.

Good, R.E. and Hamilton, P.B. (1981). Beneficial effect of reducing feed residence time in a field problem of suspected moldy feed. Poultry Sci., 60, 1403-1405.

Hamilton, P.B. (1975). Proof of mycotoxicosis being a field problem and a simple method for their control. Poultry Sci., 54, 1706-1708.

Hamilton, P.B. (1989). High Fusarium corn: A new problem for the poultry industry. Zootecnica, 6, 44-47.

Hoerr, F.J., Carlton, W.W. and Yagen, B. (1981). Mycotoxicosis caused by a single dose of T-2 toxin or diacetoxyscirpenol in broiler chickens. Vet. Pathol., 18, 652-664.

Joffe, A.Z. (1965). Toxin production by cereal fungi causing toxic alimentary aleukia in man. In: Mycotoxins in Foodstuffs. pp. 77-85 (G.N. Wogan, ed.), MIT Press, Cambridge MA.

Jones, F.T., Hagler, W.M. and Hamilton, P.B. (1982). Association of low levels of aflatoxin in feed with productivity losses in commercial broiler operations. Poultry Sci., 61, 861-868.

Jones, F.T. and Hamilton, P.B. (1986). Factors influencing fungal activity in low moisture poultry feeds. Poultry Sci., 65, 1522-1525.

Kamimura, H., Nishijima, M., Yasuda, K., Saito, K, Ibe, A., Nagayama, T., Ushiyama, H., and Naoi, Y. (1981). Simultaneous detection of several Fusarium mycotoxins in cereals, grains and foodstuffs. J. Assoc. Off. Anal. Chem., 64, 1067-1073.

Kommedahl, T. and Windels, C.E. (1981). Root-, stalk-, and ear- infecting Fusarium species on corn in the USA. In: FUSARIUM - Diseases, Biology and Taxonomy. pp. 94-103 (P.E. Nelson, T.A. Toussoun and R.J. Cook, eds.), The Pennsylvania State University Press, University Park, PA.

Marijanovic, D.R., Holt, P., Norred, W.P., Bacon, C.W., Voss, K.A., and Stancel, P.C. (1991). Immunosuppressive effects of Fusarium moniliforme cultures in chickens. Poultry Sci., Y70, 1895-1901.

Mirocha, C.J. (1983). Effect of trichothecene mycotoxins on farm animals. In: Trichothecenes - Chemical, Biological and Toxicological Aspects. pp. 177-194 (Y. Ueno, ed.), Elsevier, NY.

Rizzo, A.F., Saari, L., and Lindfurs, E. (1986). Derivatization of trichothecenes and water treatment of their trimethylsilyl ethers in an anhydrous apolar solvent. J. Chromatogr., 368, 381-386.

Shotwell, O.L., Goulden, M.L., Bothast, R.J., and Hesseltine, C.W. (1975). Mycotoxins in hot spots in grains. I. Aflatoxin and zearalenone occurrence in stored corn. Cereal Chem., 52, 687-697.

Smith, J.W. and Hamilton, P.B. (1970). Aflatoxicosis in the broiler chicken. Poultry Sci., 49, 207-215.

Trenholm, H.L., Warner, R.M., and Prelusky, D.B. (1985). Assessment of extraction procedures in the analysis of naturally contaminated grain products for deoxynivalenol. J. Assoc. Offic. Anal. Chem., 68, 645-649.

Walser, M.M., Morris, V.C., and Levander, O.A. (1988). Effect of selenium on the development of Fusarium-induced tibial dyschondroplasia in broiler chickens. Avian Dis., 32, 84-88.

Wu, W., Cook, M.E., and Smalley, E.B. (1991). Decreased immune response and increased incidence of tibial dyschondroplasia caused by growth of Fusaria on sterile corn. Poultry Sci., 70, 293-301.

Wyatt, R.D., Doerr, J.A., Hamilton, P.B., and Burmeister, H.R. (1975). Egg production, shell thickness and other physiological parameters of laying hens affected by T-2 toxin. Appl. Microbiol., 29, 641-645.

Wyatt, R.D., Hamilton, P.B. and Burmeister, H.R. (1973). The effects of T-2 toxin in broiler chickens. Poultry Sci., 52, 1853-1859.

STUDIES ON *FUSARIUM MONILIFORME*: CAUSAL AGENT OF EQUINE LEUCOENCEPHALOMALACIA

CECILE E. F. LE GRENADE, Division of Microbiology, Food and Drug Administration, Washington, D.C. 20204, USA
GEORGE A. BEAN*, Department of Botany, University of Maryland, College Park, MD 20742, USA

INTRODUCTION

For almost a hundred years, a fatal neurological disease affecting horses, donkeys and mules has been occurring (Butler, 1902). The disease, called equine leucoencephalomalacia (ELEM), is characterized by the presence of liquefactive necrotic lesions primarily in the white matter of the cerebral hemispheres of the brain of affected horses (Haliburton et al., 1979). Feeding studies have shown that mold-infected corn or corn fodder is a constant factor associated with ELEM outbreaks.

In 1971, several species of fungi were recovered from corn implicated in a naturally occurring outbreak of ELEM in Egypt (Wilson and Maronpot, 1971). Isolates of the fungus were grown on autoclaved corn and either fed or orally administered to donkeys. Only corn inoculated with *Fusarium moniliforme* produced characteristic ELEM clinical signs and lesions. Since this first pure-culture feeding experiment with *F. moniliforme*, there have been at least three additional reports of successful reproduction of ELEM in equids by feeding or orally dosing animals with corn contaminated with *F. moniliforme* (Haliburton et al., 1979; Kellerman et al., 1972; Marasas et al., 1976).

During the months of September 1986 to April 1987, an estimated 15 horses died in Maryland with symptoms of ELEM. Two of the horses were necropsied by the Maryland State Veterinary Laboratory. The first horse had

Biodeterioration Research 4, Edited by G.C. Llewellyn
et al., Plenum Press, New York, 1994

been fed a 10% protein horse feed containing corn. Necropsy revealed severe cavitation of the brain white matter whereas the liver showed mild to moderate atrophy. Pathogenic bacteria could not be isolated from lungs, spleen, liver, kidney or intestine and it was concluded that the cause of death was "moldy corn poisoning." The second horse necropsied exhibited erratic violence, loss of senses, muscle twitching and pronounced weakness prior to death. Its diet had been whole corn, hay and vitamins. Necropsy revealed multifocal areas of malacia of the cerebral hemisphere white matter indicating ELEM.

Fusarium moniliforme, believed responsible for ELEM, is one of the most prevalent fungi associated with maize intended for human and animal consumption (Booth, 1971; Marasas et al., 1979a; Trofa and Bean, 1987). Isolates of F. moniliforme are known to produce a number of compounds including fusariocin A, moniliformin, malonic acid, benzoxanthentrione pigments, fusaric acid, gibberellins, kaurane diterpeniods, phenolic derivatives, zearalenone, fusaric acid, fusarin C and most recently fumonisins. Fumonisins were produced by an isolate of F. moniliforme obtained from corn intended for human consumption in the Transkei district of Southern Africa where a high incidence of human oesophageal cancer occurs (Marasas et al., 1979b). The same isolate was shown to cause ELEM in horses (Kriek et al., 1981), and was carcinogenic to rats (Marasas et al., 1983). The structure of the fumonisin compounds has recently been elucidated (Bezudenhout et al., 1988). Marasas et al. (1988) demonstrated that both culture material and fumonisin produced by this fungus, when fed to South African horses caused ELEM. Voss et al. (1989) reported the isolation of 2 fumonisin compounds designated B_1 and B_2 as well as zearalenone at levels of 300 and 800 ng/g fumonosin, respectively, from corn samples associated with incidents of ELEM in Illinois.

The objectives of this study were: to determine the level of F. moniliforme in corn feed rations associated with horses that had died of ELEM in Maryland; to examine feed samples for the presence of toxic metabolites; and to determine toxicity of F. moniliforme isolated from feed samples. Note that a number of these studies were done prior to the availability of fumonisin standards.

MATERIALS AND METHODS

Analysis of Feed Samples

Eleven feed samples (7 corn and 4 sweet feed) associated with an outbreak of ELEM were analyzed for fungi and mycotoxins produced by F. moniliforme. Corn kernels were disinfected with 0.525% NaOCl and plated on potato dextrose agar plus NaCl to equal 7.5%. After incubation at 27 ± 2 C for 10-14 days fungi present were identified and quantitated. Single spore isolates of F. moniliforme were maintained on a sterile culture medium of 50:50 sandy loam soil at 4-6 C for longterm storage.

The toxins and the analytical methods used for their identification were fusarin C (Farber and Sanders, 1986) and moniliformin (Vesonder, 1986). The limits of detection of these toxins by thin layer chromatography (TLC) were 10 μg/g and 8 μg/g respectively. The samples were also analyzed by the multi-mycotoxin screening method of Roberts and Patterson (1979). Feed samples stored at -20 C for 18 months, were analyzed for fumonisin (Marasas et al., 1988), by a methanol/water extraction procedure.

Toxicity of Feed Samples

To test the feed samples for toxicity two bioassay methods were used, cell line cytotoxicity and a rat bioassay. The cell lines were human epithelial cells (Hep-2) and baby hamster kidney (BHK) cells. The growth medium for BHK and Hep-2 cells was a minimum essential medium (MEM) with Earl's salts, supplemented with 10% tryptose phosphate broth and 10% inactivated horse serum (Gibco Laboratories, NY). An antibiotic solution, penicillin:streptomycin (0.02%) was also added. A 1.0-ml volume of cell suspension (1×10^5 cells per ml) was added to 20 ml of the culture medium in a 75-ml tissue culture flask which was placed in a CO_2 incubator at 37 C until a monolayer formed (4-7 days). One, 2, 3, or 4 ml of feed or inoculated corn extracts dissolved in methanol was added to each cell. A toxic standard (T-2 toxin) and a methanol control were included. After incubation for 24 hr in a CO_2 incubator at 37 C, the cultures were stained with Giemsa (10%) and the level of cell death estimated using a 0-4 scoring system.

The toxicity of inoculated corn was determined using a rat bioassay developed at the University of Minnesota (C. Mirocha). A diet consisting of 1 part Purina rat chow and 1 part ground F. moniliforme inoculated corn was fed to Sprague-Dawley rats according to the procedure established by Abbas et al. (1984). Autoclaved non-inoculated corn was used as the control in these bioassays. Periodically the rats were observed for mortality and post mortem examinations performed.

Toxin Production by F. MONILIFORME

A modified method of Abbas and Mirocha (1984), was used to produce inoculum of F. moniliforme; two hundred grams of whole kernel shelled corn was placed into a 2800-ml Fernbach flask and 175 ml of water added. The flasks were allowed to sit for 6 to 8 hr. The flasks were shaken, stoppered with cotton plugs and autoclaved for 1 hour and after a 24-hr period autoclaved again. The sterile corn was then inoculated with Fusarium isolates. Flasks were incubated for 2 weeks at 27±2 C followed by 2 weeks at 4-10 C. The corn was removed from the flasks, dried in a ventilated hood for approximately 3 days, and stored in brown bags at room temperature prior to extraction for toxin analyses.

RESULTS

Mycoflora of Corn Associated with ELEM

Table 1 lists the fungi identified from surface-disinfected corn kernels obtained from 7 feed samples associated with ELEM in Maryland. The predominant fungal species isolated was F. moniliforme which agrees with previous survey results (Trofa and Bean, 1987).

Table 1. Fungal species from surface disinfected corn kernels.

Fungal Species	Sample Number						
	45	48	67	73	102	72	85
Fusarium moniliforme	90a	80	60	80	93	87	95
Aspergillus flavus	-	-	-	-	2	1	-
A. niger	-	2	-	3	-	1	-
A. glaucus	-	6	-	4	-	3	-
Penicillium citrinum	1	-	6	4	-	-	-
P. oxalicum	-	-	-	-	1	-	-
Rhizopus sp.	3	2	4	-	-	1	1
Mucor sp.	-	-	1	-	-	-	-
Trichoderma sp.	-	-	1	-	-	-	-

[a] Percentage of colonies from 100 kernels.

FUSARIUM Toxins Present in Feed Samples

Eleven feed samples from the Maryland outbreak of ELEM were examined for the presence of moniliformin and fusarin C. The results were negative for the presence of the two mycotoxins using the methods of Faber and Sanders (1986) and Vesonder (1986). The samples analysed by the Roberts and Patterson

(1979) extraction method for the presence of patulin, zearalenone, ochratoxin, citrinin, T-2 toxin and diacetoxyscirpenol were also negative.

Extracts from the 11 feed samples were next tested for their cytotoxicity to cell lines. The extracts were spotted on TLC plates and after development and spraying with p-anisaldehyde, blue spots at Rf 0.85-0.9 and 0.8-0.85 occurred which did not correspond to any of the mycotoxin standards added. The spots were removed, redissolved in methanol and added to Hep-2 and BHK cell lines. Table 2 summarizes the results of this experiment. Two TLC spots, between Rf's 0.6-0.8 from feed sample number 56; i.e., 6 and 8 and two spots from feed sample 57; i.e., 7 and 8 at Rf's between 0.7 and 0.8 were toxic to both cell lines.

Table 2. Cytotoxicity of Roberts/Patterson feed extracts.

Fraction	Hep-2 Vol[a] ml	Score[b]	BHK Vol ml	Score
56 - 6	4	1	4	1
56 - 8	4	2	4	1
57 - 7	4	2	4	3
57 - 8	4	1	4	1

[a] Concentration of extracts 2 mg/ml; volumes added to 0.2 ml cell culture media resulting in a concentration of 5, 10, 15 and 20 mg/ml.
[b] O = no cell death, 4 = complete cell death.

Frozen feed samples (approximately 18 months at -20 C) were analyzed for fumonisin using a methanol/water extraction followed by clean up on SepPak C18 cartridges. The extracts plus a fumonisin standard were spotted on reverse phase silica gel plates and after development were sprayed with p-anisaldehyde. Only 2 feed extracts were analyzed and both contained fumonisin, ie; 61 and 36 mg/g of feed. Identity of fumonisin was confirmed by GC/MS spectrometry by R. Plattner, USDA, Peoria, IL.

Toxicity of F. MONILIFORME Isolates

Two isolates of F. moniliforme, isolated at random from the feed samples were selected for further study. The fungus was grown on sterilized corn which was fed to Sprague-Dawley rats. No death occurred in rats fed the control diet which was a complete rat diet and uninoculated corn whereas 2 of 3 and 3 of

3 rats died when fed a diet containing F. moniliforme corn culture. The fermented corn samples were analyzed for fusarin C, moniliformin, zearalenone, trichothecenes, wortmannin (H-1), and fusarochromanone (TDP-1) using the methods of Mirocha et al. (1990). The results were negative, none of these mycotoxins could be detected in the inoculated corn. We next analyzed the 2 samples for fumonisin using the methanol/water extraction method developed by R. Vesonder, USDA, Peoria, IL. Corn inoculated with the 2 isolates of F. moniliforme contained 55 and 140 mg/g fumonisin based on GC/MS analysis of the extracts. Toxicity of the same F. moniliforme inoculated corn extracts was verified using the cell line bioassay. Both the methanol and the ethyl acetate extracts of the 2 corn inoculum samples were toxic to the cell lines. The cytotoxic methanol extract was confirmed to contain fumonisin whereas we were unable to identify the cytotoxic compound(s) in the ethyl acetate extracts.

DISCUSSION

The death of horses in Maryland from ELEM prompted this investigation to determine the fungi and/or their toxic metabolites responsible. Surface disinfected corn kernels from feed samples associated with ELEM contained F. moniliforme ranging from 60 to 90%. This concurs with the findings of other authors that F. moniliforme is the predominant fungus found in corn.
Analysis of the feed samples for mycotoxins that may be responsible for ELEM was not conclusive. For example, fumonisin was detected using a methanol/water extraction procedure; however, when 2 cell lines were exposed to the same feed samples extracts, at least 2 toxic fractions were present which did not correspond to any of the Fusarium toxin standards.

However, when sterilized corn was inoculated with isolates of F. moniliforme, the involvement of fumonisin in toxicity of the fungal isolates was more obvious. When inoculated corn was analyzed, fumonisin was present and toxic to the 2 cell lines used in this study. We suggest that naturally contaminated feed samples should be included in any fumonisin toxicity studies since F. moniliforme is rarely, if ever, the sole fungus found in corn. Our studies suggest that toxic compounds other than fumonisin do occur in contaminated corn, and until these compounds and their origin are identified, the fumonisin/ELEM complex can not be fully understood.

SUMMARY

An estimated 15 horses died in Maryland in 1986-87 after showing symptoms of equine leucoencephalomalacia (ELEM). The level of _Fusarium moniliforme_ in surface disinfected corn kernels from feed associated with ELEM ranged from 60-97% of the total fungi present. Feed samples were extracted and analyzed for moniliformin, fusarin C, zearalenone, and fumonisin B_1 by thin layer chromatography and gas chromatography/mass spectrometry. Only fumonisin B_1 was detected at levels of 4, 36, and 61 mg/g in 3 feed samples examined. _F. moniliforme_ isolated from feed samples and grown on corn was toxic to Sprague-Dawley rats and 2 cell lines, human epithelial cells (Hep-2) and baby hamster kidney cells (BHK). Methanol/water extracts of corn samples inoculated with 2 _F. moniliforme_ isolates contained 55 and 110 mg/g of fumonisin B_1. When grown on a chemically defined liquid medium, _F. moniliforme_ produced fusarin C and other unidentified, but cytotoxic fractions. We conclude that fumonisin was responsible for ELEM in Maryland but other toxic fractions produced by _F. moniliforme_ could be involved as well.

REFERENCES

Abbas, H.K. and Mirocha, C.J. (1984). Sensitivity of cultured human and mouse fibroplasts to trichothecenes. _J. Assoc. Off. Anal. Chem._, 67(3), 607-610.

Bezuidenhout, S.C., Gelderblom, W.C.A., Gorst-Allman, C.P., Horak, R.M., Marasas, W.F.O., Spiteller, G., and Vleggaar, R. (1988). Structure elucidation of the fumonisins, mycotoxins from _Fusarium moniliforme_. J _Chem. Soc. Chem. Commun._, 743-745.

Booth, C. (1971). _The Genus Fusarium_. Commonwealth Mycol. Inst., Kew, Surrey, England, 237 pp.

Butler, T. (1902). Notes on a feeding experiment to produce leucoencephalomalacia in a horse with positive results. _Amer. Vet. Rev._, 26, 748-751.

Farber, J.M. and Sanders, G.W. (1986). Fusarin C production by North American isolates of _Fusarium moniliforme_. _Appl. Environ. Microbiol._, 51(2), 381-384.

Haliburton, J.C., Vesonder, R.F., Lock, T.F., and Buck, W.B. (1979). Equine leucoencephalomalacia (ELEM). Study of Fusarium moniliforme as an etiologic agent. Vet. Human Toxicol., 21, 348-351.

Kellerman, T.S., Marasas, W.F.O., Pienaar, J.G., and Naude, T.W. (1972). A mycotoxicosis of equidae caused by Fusarium moniliforme Sheldon. A Preliminary Communication. Onderstepoort J. Vet. Res., 39, 305-308.

Kriek, N.P.J., Kellerman, T.S., and Marasas, W.F.O. (1981). A comparative study of the toxicity of Fusarium verticillioides (= F. moniliforme) to horses, primates, pigs, sheep and rats. Onderstepoort J. Vet. Res., 48, 129-131.

Marasas, W.F.O., Kellerman, T.S., Pienaar, J.G., and Naude, T.W. (1976). Leucoencephalomalacia: a mycotoxicosis of equidae caused by Fusarium moniliforme Sheldon. Onderstepoort J. Vet. Res., 43, 113-122.

Marasas, W.F.O., Kriek, N.P.J., Wiggins, V.M., Steyn, P.S., Towers, D.K., and Hastie, T.J. (1979a). Incidence, geographic distribution, and oxigenicity of Fusarium species in Southern African corn. Phytopathology, 69, 1181-1185.

Marasas, W.F.O., van Rensburg, S.J., and Mirocha, C.J. (1979b). Incidence of Fusarium species and the mycotoxins, deoxynivalenol and zearalenone, in corn produced in esophageal cancer areas in Transkei. J. Agric. Food Chem., 27, 1108-1112.

Marasas, W.F.O., Nelson, P.E., and Toussoun, T.A. (1983). Toxigenic Fusarium species: identity and mycotoxicology. Pennsylvania State University Press, University Park, PA. 220 pp.

Marasas, W.F.O., Kellerman, T.S., Gelderblom, W.C.A., Coeizer, J.A.W., Thiel, P.G., van der Lugt, J.J. (1988). Leucoencephalomalacia in a horse induced by fumonisin B1 isolated from Fusarium moniliforme. Onderstepoort J. Vet. Res., 55, 197-203.

Mirocha, C.J., Abbas, H.K., and Vesonder, R.F. (1990). Absence of trichothecenes in toxigenic isolates of Fusarium moniliforme. Appl. Environ. Microbiol., 56(2), 520-525.

Roberts, D.S.P. and Patterson, B.A. (1979). Mycotoxins in animal feedstuffs: sensitive thin layer chromatographic detection of aflatoxin, ochratoxin A, sterigmatocystin, zearalenone and T-2 toxin. J. Assoc. Offic. Anal. Chem., 62(6), 1265-1267.

Trofa, S. and Bean, G.A. (1987). The level of toxigenic fungi in 1985 Marylad corn before and during storage. In: Biodeterioration Research 1, pp. 119-126. (G.C. Llewellyn and C.E. O'Rear, eds.) Plenum Publishing Co., NY.

Voss, K.A., Norred, W.P., Plattner, R.D., and Bacon, C.W. (1989). Hepatotoxicity and renal toxicity in rats of corn samples associated with field cases of equine leukoencephalomalacia. Food Chem. Toxicol., 27(2), 89-96.

Vesonder, R.F. (1986). Moniliformin produced by cultures of Fusarium moniliforme var. subglutinans isolated from swine feed. Mycopathologia, 95, 149-153.

Wilson, B.J. and Maronpot, R.R. (1971). Causative fungus agent of leucoencephalomalacia in equine animals. Vet. Res., 88, 484-486.

PRODUCTION OF AFLATOXIN IN A HIGH MOISTURE MAIZE SILO IN MINNESOTA

RICHARD A. MERONUCK*, Department of Plant Pathology, College of Agriculture, University of Minnesota, St. Paul, MN 55108, USA
STEVEN P. SWANSON, Department of Drug Metabolism, Eli Lilly and Co., Indianapolis, IN 46285, USA

INTRODUCTION

Mycotoxicosis resembling aflatoxicosis in dairy cattle has been known for some time. Sipple et al. (1953) reported that cows eating moldy maize rations were depressed, off-feed, and had diarrhea containing blood. A necropsy of one of the afflicted cows revealed an enlarged liver spotted with light yellow areas. Loosmore and Markson (1961) reported that after consuming aflatoxin-contaminated peanut meal for 3-4 months, dairy cows lost their appetite and suffered reduced milk yield, especially after calving. Loosmore et al. (1964) found that cows fed aflatoxin-contaminated cottonseed meal went off-feed resulting in milk production decreases of 25 to 50% after 3-4 weeks. When the aflatoxin-contaminated cottonseed was removed from the ration, the cows were back to normal in a week. The cattle most severely affected were those which had calved 4-6 months, previously. Gopal et al. (1968) reported that peanut meal associated with deaths in a dairy herd was found to contain aflatoxin. Clinical signs included: anorexia, apathy, corneal opacity, intermittent diarrhea, and reduction or absence of milk yield. According to Guthrie (1979), cows fed aflatoxin-contaminated maize showed the following signs: diarrhea; acute mastitis; metritis; respiratory disorders; erratic milk production; abortions; hair loss; depressed appetite; and unthrifty appearance.

In addition to the above field cases, there have been a number of other cases of aflatoxicoses reported. A number of these have been reviewed (Applebaum et al., 1982; Bodine and Mertens, 1983; Raisbeck et al., 1991).

Decreased performance is a sensitive indicator of aflatoxicosis (Richard et al., 1983) and was a factor in the case reported here. In the spring of 1980 a dairy herd located near Princeton, Minnesota showed a sudden, sharp decrease in milk production. The source of the high moisture maize in the ration had changed from one stave silo to another. Because the herd was small and the silo was of a large diameter, the amount of maize removed from the exposed surface within the top unloading silo was not sufficient to prevent spoilage during warm spring weather. The sudden decrease in milk production, with no other obvious signs or symptoms, led to investigation of the high moisture maize which had obvious fungus development on the exposed surface. Therefore, the objectives of this study were: 1) to identify the fungi growing on the exposed surface of the high moisture maize in the stave silo; 2) to identify any mycotoxin involved; 3) to determine if the isolated fungi would produce a mycotoxin in the laboratory; 4) to determine if the fungal growth could be the cause of the production loss; and 5) to change the conditions of storage to prevent the fungal growth and the decreased milk production.

MATERIALS AND METHODS

Certain observations were made on the silo containing the suspect maize. A trench was dug to examine the condition of the maize below the surface. Several samples of the maize on the exposed surface of the silage and up to 0.6 M below the surface were taken for culturing and mycotoxin analysis. Cultures of the high moisture maize were made on tomato juice agar with 6% salt added (Christensen and Meronuck, 1986), and analyzed for aflatoxin (AOAC, 1980). *Aspergillus flavus* isolates obtained from the ensiled maize were incubated on autoclaved cracked maize for 7 and 10 days at 29 C and 25% moisture. These cultures were analyzed for aflatoxin.

RESULTS

The on-site observations revealed a yellow-green mold growing within the top 2.5 cm of the high moisture maize. The temperature of the maize in this layer was 29-30 C and ranged from 22-29 % moisture. The temperature of the maize down to 0.6 meters below the surface was as high as 43 C. The heating maize was shrinking away from the silo wall thus allowing for air movement, eliminating the anaerobic conditions in the top part of the silage. A. flavus was isolated from almost all the maize collected from the top 2.5 cm, but nowhere else. Aflatoxin was found in the this maize in concentrations of 80-100 ng/g.

Pure culture of A. flavus isolated from the suspect maize were inoculated on autoclaved cracked maize. The fungus grew well on this medium as exhibited by the amount of mycelial growth and sporulation, but no aflatoxin was detected in any of these cultures.

DISCUSSION

The moisture and temperature found in the top layer of maize in the stave silo was within the range previously reported for the growth of A. flavus and the production of aflatoxin (Lillehoj, 1983; Moss, 1991). Although only one set of samples was taken, it seems logical that the 1.3-2.5 cm of feed removed per day contained enough aflatoxin and possibly other mycotoxins to cause the reduction in milk yield.

The heating silage needed to be removed to eliminate the conditions necessary for the growth of A. flavus and aflatoxin production. After removing 1.3 M the temperature at the surface of the remaining silage was 20 C. After this treatment the herd consuming the maize resumed acceptable milk production. As long as the removal rates remained at 4-8 cm per day, the temperatures on the surface of the high moisture maize stayed approximately 20 C. When the recommended rate of removal was not followed, as was the case when the market for the additional feed was lost, milk production again decreased. When the faster removal rates were resumed the conditions returned to normal. These observations further implicate fungus growth and aflatoxin as the predisposing factor in the loss of milk production.

SUMMARY

Aspergillus flavus and aflatoxin were found in high moisture maize stored in a stave silo on a Minnesota dairy farm. The A. flavus infestation seemed to be within the 1-2.5 cm of the heating high moisture maize silage. Temperatures in the area containing the A. flavus infested maize were between 29 and 30 C. at the time the samples were taken. Eighty to 100 ng/g of aflatoxin was found in the layer of infested high moisture maize. This grain was part of a dairy ration which was fed to cows exhibiting a sharp and sudden loss in milk production. When the removal rate was increased so as to keep the temperature of the grain near 20 C, the milk production returned to normal. When the removal rate was changed to the original low rate, milk production again decreased. Production again returned to normal after the faster rate of removal of maize was resumed.

REFERENCES

Applebaum, R.S., Brackett, R.E., Wiseman, D.W., and Marth, E.H. (1982). Aflatoxin: toxicity to dairy cattle and occurrence in milk and milk products: a review. J. Food Prot., 45, 752-777.

Association of Official Analytical Chemists. (1980). Official Methods of Analysis, 13th ed. Secs. 26.049-26-051, p. 421.

Bodine, A.B. and Mertens, D.R. (1983). Toxicology, metabolism, and physiological effects of aflatoxin in the bovine. In: Aflatoxin and Aspergillus flavus in Corn, pp. 46-50. (U.L. Diener, R.L. Asquith, and J.W. Dickens, eds.), Southern Cooperative Series Bulletin No. 279, Ala. Agr. Exp. Sta. Rep., Auburn Univ. AL.

Christensen, C.M., and Meronuck, R.A. (1986). Quality Maintenance in Stored Grains and Seeds. p. 107, Univ. of Minn. Press, Minneapolis.

Gopal, T., Zaki, S., Narayanaswami, M. and Premlata, S. (1968). Aflatoxicosis in dairy cattle. Indian. Vet., 45, 707-712.

Guthrie, L.D. (1979). Effects of aflatoxin in corn on production and reproduction in dairy cattle. J. Dairy Sci., 62, 134.

Lillehoj, E.B. (1983). Effect of environmental and cultural factors on aflatoxin contamination of developing corn kernels, In: Aflatoxin and Aspergillus flavus in Corn, pp. 27-34 (U.L. Diener, R.L. Asquith, and J.W. Dickens, eds.), Southern Cooperative Series Bulletin No. 279, Ala. Agr. Exp. Sta., Auburn Univ., AL.

Loosmore, R.M., Markson, L.M. (1961). Poisoning of cattle by Brazilian groundnut meal. Vet. Rec., 73, 813-814.

Loosmore, T.M., Allcroft, R., Tutton, E.A., and Carnaghan, R.B.A. (1964). The presence of aflatoxin in a sample of cotton-seed cake. Vet. Res., 76, 64-65.

Moss, M.O. (1991). The environmental factors controlling mycotoxin formation, In: Mycotoxins and Animal Foods. pp. 37-56 (J.E. Smith, and R.H. Henderson, eds.), CRC Press, Boca Raton, FL.

Raisbeck, M.F., Rottinghaus, G.E., and Kendall, J.D. (1991). Effects of naturally occurring mycotoxins on ruminants. In: Mycotoxins and Animal Foods. pp. 647-688. (J.E. Smith, and R.S. Henderson, eds.), CRC Press, Boca Raton, FL.

Richard, J.L., Pier, A.C., Stubblefield, R.D., Shotwell, O.L., Lyon, R.L., and Cutlip, R.C. (1983). Effect of feeding corn naturally contaminated with aflatoxin on feed efficiency, on physiologic, immunologic and pathologic changes, and on tissue residues in steers. Am. J. Vet. Res., 44, 1294-1299.

Sipple, W.L., Burnside, J.E., and Attwood, M.B. (1953). A disease of swine and cattle caused by eating moldy corn. Proc. Am. Vet. Med. Assoc., 90, 174-181.

EFFECTS OF MOLDY COTTONSEED ON REPRODUCTION IN BEEF AND DAIRY HEIFERS

STEVEN P. WASHBURN*, Department of Animal Science, North Carolina State University, Raleigh, NC 27695-7621, USA

LON W. WHITLOW, Department of Animal Science, North Carolina State University, Raleigh, NC 27695-7621, USA

WINSTON M. HAGLER, JR., Mycotoxin Laboratory, Department of Poultry Science, North Carolina State University, Raleigh, NC 27695-7636, USA

INTRODUCTION

Weather conditions before harvest and storage conditions after harvest contribute to growth of molds and subsequent production of mycotoxins in various commodity feeds. Although numerous mycotoxins have been identified, information on adverse effects on farm animals is often speculative, or based only on case studies and field observations. This is particularly true for suspected mycotoxicoses in ruminants.

Cottonseed and other cotton by-products are important feedstuffs in beef and dairy production. The cellulosic, lipid, and protein components of cottonseed are efficiently digested and utilized by these ruminants. In a region of increasing cotton production such as North Carolina, beef and dairy production are important outlets for these products.

Two potential disadvantages of feeding cottonseed to beef and dairy cattle are: 1) gossypol, a natural component of cottonseed which, if present in excessive amounts, will increase osmotic fragility of red blood cells and lower bull fertility, but any negative effects on female reproduction are not well documented (Calhoun and Holmberg, 1991.) and 2) cottonseed from cotton grown under irrigation in the arid Southwestern United States are highly susceptible to field contamination with aflatoxin (Diener, 1989). However, Southwestern cottonseed are routinely treated with ammonia to destroy the aflatoxin before feeding to cattle.

Cottonseed grown in the Southeastern United States are not likely to contain significant contamination with aflatoxin (Deiner, 1989). Supporting this supposition, Hagler (1990) reported that of 35 samples of cottonseed submitted for aflatoxin analysis to his laboratory over several years, 11 were contaminated, and contained a mean of only 23 ng/g of aflatoxin B_1. However, being rather bulky, undelinted cottonseed pose a problem for safe storage in the Eastern United States, and storing outdoors or with exposure to moisture is risky.

One of the most costly problems for cattle farmers is reproductive failure. These problems are often associated with relatively low-level mycotoxin contamination of feedstuffs (Whitlow et al., 1986). The cottonseed used in this study were associated with a severe case of reproductive failure in a well-managed herd of beef heifers. The cottonseed involved contained visible mold growth. However, aflatoxin analysis revealed no aflatoxin and extensive chemical analysis for other mycotoxins revealed no contamination except that the suspect cottonseed tested positive for deoxynivalenol (DON) by a commercially available enzyme-linked immunosorbent assay (ELISA).

The current study was possible because sufficient moldy cottonseed was available to conduct a controlled study in dairy heifers. The objectives of this study were to: 1) document the reproductive performance of the beef heifers on the problem farm; and 2) to use the cottonseed from that farm in a controlled study to determine effects on growth and reproduction of dairy heifers.

MATERIALS AND METHODS

Collection and Analysis of Field Data

A request for assistance with a reproductive problem in beef heifers was received by the first author. Feeding, vaccinations, estrus detection, synchronization of estrus, insemination, and other management procedures were reviewed with the producer and the local extension agent. Procedures used on the farm were consistent with established recommendations and were similar to procedures used in previous years when reproductive performance was acceptable. Although not weighed, the beef heifers appeared to be well-grown and almost all should have reached puberty. Reproductive performance of the lactating cows on the farm was normal during this year of low fertility among the heifers. The difference in the feeding program between the affected heifers and the cows was that the heifers were being fed a supplement of corn and whole cottonseed. Feeding of heifers included ad libitum coastal bermudagrass hay supplemented with a grain ration (approximately 2 kg corn and 1 kg cottonseed/head/day).

Reproductive records on crossbred heifers from the beef farm were examined over three years. Whole cottonseed was fed during the second and third years. Reproductive data included the two years before (21 heifers each year) and the year of (30 heifers) the reproductive problem. During all three years, heifers had estrus synchronized using Synchro-Mate-B and were inseminated by the same technician. Estrous synchronization using this product included an injection (IM) of 3 mg of a synthetic progesterone, norgestomet, with 5 mg of estradiol valerate, followed by a subcutaneous ear implant (6 mg norgestomet) for 9 days. After removal of the implants, heifers were observed for estrus twice daily and artificially inseminated (AI) 8 to 18 hours after onset of estrus. Estrous detection and insemination were continued for 7 days. Heifers were then exposed to a bull for natural service. Although exact times were not available, the bull was allowed approximately 20 days to breed during the problem year in an attempt to get more heifers pregnant.

Reproductive data are categorical, and were analyzed using Chi-Square analysis (Steel and Torrie, 1960). Two comparisons were made for each variable: 1) comparison of the first two years to the problem year; and 2) comparisons of year 1 to year 2. Variables examined included estrous response to synchronization, pregnancy rate among heifers receiving AI, pregnancy rate to AI among all heifers in a group, and overall pregnancy rate (AI plus natural service).

<u>Mycotoxin Analysis</u>

Because the cottonseed were visibly discolored, a potential mycotoxin problem was suspected and samples of the corn and cottonseed were screened for common mycotoxins. Samples were tested for the presence of aflatoxin and T-2 toxin by the antibody-based EZ-SCREEN Quick Card tests (Environmental Diagnostics, Inc. Burlington, NC 27215). Zearalenone and DON were determined by ELISA (AGRI-SCREEN, Neogen, Inc. Lansing, MI 48912). Aflatoxin, T-2 toxin, zearalenone, tested negative by the antibody assays. Aflatoxin analysis by high performance liquid chromatography (Hutchins and Hagler, 1983) was also negative. Zearalenone analysis by thin layer chromatography (Swanson et al., 1984) was also negative. However, using the AGRI-SCREEN test for DON, both the corn (1,362 ng/g) and the moldy cottonseed (2,618 ng/g) had sufficient apparent concentrations of DON to suspect an association with impaired fertility. Although most of the corn involved had been fed, approximately 7 metric tons of moldy cottonseed remained and were donated by the producer for use in a controlled study.

Control and treatment composite ration samples used in the controlled study with dairy hiefers were analyzed for presence of DON by electron capture gas chromatography (Trenholm et al., 1985; Rizzo et al., 1986). Further analyses for four aflatoxins (B_1, B_2, G_1, G_2), ochratoxin A, citrinin, sterigmatocystin, patulin, T-2 toxin, HT-2 toxin, diacetoxyscirpenol, neosolaniol, fusarenon-X, deoxynivalenol, nivalenol, GY25, zearalenone, and zearalenol were done on composite samples by Romer Labs, Inc., Washington, MO. Dr. Howard Casper, North Dakota State University, Fargo, ND performed an exhaustive mass spectral screen for trichothecene mycotoxins in the control and treatment diets.

Controlled Study Using Dairy Heifers

Rations were formulated using either moldy cottonseed from the problem herd or control cottonseed (not visually mold-damaged and mycotoxin-tested). Cottonseed was included in these rations at 31% of dry matter (Table 1), so that heifers would consume approximately 2.7 kg of cottonseed per day. The nutrient composition from composite samples of the control and treatment rations is included in Table 2.

Table 1. Feed Ingredients in Mixed Rations of Controlled Study.[a]

Ingredient	Percent of Ration Dry Matter
Whole Cottonseed	31.0
Cottonseed Hulls	61.3
Ground Corn	2.0
48% Soybean Meal	3.8
Mineral and Vitamin Supplement	2.0

[a]Rations differed only in source of whole cottonseed. Rations were fed ad libitum to two groups of heifers.

Twenty-four cyclic Holstein heifers (weight = 339 ± 14 kg) were grouped in eight sets of three according to weight. One heifer in each of the eight sets was randomly assigned to be fed the control ration while the other two heifers within each group were assigned to be fed the ration containing moldy cottonseed. These rations were fed ad libitum as total mixed rations for a 70-day period. Heifers on respective rations were housed in two groups in adjacent pens with an open-fronted shelter. Weights of heifers were recorded weekly. Beginning and ending weights were after an overnight shrink. Weekly blood samples were collected to monitor serum progesterone (Archbald et al., 1984) and estrous cycles. Serum from blood samples collected at initiation and end of the trial was analyzed for albumin, alkaline phosphatase, bilirubin, urea nitrogen, calcium, cholesterol, chloride ion, glutamic-oxaloacetic transaminase, total protein, creatinine, globulin, gamma-glutamyl transpeptidase, glucose, potassium, lactate dehydrogenase, glutamic-pyruvic transaminase, triglycerides, uric acid, and ratios of urea nitrogen to creatinine and albumin to globulin. These blood chemistry analyses were conducted by Roche Biomedical Laboratories, 1447 York Court, Burlington, NC. Fecal samples collected at the beginning and end of the trial were examined for presence of occult blood by Roche Biomedical Laboratories.

Table 2. Nutrient Composition of Treated and Control Rations Expressed as a Percentage of Dry Matter.

Composition	Control	Treated
Dry Matter, %	93.6	93.1
Crude Protein, %	11.7	12.2
Unavailable Protein, %	2.8	2.3
Adjusted Crude Protein, %	10.1	11.1
Acid Detergent Fiber,%	51.9	52.0
TDN, %	56.5	56.5
Net Energy (Lactation), Mcal/kg	1.25	1.25
Crude Fat, %	6.54	6.42
Sodium, %	.08	.08
Potassium, %	1.21	1.22
Calcium, %	.63	.57
Magnesium, %	.24	.24
Phosphorus, %	.22	.24
Sulfur, %	.14	.15
Iron, ppm	120	325
Zinc, ppm	52	57
Copper, ppm	19	22
Manganese, ppm	51	66

All heifers received two injections (IM) of prostaglandin $F_2\alpha$ to synchronize estruses before breeding. Injections were at an interval of 13 days with the second injection occurring on day 41 of treatment. Heifers were observed twice daily for estrous behavior and were inseminated 8 to 18 hours after onset of estrus. Heifers returning to estrus during or after the treatment period were reinseminated. Pregnancy diagnoses were conducted by palpation of uterine horns via the rectum.

Analyses of data included use of GLM procedures of SAS (SAS, 1985) and Chi-Square for categorized data (Steel and Torrie, 1960). Variables examined included blood chemistry analyses, growth rates of heifers, response to estrous synchronization, apparent pregnancy rates (based on progesterone profiles), actual pregnancy rate (based on palpation) and net pregnancy rate (including breedings after treatment corrected for abortions).

RESULTS AND DISCUSSION

Field Data

Heifer reproductive data from the beef farm over a three-year period are included in Table 3. Estrous responses and subsequent fertility in year three was significantly lower than in the two previous years. However, the overall pregnancy rate did not differ significantly probably because those heifers were exposed to a bull for natural service for a longer period in the third year than in either of the previous years. Although the data were not available, the producer did not observe lower reproductive rates in the third year among the rest of his cattle. There were no significant differences between year one and year two in estrous responses or pregnancy rates. These data are not from a controlled study but do provide strong evidence that either (or both) the corn and cottonseed fed to heifers but not to cows in year three was associated with the reproductive failure.

Controlled Study with Dairy Heifers

Estrous synchronization and fertility data from the controlled study with Holstein heifers are included in Table 4. The proportion of heifers with synchronous estrual responses did not differ significantly, but those consuming moldy cottonseed had numerically lower responses. This response to prostaglandin $F_2\alpha$ is consistent with low responses to Synchro-Mate B for the beef heifers in the field study. Of particular interest in this controlled study was the fact that apparent conception among heifers actually inseminated was not different and was high (>70%) in both groups. However, eight heifers apparently had pregnancies initiated while consuming the moldy cottonseed, but four of those could not be confirmed by palpation.

Table 3. Responses to Estrous Synchronization and Breeding Among Cross-bred Beef Heifers Over Three Years on a Commercial Farm.

Responses	Year 1	Year 2	Year 3
No. Heifers	21	21	30
Estrous Responses	17	18	15
% of Group	81	86	50[a]
AI Pregnancies	10	9	2
% of Inseminations	59	50	13[a]
Total Pregnancies	19	18	22
% of Group (AI + Natural)	90	86	73[b]

[a]Year 3 was significantly lower ($P<.01$) than first two years.
[b]Overall pregnancy rates tended to be lower ($P=.11$) in year three, but natural service was allowed for additional time during that year.

Table 4. Response of Holstein Heifers to Estrous Synchronization and Breeding after Exposure to Diets with Moldy (Treated) or Control Cottonseed.

	Ration		Level of Significance
	Control	Treated	
Number of Heifers	8	16	---
Synchronized Estrus (≤ 5 days)	7 of 8 (88%)	11 of 16 (69%)	N.S.
Apparent Pregnancies[a] (During Synchrony)	5 of 8 (62%)	8 of 16 (50%)	N.S.
Actual Pregnancies[b] (During 29-day Breeding)	6 of 8 (75%)	6 of 16 (38%)	$P<.1$
Net Pregnancies[c]	8 of 12 (67%)	13 of 32 (41%)	$P\approx.13$

[a]Apparent pregnancies of total group determined by serum progesterone ≥ 1ng/ml through 28 days after breeding. Note: 5/7 (71%) vs 8/11 (73%) heifers actually inseminated were apparently pregnant.
[b]Actual pregnancies of total group based on palpation ≥ 35 days after breeding. Note: 4 apparent pregnancies lost among treated heifers from synchronized breeding.
[c]Net pregnancies are expressed as total heifers maintaining pregnancies as a proportion of all inseminations before and after treatment. Note: Two treated heifers inseminated within 5 weeks after end of treatment were diagnosed pregnant and later aborted.

This suggests either a disruption of normal estrous cycles with the presence of a persistent corpus luteum or probable loss of early pregnancies. This may explain a strong trend ($P<.1$) for lower pregnancy rates during treatment among heifers consuming the moldy cottonseed. Eventually, all but one of the heifers consuming moldy cottonseed conceived, but two of those heifers that conceived within five weeks after treatment aborted. Overall net pregnancy rates, as a percentage of inseminations attempted, tended to favor control heifers (67% vs 41%, $P=.13$).

Although the ration used in the control study was quite different than that used on the beef cattle farm, the results implicate the moldy whole cottonseed as the source of the problem at the farm. Although daily gains tended to favor heifers on the control diet (1.07 vs .91 $\pm$ .05 kg/day, $P=.06$), all heifers gained quite well and appeared normal. Again this was consistent with the perceived situation on the farm case study.

The analysis of feces for the presence of occult blood was negative and there were no consistent differences in the concentrations of various blood chemistry constituents. Therefore, the moldy cottonseed appeared to affect reproduction greatly, gains moderately, and had little effect on other variables.

Concentrations of free (.147 vs .122%) and total (.197 vs .203%) gossypol in samples of problem and control cottonseed lots, respectively, were similar and indicates that gossypol was not the cause of the observed reproductive problems.

Although the initial ELISA test indicated the presence of DON in the problem cottonseed, subsequent analyses using gas chromatography by the Mycotoxin Laboratory were negative. Analyses by Romer Labs, Inc. failed to detect DON (detection limits - 125 to 500 ng/g) or any of the other mycotoxins listed in the methods section. Further analysis by Dr. Howard Casper of North Dakota State University also revealed no trichothecene mycotoxins.

Although substantial evidence from the combined field data and the controlled study implicates the mold-damaged cottonseed in the observed reproductive failure, a specific causative agent was not identified.

An additional observation made during the course of these studies needs clarification. The fact that the moldy cottonseed repeatedly tested strongly positive for DON by Neogen Inc. ELISA test, but negative for DON, or DON derivatives cross reactive with Neogen's antibody, by chemical analysis in the

North Carolina State University Mycotoxin Laboratory, Romer Labs, Inc., or in the laboratory of Dr. Howard Casper at North Dakota State University is important. These results imply that there was a cross reactive compound in the sample extract, an enzyme in the extract that interfered with the color reaction in the test, or some extract component denatured the antibody. We have more recently seen this same type of divergence in results on analysis of various silages with ELISA test for DON compared to chemical analysis (Hagler and Whitlow, Unpublished data).

SUMMARY

Field data on reproductive performance of beef heifers over a three-year period were indicative of a problem with moldy whole cottonseed in the third year. Cottonseed from that farm was used in a complete ration and compared to a similar ration containing another source of cottonseed using dairy heifers. These rations were fed ad libitum in a controlled study to two groups of Holstein heifers. Reproductive rates were lower among heifers consuming the moldy cottonseed. Attempts to identify a specific mycotoxin as the cause of reproductive problems were unsuccessful.

Aknowledgements

The technical assistance of Faye Suggs, Jim Brewer, Jennifer Lee and Karl Hedrick, Jr. is appreciated. Dr. Howard Casper's (North Dakota State University, Fargo, ND) willingness to exhaustively test the diets made from the control and moldy cottonseed lots for DON and other trichothecenes is greatly appreciated. Cooperation from beef producers Roy and Richard Wood, Extension Agent Richard Melton, and the staff at the Piedmont Research Station in Salisbury, NC is appreciated. The use of trade names in this publication does not imply endorsement by the North Carolina Agricultural Research Service, nor criticism of similar products not mentioned.

REFERENCES

Archbald, L.F., Ingraham, R.H. and Godke, R.A. (1984). Inability of progesterone pretreatment to prevent premature luteolysis of induced corpora lutea in the anestrous bitch. Theriogenology, 21, 419-426.

Calhoun, M., and Holmberg, C. (1991). Safe use of cotton by-products as feed ingredients for ruminants: A review. In: Cattle Research with Gossypol Containing Feeds, pp. 97-129 (L.A. Jones, D.H. Kinard, and J.S. Mills, eds.) National Cottonseed Products Assn., Memphis, TN.

Diener, U. L. (1989). Preharvest aflatoxin contamination of corn, peanuts, and cottonseed: A review. In: Biodeterioration Research 2, pp. 217-244 (G.C. Llewellyn and C. E. O'Rear, eds.) Plenum Press, NY.

Hagler, W. M., Jr. (1990). Aflatoxin in field crops in the Southeastern United States. In: A Perspective on Aflatoxin in Field Crops and Animal Food Products in the United States. ARS-83, pp. 35-42. U.S.D.A., A.R.S., Washington, DC.

Hutchins, J. E., and Hagler, W.M., Jr. (1983) Rapid liquid chromatographic determination of aflatoxins in heavily contaminated corn. J. Assoc. Offic. Anal. Chem., 66, 1458-1465.

Rizzo, A.F., Sarri, L., and Lindfors, E. (1986). Derivatization of trichothecenes and water treatment of their trimethylsilyl ethers in an anhydrous apolar solvent. J. Chrom., 368, 381-386.

SAS (1985). SAS User's Guide: Statistics. SAS Institute, Inc. Cary, NC.

Steel, R.G.D, and Torrie, J.H. (1960). Principles and Procedures of Statistics. McGraw-Hill Book Company, Inc. pp. 41-43.

Swanson, S.P., Corley, R.A., White, D.G., and Buck, W.B. (1984). Rapid thin layer chromatographic method for determination of zearalenone and zearalenol in grains and animal feeds. J. Assoc. Offic. Anal. Chem., 67, 580-582.

Trenholm, H. L., Warner, R.M., and Prelusky, D.B. (1985). Assessment of extraction procedures in the analysis of naturally contaminated grain products for deoxynivalenol (Vomitoxin). J. Assoc. Offic. Anal. Chem., 68, 645-649.

Whitlow, L.W., Nebel. R.L., Behlow, R.F., Hagler, W.M., Jr., and Brownie, C.F.-G. (1986). Mycotoxins in North Carolina dairy feeds - A survey of 100 dairy farms. J. Dairy Sci. 69 (Suppl. 1), 223.

THE ASSOCIATION OF DEOXYNIVALENOL IN GRAIN WITH MILK PRODUCTION LOSS IN DAIRY COWS

LON W. WHITLOW*, Department of Animal Science, North Carolina State University, Raleigh, NC 27695, USA

RAY L. NEBEL, Department of Dairy Science, Virginia Polytechnic Institute and State University, Blacksburg, VA 24061, USA

WINSTON M. HAGLER, JR., Mycotoxin Laboratory, Department of Poultry Science, North Carolina State University, Raleigh, NC 27695, USA

INTRODUCTION

Reports have indicated that in certain geographical locations and in certain years there can be a high incidence of feedstuffs contaminated with aflatoxin (Nichols 1983), deoxynivalenol (DON) (Vesonder et al., 1978; Mirocha, 1974), zearalenone (Shotwell et al., 1980) or ochratoxin A (Hamilton et al., 1982). The practical importance of these and other mycotoxins in feedstuffs which have the potential of affecting the health and productivity of farm animals is a topic of intensive investigation. The significance of DON-contamination of poultry and dairy feedstuffs is particularly controversial.

When we began to analyze samples of suspect dairy feedstuffs for DON in 1983, there seemed to be an association between DON and productivity problems. Milk production losses and decreased reproductive efficiency were of particular note.

In individual case studies in North Carolina, production and/or health problems in dairy cows have been attributed to mycotoxins by diagnosticians and extension specialists. Aflatoxin, DON, and zearalenone are commonly detected mycotoxins and seem to be related to production and health problems even at fairly low concentrations (Whitlow and Hagler, unpublished observations).

This study was initiated to determine the prevalence of mycotoxins in grain mixes collected from NC dairy farms and to determine if changes in milk production were associated with mycotoxins, DON in particular, in the grain mix component of dairy rations.

MATERIALS AND METHODS

Grain samples were collected from selected NC dairy farms by county extension agents during July and August each year of a two-year survey. In the first year, 100 farms were selected on the basis of milk production performance in the previous six months. Herds maintaining or increasing in milk production; i.e., those having an above average production and low average number of days not pregnant, were designated as control herds. Problem herds were below average in milk production; i.e., those showing a loss in production and a high average number of days not pregnant. Official dairy herd improvement association (DHIA) records were used to provide rolling herd average milk (RHAM) production data. Change in RHAM was designated as the difference from May to September.

First-year grain mix samples were analyzed for aflatoxins B_1, B_2, G_1, and G_2 (Hutchins and Hagler, 1983; Hutchins et. al., 1988), ochratoxin (Tyczkowska et al., 1984), zearalenone (Swanson et al., 1984), and DON (Scott et al., 1981; or Trenholm et al., 1985b; Rizzo et al., 1986). Aflatoxin concentrations are reported as the sum of the four major aflatoxins tested.

Based on first-year results, an additional 200 grain mixes were collected in the second year from 200 randomly selected herds and analyzed for DON only. Change in RHAM was calculated in a similar manner. Analysis of data included use of general linear models (SAS, 1985) and Chi-Square for categorical data (Steel and Torrie, 1960).

RESULTS

Of the 100 samples collected in year one, 56% contained one or more of the seven mycotoxins tested. No sample contained detectable ochratoxin and only one contained zearalenone at 12 ng/g. Eight samples contained both aflatoxin and DON. Incidence and concentrations of aflatoxins and DON are summarized in Table 1. There were no significant differences between control and problem herds in the incidence or concentration of aflatoxins or DON or incidence of mycotoxin positive samples, although control herds tended to have a lower incidence and concentration of DON. Although 37% of the grain samples contained aflatoxins, the highest concentration was 70 ng/g. Incidence of aflatoxins was not statistically related to milk production changes, thus was not considered in more detail. Twenty-three of the 100 herds in the first year survey participated in the USDA-administered Milk Diversion Program (Public Law # 98-180). Under this program dairymen were paid to reduce amount of milk marketed. Since many dairymen reduced production per cow to achieve a reduction in marketed milk, all diversion herds were removed from the data set, leaving only nondiversion herds for evaluating the association of DON with change in RHAM. Incidence and concentrations of DON in grain mixes are presented for all herds and nondiversion herds in Table 2. Within the nondiversion herds, those with grain mixes testing negative for DON had less ($P<.03$) RHAM loss than did herds with grain mixes testing positive for DON. In addition, a negative change in RHAM was related ($p<.02$) to an increased concentration of DON in the grain mix.

Of the 200 grass mix samples from the second year, 127 contained DON. Data of both years were combined, and change in RHAM evaluated with respect to DON level in the grain mixes. Results indicated a significant ($p<0.01$) year effect, but no year by DON interaction. Therefore, combined data showing the relationship of DON with RHAM are presented in Table 3. Downward stepwise regression indicated that a cubic model was most appropriate. Based on the cubic model in Table 3, predicted changes in RHAM and daily milk production over a four-month period are presented in Table 4. This model suggested that milk production losses were not associated with grain mix DON concentrations below 300 to 400 ng/g. Above this range, higher levels of DON are associated with a loss in milk production. Personal observations of field cases support this association (L.W. Whitlow, unpublished observations).

DISCUSSION

Aflatoxin levels in commercial feed products are regulated and both the feed industry and dairy farmers recognize that milk can be contaminated with aflatoxin M_1 as a result of feeding aflatoxin-contaminated feeds. Regulatory efforts to minimize aflatoxin contamination of milk have been reasonably effective (GAO/RCED-91-109). The result is that aflatoxin contamination of dairy feeds is apparently minimal, a conclusion supported by the data in the present study.

Table 1. Incidence of Mycotoxins and Mean Concentration of Aflatoxin and Deoxynivalenol (DON) in Positive Grain Mix Samples Collected from NC Dairy Farms over Two Years.

	Total Samples n	Mycotoxin-Positive n	Mean ng/g	Standard Error ng/g	Range ng/g
Any Mycotoxin[a]	100[c]	56	---	---	---
Aflatoxin[b]	100[c]	37	16	3	4-70
Deoxynivalenol	300[d]	153	105	35	10-5250

[a]Includes samples positive for aflatoxin B_1, B_2, G_1, G_2, deoxynivalenol, zearalenone, or ochratoxin.
[b]Aflatoxin values are the sum of the four aflatoxins tested.
[c]Year one samples only.
[d]Year one and two samples.

Table 2. Least Squares Means[a] for Grain Mix Deoxynivalenol (DON) Concentrations and Change in Rolling Herd Average Milk (RHAM)[b] for All Year One Dairy Herds and Nondiversion[c] Dairy Herds Grouped by DON Presence[d] or by Herd Treatment[e] Group.

	All Herds			Nondiversion Herds		
Herd Type	n	DON (ng/g)	RHAM Change	n	DON (ng/g)	RHAM Change
DON -	74	0 ± 60	-67 ± 33	58	0 ± 77	-33 ± 39
DON +	26	299 ± 101	-218 ± 55	19	385 ± 134	-204 ± 69
P Value[f]		0.02	0.03		0.02	0.04
Control	22	4 ± 114	-11 ± 61	16	5 ± 139	8 ± 70
Problem	78	99 ± 60	-133 ± 32	61	+ 125 ± 80	-102 ± 40
P Value[f]		N.S.	N.S.		N.S.	N.S.

[a]Means ± standard error.
[b]Rolling herd average milk from official records measured from May to September (Feed samples from July and August).
[c]Nondiversion herds include those not participating in government milk reduction program (Public Law #98-180).
[d]Herds are grouped by presence of DON in grain mixes, where 0 Don = negative and >0 DON = positive.
[e]Control or problem herds were identified prior to sample collection based on herd performance for the previous six months.
[f]Probability that means are statistically different. N.S. = not significantly different.

Deoxynivalenol (DON) has not been recognized as a mycotoxin causing major detrimental effects in ruminants. It is generally thought that the rumen microflora provide some protection to the ruminant by metabolizing DON and some other mycotoxins to less toxic or nontoxic metabolites. Côté et al. (1986)

demonstrated that DON is metabolized by rumen microorganisms to the less toxic metabolite, deepoxydeoxynivalenol, referred to as DOM-1. Only very limited research has investigated possible DON effects in ruminants. Some field observations have associated DON with general unthriftiness in cattle and suggested that DON may be of concern (Mirocha et al., 1974). Harvey et al. (1986) fed seven lambs a diet containing approximately 9.36 μg/g DON for 28 days. Compared with eight lambs on DON-free diets, no significant differences were noted for feed intake, weight gain, feed efficiency, or in an array of serum biochemical or hematologic variables. Trenholm et al. (1985a) fed a group of nonlactating dairy cows a grain mix containing 1.5 μg/g DON for three weeks, fed a grain mix containing 6.4 μg/g DON for six weeks, and then fed a grain mix containing 1.5 μg/g DON for the final three weeks. Consumption of the high DON grain mix was less than for the low DON grain mix. No clinical signs of illness were observed and average weight gain was apparently not affected; however, feeding periods were short and clean grain was not fed. Therefore, the experiment appears to compare two fairly high levels of DON contamination without a toxin-free feed control. Noller et al. (1979) fed lactating dairy cows diets containing either 1) "normal" corn, 2) Gibberella zeae-infected corn, or 3) equal parts of 1 and 2. The infected grain contained 500 ng/g zearalenone. Concentration of DON was not determined; however, corn grown in the same field but harvested earlier contained 12 to 13 μg/g DON. The experimental design was a 3X3 Latin square with three-week periods. Infected corn did not affect milk or milk fat production but did reduce feed intake resulting in lower weight gains. Unfortunately, these studies did not address the question of long-term feeding effects.

Table 3. Statistical Model for the Association of Grain Don Level, ng/g on Change in Rolling Herd Average Milk Production (RHAM) over a Four Month Period Based on Data from Years 1 and 2[a,b].

Source	Estimate ± standard error
Intercept	0-446 ± 119
Year	272 ± 61
DON	0.87 ± 0.51
DON * DON	$-2.86 \times 10^{-3} \pm 0.81 \times 10^{-3}$
DON * DON * DON	$5.99 \times 10^{-7} \pm 1.36 \times 10^{-7}$

[a]Coefficient of determination (R^2) = .29, P<.0001.
[b]DON x year interaction was not significant.

Table 4. Model Predictions for Change in Rolling Herd Average Milk (RHAM) and Changing Daily Milk Relative to Zero Deoxynivalenol (DON) in the Grain Mix[a,b].

Grain Mix DON ng/g	Predicted Change in RHAM Kg	Predicted Change in Daily Milk Kg
0	-38	0
100	21	0.5
200	25	0.5
300	-21	0.1
400	-116	-0.6
500	-256	-1.8
600	-439	-3.3
700	-660	-5.1
800	-919	-7.2
900	-1210	-9.5
1000	-1532	-12.1

[a]Change RHAM = -446 + 272 * Year + .87 DON - .00286 DON^2 + .000000599 DON^3 where year is assumed equal +1.5 and change is for a four-month period.

[b]Predicted change in daily milk at X concentration of DON is calculated as change in RHAM at 0 DON and divided by 123 days.

In contrast to ruminants, swine are known to respond more adversely to consumption of DON (Vesonder et al., 1976). An Illinois study of field problems by swine producers (Côté et al., 1984) found that DON was associated with problems in 80% of the cases. Problems reported and frequency of the problems were: reproductive problems, 50%; feed refusal, 43%, reduced weight gains, 25%; diarrhea, 17%; deaths, 14%; and emesis, 11%. Range of DON concentrations was 0.1 to 8.0 μg/g.

Swine data (Foster et al., 1986) would suggest that grain mixes naturally contaminated with DON are more detrimental to performance than are grain mixes with the same concentration of pure DON added. This infers that other toxic substances may, at times, be present in feeds naturally contaminated with DON. Thus, DON may serve as a marker for toxic, moldy feed. The idea of using DON (and/or zearalenone) as a marker for the potential presence of other mycotoxins has been discussed in more detail by Romer (1990). DON may be a better marker for problem feeds than other commonly used measures such as mold count. Wyatt (1991) makes an argument that mold spore counts probably have little utility in measuring the moldiness or toxicity of a feedstuff.

In our study, T-2 toxin was not measured. In a separate group of 344 feed samples taken from dairy farms, T-2 was found in 4.6%. Of these, 63% (or 2.9% of the total) also contained DON. There could be some question about the potential presence of T-2, but results from this separate group of samples suggests that the incidence of T-2 may be minimal (L. W. Whitlow, unpublished data). However, Romer (1990) points out there are about 45 trichothecene mycotoxins produced by Fusarium and practical analytical methods exist for only about ten of these.

SUMMARY

The presence of DON in dairy feeds may indicate that the feed was subjected to a situation conducive to mold growth and mycotoxin formation. It is possible that DON may affect dairy cattle, or other associated but unidentified mycotoxins may be present and toxic to dairy cows, or other mycotoxins may interact with DON to produce detrimental effects. The presence of DON may thus serve as a marker for feeds with the potential to cause production losses in dairy cattle. The association of change in RHAM with grain mix DON concentrations does not imply a cause and effect, but suggests that DON effects on dairy cattle should be further investigated. This is especially true in light of the prevalence of DON and the few studies which have been conducted with dairy cattle. These results also suggest the need to investigate the potential presence of other mycotoxins in DON-contaminated feeds.

Acknowledgments

Appreciation is expressed to Faye Suggs, Jim Brewer, Jennifer Lee, and Janice Wheeling for technical assistance, to Donna Suggs for assistance with manuscript preparation, and to Robert Behlow and Cecil Brownie for their contributions. The use of tradenames in this publication does not imply endorsement by the North Carolina Agricultural Research Service nor criticism of similar ones not mentioned.

REFERENCES

Côté, L.M., Nicoletti, J., Swanson, S.P., and Buck, W.B. (1986). Production of deepoxydeoxynivalenol (DOM-1), a metabolite of deoxynivalenol, by in vitro rumen incubation. J. Agric. Food Chem., 34, 458-460.

Côté, L.M., Reynolds, J.D., Vesonder, R.F., Buck, W.B., Swanson, S.P., Coffey, R.T., and Brown, D.C. (1984). Survey of vomitoxin - contaminated feed grains in Midwestern United States, and associated health problems in swine. J. Am. Vet. Med. Assoc., 184, 72-192.

Foster, B.C., Trenholm, H.L., Friend, D.W., Thompson, B.K., and Hartin, K.E. (1986). Evaluation of different sources of deoxynivalenol (vomitoxin) fed to swine. Can. J. Anim. Sci., 66, 1149-1154.

Hamilton, P.B., Huff, W.E., Harris, J.R., and Wyatt, R.D. (1982). Natural occurrence of ochratoxicosis in poultry. Poultry Sci. 61, 1832-1841.

Harvey, R.B., Kubena, L.F., Corrier, D.E., Witzel, D.A., Phillips, T.D., and Heidelbaugh, N.D. (1986). Effects of deoxynivalenol in a wheat ration fed to growing lambs. Am. J. Vet. Res., 47, 1630-1632.

Hutchins, J.E. and Hagler, W.M., Jr. (1983). Rapid liquid chromatographic determination of aflatoxins in heavily contaminated corn. J. Assoc. Offic. Anal. Chem., 66, 1458-1465.

Hutchins, J.E., Lee, Y.J. Tyczkowska, K., and Hagler, W.M., Jr. (1988). Evaluation of silica cartridge purification and hemiacetal formation for liquid chromatographic determination of aflatoxins in corn. Arch. Environ. Contam. Toxicol., 18, 319-326.

Mirocha, C.J., Schauerhamer, B., Pathre, S.V. (1974). Isolation, detection and quantitation of zearalenone in maize and barley. J. Assoc. Offic. Anal. Chem., 57, 1104-1110.

Nichols, T. E. Jr. (1983). Economic impact of aflatoxin in corn. In: Aflatoxin and Aspergillus flavus in Corn: Southern Cooperative Series Bulletin 279, pp 67-71 (U.L. Diener, R.L. Asquith and J.W. Dickens, eds.), Alabama Agricultural Experiment Station, Auburn University, Auburn, AL.

Noller, C.H., Stob, M., and Tuite, J. (1979). Effects of feeding Gibberella zeae - infected corn on feed intakes, body weight gain, and milk production of dairy cows. J. Dairy Sci., 62, 1003-1006.

Rizzo, A.F. Saari. L., and Lindflors, E. (1986). Derivatization of trichothecenes and water treatment of their trimethylsilyl ethers in an anhydrous apolar solvent. J. Chromatogr., 368, 381-386.

Romer, Tom (1990). Advances in mycotoxin analysis: Marking Fusarium. Feed Management, 41, 30-33.

SAS Institute, Inc. SAS User's Guide, Version 5. Statistical Analysis Systems, Inc., Cary, NC.

Scott, P.M., Lan, P., and Kanhere, S.R. (1981). Gas chromatography with electron capture and mass spectrometric detection of deoxynivalenol in wheat and other grains. J. Assoc. Offic. Anal. Chem., 64, 1364-1371.

Shotwell, O.L., Bennett, G.A., Goulden, M.L., Plattner, R. D., and Hesseltine, C.W. (1980). Survey for zearalenone, aflatoxin and ochratoxin in U.S. grain sorghum from 1975 and 1976 crops. J. Assoc. Offic. Anal. Chem., 63, 922-926.

Steel, R.G.D., and Torrie, J.H. (1960) Principles and Procedures of Statistics, pp. 346-350. McGraw-Hill Book Company, Inc. New York.

Swanson, S.P., Corley, R.A. White, D.G., and Buck, W.B. (1984). Rapid thin layer chromatographic method for determination of zearalenone and zearalenol in grains and aninal feeds. J. Assoc. Offic. Anal. Chem., 67, 580-582.

Trenholm, H.L., Thompson, B.K., Hartin, K.E., Greenholgh, R., and McAllister, A.J. (1985a). Ingestion of vomitoxin (deoxynivalenol) contaminated wheat by nonlactating dairy cows. J. Dairy Sci., 68, 1000-1005.

Trenholm, H.L., Warner, R.M., and Prelusky, D.B. (1985b). Assessment of extraction procedures in the analysis of naturally contaminated grain products for deoxynivalenol (vomitoxin). J. Assoc. Offic. Anal. Chem., 68, 645-649.

Tyczkowska, K., Hutchins, J.E., and Hagler, W.M., Jr. (1984). Determination of ochratoxin A in feeds and biological fluids using pre-packed Celite colums. Assoc. Offic. Anal. Chem. Abstract 134, p. 13.

Vesonder, R.F., Ciegler, A., Jensen, A.H., Rohwedder, W.K., and Weisleder, D. (1976). Co-identity of the refusal and emetic principle from Fusarium-infected corn. Appl. Environ. Microbiol., 31, 280-285.

Vesonder, R.F., Ciegler, A., Rogers, R.F., Barbridge, K.A., Bothast, R.J., and Jensen, A.A. (1978). Survey of 1977 crop year preharvest corn for vomitoxin. Appl. Environ. Microbiol., 36, 885-888.

Wyatt, R.D. (1991). Measurement of mold growth and mycotoxins in feed: fallacies and innovations, In: Proc. 1991 Georgia Nutrition Conference for the Feed Industry, pp. 73-82, Atlanta, GA.

ULTRASTRUCTURAL ANALYSIS OF WHETHER DIETARY MANGANESE PROTECTS AGAINST AFLATOXIN-INDUCED LIVER DAMAGE IN RATS

GERALD C. LLEWELLYN*, Department of Forensic Science, The George Washington University, Washington, DC 20052, USA and The Delaware Department of Natural Resources and Environmental Control, New Castle, DE 19720, USA

BETTY A. FLOYD, Department of Biology, Virginia Commonwealth University, Richmond, VA 23284, USA

ROBERT L. KNIGHT, Electron Microscopy Section, Charleston Area Medical Center, Charleston, WV 25301, USA

WILLIAM V. DASHEK, Department of Biology, Clark Atlanta University, Atlanta, GA 30314, USA

INTRODUCTION

Aflatoxins (AFTs) are secretions of both Aspergillus parasiticus and A. flavus (Nesbit et al., 1962), two fungi, which can grow upon economically-important crop plants (Wogan, 1964). These secretions are bifuranocoumarins belonging to the family of compounds know as mycotoxins. Of interest from this group are the aflatoxins (AFTs), especially aflatoxin B_1 (AFB_1). The AFTs are heptocarcinogenic, mutagenic, and teratogenic. Perhaps, the most widely studied aspect of the AFTs is their carcinogenic property but only a few investigators have concerned themselves with preventing AFT-induced heptocarcinogenesis. For example, Novi (1981) reported that glutathione can alter the development of AFB_1-induced hepatocellular carcinomas. One under-explored prevention possibility involves the employment of trace elements, e.g.,

Biodeterioration Research 4, Edited by G.C. Llewellyn *et al.*, Plenum Press, New York, 1994

copper, manganese, and zinc as potential protectors against aflatoxicosis. Several studies regarding this topic have emanated from a single laboratory (Kimbrough et al., 1981; Lalor and Llewellyn, 1981; Llewellyn et al., 1980; 1981; 1983; 1985; Weekley et al., 1981). Both comprehensive (Llewellyn et al., 1982) and critical (Llewellyn et al., 1984) reviews evaluating studies regarding trace element-induced protection against aflatoxicosis have appeared.

The basis to employ trace elements had its origins in the proposed role of these elements as possible protectors against carcinogenesis (Fare and Howell, 1964; Carlton and Price, 1967; Kamamato et al., 1973). Indeed, recent investigations have revealed that the distribution of trace elements differs in normal and malignant human tissues (Drake and Sky-Peck, 1989). Various reviews have been published regarding the role of trace elements and health (Underwood, 1979) as well as their possible role in cancer prevention (Schwartz, 1975) or therapy (Bertino, 1979; Taylor, 1985). These reviews also include those connected with carcinogenesis as a result of the deficiency of certain essential trace elements (Marczynski, 1988) and the distribution of trace elements in malignant neoplasms. However, Watson (1986) cautioned that it is premature for investigators to conclude that dietary Mn^{++}, Mo^{++}, and Cu^{++} are useful in cancer prevention.

Copper can inhibit neoplastic growth thereby shielding rat liver from damage resulting from 2-acetminofluorene, α-napthylaminobenzene, DL-ethionine, thioacetamide, 3-methoxy-4-aminobenzene or its N-methyl derivatives (Fare and Howell, 1964; Kamamoto et al., 1973; Carlton and Price, 1967). This sparing action might result from the competitive binding of copper with carcinogens for available protein sites within the liver (Fare and Howell, 1964). Barber et al. (1968) maintained swine upon AFB_1 contaminated diets with and without supplemental copper sulfate. These investigators found that the addition of 250 ppm copper resulted in an improved weight gain and feed efficiency over a diet containing only AFB_1. In contrast, Goodall (1964) demonstrated that the inhibition of azo-dye-hepatocarcinogenesis by excess copper was not a general phenomenon.

The investigations of Llewellyn and his associates (Kimbrough et al., 1981;, Llewellyn et al., 1981; Weekley et al., 1981) regarding whether trace elements can forestall aflatoxicosis have centered about blood chemistry, body and organ weights, as well as histological examination of excised organs as monitors of prevention. Here is reported an ultrastructural analysis of whether Mn^{++} can alter aflatoxin-induced liver damage in rats.

MATERIALS AND METHODS

Aflatoxin Preparation

Mixed AFTs were prepared using coconut (Baker Angel Flake, General Foods, White Plains, NY and Durkee Sweetened Flaked, Cleveland, OH) as a substrate (Samarajeewa, 1972). Twenty g coconut and 10 ml distilled water within cotton stoppered 500 ml flasks were sterilized by autoclaving (20 min, 15 psi, 12°C). Eighty flasks were inoculated with A. parasiticus strain NRRL 2999 spores (Shotwell et al., 1966; Llewellyn et al., 1973) using sterile procedures within a hood. The flasks, which were maintained at 24 ± 4°C in the dark, were shaken daily to enhance mycelial growth. After two weeks, the media became yellow to orange in color whereupon the AFTs were extracted through the addition of 50 ml chloroform to each flask.

The flasks were shaken upon a gyrotory shaker for at least 0.5 hr. A vegetable press was utilized to obtain the chloroform extract which was filtered by vacuum filtration through a Buchner funnel lined with medium porosity Whatman filter paper. The filtrate containing the AFTs was maintained within an amber bottle wrapped with aluminum foil to eliminate AFT photodestruction. When utilizing chloroform, a Norton Protex (series 7500, Cranston, RI) respirator equipped with pesticide type filters was used. Analysis of the extract for AFTs content using an Association of Official Analytical Chemists' procedure (Horwitz et al., 1975a) yielded 165. 1 μg/ml AFB_1, 4.3 μg/ml AFB_2, 736. 1 μg/ml AFG_1, and 13.8 μg/ml AFG_2.

Diet Preparation

The control (C) diet consisted of Purina Laboratory Meal for rats, mice and hamsters (Ralston Purina, St. Louis, MO). For construction of AFT (A) diets, a stock was prepared by spraying 20 ml aliquots of the chloroform extract under controlled conditions onto 500 g of powdered meal until 150 ml was added. For preparation of the A diet, 83.3 g of the stock was mixed with 2916.7 g of regular meal within a closed container upon a roller type mixer for two to three hr. The AFTs plus Mn^{++} diet (AMn) had 27.0 g $MnSO_4$ H_{2O} added. This addition yielded an A diet containing 0.92 μg/g AFB_1, 0.03 μg/g AFB_2, 4.64 μg/g

AFG_1 and 0.08 $\mu g/g$ AFG_2, and a Mn^{++} diet of 2995 ppm since there was 71 ppm of the element within the feed (Horwitz et al., 1975b).

Animals

Weanling male Fischer rats, approximately five weeks old, were purchased from the Charles River Breeding Laboratories, Inc. (Wilmington, MA). The rats were housed upon a Wahmann (Baltimore, MD) wire-bottomed animal rack consisting of five shelves with six cages/shelf on both sides, back to back. The animals were placed upon the upper four shelves of one side and the third and fourth shelf of the opposite side so that animals fed an A diet were not adjacent to either C or Mn^{++} diet groups. To avoid the influence of estrogen on the AFTs response, only males were employed. Nine animals each were assigned to the C and treatment groups.

Purina Laboratory Meal was provided ad libitum within 125 ml glass feeders which were bolted to the cages' bottoms to eliminate spillage. Distilled water was provided ad libitum within standard glass cage water bottles fitted with a metal shield between the rubber stopper and the cage in order that the animals would not chew upon the rubber. The animal quarters were maintained at 25 $\pm 3°C$ and a photoperiod of 14 hr light/10 hr darkness.

For 19 days the animals were allowed to both acclimate and stabilize their body weights. Then, they were divided into four groups of nine animals possessing equivalent weights. Both body weight and feed consumption were recorded every two to three days for 122 days with a Mettler PC 4400 balance (Mettler Instrument Corporation, Hightstown, NJ).

Animal Sacrifice

One hundred twenty-two days after being transferred to the above diet groups, rats were sacrificed by exsanguination.

Preparation of Samples for Electron Microscopy

Freshly excised livers were placed into 6.25% glutaraldehyde buffered with pH 7.2, 0.1 M KH_2PO_4 for 12 hr (Sabatini et al., 1963). Six hr after being placed into the glutaraldehyde, the livers were cut into small pieces (1-5 mm) and pre-fixation was permitted to continue for 6 additional hr. Subsequent to pre-fixation, the liver pieces were washed twice with 0.1 M phosphae buffer (5 min each time) and then post-fixed for 1 hr in 2% aqueous $KMnO_4$. Following post-fixation, the pieces were washed twice with 0.1 M phosphate buffer (5 min each time), dehydrated through a graded ethanol series, washed twice with propylene oxide (30 min each time) and progressively embedded in propylene oxide-Epon (Luft, 1961). Thin sections were viewed with an electron microscope.

RESULTS

Figure 1 A-D presents electron micrographs of hepatocytes from control (A), AFTs (B), Mn^{++} (C), and AFTs plus Mn^{++} (D) fed animals. Except for the AFTs-treated animals, the intrastructural appearance of the cytoplasm of the hepatocytes from both the control and treatment groups was quite similar, i.e., an abundance of mitochondria and stacked smooth membranes and few electron dense granules. In contrast, the AFTs-treated animals exhibited a marked diminution in the numbers of stacked smooth membranes and an enhancement in the quantity of the electron dense particles.

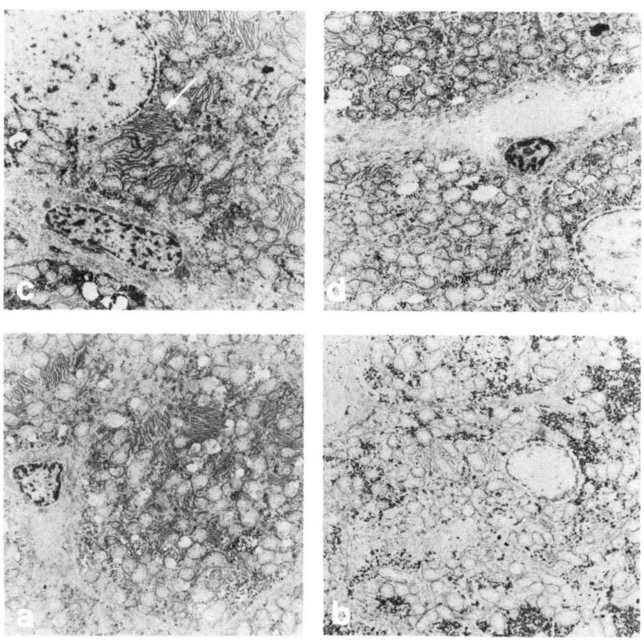

Figure 1A-D. Survey Electron Micrographs of Hepatocytes From Rats Fed Control (A), AFTs (B), Mn^{++} (C), and AFTs plus Mn^{++} (D) Diets. Counterstained with Lead Hydroxide (X 4000-7000).

DISCUSSION

Mn^{++} Altered AFTs-Induced Damage

The electron micrograph presented within Figure 1B demonstrates that dietary AFTs reduced the quantity of the stacked smooth membranes as well as increased the amount of electron dense particles whose origin is uncertain. In contrast, diets containing Mn^{++} either singly (Figure 1C), or in combination with AFTs (Figure 1D) yielded an image similar to that of the control, i.e., an abundance of the stacked smooth membranes and a few electron dense particles (Figure 1A). These observations suggest that Mn^{++} can alter AFT-induced ultrastructural hepatocyte damage. This suggestion is supported by both growth and biochemical investigations utilizing trace elements and a variety of environmental carcinogens.

Limitations

Both the identity and function of the stacked smooth membranous structure which appeared to be affected by dietary AFTs need to be established with certitude. This could be accomplished through isolation and purification of the membranes by combined differential and sucrose gradient centrifugations coupled with the utilization of marker enzymes and/or histochemical substrate methods for detecting enzymes within the hepatocyte organelles. Substrate and immunoelectronmicroscopy localizations of enzymes within the stacked smooth membranous component might provide keys to its function. The morphological appearance of the stacked smooth membranes suggests that they are either smooth endoplasmic reticulum or dictyosomes. However, since $KMnO_4$ was employed as the post-fixative, the membranes could be endoplasmic reticulum since the oxidant usually does not preserve ribosomes. In this connection, Svoboda and Reddy (1975) pointed out that one of the effects of AFTs is to detach ribosomes from the rough endoplasmic reticulum. Similarly, the identity of the electron dense particles remains to be established. While these particles are too large to be ribosomes, it is conceivable that they may represent an iron-containing constituent since infiltration of the liver with RBCs is characteristic of AFTs-treated animals.

Although the concentration of Mn^{++}, which was employed herein, was elevated it did not appear to produce toxic effects as measured by serum cholesterol and 'crude' liver lipid contents (Floyd and Llewellyn, unpublished).

Nevertheless, it is desirable to repeat the experiment reported here but with diets containing other levels of Mn^{++} (100-2000 ppm).

Another limitation is that only a single fixation image has been presented. Although it is replicated in triplicate, it would be useful to post-fix segments of livers from the various dietary groups in buffered OsO_4. Should the stacked smooth membrane complex which was affected by AFTs be present following OsO_4 post-fixation, evidence would be provided that the membrane complex is present in situ.

Finally, the mechanism by which Mn^{++} may prevent AFTs-induced liver damage should be explored. Since Mn^{++} can substitute for either Ca^{++} or Mg^{++} as a co-factor for certain enzymes, the mechanism may involve an alteration in the activity of a critical enzyme. In this connection, glutathione S-transferase activity can be induced by dietary AFTs (Maingi and Campbell, 1981). Glutathione has been reported to inhibit the development of hepatocellularcarcinomas in Wistar rats (Novi, 1981). The precise mechanism by which glutathione could reverse hepatocellularcarcinogenesis is unclear. Glutathione may be involved in amino acid transport at the membrane level via the action of δ-glutamyl transferase. Whether there is a relationship between this purported role for glutathione in its reversal of hepatocellular carcinogenesis requires further confirmation but the positive response of Mn-supplements on the livers of AFT-treated Fischer rats was definitely evident at this biolevel.

An alternative explanation is that trace elements may enhance immune functions. This possibility cannot be ignored (Watson, 1986). Whatever the case, human cancer development seems to be linked to possible deficiencies or even excesses in dietary trace elements (Schrauzer et al., 1977).

SUMMARY

This report provides insight into the potential for Mn^{++} treatment to aflatoxin-induced liver damage in rats to alter in a positive fashion some of the toxic aspects. These probable protective responses to aflatoxicosis complement previous companion changes indicating that supplemental Mn^{++} is indeed interacting in some fashion with the ultrastructural changes associated with this carcinogenic mycotoxin. The exact mode of action and specifics will require further clarification. Possibly the Mn^{++} acts more in the regenerative or healing process and less in the protective mode. What ever be its mode of action, there is little doubt that there is an observable and positive influence noted in this study. Hopefully such aspects would be found in other animal systems too.

Acknowledgments

The authors are grateful to Marcella Gower, Pamela Markley, Pamela Caldwell and Gladys Llewellyn for assistance. This work was supported in part by an NIH-RCMI grant No. RRO-3062002 and Virginia Commonwealth University. The views expressed in this paper are those of its authors and do not necessarily represent the official positions or policies of the departments included.

REFERENCES

Barber, R.S., Braude, R., Mitchell, K.G., Harding, J.D., Lewis, G. and Loonsmore, R. M. (1968). The effects of feeding toxic groundnut meal to growing pigs and its interaction with high copper diets. Brit. J. Nutr. 22:535.

Bertino, J.R. (1979). Nutrients, vitamins and minerals as therapy. Cancer. 43:2137.

Carlton, W.M. and Price, P.S. (1967). Dietary copper and the induction of neoplasms in the rat by acetylaminofluorene and dimethylnitrosamine. Fd. Cosmet. Toxicol. 2:827.

Drake, E.N. and Sky-Peck, H.H. (1989). Discriminant analysis of trace element distribution in normal and malignant human tissues. Cancer Res. 49:4210.

Fare, G. and Howell, J.S. (1964). The effect of dietary copper on rat carcinogenesisby 3-methoxy dyes. I. Tumors induced at various sites by feeding 3-methoxy-4-aminoazobenzene and its N-methyl derivative. Cancer Res. 24:1279.

Goodall, C.M. (1964). Failure of copper to inhibit carcinogenesis by 2-aminofluorene. Br. J. Cancer. 18:777.

Horwitz, W., Sensel, A., and Reynolds, H. (1975a). Natural poisons. In: Official Methods of Analysis, p. 465. Washington, D.C.: Assoc. Offic. Anal. Chem. Soc.

Horwitz, W., Sensel, A., and Reynolds, H. (1975). Metal analyses. In: Official Methods of Analysis, p. 18. Washington, D.C.: Assoc. Offic. Anal. Chem. Soc.

Kamamoto, Y., Makiura, S., Hiasa, Y., Arai, M., and Nobuyuki, I. (1973). The inhibitory effect of copper on DL-ethionine carcinogenesis. Cancer Res. 23:1129.

Kimbrough, T.D., Weekley, L.B., Lawson, C.J., O'Rear, C.E. and Llewellyn, G.C. (1981). Mycotoxin and copper induced gastrointestinal disturbances in male Holtzman rats. Dev. Ind. Microbiol. 22:771.

Lalor, J.H. and Llewellyn, G.C. (1981). Biointeraction of sodium selenite and of aflatoxin B_1 in the Mongolian gerbil. J. Toxicol. Envir. Health. 8:387.

Llewellyn, G.C., Carlton, W.W., Robbers, J.E., and Hansen, W.G. (1973). A rapid and simplified method for the preparation of pure aflatoxin B_1. Dev. Ind. Microbiol. 14:325.

Llewellyn, G.C., Hastings, C.E., and Hoffman, J.W. (1980). Biointeraction of aflatoxin and zinc in Mongolian gerbils. Meriones unguiculatus. Toxicon. 18:107.

Llewellyn, G.C., Thomen, L.E., and Katzen, J.S. (1981). Effects of dietary copper on developing aflatoxicosis in Syrian hamsters. J. Envir. Sci. Health. B16:211.

Llewellyn, G.C., (1982). Biointeraction of aflatoxins and metals. A review. Dev. Ind. Microbiol. 27:237.

Llewellyn, G.C., O'Rear, C.E,. and Dashek, W.V. (1984). Do dietary trace elements reverse aflatoxicosis in selected rodents? Dev. Ind. Microbiol. 23:779.

Llewellyn, G.C., Morgan, J.D., and Dashek, W.V. (1985). Evaluation of the response of Syrian hamsters exposed to copper acetate and mixed aflatoxin diets. Proceedings of the 4th International Congress of Plant Pathology and the 22nd World Veterinary Congress, Sydney, Australia. Chapt. 43, pp. 469-485.

Luft, J.H. (1961). Improvements in Epoxy resin embedding methods. J. Biophy. Biochem. Cytology. 9:409.

Mainigi, K.D. and Campbell, T.C. (1981). Effects of low dietary protein and dietary aflatoxin on hepatic glutathione levels in F-344 rats. Toxicol. Appl. Pharmacol. 59:196.

Marczynski, B. (1988). Carcinogenesis as the result of the deficiency of some essential trace elements. Med. Hypotheses. 26:239.

Nesbitt, B.F., O'Kelly, J., Sargeant, K., and Sheridan, A. (1962). Toxic metabolites of Aspergillus flavus. Nature. 195:106.

Novi, A.M. (1981). Regression of aflatoxin B_1-induced hepatocellular carcinomas by reduced glutathione. Science. 212:421.

Sabatini, D.D., Bensch, K., and Barrnett, R.J. (1963). Cytochemistry and electron microscopy. J. Cell Biol. 17:19.

Samarajeewa, U. (1972). Aflatoxins in coconut products. Ceylon Coconut Quart. 23:108.

Schrauzer, G.N., White, D.A., and Schneider, C.J. (1977). Cancer mortality correlation studies - IV: association with dietary intakes and blood levels of certain trace elements, notably Se-antagonists. Bioinorg. Chem. 7:35.

Schwartz, M.K. (1975). Role of trace elements in cancer. Cancer Res. 35:3481.

Shotwell, O.L., Hesseltine, C.W., Stubblefield, R.D., and Sorenson, W.D. (1966). Production of aflatoxin on rice. Appl. Microbiol. 14:425.

Svoboda, D. and Reddy, J. (1975). Some effects of chemical carcinogens on cell organelles. In: Cancer. A comprehensive Treatise. I. Etiology and Physical Carcinogenesis (F.F. Becker, Eds.), Plenum Press, NY.

Taylor, A. (1985). Therapeutic uses of trace elements. Clin. Endocrinol. Metab. 14:703.

Underwood, E.J. (1979). Trace elements and health: an overview. Philos. Trans. R. Soc. London 288:5.

Weekley, L.G., Kimbrough, T.D., and Llewellyn, G.C. (1981). Dietary aflatoxin and cooper acetate effects on various blood parameters in rats. Drug Chem. Toxicol. 4:113.

Watson, R.R. (1986). Immunological enhancement by fat-soluble vitamins, minerals and trace metals: a factor in cancer prevention. Cancer Detect. Prev. 9:67.

Wogan, G.N. (1964). Mycotoxins in Foodstuffs. MIT Press, Cambridge, MA.

THE EFFECT OF FUNGAL TOXINS ON THE GERMINATION OF MITOSPORES FROM THERMOPHILIC FUNGI

JAMES J. DEPLOEY[*], Department of Biology, The Pennsylvania State University, York Campus, York, PA 17403, USA.
CHARLES E. O'REAR, Department of Forensic Sciences, The George Washington University, Washington, D.C. 20052, USA
GERALD C. LLEWELLYN, Department of Forensic Sciences, The George Washington University, Washington, D.C. 20052, and Delaware Department of Natural Resources and Environmental Control, New Castle, DE 19720, USA

INTRODUCTION

Fungi in several genera produce toxins that vary in toxicity to humans and animals. The most prominent fungal toxins are those that are used as antibiotics in human and veterinary medicine, and their effect on pathogenic microbes is well known. The lesser known fungal toxins include those produced by *Penicillium* species (citreoviridin, penicillic acid, gliotoxin, patulin), *Aspergillus* species (aflatoxin, citrinin, gliotoxin, helvolic acid, ochratoxin, patulin), *Trichoderma* species (gliotoxin), *Stemphilium* (radicinin), *Gibberella* and *Fusarium* (zearalanone). Some toxigenic fungi (*A. fumigatus*, *A. flavus*, *A. parasiticus*) grow on stored vegetation and compost contributing to their decay while releasing toxin. Since both mesophilic and thermophilic fungi grow on these substrates their spores would be exposed to fungal toxins and spore germination could be affected. A few studies have been done on the effects of fungal toxins on the germination of spores from mesophilic fungi (Brian and Hemming, 1945; Reiss, 1973; Deshmukh and Agrawal, 1984), and only one report (Thakre and Johri, 1973) was found dealing with the effects of fungal toxins on the germination of spores from thermophilic fungi. Furthermore, several studies have been done concerning the effects of fungal toxins on the germination of tracheophyte spores, pollen, and seeds (Reynolds et al., 1978; Jones et al., 1980; Dashek et al., 1981a; Dashek et al., 1981b; Dashek et al., 1982; Llewellyn et al., 1982; Llewellyn et al., 1985) to determine the effects of toxins on the germination of these structures or to use their germination in toxin solutions as a bioassay for toxin.

Biodeterioration Research 4, Edited by G.C. Llewellyn *et al.*, Plenum Press, New York, 1994

Since so little is known of the effects of fungal toxins on fungal spores, especially spores from thermophilic fungi, it was believed that such a study would be worthwhile. The results of this study are presented here.

METHODS

Two thermophilic fungi were studied: R. pusillus (ATCC 22074), and Dactylomyces thermophilus (NRRL 5208). The fungi were cultured in tubes containing slanted glucose-yeast extract agar (Adams, 1981), and were at 40C for 3 days. Spores were obtained from the tube cultures and washed several times with sterile distilled water using a method previously described (Deploey, 1985). The spore concentration of each solution was adjusted to 1,000,000/ml (double the required concentration) by adding sterile, glucose-basal (GB) (Deploey and Fergus, 1975) liquid medium prepared at double strength.

Five toxins were studied: helvolic acid, patulin, penicillic acid, rubratoxin B, and secalonic acid. Three concentrations of each toxin were prepared (at double strength) 100, 50, and 20 μg/ml. The spore suspensions, the GB liquid medium used to make the final spore suspension, and the 3 concentrations of each toxin solution were prepared at double strength so that when equal amounts of the spore suspension and a toxin solution were mixed they diluted each other to the desired concentrations.

Fifteen sterile test tubes (4 ml) were prepared and labeled (five toxins, three concentrations) for each fungus studied. To each test tube were added equal amounts of a GB medium/spore suspension and the appropriate concentration of a toxin.

Two types of controls were prepared for each fungus. One control contained equal amounts of GB medium/spore suspension and sterile distilled water. The second type of control contained equal amounts of GB medium/spore suspension and 0.8% sterile, aqueous, acetone solution to simulate the concentration of acetone used to dissolve four of the toxins (helvolic acid, penicillic acid, rubratoxin B, secalonic acid).

All of the tube openings were sealed with sterile parafilm, and incubated in a water bath (45C) for 6 hr. Percent spore germination was determined by examining (450 X) 300 spores from each tube. The average germ tube length of at least 12 randomly selected spores was also determined.

Experiments were repeated three times for R. pusillus, and twice for D. thermophilus. All materials used were of reagent grade.

RESULTS

Some toxins appeared to stimulate spore germination (Table 1, 1a). The percent germination of D. thermophilus and R. pusillus spores was higher than that of the controls when they were incubated in one or more of the three solutions (10, 20, 50 μg/ml) of helvolic acid, penicillic acid, and rubratoxin B. However, other toxins appeared to inhibit spore germination. Spores of D. thermophilus and R. pusillus germinated at a lower percent than the controls when they were incubated in the three solutions (10, 25, 50 μg/ml) of patulin and secalonic acid. When these data were analyzed statistically (analysis of variance; Duncan's multiple range test), there was no significant difference (at the 5% level) in percent germination between the controls and the three toxin concentrations studied. Furthermore, no significant difference in percent germination was evident (at the 5% level) when the spores were incubated in the three concentrations of the five toxins.

The pattern of fungal toxin effects on germ tube elongation was similar to their effect on percent spore germination (Table 1). Germ tube elongation by D. thermophilus spores was greater than that of the germ tube elongation by control spores when they were incubated in at least one of the three solutions of the five toxins. Germ tube elongation by spores of R. pusillus was greater than that of germ tube elongation by control spores when they were incubated in at least one of the three solutions of helvolic acid, penicillic acid, and rubratoxin B. When incubated in solutions of patulin and secalonic acid, germ tube elongation by spores of R. pusillus was less than that observed in the controls. Based on the statistical analysis of these data, when spores of D. thermophilus and R. pusillus were incubated in solutions of penicillic acid and rubratoxin B, respectively, a significant difference (at the 5% level) in germ tube length was observed between spores incubated in the control solutions lacking these toxins and solutions containing 10 μg/ml of these two toxins. Also, a significant difference (5% level) in germ tube length was observed when spores of D. thermophilus and R. pusillus were incubated in solutions (10 and 50 μg/ml) of penicillic acid and rubratoxin B. Differences in germ tube length resulting when spores of D. thermophilus and R. pusillus were incubated in the three concentrations of the other three toxins (helvolic acid, patulin, secalonic acid) were not significantly different (at the 5% level) when comparisons were made between the three concentrations of each toxin or between the controls and the three concentrations of each toxin.

DISCUSSION

Of the 5 toxins tested here only patulin is soluble in water. Aqueous acetone (0.8%) was used to dissolve the remaining toxins. This required preparing controls containing and lacking acetone to determine its effect on

Table 1. Effect of Fungal Toxins on the Percent Spore Germination and Germ Tube Elongation of Thermophilic Fungal Mitospores Incubated in Liquid Glucose-Basal Medium at 45C for 6 hr.

Toxin Conc.µg/ml	Helvolic Acid		Patulin		Penicillic Acid		Rubratoxin B		Secalonic Acid	
	% Germ	Germ Tube Length (µ)	% Germ	Germ Tube Length (µ)	% Germ	Germ Tube Length (µ)	% Germ	Germ Tube Length (µ)	% Germ	Germ Tube Length (µ)
				Dactylomyces thermophilus						
10	26.9	49.5	29.0	49.4	32.9	89.2	34.4	14.3	0.2	5.5
25	34.3	36.4	36.8	38.9	34.6	34.8	0.4	29.7	0.3	11.0
50	32.4	26.1	32.9	35.2	22.0	11.8	0.2	7.7	0.3	9.9
				Rhizomucor pusillis						
10	30.5	55.1	0.9	53.1	4.9	67.2	23.6	52.2	0.5	4.8
25	20.7	49.7	0.1	0.7	3.3	115.5	4.1	23.5	0.2	2.9
50	29.4	45.9	0.0	0.0	1.0	39.5	0.5	9.5	0.4	11.0

Table 1a. Control effects

	Control GB only no acetone		Control - GB Acetone	
	% Germ	Germ Tube length (µ)	% Germ	Germ tube length (µ)
D. thermophilus	41.8	29.8	27.7	7.1
R. pusillus	12.0	62.3	1.6	13.9

percent germination and germ tube elongation. Percent germination and germ tube elongation of *R. pusillus* spores were 7.5 and 4.5 times greater, respectively, in the controls lacking acetone compared to controls containing acetone. The percent germination and germ tube length of *D. thermophilus* spores incubated in the control solutions lacking acetone were also more (1.5 fold and 4.2 fold, respectively) than that observed in the controls containing acetone.

When the control data for both fungi were analyzed statistically (analysis of variance), the difference between the percent germination (observed $P = 0.076$) and germ tube length (observed $P = 0.059$) of R. pusillus "GB only" controls and the "GB-acetone" controls was significant at the 10% level. The difference between the percent germination of D. thermophilus "GB only" controls and the "GB-acetone" control was not significant (observed $P = 0.7813$), but the difference in the length of the germ tubes of the 2 controls proved to be statistically significant at the 5% level (observed $P = 0.0323$). Accordingly, when comparisons are made between controls and experimentals herein the "GB-acetone" controls should be used. The exception to this would be the study of spore germination using solutions containing patulin. Since patulin is water-soluble, acetone was not added to the spore germination medium.

In an earlier study by Reiss (1973) of the effect of patulin on spore germination of mitospores from several species of mesophilic fungi dimethylformamide was used to dissolve the toxin; however, no reduction in spore germination was observed in the control solutions containing dimethylformamide. If it can be demonstrated that dimethylformamide dissolves the other toxins studied herein and doesn't inhibit germination of spores from thermophilic fungi, future studies of insoluble toxins on thermophilic spore germination should be done using this solvent.

One aspect of the report by Brian and Hemming (1945) centered on measuring gliotoxin activity using percent germination of Botrytis allii spores as a measure of toxicity. Serial dilutions were made of the gliotoxin solution to be assayed. Equal amounts of the diluted gliotoxin solutions and a spore suspension of B. allii spores were mixed and several drops of each mixture were placed on microscope slides. Then, the slides were incubated for 16-18 h in a moist chamber. Evaluation of spore germination was semiquantitative in that the degree of spore germination was estimated and not actually counted. Six levels of spore germination were established (0%, less than 1%, 1-5%, 5-50%, 50-90%, 90-100%). This system proved to be sufficiently accurate to allow for rapid evaluation of the toxicity of the large number of samples they evaluated. However, such data could not be analyzed statistically and thus would be of limited value.

Later, Reiss (1973) reported that the effect of patulin on spore germination by Thamnidium elegans was dose related. He observed no germination by T. elegans spores incubated in solutions containing 100 μg of patulin/ml, but spore germination occurred at a reduced level in solutions containing 10 μg/ml (compared to the controls). Spores of R. pusillus responded in a fashion similar (Table 1) to T. elegans when incubated in solutions of patulin. The percent spore germination and germ tube elongation of R. pusillus spores declined markedly as patulin concentration increased, but these declines were not linear, and there was no significant difference (at the 5% level) in percent spore germination or germ tube elongation. Percent germination and germ tube length of D. thermophilus spores incubated in the three patulin solutions (Table 1) were virtually identical to the controls (GB only).

The apparent stimulation of spore germination and germ tube elongation by some of these toxins (Table 1) was unexpected since all of the reports in the literature concerning the effect of fungal toxins on fungal spore germination stated that the toxins either inhibited spore germination or had no effect on germination. Especially striking in this regard was the effect of helvolic acid on spore germination and germ tube elongation by R. pusillus where as much as a 19-fold increase in spore germination and a 4-fold increase in germ tube elongation was observed compared to the controls (GB-acetone), but these differences proved to be statistically insignificant at the 5% level. Rubratoxin B also stimulated spore germination and germ tube elongation of R. pusillus spores by 14.8 fold and 3.8 fold, respectively, at the lowest concentration tested (10 μg/ml), but at increasing concentrations of the toxin the percent spore germination and length of the germ tube declined. When analyzed statistically, a significant difference (5% level) was observed in germ tube elongation when R. pusillus spores incubated in the control solution (containing acetone) and those cultured in solutions containing 10 μg/ml of rubratoxin B were compared. Also, a statistically significant difference (5% level) in germ tube elongation was recorded when R. pusillus spores, incubated in solutions containing 10 and 50 μg/ml of rubratoxin B, were compared.

Other investigators have reported that fungal spore germination could be inhibited by fungal metabolites. For example, Wadia et al. (1986) found that the culture filtrates of Colletotrichum gleosporioides and Penicillium expansum inhibited the germination of conidia from the fruit rot fungus Pestalotiopsis palmarum. Furthermore, growth of P. palmarum was suppressed by the culture filtrates of Aspergillus terreus, C. gleosporioides, Curvularia lunata, Fusarium semitectum, and P. expansum, but the factors inhibiting spore germination and growth by P. palmarum were not identified.

The only study of the effects of fungal toxins (antibiotics) on the germination of spores from thermophilic fungi was the report by Thakre and Johri (1973). They observed that three antibiotics (aureofungin, amphotericin, hamycin) reduced the percent germination and the degree of spore swelling of spores from A. fumigatus, Humicola lanuginosa, Sporotrichum thermophile, and Rhizopus sp. It would be interesting to see how these 3 antibiotics affected the germination of spores from R. pusillus and D. thermophilus.

A number of studies have been done to determine the effect of fungal toxins on the germination of spores, pollen, and seeds from tracheophytes. Reynolds et al. (1978) found that aflatoxin B_1 reduced or completely inhibited the germination of fern (Athyrium asplenoides) spores as the concentration of toxin increased. They also reported that A. asplenoides spores were more sensitive to aflatoxin than other tracheophyte structures (angiosperm pollen and seeds, fern sporocarps), but were not more sensitive to aflatoxin than the chick embryo assay (sensitive to 0.025 μg/ml). Later, Jones et al. (1980) reported that germination of soybean seeds was inhibited by exposure to various concentrations of aflatoxin for 18 hr. Also, inhibition of germination increased as concentration of aflatoxin increased but aflatoxin B_1 stimulated pollen tube elongation (Lilium longiflorum) and the degree of elongation increased as toxin concentration increased. However, Dashek et al. (1982) concluded that lily pollen was not as sensitive as other chemical and biological assays although they believed it could function as a confirmatory assay for aflatoxin. Studies to determine if the percent germination and/or germ tube elongation of spores from thermophilic fungi could be used to assay for aflatoxin would be worth undertaking since no one has attempted to assay for aflatoxin using these organisms.

SUMMARY

The effects of five fungal toxins (helvolic acid, patulin, penicillic acid, rubratoxin B, and secalonic acid) on the percent germination and germ tube elongation of spores from Rhizomucor pusillus and Dactylomyces thermophilus were studied. Spore suspensions were prepared from tube slant cultures of R. pusillus and D. thermophilus. The spore concentration of each solution was adjusted to 1,000,000/ml by adding sterile glucose-basal medium. Three concentrations of each toxin were prepared (10, 25, 50 μg/ml). Fifteen sterile test tubes were prepared (plus controls) for each fungus (five toxins, three concentrations) containing the spore suspension and toxin.

The tubes were sealed and incubated in a water bath at 45C for 6 hr. Percent germination and germ tube length were determined by examining 300 randomly selected spores from each tube.

The percent germination of D. thermophilus spores incubated in solutions of helvolic acid, patulin, and penicillic acid did not differ markedly from the controls, but when incubated in some or all of the solutions of rubratoxin B and secalonic acid percent germination declined noticeably compared to the controls. The percent germination of R. pusillus spores increased (compared to the controls) when incubated in some or all of the solutions of helvolic acid, penicillic acid, and rubratoxin B, but percent germination was lower than the controls when they were incubated in solutions of patulin and secalonic acid. However, when these data were analyzed statistically (analysis of variance, Duncan's multiple range test) no significant difference in percent germination of D. thermophilus or R. pusillus spores was evident when compared to the controls or when comparisons were made between spores incubated in the three concentrations of these five toxins.

The elongation of D. thermophilus germ tubes appeared to increase when incubated in some or all of the solutions of these five toxins. Also, the germ tube elongation of spores from R. pusillus seemed to be greater than that of the controls when incubated in some or all of the solutions of helvolic acid, penicillic acid, and rubratoxin B, but spores incubated in solutions of patulin and secalonic acid produced germ tubes that were shorter than those of the control spores. With only a few exceptions, no statistically significant difference (5% level) was observed in germ tube elongation when the germ tubes of D. thermophilus and R. pusillus were compared to the control spores or to spores incubated in the three concentrations of the five toxins. The exceptions to this were seen when D. thermophilus spores incubated in solutions of 10 and 50 μg/ml of penicillic acid were compared, and when spores incubated in 0 and 10 μg/ml of penicillic acid were compared. Also, a statistically significant difference (5% level) in germ tube elongation was observed when spores of R. pusillus incubated in solutions of 10 and 50 μg/ml of rubratoxin B were compared, and when those incubated in 0 and 10 μg/ml of rubratoxin B were compared.

Acknowledgments

We wish to express our sincere appreciation to Mrs. Linda Kline for typing the manuscript, and to Mr. Roger Twitchell for suggestions concerning its content.

REFERENCES

Adams, P.R. (1981). Amylase production by Mucor pusillus and Humicola lanuginosa as related to mycelial growth. Mycopathologia, 76, 97-101.

Brian, P.W. and Hemming, H.G. (1945). Gliotoxin, a fungistatic metabolic product of Trichoderma viride. Ann. Appl. Biol., 32, 214-220.

Dashek, W.V., Bragg, C.L., and Llewellyn, G.C. (1981a). Comparative responses of Lilium longiflorum, cv, "ace" pollen to aflatoxin B_1, B_2, and G_1 Bull. Environm. Contam. Toxicol., 27, 738-751.

Dashek, W.V., Harman, R.L., Adlestein, L.B., Morton, W.A., Rapisarda, B.M., Chancey, J.C., and Llewellyn, G.C. (1981b). Use of Lilium longiflorum, cv. ace pollen germination and tube elongation as a bioassay for the hepatocarcinogens, aflatoxins, Environ. Hth. Perspect., 40, 267-278.

Dashek, W.V., Boggs, B.W., Gamber, R.B., Hrycyk, M., Schell, S.L., and Llewellyn, G.C. (1982). Evaluating the potency of the mycotoxin aflatoxin G_1 by lily pollen germination and tube elongation bioassays. Dev. Ind. Microbiol., 23, 273-278.

Deploey, J.J. (1985). The influence of atmosphere composition and nutrients on the germination of Rhizomucor pusillus sporangiospores. Mycologia, 77, 97-102.

Deploey, J.J. and Fergus, C.L. (1975). Growth and sporulation of thermophilic fungi and actinomycetes in O_2-N_2 atmospheres. Mycologia, 67, 780-797.

Deshmukh, S.K. and Agrawal, S.C. (1984). Volatile metabolites of aspergilli in relation to spore germination of some keratinophilic fungi. Experientia, 40, 583-584.

Jones, H.C., Chancey, J.C., Morton, W.A., Dashek, W.V., and Llewellyn, G.C.(1980). Toxic responses of germinating pollen and soybeans to aflatoxins. Mycopathologia, 72, 67-73.

Llewellyn, G.C., Gentry, C.L., Mayo, E.S., and Dashek, W.V. (1982). Aflatoxin effects on spice seed germination and root elongation. Z. Lebensm. Unters. Forsch., 174, 18-22.

Llewellyn, G.C., Reynolds, J.D., and Dashek, W.V. (1985). Water fern (Marsilea quadrifolia) reproduction as a bioassay for aflatoxin B_1 and T-2 trichothecene toxin. Proceedings for Sexual Reproduction In Plants. The Netherlands, 8, 127-130.

Reiss, J. (1973). Influence of the mycotoxins patulin and diacetoxyscirpenol on fungi. J. Gen. Appl. Microbiol., 19, 415-420.

Reynolds, J.D., Cahill, S.G., and Llewellyn, G.C. (1978). Reduced fern spore germination associated with aflatoxin B_1 treatment. Incompat. Newsletter, 9, 60-65.

Thakre, R.P. and Johri, B.N. (1973). Influence of antibiotics on the swelling phase and spore germination of thermophilic fungi. Hindustan Antibiotics Bull., 16, 109-114.

Wadia, K.D.R., Manoharachary, C., and Prakash, P. (1986). Effect of fungal metabolites on some pathogenic fungi. Indian J. Bot. 9, 180-182.

LACK OF FUNGICIDAL CONTROL OF *ASPERGILLUS FLAVUS* IN FIELD CORN

H.E. DUNCAN*, Department of Plant Pathology, North Carolina State University, Raleigh, NC 27695-7616, USA
A.R. AYERS, Rhone-Polenc Ag. Company, Research Triangle Park, NC 27709-12014, USA
G.A. PAYNE, Department of Plant Pathology, North Carolina State University, Raleigh, NC 27695-7608, USA
W.M. HAGLER, JR., Mycotoxin Laboratory, Department of Poultry Science, North Carolina State University, Raleigh, NC 27695-7608, USA

INTRODUCTION

Aflatoxins B_1 and B_2, produced by Aspergillus flavus Link ex. Fries, in corn have received considerable attention. Aflatoxins were a serious problem in the 1977, 1980, and 1983 southeastern corn crops. For many years, it was thought that aflatoxins were produced only in storage. However, surveys done in South Carolina in the early 1970's clearly demonstrated that aflatoxins can be produced in corn prior to harvest (Anderson et al., 1975). The problem of aflatoxin contamination in the field, in addition to the hazard of aflatoxin buildup in storage, accentuated the need for effective control measures.

Infection of corn by A. flavus and subsequent production of aflatoxins before harvest have been well documented (Anderson et al., 1975; Jones et al., 1980; Jones et al., 1981; Marsh and Payne, 1984; Payne, 1983). Taubenhaus (1920) first reported the occurrence of A. flavus on Texas field corn in 1920. Since that time, insects have been implicated in transporting inoculum to the developing ear, moving inoculum from silks into the kernel region, and providing wounds for establishment of the fungus. Insecticides and various fumigants

Biodeterioration Research 4, Edited by G.C. Llewellyn *et al.*, Plenum Press, New York, 1994

have been used to control insects in stored grain. They have also been used in some instances in an attempt to prevent the buildup of aflatoxins. The role of insects in epidemiology of A. flavus on corn was reviewed by Widstrom (1979).

Other chemicals have been used in some cases to limit aflatoxin production in postharvest corn. The use of ammonia as a fumigant has been studied extensively. Wales and Somers (1968) screened fifteen fungicides against in vitro growth of three strains of A. flavus. Maneb prevented growth of all three strains of A. flavus. Jones and Mulkey (1982) demonstrated significant reduction in percent infection and aflatoxin B_1 concentrations in corn in the field with benomyl (Benlate) applied at 90% silk. Spray and whorl applications of triazole derivatives (CGA 64250 2.5G and CGA 64250 3.6EC) significantly reduced aflatoxin B_1 concentrations, but soil applications were ineffective. Barnes (1983) showed that foliar applications of benomyl reduced aflatoxin levels in mechanically inoculated corn in 1979, and in both mechanically inoculated and silk inoculated corn in 1980. No effect was noted in 1981. Draughon (1983) has reviewed the control or suppression of aflatoxin production with pesticides.

Work in North Carolina has demonstrated that A. flavus can infect the corn kernel through the silk in the absence of insect injury (Jones et al., 1980; Marsh and Payne, 1984; Payne, 1983). This led to the objectives of this study, which were to further elucidate the infection process of A. flavus and to see if selected fungicidal treatments would control A. flavus and thus reduce the levels of aflatoxin contamination.

MATERIAL AND METHODS

Experimental Design

Fungicidal trials were conducted at the Central Crops Research Station (CCRS), Clayton, NC, and at the Upper Coastal Plain Research Station (UCPRS), Rocky Mount, NC, in 1981 and 1982. Corn (Zea mays L.) cv "Pioneer Brand 3147" was used in all of the experiments. Seeds were planted 15 April and 22 April, 1981 at CCRS and UCPRS, respectively. The plant population was 5.4 X 10^4 plants/ha at UCPRS. The experiments were fertilized according to soil test recommendations and other good agronomic practices were followed.

The experimental design was split-plot within a randomized complete block design with four replications. Fungicides were the main plot and A. flavus inoculations were the subplots. One-half of the main plot was inoculated with a spore suspension of A. flavus, whereas the other subplot was exposed only to natural inoculum. Main plots were 9.4m long and the subplots 4.7m long. Each plot consisted of two rows spaced 91.4 cm apart, and was separated by two buffer rows on one side and two unplanted rows on the other. The unplanted rows provided an alley for the tractor and sprayer.

Inoculation Procedures

A. flavus NRRL 3357 was grown on potato dextrose agar at 28°C. The culture plates were flooded with sterile, distilled water and the spores were dislodged with a glass rod. Silks on the dominant ear were inoculated when they were yellow-brown in color by atomizing 1.0 ml/ear of a suspension containing 1×10^6 conidia/ml in 1981 and 1.5 ml/ear of a suspension containing 7.5×10^4 conidia/ml in 1982. Inoculated ears were covered with a plastic bag and then covered with a paper bag. Both bags were removed after 3-4 days. Plants were inoculated in 1981 on 15 and 16 July at CCRS and on 17 July at UCPRS. In 1982, plants were inoculated 16 July and 22 July at CCRS and UCPRS, respectively. Inoculations were done 2-3 days after the first fungicide spray applications were made.

Fungicide Applications

Various fungicidal formulations, rates and methods of application were used in these tests (Tables 1, 2). The experimental fungicide CGA 64250 (Ciba-Geigy Corp., Tilt 2.5G and/or 3.6EC) was applied as a layby treatment, tassel treatment, or a spray. The other fungicides, benomyl (Benlate 50WP), maneb + zinc (Manzate D), thiabendazole (Mertect 340F) and liquid maneb + zinc (Manex) were applied as sprays. In 1982, the experimental fungicide DPX 3866 75G (E. I. DuPont de Nemours & Co., Inc.) was applied as a layby treatment. The layby treatments were applied in an 18-cm band on both sides of the row and incorporated to a depth of 5-8 cm with a rolling cultivator in 1981 and manually with a hoe in 1982. Tassel treatments were applied manually by shaking the fungicide onto newly emerged tassels from glass jars with perforated lids designed to deliver the chemical at the specified rate. Spray treatments were applied with a tractor-mounted, boom sprayer operated at 21 kg/cm^2 pressure. The boom contained seven nozzles per row arranged so that the spray pattern gave complete coverage of the corn ear silk, the tassel and the upper 50% of the foliage.

Layby and tassel treatments for CGA 64250 were applied on 10 June and 29 June, respectively in 1981. In 1982, the layby treatments for CGA 64250 and DPX 3866 were applied on 14 June. The tassel treatment for CGA 64250 was applied on 6 July.

Spray schedules began at silking (90% silk) and continued on weekly intervals for 1 and/or 2 weeks, thereafter. The spray schedule in 1981 started on 13 and 14 July at CCRS and UCPRS, respectively. In 1982, the schedule started 13 and 19 July at CCRS and UCPRS, respectively.

Aflatoxin Analysis

Every dominant ear in each plot was hand harvested in 1981 on 16 and 22 September at CCRS and UCPRS, respectively, and in 1982, 14 and 21 September at CCRS and UCPRS, respectively. Each ear was rated for visible growth of *A. flavus* and *Fusarium spp.*, and insect damage (European corn borer and corn earworm). After the ears were rated, they were bulked and machine shelled. The grain was weighed, and the moisture content was determined using a Dicky-John moisture meter (Model 15-C, Dicky-John Corp., Auburn, IL 62615). A 4.6 kg subsample of grain was randomly taken for aflatoxin analysis. Each subsample was course ground through a Dickens subsampling mill (Dickens and Satterwhite, 1969). Subsamples were then riffle divided and approximately 250g were finely ground in a Wiley Mill (Model 4 with a 0.84 mm, 20-mesh screen). Samples were dried at 70°C for at least 48 hours in a forced air oven. Fifty grams from each subsample were stored at 2 - 5°C until analyzed for aflatoxin. The procedures of the CB method and thin layer chromatography were used for extraction and detection of aflatoxins (Association of Official Analytical Chemists, 1980).

Data on plant populations, total numbers of ears, number of ears showing visible growth of *A. flavus* and *Fusarium spp.*, insect damage, yields, and aflatoxin B_1 and B_2 concentrations were subjected to an analysis of variance. Mean comparisons were made with the Waller-Duncan K-ratio t-test. Correlation analyses were also used to determine the relationships among aflatoxin B_1 and B_2 concentrations, number of ears showing visible growth of *A. flavus* and *Fusarium spp.*, insect damage, and yields.

Table 1. Effects of Fungicides on Aflatoxin Concentrations, Central Crops Research Station (CCRS) and Upper Coastal Plain Research Station (UCPRS) 1981.

Fungicide	Formulation	Rate (kg a.i./ha)	Method of Application		Aflatoxin B_1 Concentrations ($\mu g/kg$)			
					CCRS		UCPRS	
					Inoc.	Non-Inoc.	Inoc.	Non-Inoc.
CGA-64250	2.5G	0.6	Layby		2896	288	1925 BCD	282
CGA-64250	2.5G	0.9	Layby		4459	390	1760 CD	388
CGA-64250	2.5G	1.2	Layby		5714	646	2246 BCD	291
CGA-64250	2.5G	0.6	Tassel		6741	762	1388 D	991
CGA-64250	2.5G	1.2	Tassel		4668	784	2435 BCD	723
CGA-64250	3.6EC	0.1	Spray-A		3585	586	2106 BCD	1929
CGA-64250	3.6EC	0.2	Spray-A		3686	611	5111 A	438
benomyl	50WP	0.3	Spray-B		2423	473	1824 BCD	364
benomyl	50WP	0.6	Spray-B		3436	2693	1914 BCD	657
benomyl	50WP	1.2	Spray-B		1546	267	2263 BCD	858
maneb + zinc	80WP	2.4	Spray-B		4839	471	4100 AB	418
thiabendazole	340-F	1.2	Spray-B		3825	776	1444 D	607
thiabendazole	340-F	1.2, 0.6	Spray-C		3996	333	1966 BCD	514
maneb + zinc	37-F	1.9	Spray-B		5394	1462	------	----
control	----	----	----		3054	324	3980 ABC	485
					NS	NS	P=0.01	NS
				$\bar{x}$..	4017	722	2462	639

[a]Sprays-A,B and C represent split-fungicide applications. Spray-A=fungicide applied at time of silk, and 2 weeks later; Spray-B=fungicides applied at time of silk, 1 week later, and 2 weeks later. Spray-C=1.2 kg a.i./ha applied at time of silk and 0.6 kg a.i./ha applied 1 week later.
[b]Data are means of 4 replications.
[c]Means with the same letter are not significantly different according to Waller-Duncan's K-ratio t-test.

Table 2. Effects of Fungicides on Aflatoxin Concentrations, Central Crops Research Station (CCRS) and Upper Coastal Plain Research Station (UCPRS), 1982.

Fungicide	Formulation	Rate (kg a.i./ha)	Method of Application		Aflatoxin B_1 Concentrations ($\mu g/kg$) CCRS Inoc.	CCRS Non-Inoc.	UCPRS Inoc.	UCPRS Non-Inoc.
CGA-64250	2.5G	0.6	Layby		67	12 BCDE	30	26
CGA-64250	2.5G	0.9	Layby		19	6 E	100	6
CGA-64250	2.5G	1.2	Layby		73	4 E	56	24
CGA-64250	2.5G	0.6	Tassel		87	39 AB	104	12
CGA-64250	2.5G	1.2	Tassel		186	21 ABCDE	37	33
CGA-64250	3.6EC	0.1	Spray-A		59	35 ABCD	73	8
CGA-64250	3.6EC	0.2	Spray-A		34	27 ABCDE	105	38
DPX-3866	75G	1.2	Layby		27	37 ABC	87	4
benomyl	50WP	0.3	Spray-B		78	46 A	63	7
benomyl	50WP	0.6	Spray-B		70	4 E	130	26
benomyl	50WP	1.2	Spray-B		23	6 E	45	6
maneb	80WP	2.4	Spray-B		84	6 E	106	72
thiabendazole	340-F	1.2	Spray-B		27	14 BCDE	133	10
thiabendazole	340-F	0.6, 1.2	Spray-C		10	15 BCDE	56	12
maneb + zinc	37-F	1.9	Spray-B		92	8 DE	84	15
control	----	----	----		66	10 CDE	80	9
					NS	P=0.01	NS	NS
				$\bar{x}$..	63	18	81	19

[a]Sprays-A,B and C represent split-fungicide applications. Spray-A=fungicide applied at time of silk, and 2 weeks later; Spray-B=fungicide applied at ime of silk, 1 week later, and 2 weeks later. Spray-C=0.6 a.i./ha applied at time of silk and 1.2 kg a.i./ha applied 1 week later.
[b]Data are means of 4 replication.
[c]Means with the same letter are not significantly different according to Weller-Duncan's K-ratio t-test.

Table 3. Effects of Fungicide Treatments on Yields, Central Crops Research Station (CCRS) and Upper Coastal Plain Research Station, (UCPRS), 1981.

				Yields (kg/ha)			
				CCRS		UCPRS	
Fungicide	Formulation	Rate (kg a.i./ha)	Method of Application	Inoc.	Non-Inoc.	Inoc.	Non-Inoc.
CGA-64250	2.5G	0.6	Layby	9070[b]	8682	8183 A	7845 AB
CGA-64250	2.5G	0.9	Layby	8294	8851	7313 AB	8235 A
CGA-64250	2.5G	1.2	Layby	8234	8343	7806 A	7599 AB
CGA-64250	2.5G	0.6	Tassel	8476	7798	7222 AB	7287 ABCD
CGA-64250	2.5G	1.2	Tassel	9130	9614	7689 A	7209 ABCD
CGA-64250	3.6EC	0.1	Spray-A	8645	9566	6910 AB	7170 ABCD
CGA-64250	3.6EC	0.2	Spray-A	8839	9214	6832 AB	7339 ABC
benomyl	50WP	0.3	Spray-B	9263	9541	7910 A	7793 AB
benomyl	50WP	0.6	Spray-B	8415	8076	7702 A	6897 ABCD
benomyl	50WP	1.2	Spray-B	8791	8561	7248 AB	6183 CD
maneb	80WP	2.4	Spray-B	9093	8488	7430 AB	7676 AB
thiabendazole	340-F	1.2	Spray-B	8500	8209	7430 AB	6598 BCD
thiabendazole	340-F	1.2, 0.6	Spray-C	8536	8815	5988 B	5936 D
maneb + zinc	37-F	1.9	Spray-B	8125	8306	------	----
control	----	----	----	8754	8912	6676 AB	7443 ABC
				NS	NS	P=0.05[c]	P=0.01
			$\bar{x}$..	8678	8732	7310	7229

[a]Sprays-A,B and C represent split-fungicide applications. Spray-A=fungicide applied at time of silk, and 2 weeks later; Spray-B=fungicides applied at time of silk, 1 week later, and 2 weeks later. Spray-C=1.2 kg a.i./ha applied at time of silk and 0.6 kg a.i./ha applied 1 week later.
[b]Data are means of 4 replication.
[c]Means with the same letter are not significantly different according to Waller-Duncan's K-ratio t-test.

Table 4. Effects of Fungicide Treatments on Yields, Central Crops Research Station (CCRS) and Upper Coastal Plain Research Station (UCPRS), 1982.

Fungicide	Formulation	Rate (kg a.i./ha)	Method of Application		Yields (kg/ha) CCRS Inoc.	CCRS Non-Inoc.	UCPRS Inoc.	UCPRS Non-Inoc.
CGA-64250	2.5G	0.6	Layby		12,466[b]	13,238	11,598	11,227
CGA-64250	2.5G	0.9	Layby		12,656	13,048	11,086	10,778
CGA-64250	2.5G	1.2	Layby		13,072	12,929	10,022	10,919
CGA-64250	2.5G	0.6	Tassel		13,095	13,451	11,573	11,650
CGA-64250	2.5G	1.2	Tassel		12,644	13,748	11,893	11,124
CGA-64250	3.6EC	0.1	Spray-A		12,917	13,226	10,022	9,945
CGA-64250	3.6EC	0.2	Spray-A		13,618	13,273	10,855	10,599
DPX-3866	75G	1.2	Layby		13,701	12,905	10,612	10,317
benomyl	50WP	0.3	Spray-B		13,238	12,953	10,970	11,611
benomyl	50WP	0.6	Spray-B		12,205	13,167	11,342	11,252
benomyl	50WP	1.2	Spray-B		12,359	13,653	11,509	12,137
maneb + zinc	80WP	2.4	Spray-B		13,725	13,511	10,676	10,189
thiabendazole	340-F	1.2	Spray-B		12,074	11,682	11,047	10,125
thiabendazole	340-F	0.6, 1.2	Spray-C		13,677	13,155	9,817	11,163
maneb + zinc	37-F	1.9	Spray-B		12,703	13,891	9,868	11,137
control	----	----	----		12,656	12,620	11,137	11,393
					NS	NS	NS	NS
				$\bar{y}_{..}$	12,925	13,153	10,877	10,973

[a]Sparys-A,B and C represent split-fungicide applications. Spray-A=fungicide applied at time of silk, and 2 weeks later; Spray-B=fungicides applied at time of silk, 1 week later, and 2 weeks later. Spray-C=0.6 a.i./ha applied at time of silk and 1.2 kg. a.i./ha applied 1 week later.
[b]Data are means of 4 replication.

Table 5. Correlation Coefficients Between Aflatoxins, Fungi, Insects, and Yields, Central Crops Research Station (CCRS) and Upper Coastal Plain Research Station (UCPRS), 1981[a].

Location	Inoculation Method	Variables	*A. flavus*	*Fusarium* spp.	Insect: Cornborer	Insect: Earworm	Yields
CCRS	Artifically Inoculated	$[B_1]$	.57**	-.21NS	-.23NS	-.07NS	-.67**
		$[B_1+B_2]$	.59**	-.21NS	-.22NS	-.07NS	-.66**
		A. flavus	. . .	-.22NS	.02NS	-.16NS	-.46**
	Naturally Inoculated	$[B_1]$	.10NS	.13NS	-.16NS	-.12NS	-.11NS
		$[B_1+B_2]$	.08NS	.13NS	-.16NS	-.12NS	-.10NS
		A. flavus	. . .	.07NS	-.13NS	-.19NS	-.26*
UCPRS	Artifically Inoculated	$[B_1]$	.50**	-.10NS	-.11NS	-.18NS	-.32*
		$[B_1+B_2]$	.49**	-.08NS	-.10NS	-.17NS	-.31*
		A. flavus	. . .	-.03NS	-.16NS	-.34NS	-.20NS
	Naturally Inoculated	$[B_1]$	.20NS	-.18NS	-.11NS	.02NS	-.23NS
		$[B_1+B_2]$	.19NS	-.18NS	-.11NS	.03NS	-.22NS
		A. flavus	. . .	-.12NS	.09NS	-.03NS	-.44**

[a]Correlation analysis performed across treatments. Astrisks *a and ** indicate significance at P=0.05 and P=0.01, respectively.

Table 6. Correlation Coefficients Between Aflatoxins, Fungi, Insects, and Yields, Central Crops Research Station (CCRS) and Upper Coastal Plain Research Station (UCPRS), 1982[a].

Location	Inoculation Method	Variables	*A. flavus*	*Fusarium* spp.	Insects	Yield
CCRS	Artifically Inoculated	[B_1]	.71**	-.03NS	.19NS	-.38**
		[B_1+B_2]	.73**	-.04NS	.18NS	-.38**
		A. flavus	. . .	.04NS	.04NS	-.24NS
	Naturally Inoculated	[B_1]	-.06NS	-.17NS	-.10NS	-.12NS
		[B_1+B_2]	-.06NS	-.18NS	-.10NS	-.11NS
		A. flavus	. . .	.09NS	.18NS	-.05NS
UCPRS	Artifically Inoculated	[B_1]	.17NS	.08NS	-.02NS	-.02NS
		[B_1+B_2]	.17NS	.08NS	-.03NS	-.02NS
		A. flavus	. . .	.19NS	.19NS	.29*
	Naturally Inoculated	[B_1]	-.04NS	.06NS	.00NS	-.09NS
		[B_1+B_2]	-.04NS	.06NS	.00NS	-.09NS
		A. flavus	. . .	.16NS	.00NS	.25*

[a]Correlation analysis performed across treatments. Astrisks * and ** indicate significance at P=0.05 and P=0.01, respectively.

RESULTS

Much more aflatoxin developed in the corn in 1981 than in 1982. Aflatoxin B_1 concentrations (averaged across treatments and locations) were 41-fold greater in inoculated and 33-fold greater in naturally infected corn in 1981 than in 1982. Aflatoxin B_2 levels were 36-fold greater in inoculated and 38-fold greater in naturally infected corn, respectively. Aflatoxin concentrations were greater at CCRS than at UCPRS in 1981, but the opposite was true in 1982. Artificially inoculated corn consistently had greater aflatoxin concentrations than naturally infected corn. Over all fungicide treatments, artificially inoculated plots had 4 - 6 times more aflatoxins than uninoculated plots in 1981 and 1982 (Tables 1, 2).

Fungicides did not prevent infection of corn by *A. flavus* nor did they consistently reduce aflatoxin contamination at harvest (Tables 1, 2). Artificially inoculated corn treated with CGA 64250 and thiabendazole at UCPRS in 1981 had less aflatoxin B_1 than the control, but in naturally infected corn there was no difference in aflatoxin levels between treatments (Table 1). None of the chemical treatments significantly affected aflatoxin concentrations at CCRS in 1981. Naturally infected corn treated with CGA 64250 and benomyl at CCRS in 1982 had higher concentrations of aflatoxin B_1 than the control in 1982 (Table 2). None of the fungicides were significantly better than the control. Chemical treatments also were not effective in significantly reducing the aflatoxin contamination level at UCPRS in 1982 (Table 2). Corn yields were not increased with fungicides (Tables 3, 4). In fact, in naturally infected plots at UCPRS in 1981, thiabendazole decreased yields compared to the control. Corn yields were greater at both locations in 1982 than 1981. Yields were generally less at UCPRS than at CCRS.

Aflatoxin B_1 was positively correlated with numbers of ears showing visible growth of *A. flavus* in artificially inoculated plots at CCRS in 1981 and 1982, and UCPRS in 1981 (Tables 5, 6). Aflatoxin B_2 was highly correlated with numbers of *A. flavus* infected ears in artificially inoculated corn at CCRS in 1981 and 1982. There was no association between these factors at UCPRS. No correlation existed between aflatoxins and visible growth of *A. flavus* in naturally infected corn.

Aflatoxin B_1 was negatively correlated with yields in artificially inoculated corn at CCRS in 1981 and 1982 and at UCPRS in 1981 (Tables 5, 6). There was no correlation between aflatoxins and yields in naturally infected corn.

Numbers of A. flavus infected ears were inversely related with yields at CCRS in 1981, but not 1982 (Table 5). At UCPRS, numbers of A. flavus infected ears were inversely correlated with yields in naturally infected corn in 1981, but positively correlated with yields in 1982 in both artificially, inoculated, and naturally infected corn (Tables 5, 6).

There were no correlations between either A. flavus or aflatoxins and insect damage (Tables 5, 6). Fusarium spp. were not significantly related to any of the variables.

DISCUSSION

Fungicides did not reduce the number of ears showing growth of A. flavus or reduced levels of aflatoxins. There may be several reasons for lack of control,but it was not related to coverage of the silks since the materials were applied at high pressure with nozzles arranged in such a manner that the silks were completely soaked with the materials. The timing of the application may have accounted for the poor performance of the fungicides applied as sprays as well as those applied either as tassel or layby treatments. We assumed that the materials applied as tassel and layby treatments were systemic enough to move into the silk tissues and provide protection against infection. However, this probably was not the case. It would appear that fungicides cannot be used to control infection by A. flavus and contamination with aflatoxins and that resistant cultivars may be necessary in the future.

Inoculations with spore suspensions of A. flavus resulted in an increased amount of aflatoxins at both locations in both 1981 and 1982 when compared to natural infection. This has been demonstrated before and supports other evidence that A. flavus can cause infection through silk tissues. Silk inoculation may be a useful technique for breeders to consider in testing for resistance to infection by A. flavus and aflatoxin contamination.

The fungicides also did not have an influence on the Fusarium population. Again, this was surprising since several of the materials used in these trials would be expected to control Fusarium. The 1982 growing season was cooler and wetter than the 1981 growing season, but the Fusarium spp. were frequently observed during both years. Thus, the population was high enough to get an accurate rating on the effectiveness of these treatments against

Fusarium. It would appear that fungicides are not a good method of controlling Fusarium on corn.

The fungicides had very little influence on yields. Although thiabendazole significantly reduced the yields in 1981 at the UCPRS, it did not cause any reduction in 1982 at the UCPRS or either year at the CCRS. The reduced yield at UCPRS in 1981 cannot be explained, since no phytotoxicity was noted with this, or any other material. Yields were much higher and aflatoxin levels much lower in 1982 than in 1981. The association with high yields and low aflatoxin contamination has been noted on several occasions. This association appears to be strong enough to allow the development of a predictive system for aflatoxin problems in field corn based upon the potential yield and weather conditions at time of pollination.

The lack of correlation between insect damage and aflatoxin contamination in this study raises serious questions about the role of insects in the aflatoxin problem, particularly in North Carolina. This lack of any association between insects and aflatoxin in corn has been consistent over all of the field experiments conducted in North Carolina on aflatoxins in corn. This was why an insecticide was not included in these fungicidal trials, and it would appear that if future fungicidal trials are conducted, an insecticide would not be necessary.

SUMMARY

The fungicides CGA 64250, benomyl, maneb + zinc (both liquid and WP formulations), thiabendazole, and DPX 3866 where ineffective in controlling Aspergillus flavus and limiting aflatoxin production when applied either as a spray, tassel, or a layby (soil) treatment at two locations in 1981 and 1982. The fungicidal treatments also had no effect on the number of corn ears infected with Fusarium. There were no associations between the European corn borer or the corn earworm feeding injury and aflatoxin contamination. Yields were lower and aflatoxin contamination was greater in 1981 than in 1982. Fungicidal treatments had no significant effect on yield of corn. Silk inoculation with conidial suspension of A. flavus significantly increased the aflatoxin contamination in corn grain when compared to natural inoculation.

Acknowledgments

The authors would like to express their appreciation to the personnel at the

Central Crops Research Station and Upper Coastal Plain Research Station for their technical assistance. We also thank the technicians, summer employees, and secretaries in the two departments for their very able assistance. We also thank the Corn Growers Association of North Carolina, Inc. and the various chemical companies for their financial assistance. Use of trade names in this article does not imply endorsement of the products named by the North Carolina Agricultural Research Service nor criticism of similar ones not mentioned.

REFERENCES

Anderson, H.W., Nehring, E.W., and Wichser, W.R. (1975). Aflatoxin contamination of corn in the field. J. Agric. Food Chem., 23, 775-782. Association of Official Analytical Chemists. (1980). Natural poisons. Chapter 12. Official Methods of Analysis, 13th ed. Assoc. Off. Anal. Chem., Washington, D. C. 1018 pp.

Barnes, L.W. (1983). Aflatoxin development in preharvest corn. Ph.D. Dissertation. Texas A & M University. 64 pp.

Dickens, J.W., and Satterwhite, J.B. (1969). Subsampling mill for peanut kernels. Food Technol. 23, 190-92.

Draughon, F.A. (1983). Control or suppression of aflatoxin production with pesticides. In: Aflatoxin and Aspergillus flavus in Corn. pp. 81-86 (U.L. Diener, R.A. Asquith, and J.W. Dickens, eds.) Southern Cooperative Series Bulletin No. 279. Auburn University, Auburn, AL.

Jones, R.K., Duncan, H.E., Payne, G.A. and Leonard, K.J. (1980). Factors influencing infection by Aspergillus flavus in silk-inoculated corn. Plant Disease, 64, 859-863.

Jones, R.K., Duncan, H.E., and Hamilton, P.B. (1981). Planting date, harvest date, and irrigation effects on infection and aflatoxin production by Aspergillus flavus in field corn. Phytopathology, 71, 810-816.

Jones, R.K., and Mulkey, J.R. (1982). Evaluation of benomyl and triazole in reducing aflatoxin B_1 concentrations in field corn (Abstr.). Phytopathology, 72, 970.

Marsh, S.F., and Payne, G.A. (1984). Infection of silks and kernels of preharvest corn by Aspergillus flavus. Phytopathology, 74, 1284-1289.

Payne, G.A. (1983). Nature of field infection of corn by Aspergillus flavus. pp. 16-20 In: Aflatoxins and Aspergillus flavus in Corn. (U.L. Diener,

R.A. Asquith, and J.W. Dickens, eds.) Southern Cooperative Series Bulletin No. 279. Auburn University, Auburn, Alabama.

Taubenhaus, J.J. (1920). A study of the black and yellow molds on ear corn. Tex. Agric. Exp. Sta. Bull., No. 270. 38 pp.

Wales, P., and Somers, E. (1968). Susceptibility of aflatoxin-producing strains of *Aspergillus flavus* to a range of fungicides. Can. J. Plant Sci., 48, 377-379.

Widstrom, N.W. (1979). The role of insects and other plant pests in aflatoxin contamination of corn, cotton, and peanuts: A review. J. Environ. Qual., 8, 5-11.

MOLECULAR GENETIC CRITERIA FOR THE IDENTIFICATION OF ATYPICAL TOXIGENIC *FUSARIUM* STRAINS

ANTONIO LOGRIECO, Istituto Tossine e Micotossine da Parassiti Vegetali, Consiglio Nazionale delle Ricerche, 70126 Bari, Italy

ANTONIO BOTTALICO, Istituto di Patologia vegetale dell'Università degli Studi, 07100 Sassari, Italy

STEPHEN W. PETERSON, Northern Regional Research Center, Peoria, IL 61604, USA

ANTONIO EVIDENTE, Dipartimento di Scienze Agrarie dell' Università degli Studi, 80055 Portici, Italy

GIUSEPPINA MULE', Istituto Tossine e Micotossine da Parassiti Vegetali, Consiglio Nazionale delle Ricerche, 70126 Bari, Italy

INTRODUCTION

Several systems for Fusarium taxonomy can be adopted to species identification, and to define different numbers of taxa (Wollenweber and Reinking, 1935; Snyder and Hansen, 1945; Booth, 1971; Gerlach and Nirenberg, 1982; Nelson et al., 1983). However, these taxonomic systems not always agree on species limits, and character variability in species as well as the loss of characters which sometimes occurs during laboratory cultivation of strains, also make identification difficult. Correct identification is important because many species are plant pathogens and certain isolates produce potent mycotoxins (Booth, 1971; Nelson et al., 1981). Moreover, culture conditions and frequent subculturing of the strains can lead to degeneration of morphological features or loss of toxigenic capability. In fact, Marasas et al. (1984) examined several

Biodeterioration Research 4, Edited by G.C. Llewellyn *et al.*, Plenum Press, New York, 1994

reported toxigenic isolates of Fusarium species todetermine the relationship among species identity and mycotoxin production, and found a number of strains with atypical morphological features which prevented the identification of those strains. To this regard, molecular genetic and chemotaxonomic methods of systematic analysis can provide useful data concerning the relatedness of fungal strains (Kurtzman, 1985; Frisvad and Filtenborg, 1983).

In the course of a screening of toxigenic Fusarium strains from several geographical areas a highly toxic, but morphologically atypical and unclassified strain, was encountered. It appeared an useful occasion to investigate the application of molecular genetic methods in order to support, in a more general approach, the morphotaxonomic identification of atypical Fusarium strains.

MATERIALS AND METHODS

Fungal strain

Strain KF-302, come from England (Dr. C. Booth), was received from Prof. J. Chelkowski (Department of Plant Pathology, Warsaw, Poland). Single conidial culture of the fungus was deposited as lyophilized cultures in the collection of the Institute of Toxins and Mycotoxins, Bari, Italy under the accession number of ITEM-709. The strain was also deposited in the ARS culture collection, Peoria, Illinois, USA as NRRL-13772.

Chemical analysis

Two hundred grams of polished rice var. "Flora", durum wheat var. "Appulo" or yellow maize kernels var. "Plata" were brought to 45% moisture in 500 ml Erlenmeyer flasks and left overnight. Additionally, 150 ml of Czapek and Nachmias liquid media (Nachmias et al., 1977) were prepared. The substrates were inoculated (after sterilization) with ca. 3 x 10^3 conidia suspended in 0.05% Triton X 100. Solid substrate cultures were incubated in the dark for 3 weeks at 25 °C and shaken by hand daily. Liquid cultures were shaken at 200 rpm for 1 week. The culture materials were extracted and analyzed for the principal Fusarium-toxins zearalenone, zearalenols, moniliformin, deoxynivalenol, nivalenol,diacetoxyscirpenol, neosolaniol, and T-2 toxin, using published procedures (Bottalico et al., 1989). The T-2 toxin production wasconfirmed by ^{1}H NMR and ^{1}NMDR spectra. The samples of T-2 toxin and NEOS used as reference compound, were purchased from Sigma Chemical Co., St. Louis, U.S.A.

Morphological analysis

Pigmentation and growth of ITEM-709 from single-conidium cultures were tested on fresh prepared PDA (800 ml filtrate from 200 g peeled, sliced and autoclaved potatoes, 20 g dextrose, 15 g Bacto agar (Difco), up to 1000 ml with distilled water). Small square (ca 2 mm) of water agar containing 12-h-old single germinating conidia were cut out with a dissecting needle and inoculated into the centre of 90 mm Petri dishes containing 20 ml of PDA. The growth rate of ITEM-709 was calculated as the mean of four colony diameters after 7 d in the dark at 25 °C. Microscopic features were observed on subcultures grown on liquid with sterile carnation leaf (CLA) (Nelson et al., 1983). The cultures were incubated at 24 °C under fluorescent and black (ultraviolet) lamps (2700 lux) with 12 hr photoperiod. A subculture was sent to Dr. W. Gams (Centraalbureau voor Schimmelcultures, Baarn, The Netherlands) and to Prof. P.E. Nelson (The Pennsylvania State University, University Park, PA, U.S.A.).

Molecular Genetics Analysis - Ribosomal DNA Comparison

A sporulating subculture was used to inoculate a flask containing 100 ml of YM (Wickerham, 1951). The flask was incubated at 25 °C on a rotary shaker (200 rpm) until the culture was in log phase growth (ca. 36 hr). Ribosomal RNA was extracted and purified by the guanidinium thiocyanate procedure of Chirgwin et al. (1979) after cell breakage in a Braun cell homogenizer. Nucleotide sequences were determined from a highly variable region of large subunit (25-28S) rRNA [D2 region (Hassouna et al., 1984), ca 220 nucleotides]. Direct rRNA sequencing was performed using the primer-extension, dideoxynucleotide method of Sanger et al. (1977) as modified by Lane et al. (1985) using synthetic oligonucleotide primer 25S-635 (Guèho et al., 1990). Sequences were aligned manually. Where the proper base could not be determined (dark bands on the autoradiographs of the sequencing gel), N was assigned. Unrelated fungus (Saccharomices cerevisiae Hansen, strain NRRL Y-12632) was used as control.

Nuclear DNA Comparison

In accordance with our chemical results we chosed a typical strain of F. sporotrichioides Sherb. T-2 producer (NRRL-3299) (Marasas et al., 1984) to study the DNA/DNA relatednesss with ITEM-709. Unrelated fungus (Rhizopus arrhizus Fischer var.delemar (Wehmer & Hanz) Ellis, strain NRRL-1472) was used as control. The strains were grown to stationary phase (ca. 42 hr) in 2.8 L Fernbach flasks with 1.5 L YM broth, at 25 °C on a rotary shaker at 200 rpm.

Whole cell DNA was extracted and purified as previously described (Ellis, 1988; Kurtzman et al., 1980). Purity of the DNA preparations was determined from UV-absorbance ratios and thermal denaturation profiles. Renaturation was performed using a Gilford Response II spectrophotometer with a thermal programmer (Kurtzman et al., 1980; Seidler and Mandel, 1971; Seidler et al., 1975).

RESULTS

Chemical analysis

The main toxic metabolite produced in culture from ITEM-709 was T-2 toxin. Neosolaniol was also synthesize. Both toxins were found in all tested media (Table 1), and are typical metabolites of the genus Fusarium (Cole and Cox, 1981) and particularly of F. sporotrichioides (Marasas et al., 1984).

Table 1. Production of Mycotoxins by Strain ITEM-709 on Different Substrates.

	Trichothecene Yield (ppm)	
Substrate	T-2 Toxin	Neosolaniol
Maize	200	80
Wheat	60	40
Rice	20	100
Czapek	15	20
Nachmias	80	40

Morphological examination

The colonies on PDA initially formed a white, floccose mycelium which became carmine red to reddish brown in older cultures. The growth rate of ITEM-709 was 8.3 cm. On CLA, smooth-walled conidia formed abundantly on aerial mycelium with no clear subdivision between micro- and macro- conidia (Figure 1). Conidia are thick walled, clavate, fusiform, usually with a flat base and sometimes pedicellate with tapering or rounded apical cell; 0-septate (6-7.5 μm x 3-4.2 μm); 1-septate (12.5-20 μm x 3-5 μm); 2-septate (20-30 μm x 3-4.2 μm); 3-septate (20-35 μm x 3-4.2 μm); 4-septate (27.5-35 μm x 3-4.2 μm). Conidia are solitary, borne directly on the terminal hyphal segments

(Figures 1A,B). Chlamydospores formed in 2-4 weeks and were usually more abundant on hyphae (both aerial and submerged) than conidia. They form singly in chains or in clusters, are smooth or rough walled, and hyaline to pale yellow.

On the basis of the morphological features, this fungus prove to be a atypical strain. Dr. W. Gams (personal communication) regarded ITEM-709 as an unusual strain and Prof. P.E. Nelson (personal communication) consider it as not clearly referable to F. sporotrichioides or any other Fusarium species.

Molecular comparisons

The ribosomal RNA sequence of ITEM-709 (NRRL-13772) (Figure 2) was compared to sequence from the homologous regions of two well documented strains T-2 toxin producers of F. sporotrichioides (NRRL-3299) and F. acuminatum Ell. & Ev. sensu Gordon (NRRL-6227) (Marasas et al., 1984). The sequence of ITEM-709 is identical to that of F. sporotrichioides NRRL-3299 (Logrieco et al., 1991), suggesting conspecificity of the isolates (Peterson and Logrieco, 1991). On the other hand, the other T-2 toxin species (F. acuminatum) show three base differences. Saccharomices cerevisiae used as unrelated microrganism, show many base differences. DNA/DNA hybridization between the atyical strain (ITEM-709) and F. sporotrichioides (NRRL-3299) confirmed their high relatedness. Infact the nuclear DNA similarity of these two strains was found to be 90% (average of two determinations) showing conspecificity of the isolates, while the similarity of these DNA's to that from R. arrhizus var. delemar was 4.5 %, a typical background level of hybridization seen in unrelated organisms (Ellis, 1988).

DISCUSSION

Conidiogenesis and conidial form are critically important characters in mycological taxonomy (Hughes, 1953). Booth (1971) regarded the Fusarium conidiogenous cell (mono or poly-phialide) as a fundamental taxonomic character and on that basis modified and discarded some sections of the genus. However, this fundamental taxonomic character sometimes mutates or becomes degenerate in culture and consequently the correct identification of the strain is difficult to determine. Several toxigenic Fusarium strains are reported that have unusual or atypical conidiogenesis that makes them difficult or impossible to identify. For example "F. nivale" Fn 2B (= NRRL 6490) produces mycotoxins, but yet it is unclassifiable morphologically (Marasas et al., 1984).

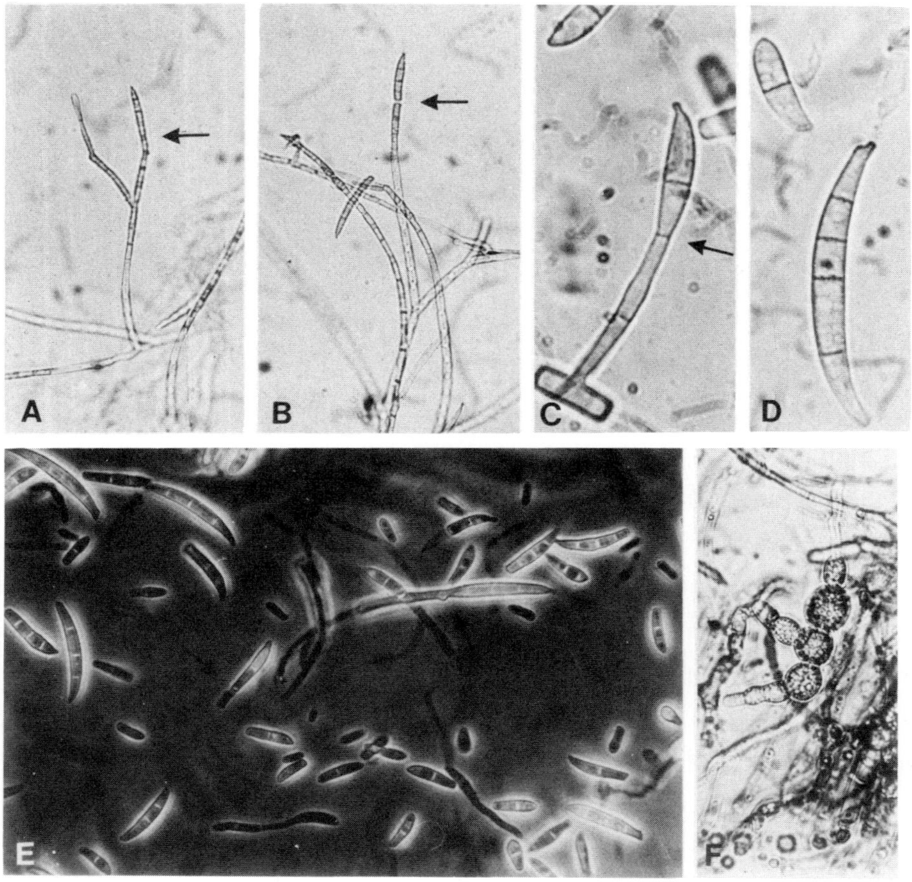

Figure 1. Conidiogenesis of Strain ITEM-709. A-C Conidia borne on the terminal hyphal segment, arrows show the point of detachment of the spore [A, B (x360), C (x1000)]. D. Conidia (x1000). E. Conidia (x500), arrows showing the strong connection of the Conidia to the hyphae. F. Chlamydospores (x500).

```
(375-470)
F. sporotrichioides   NRRL-13772        UUUGAA AAGAGAGUUA AAAAGUACGU GAAAUUGUUG AAAGGGNAGU GUUUAUGACC AGACUUGGGC UUGGUUAAUC AUCUGGGGUU CUCCCNAGUG
F. sporotrichioides   NRRL-3299         ...... .......... ......N... .......... ......N..N .......... .......... .......... ..N....... .....C....
S. cerevisiae         NRRL Y-12632      ...... ........G. .......... .......... ......A..G .CA.U...U. ....A...UG ..UUG.GCC. UCUGCUCC.. G.GGGU..G.

(471-570)
F. sporotrichioides   NRRL-13772    CNCUUU-UCC -AGUUC-AGG CCAGCAUCAG UUUUCGC-CG GGGGAUAAAG ACUUCGGGAA UGUGGCUCCU CUCGGGGAGU GUUAUAGCCC NUUGUGUNAU
F. sporotrichioides   NRRL-3299     .A....-... -.....-... .......... .......-.. .......... .......... .......... .......... .......... N......A..
S. cerevisiae         NRRL Y-12632  GAA.C.CG.A UUUCA.UG.. .......... ....G.U-G. CA.......- U.CAUA.... ...A...UGC .....UA... A........U --GUG.GA..

(571-624)
F. sporotrichioides   NRRL-13772    ACC-CUGGCG GGGACUGAGG UUCGCGCUUC U-G-CAAGGA UGCUGGCGUN AUGG
F. sporotrichioides   NRRL-3299     ...-...... .......... .......... .-.-...... .........A ....
S. cerevisiae         NRRL Y-12632  ..UG.CA..U .......... AGU...ACGU AA.U...... .......A.A ....
```

Fig. 2. 25S-635 Region (Guèho et al., 1990) rRNA Sequences of ITEM-709 and F. sporotrichioides NRRL 3299 (Logrieco et al., 1991) Aligned with the homologous sequences from Saccharomyces cerevisiae. Nucleotide numbering is based on the complete sequence of S. cerevisiae (Georgiev et al., 1981). The complete sequence is given for the first strain, dots in subsequent strains indicates the same base, N indicates an unreadable nucleotide position and a dash indicates a gap added to a sequence to account for sequence length differences.

Secondary metabolite patterns of Fusarium species prove to be a potential useful tool for fungal characterization, but they are not still clearly categorized at present (Thrane, 1989). To this regard, a major impediment has been in turn the difficulty to correctly identify the strains whose metabolite profiles have been studied (Marasas et al., 1984; Thrane, 1989). In this note, the finding of T-2 toxin and NEOS production by studied strain is suggestive of its placement in one of several species, but does not permit, without any doubts, to identify the species.

Molecular analyses of nucleotide sequences from rRNA (25-28S) obtained from Fusarium atypical strain (ITEM-709) was identical to the homologous sequence from the morphologically typical F. sporotrichioides (NRRL-3299). Moreover, DNA complementarity confirmed the identity of atypical ITEM-709 as F. sporotrichioides (90% similarity). These combined findings from rRNA sequence analysis and DNA similarity studies indicates that the atypical strain is a variant of F. sporotrichioides. Whole genomic DNA complementarity is a sensitive measurement of genetic distance between strains and can be used to confirm the identification of strains (Ellis, 1988). However, if a presumptive identification cannot be made morphologically or on another basis, DNA from each species of Fusarium would have to be tested against the unknown until a positive result was achieved because DNA hybridization has no resolution beyond the sibling species (Kurtzman et al., 1980).

An easier way to identify morphologically degenerate strains is through rRNA (or rDNA) sequence comparison. Ribosomal RNA is very slowly mutating molecule which shows structural and sequence homology among all cellular organisms (Woese, 1987). However, not all parts of these large molecules are so highly conserved. The 25S-635 region (Guèho et al., 1990) of rRNA appears to have little variability at the species level, but even very closely related species can be distinguished based on sequences from this part of the rRNA. Sequence determination of the complete rRNA molecule or its gene (rDNA) can be obtained rapidly and on the basis of comparison to known sequences (Guadet et al, 1989; Logrieco et al., 1991; Peterson and Logrieco, 1991), a presumptive strain identification can be made. For several Gibberella species that have been examined, there is little or no sequence variation among conspecific strains, but differences between closely related species are seen (Peterson and Logrieco, 1991).

The relative ease of obtaining ribosomal RNA or rDNA (by polymerase chain reaction) sequences, together with the low level of interspecific variation of this D2 region makes these techniques suitable for the presumptive identification of fungal strains and will be a useful tool in identifying unusual or morphologically degenerate isolate as well as their potential secondary metabolites.

SUMMARY

In the course of our studies on toxigenic fungi, a morphologically atypical Fusarium strain (ITEM-709 = NRRL 13772) was encountered. Secondary matabolite analysis revealed the production of T-2 toxin and neosolaniol, which are toxins typically produced by Fusarium genus and particularly by F. sporotrichioides. In order to determine the correct identity of the strain, molecular genetics analyses were performed. Nucleotide sequences from a highly variable region of large subunit (25-28S) rRNA (D2 region, ca 220 nucleotides) obtained from this strain was identical to the homologous sequence from a morphologically typical and T-2 producer F. sporotrichioides. DNA complementarity confirmed the identity of the strain as F. sporotrichioides (90 % similarity). Secondary metabolite analysis, DNA hybridization and D2 region sequence comparison each has value for identification of the atypical strain, but the ease of obtaining partial rRNA or rDNA sequences and the sequence uniformity within species shows that D2 region sequence will be a particularly valuable screening tool for the identification of morphologically unusual or degenerate Fusarium strains.

Acknowledgments

The authors are grateful to Dr. W. Gams, Centraalbureau voor Schimmelcultures, Baarn, The Netherlands and Prof. P.E. Nelson, The Pennsylvania State University, University Park, PA, U.S.A. for their taxonomic contribution.

REFERENCES

Booth, C. (1971). The Genus Fusarium. Commonwealth Mycological Institute, Kew, Surrey, England.

Bottalico, A., Logrieco, A., and Visconti, A. (1989). Fusarium species and their mycotoxins in infected cereals in the field and in stored grains, In: Fusarium - Mycotoxins, Taxonomy and Pathogenicity, pp. 85-119 (J. Chelkowski, ed.), Elsevier, Amsterdam.

Chirgwin, J.M., Przybyla, A.E., MacDonald, R.J., and Rutter, W.J. (1979). Isolation of biologically active ribonucleic acid from sources enriched inribonuclease. Biochemistry, 18, 5294-5299.

Cole, R.J., and Cox, R.H. (1981). Handbook of Toxic Fungal Metabolites, 937 p. Academic Press, New York.

Ellis, J.J. (1988). Section Liseola of Fusarium. Mycologia, 80, 255-258.

Frisvad, J.C. and Filtenborg, O. (1983). Classification of terverticillate penicillia based on profiles of mycotoxins and other secondary metabolites. Appl. Environ. Microbiol., 46, 1301-1310.

Guadet, J., Julien, J., Lafay, J.-F., and Brygoo, Y. (1989). Phylogeny of some Fusarium species determined by large-subunit rRNA sequence comparison. Mol. Biol. Evol., 6, 227-242.

Georgiev, O.I., Nikolaev, N., Hadjiolov, A.A., Skryabin, K.G., Zakharyev, V.M., and Bayev, A.A. (1981). The structure of yeast ribosomal RNA genes. 4. Complete sequence of the 25S rRNA gene from Saccharomyces cerevisiae. Nucleic Acid Res., 9, 6953-6958.

Gerlach, W., and Nirenberg, H. (1982). The genus Fusarium- a pictorial atlas. Mitt. Biol. Bundesanst. Land-Forstwirt. Berlin-Dahlem 209, 1-406.

Guèho, E., Kurtzman, C.P., and Peterson, S.W. (1990). Phylogenetic relationships among species of Sterigmatomyces and Fellomyces as determined from partial rRNA sequences. Int. J. Syst. Bacteriol., 40, 60-65.

Hassouna, N., Michot, B., and Bachellerie, J.-P. (1984). The complete nucleotide sequence of mouse 28S rRNA gene. Implications for the process of size increase of the large subunit rRNA in higher eukaryotes. Nucl. Acid. Res., 12, 3563-3583.

Hughes, S.J. (1953). Conidia, conidiophores and classification. Canad. J. Bot., 31, 577-659.

Kurtzman, C.P. (1985). Molecular taxonomy of the fungi. In: Gene Manipulations in Fungi, pp. 35-63 (J.W. Bennett and L.L. Lasure, eds.), Academic Press, Orlando.

Kurtzman, C.P., Smiley, M.J., Johnson, C.J., Wickerham, L.J., and Fuson, G.B. (1980). Two new and closely related heterothallic species, Pichia amylophyla and Pichia mississippiensis: Characterization by hybridization and deoxyribonucleic acid reassociations. Int. J. Syst. Bacteriol., 30, 208-216.

Lane, D.J., Pace, B., Olsen, G.J., Stahl, D.A., Sogin, M.L. and Pace, N.R. (1985). Rapid determination of 16S ribosomal RNA sequences for phylogenetic analyses. Proc. Nat. Acad. Sci. USA, 82, 6955-6959.

Logrieco, A., Peterson, S.W., and Bottalico, A. (1991). Phylogenetic affinities of the species in Fusarium section Sporotrichiella. Exper. Mycol., 15, 174-179.

Marasas, W.F.O., Nelson, P.E., and Toussoun, T.A. (1984). Toxigenic Fusarium species: Identity and Mycotoxicology, 328 p. Pennsylvania State University Press, University Park.

Nachmias, A., Solel Barash, Z., and Strobel, G.A. (1977). Purifications and characterization of phytotoxin produced by Phoma tracheiphila, the causal agent of mal secco of citrus. Physiol. Pl. Path., 10, 147-157.

Nelson, P.E., Toussoun, T.A., and Cook, R.J. (1981). Fusarium: Disease, Biology, and Taxonomy, 459 p. Pennsylvania State University Press, University Park.

Nelson, P.E., Toussoun, T.A., and Marasas, W.F.O. (1983). Fusarium Species: An Illustrated Manual for Identification, 193 p. Pennsylvania state University Press, University Park.

Peterson, S.W., and Logrieco, A. (1991). Ribosomal RNA sequence variation among interfertile strains of some Gibberella species. Mycologia, 83(4), 397-402.

Sanger, R., Nicklen, S., and Coulson, A.R., (1977). DNA sequencing with chain-terminating inhibitors. Proc. Natl. Acad. USA, 74, 5463-5467.

Seidler, R.J. and Mandel, M. (1971). Quantitative aspects of DNA renaturation: DNA base composition, state of chromosome replication, and polynucleotide homologies. J. Bacteriol., 106, 608-614.

Seidler, R.J., Knittel, M.D., and Brown, C. (1975). Potential pathogens in the environment. Cultural reactions and nucleic acid studies on Klebsiella pneumoniae from chemical and environmental sources. Appl. Microbiol., 29, 819-825.

Snyder, W.C., and Hansen, H.N. (1945). The species concept in Fusarium with reference to discolor and other sections. Amer. J. Bot., 32, 657-666.

Thrane, U. (1989). Fusarium species and their specific profilesofsecondary metabolites, In: Fusarium Mycotoxins, Taxonomy, and Pathogenicity, pp.199-225 (J. Chelkowski, ed.), Elsevier, Amsterdam.

Wickerham, L.J. (1951). Taxonomy of yeasts. U.S. Dept. Agric. Tech. Bull., 1029, 1-56.

Woese, C.R. (1987). Bacterial evolution. Microbiol. Rev., 51, 221-271.

Wollenweber, H.W., and Reinking O.A. (1935). Die Fusarien, ihre Beschreibung, Schadwirking und Bekämpfung, 355 p. Paul Parey, Berlin.

THE EFFECTS OF SOIL AND POPLAR TREES ON AFLATOXIN B_1

UFORD A. MADDEN* and HENRY M. STAHR, Veterinary Diagnostic Laboratory, College of Veterinary Medicine, Iowa State University, Ames, IA 50011, USA

INTRODUCTION

Mycotoxins which are associated with several crops are becoming more important to the public and environmental health workers because of their potential for causing health problems. Contaminated crops with high levels of aflatoxin are unfit for consumption by humans and animals and are usually disposed of by plowing under soil (Angle 1986; Angle and Wagner 1981; and Goldberg and Angle, 1985), thereby creating the potential for contamination of groundwater. Aflatoxin contamination of the water could be detrimental to the health and welfare of the population since aflatoxin B_1 is both toxic and carcinogenic (Angle, 1986, Goldberg and Angle, 1985).

Leaching and absorption studies have shown that aflatoxin B_1 and its metabolites were retained in different types of soil (Angle,1986; Angle and Wagner, 1981). Other studies have also shown that soil microorganisms affect the concentration and the movement of aflatoxin in the soil (Angle, 1986, Goldberg and Angle, 1985). Poplar trees were found to be effective in removing nitrate-nitrogen from the soil, thereby reducing the amount of nitrates in groundwater. Trees were planted in the area, providing rooting systems which bind the soil, preventing soil erosion, and removing excess nitrogen from the water-table (Litch, 1990).

The purpose of this study was to investigate the effectiveness of soil and poplar trees in the detoxification of aflatoxin B_1 in soil. The objectives of this study were to determine the fate of aflatoxin in Iowa soil, to determine procedures for the disposition of mycotoxin-contaminated commodities and to utilize this control system to successfully prevent the contamination of municipal and private wells. Trees will prevent erosion of the soil which will prevent the transport of aflatoxins and other contaminants in the soil from gaining access to the groundwater from which a portion of the potable water that the public consumes is derived.

MATERIALS and METHODS

Three tubs of soil were used in this study. One tub was filled with soil taken from a soybean field while the other two were filled with soil taken from a corn field. Mini-wells were made from a one and one half inch PVC pipe tubing with aluminum cloth screen covering the bases of the wells. The deep wells were placed at a depth of about 30 cm into the soil in the tubs while the shallow wells were at a depth of about 15 cm. In one tub, containing soybean soil and the second containing corn soil, 150 g of aflatoxin-contaminated corn was buried into the soil equidistant from each well, which was about 7.5 cm. In the third tub containing corn soil, four cuttings of propagated poplar trees were planted approximately 7.5 cm apart and 10 cm away from the deep well; then 150 g of aflatoxin-contaminated corn was buried into the soil approximately 5 cm from the trees and 10 cm away from the 15 cm well.

Water samples were collected from the mini-wells after rainfall during the first, second, fourth, seventh, and ninth week and analyzed for the presence of aflatoxin. After twelve weeks samples of buried aflatoxin-contaminated corn and soil surrounding the aflatoxin-contaminated in tubs were taken and analyzed, along with samples from the original soil and original aflatoxin-contaminated corn for the presence of aflatoxin. Extraction of aflatoxin from water samples was performed by using methylene chloride ($MeCl_2$) in a 1:1 ratio to water (H_2O), and soil samples were extracted using 90:10 acetonitrile:water, then with $MeCl_2$:H_2O (1:1). The corn samples were extracted according to Analytical Methods in Toxicology procedure (Stahr, 1991). All extracts were cleaned up by C-18 Sep-Pak^R (Waters, Division of Millipore) and dried under nitrogen gas, then redissolved in 100 ul of $MeCl_2$. Detection and identification of aflatoxin was by thin layer chromatography (TLC). Ten ul of each sample were spotted on TLC plates along with appropriate amounts of standards. The plates were developed in 90:10 chloroform:acetone, 1-2 cm from the top of each plate, and examination of plates was done under longwave UV light for positive identification of aflatoxins. Each fluorescent spot was scraped from the plates and placed in a two dram vial; 3 ml of 90:10 HPLC-methanol:water was added to each vial in order to elute the aflatoxin. Quantitation of aflatoxins was performed by spectrophotofluorometry with excitation set at 360 nm, and emission set at 430 nm, with sensitivity set at 0.1% scale (Stahr, 1991).

RESULTS

The total aflatoxins present in the corn sample buried in the soil in each tub was 208 ng/g B_1 and 12 ng/g B_2. No aflatoxin was detected in water samples collected from the mini-wells after rainfall during weeks 1, 2, 4, 7 and 9 nor from original soil samples. After three months, no aflatoxin B_1 was detected in soil surrounding the aflatoxin-contaminated corn, but aflatoxin B_2 was present. Both aflatoxin B_1 and B_2 were present in the aflatoxin-corn samples that were buried into the soil (Table 1). Aflatoxin B_1 was decreased in all aflatoxin-contaminated corn samples buried in soil while aflatoxin B_2 was increased (Table 2).

Table 1. Detection of Aflatoxin B_1 and B_2 in Original Aflatoxin-Contaminated Corn, Original Soil, Buried Aflatoxin-Contaminated Corn and Soil Surrounding Aflatoxin-Contaminated Corn After Three Months in Corn Soil with Poplar Trees, Soybean Soil and Corn Soil[a].

Samples	Orig. AFB_1-Corn		Orig. Soil		Buried AFB_1-Corn		Soil/Corn	
	B_1	B_2	B_1	B_1	B_1	B_2	B_1	B_2
Corn Soil Trees	+	+	-	-	+	+	-	+
Soy Soil	+	+	-	-	+	+	-	+
Corn Soil	+	+	-	-	+	+	-	+

\+ = positive detection of aflatoxin
\- = negative detection of aflatoxin
[a]Positive detection and identification of aflatoxins was done by thin layer chromatography with longwave UV light at Rf values corresponding to that of aflatoxin standards.

Table 2. Concentrations aflatoxin B_1 and B_2 in Original Aflatoxin-Contaminated Corn, Original Soil, Buried Aflatoxin-Contaminated Corn and Soil Surrounding Aflatoxin-Contaminated Corn After Three Months in Corn Soil with Poplar Trees, Soybean Soil and Corn Soil[a].

Samples	Orig. AFB$_1$ Corn ng/g		Orig. Soil ng/g		Buried AFB$_1$ Corn ng/g		Soil/Corn ng/g	
	B_1	B_2	B_1	B_2	B_1	B_2	B_1	B_2
Corn Soil Trees	208	12	-	-	37	21	-	10
Soy Soil	208	12	-	-	3	25	-	7
Corn Soil	208	12	-	-	13	17	-	11

[a]Concentrations of aflatoxins was determined by spectrophotofluorometry. Quantitation of aflatoxins was done by linear regression from aflatoxin standard curves.

DISCUSSION

The results from this study have shown that no aflatoxin was detected in any water sample taken from the mini-wells, which suggests that no aflatoxin from aflatoxin-contaminated corn leached through soil with water. These results agree with previous reports that aflatoxin B_1 and its metabolites were retained in soil. Concentrations of aflatoxin determined by spectrophotofluorometry revealed that aflatoxin B_1 was present in all samples from the aflatoxin-contaminated corn buried in soil in the tubs. Corn soil with trees contained 37 ng/g, soybean soil had 3 ng/g, and corn soil had 13 ng/g, were significantly lower than the original aflatoxin-contaminated corn (208 ng/g). Aflatoxin B_2 was higher in all samples from the aflatoxin-contaminated corn buried in soil suggesting that all soils were effective in reducing the amount of aflatoxin B_1 present. The concentration of aflatoxin B_1 detected in aflatoxin-contaminated corn buried in soybean soil (3 ng/g) was significantly less than the amount present in the aflatoxin-contaminated corn buried in corn soil, corn soil with trees and the original aflatoxin-contaminated corn sample. The amount of aflatoxin B_2 present in the aflatoxin-contaminated corn buried in soybean soil (25 ng/g) was greater than the amount in the tubs with the corn soil (corn soil with trees 21 ng/g, corn soil 17 ng/g) and also the original aflatoxin-contaminated corn (12 ng/g). Therefore, soybean soil was more effective than corn soil with poplar trees and corn soil in reducing aflatoxin B_1 present in the aflatoxin-contaminated corn and converting it to aflatoxin B_2 (Figure 1). No aflatoxin B_1 was detected in any soil samples surrounding the aflatoxin-contaminated corn which suggests that aflatoxin B_1 in soil is readily degraded to aflatoxin B_2. The concentrations of aflatoxin B_2 present in all soil samples surrounding the aflatoxin-contaminated corn in tubs (corn soil/trees 10 ng/g, soybean soil 7 ng/g and corn soil 11 ng/g) were lower than

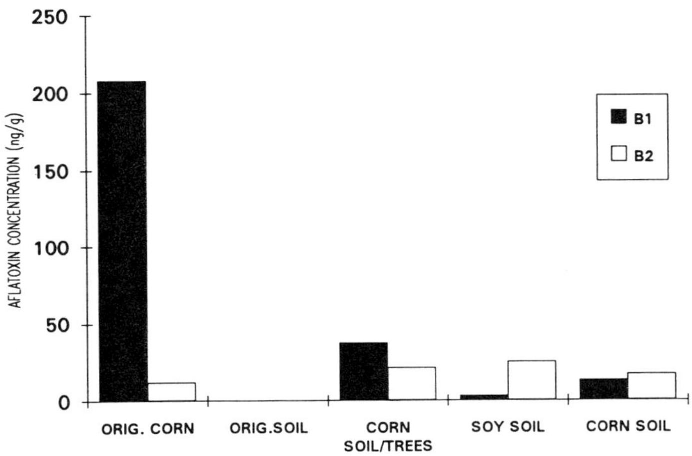

Figure 1. Concentrations of Aflatoxin B_1 and B_2 in Original Aflatoxin Contaminated Corn, Original Soil, Buried Aflatoxin-Contaminated Corn in Corn Soil with Trees, Soybean Soil and Corn Soil After Three Months.

the amount present in the original aflatoxin-contaminated corn sample (12 ng/g) and the buried aflatoxin-contaminated corn samples (Figure 2). This suggest that aflatoxin B_2 like aflatoxin B_1 in soil is readily degraded to other compounds. The concentration of aflatoxin B_2 present in soybean soil sample surrounding the aflatoxin-contaminated corn was lower than the samples from the other tubs. These results show that soybean soil was more effective in reducing both aflatoxin B_1 and B_2 than the corn soil (Table 2).

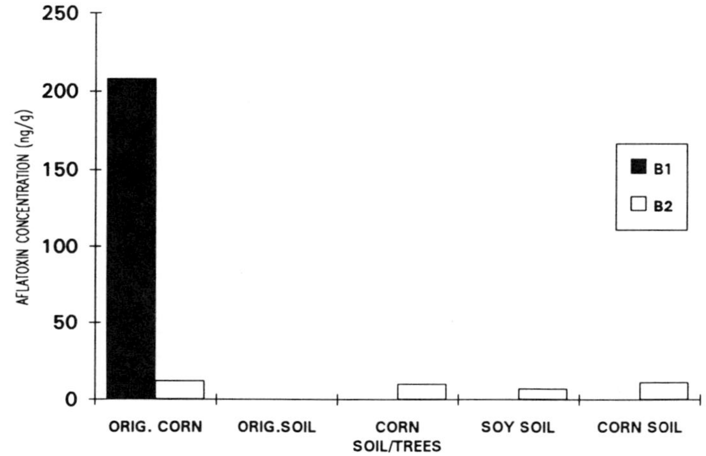

Figure 2. Concentrations of Aflatoxin B_1 and B_2 in Original Aflatoxin-Contaminated Corn, Original Soil, and Soil Surrounding Aflatoxin-Contaminated Corn Buried in Corn Soil with Trees, Soybean Soil and Corn Soil After Three Months.

SUMMARY

Aflatoxin B_1 in corn samples was reduced significantly when buried in soil for a period three months. Aflatoxin B_1 did not remain in the soil and was converted to B_2, since only aflatoxin B_2 was detected in the soil surrounding the buried aflatoxin-contaminated corn. The amount of aflatoxin B_2 present in the aflatoxin-contaminated corn buried the soil was significantly higher than that which was present in the original aflatoxin-contaminated corn and the soil samples surrounding the aflatoxin-contaminated corn. This result suggests that if aflatoxin B_1-contaminated commodities are buried in the soil, aflatoxin B_1 is readily converted to aflatoxin B_2 which is a less toxic compound and non mutagenic. The amount of aflatoxin B_2 detected in soil samples surrounding the aflatoxin-contaminated corn was lower that the amount present in the original aflatoxin-contaminated corn and the buried aflatoxin-contaminated corn which suggests that aflatoxin B_2 in soil is readily degraded to other compounds.

There was no detectable amount of aflatoxin present in water samples taken from mini-wells around buried aflatoxin-contaminated corn which indicates that no aflatoxin leached into the mini-wells after rainfall. This result further indicates that aflatoxin B_1 is degraded to aflatoxin B_2 which when further degraded is not detected as a leachate in the water samples taken from the mini-wells.

The recommendation that could be given concerning aflatoxin-contaminated commodities is that, plowing aflatoxin-contaminated commodities under soil will allow for successful detoxification of aflatoxin B_1 without potential contamination of groundwater. Present farm practices of discing aflatoxin-contaminated corn into the soil should result in successful detoxification of aflatoxin-contaminated land if the land is allowed enough time for reduction of aflatoxin to occur. If proper management is instituted and practiced then public concerns about the safety of potable water will be eliminated or reduced.

Acknowledgments

The computational assistance of Mr. Bruce Eveland and the editorial suggestions of Dr. Gary D. Osweiler and Dr. Yosiya Niyo are sincerely appreciated.

REFERENCES

Angle, J.S. (1986). Aflatoxin decomposition in various soils. J. Environ. Sci. Hlth., B21(4), 277-288.

Angle, J.S. and Wagner, G. H. (1981). Aflatoxin B_1 effects on soil micro-organisms. Soil Biol. and Biochem., 13, 381-384.

Goldberg, B.S. and Angle, J.S. (1985). Aflatoxin movement in soil. J. Environ. Qual., 14(2), 224-228.

Litch, Louis A. (1990). Poplar tree roots for water quality improvement. Natl. Con. on Enhg. Sts. La. Mgmt. Pgms, 55-61.

Stahr. H.M. (1991). Analytical Methods in Toxicology. pp. 93-104, A Wiley Interscience Publication, John Wiley and Sons Inc., New York, Chichester, Brisbane, Toronto, Singapore.

ALLERGENICITY OF CERTAIN AIRBORNE FUNGAL SPORES

EMMANUEL MBA* and **STEPHANIE HARRISON**, Department of Biology, North Carolina Central University, Durham, NC 27707, USA
SUMANA BANERJEE, Durham Technical Community College, Durham NC 27703, USA
JOHN E. MAYFIELD, Department of Biology, North Carolina Central University, Durham, NC 27707, USA

INTRODUCTION

Fungi have been implicated in hypersensitivity reactions in sensitive humans. These allergic reactions can be related to exposure to either spores, mycelia or metabolites of fungi (Burge, 1989) and are more often experienced in the lower respiratory passages than in the nasal tissues due to the small spore sizes (Hjelmroos, 1988). Fungal spores, present in both indoor and outdoor air samples, vary in both quantity and type according to the season of the year, time of day, geographical location and nearness to the spore sources (Lacey, 1981). Aspergillus and Penicillium are usually non-seasonal indoor molds. Fungi such as Alternaria and Cladosporium are universally the most dominant among the outdoor molds (Hoffman, 1984).

Both the spores and mycelia of fungi contain certain proteins or glycoproteins that are capable of stimulating the production of and binding to IgE antibodies (Budd, 1986). In 1981, Hoffman and his associates compared preparations which were either 96% pure spores or 97% pure mycelium; the results indicated that the spore extracts contained all antigens found in the mycelial extracts as well as eight additional antigens.

Biodeterioration Research 4, Edited by G.C. Llewellyn *et al.*, Plenum Press, New York, 1994

Significant progress has been made in characterizing allergens through application of immunochemical and physiochemical technology such as polyacrylamide gel electrophoresis (PAGE), immunoelectrophoresis and radioallergosorbent test (RAST) (Yunginger, 1984). Western blotting or protein blotting has proven to be less intricate in characterizing allergens. It combines the excellent resolution ability of PAGE with electrophoretic transfers of these components to nitrocellulose paper where they are immobilized (Towbin et al., 1979). This highly successful approach has been used in identifying important allergens in a variety of allergen sources such as ticks (Gauci et al., 1990), house dust mites (Tovey and Baldo, 1985) grasses (Ford et al., 1985) and molds (Ruffin et al., 1988).

Progress has been made in analyzing the allergenic components of the most dominant universal molds, such as, Cladosporium and Alternaria. There are other genera of fungi which deserve equal attention because they produce respirable airborne spores and are collected along with Cladosporium and Alternaria.

The objectives of this investigation were to extract and separate Coomassie blue-positive components of airborne spores of Bispora, Candida. Epicoccum and a yeast and to determine the allergenic components of these molds by enzyme-immunoassay after electrophoretic transfer to nitrocellulose paper.

MATERIALS AND METHODS

Airborne fungal spores were collected on Potato dextrose agar (PDA) plates with the Andersen two-stage sampler. Colony-forming units were counted and isolated to culture tubes and maintained as stock cultures. These stock cultures were the sources for the inoculation of both fresh liquid or solid plates of PDA. All fungi except Bispora were grown in either liquid or solid potato dextrose media. Bispora was grown in both media.

After growth and sporulation on PDA or five days of constant agitation in liquid medium, spores and mycelial fragments or mycelial pellets, respectively, were aseptically collected in phosphate buffered saline (PBS) pH 7.2 and continuously stirred for 4 hr. Extraction was continued at 4° C overnight and clarified by centrifugation. Supernatants were frozen, lyophilized and stored at -10° C. Protein concentration was determined by a Bio-Rad dye-binding protein assay (Bradford, 1976).

Sodium dodecyl sulphate polyacrylamide gel electrophoresis (SDSPAGE) was carried out at 35 mA in the Mini Protein II cell (Bio-Rad) with 0.75 mm thick gels utilizing a discontinuous buffer system (Laemmli, 1970). Approximately 45 ug of protein extract was loaded into each well. After electrophoresis the gels were stained with Coomassie brilliant blue (CBB).

Protein bands separated by SDS-PAGE were electrophoretically transferred to nitrocellulose membrane (Towbin et al., 1979) with the BioRad transblot cell. Transfer was conducted at 30 V overnight at room temperature. The transfer buffer contained 0.025 Tris, 0.019 M glycine, 20% (v/v) methanol and 0.1% SDS, pH 8.3. After the transfer, the unbound membrane surfaces were blocked by 3% gelatin, 0.5% Tween-20 in PBS at room temperature.

IgE binding allergens on the nitrocellulose were detected with the immunoblot assay of Tovey and Baldo (1984). Nitrocellulose membranes were incubated with sera (1:2 v/v dilution with TPBS buffer) overnight and washed three times with TPBS buffer. Membranes were then incubated with goat-anti-human IgE (1:100 [v/v]) dilution for 3 h and washed three times prior to incubation with rabbit-anti-goat IgG conjugated with horseradish peroxidase (1:200 [v/v]) for 1.5 h. Membranes were washed three times and incubated in the dark with a substrate solution containing 3,3 diaminobenzidine and 0.003% hydrogen peroxide which developed as dark brown bands, indicating antigen-antibody complex.

RESULTS AND DISCUSSION

The extracts of Bispora, Epicoccum, Candida and yeast contained certain allergenic components. But, the number and molecular weights of these allergenic proteins or glycoproteins varies and range from 25.5 to 74 KD, respectively (Fig. 1 and 2). The diversity of the IgE antibody response to the allergenic components is not surprising because individual IgE responds differently to prevailing allergens. Some individuals will react to only one allergen while others may react to as many as 30 allergens in the same environment (Potter et al., 1991).

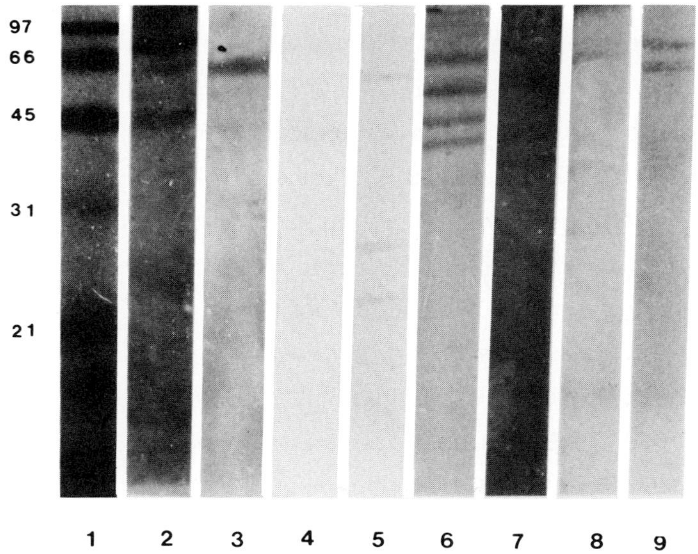

Figure 1. Identification of IgE-binding components using immunoenzyme staining in extracts of Candida, Epicoccum, yeast and Bispora after transfer from SDS-PAGE to nitrocellulose membrane and incubated with pooled sera of atopic patients. Lane 1 is SDSgel of marker proteins. Lanes 2, 4, 6 and 8 show Candida, Epicoccum, yeast and Bispora components, respectively, after SDS-PAGE. Lanes 3, 5, 7 and 9 show the detection of IgE binding in Candida, Epicoccum, yeast and Bispora extracts, respectively.

KD	Candida		Epioccum		Yeast		Bispora	
97.4			I:::::I		I:::::I			
	I:::::I						I:::::I	IIIIII
66.2								
			I:::::I					
	I:::::I	IIIII			I:::::I	IIIIII		
			I:::::I		I:::::I			IIIIII
			I:::::I	IIIII				
45.0								
	IIIIII	I:::::I		I:::::I	IIIIII			
		I:::::I						
	I:::::I						I:::::I	
						IIIIII		
31.0								
			I:::::I					
				IIIIII				
							I:::::I	
			I:::::I					
				IIIIII			I:::::I	
21.5								

I:::::I = Coomassie blue-stained bands
IIIIII = Immunoenzyme labeled bands

Figure 2. Graphic comparison of Coomassie blue-stained bands with immunoenzyme labeled bands of each fungal extract.

The SDS-PAGE pattern of the four fungal extracts showed that some bands are more intense than others when stained with Coomassie-brilliant blue (Figs. 1 and 2). Some bands were not detected by CBB but were identified as allergens by sensitive immunoblot assay. The color intensity of the bands detected by the immunoblot also varies depending on the allergenic contents present in the bands, and the concentration of specific IgE antibodies in the pooled serum used in the immunoblotting.

To our knowledge, there has been no previous report on the allergenicity of Bispora. This fungus is collected along with known moldcontaining allergens such as Cladosporium and Alternaria. Resolution of proteins of Bispora grown in both solid and liquid media by SDS-PAGE revealed the same number of bands when stained with coomassie brilliant blue R-250. The two allergenic proteins detected by immunoblotting were very light (Fig. 1, lane 9). This may be due to low allergen contents in the resolved bands and low concentration of specific IgE antibodies in the pooled serum (Axelsson et al., 1990). This assumption was supported by the fact that when undiluted serum was used in immunoblotting the allergenic bands were more conspicuous. The Bispora grown in the liquid media showed more intense bands than when grown on solid media. The difference in color intensity suggests that the Bispora grown in liquid medium contains more allergens than the one grown in the solid medium. This difference may be a result of the antigens that leached out and were entrapped in the solid phase; causing their exclusion during extraction. It was confirmed when the PDA medium used in culturing Epicoccum showed that both the Epicoccum and the medium contain the same allergenic components.

Epicoccum has been reported to be among the most prevalent aeroallergens (Burge, 1986). Little is known about its allergenic components. Based on our immunoblotting results, Epicoccum has three proteins that reacted with IgE antibodies. The immunoband of IgE binding protein 55 KD was very strong but the proteins 29 and 25 KD were very weak as indicated by color intensity (Fig. 1, lanes 4-5, Fig. 2).

Candida extract and yeast shared two common allergenic bands at 63 and 46 KD (Fig. 1, lanes 3 and 7, Fig. 3). Candida is dimorphic and as such can change from yeast-like to mycelial fungus depending on the environmental condition. According to the studies conducted on the allergenicity of Candida ablicans (yeast phase) by Savolainen and his associates (1989), two bands at 46 and 29 KD were identified. The 46 KD allergen was believed to be the major allergen because it was noticed in all six strains of Candida ablicans studied.

Based on the number of allergenic bands detected and the number of proteins resolved by SDS-PAGE in each fungal extract, Candida and yeast have a higher percentage of allergenic bands (Fig. 2). For instance, Candida had two out of three bands that reacted with IgE antibodies. Also, three out of four bands reacted with IgE antibodies in yeast. In Bispora, two out of four bands reacted while Epicoccum revealed three allergenic bands out of eight bands. This may be due to presence of melanin found in dark spore fungi such as Bispora and Epicoccum. This pigment melanin precipitates with protein during extraction and thereby tends to reduce the number of allergens recovered in an extract (Paris et al., 1990).

It is evident that these fungi contain components that are capable of binding IgE immunoglobulins. However, further study is needed to further determine the exact chemical nature of these components. Also, the use of monoclonal antibodies maybe considered to provide a more definitive determination of allergenicity. Improved characterization of fungal allergens is essential to both the development of effective diagnosis and the establishment of spore prevalence with allergenic reactions.

SUMMARY

The frequency of IgE-binding components within the crude extracts of airborne spore producing fungi (Epicoccum, Candida. Bispora and unidentified yeast) was determined by enzyme-linked immunoassay. The extracts were electrophoretically separated and transferred by the Western blot technique to nitrocellulose membranes. The allergenic components were detected using pooled sera from allergic patients. The allergen-antibody complexes were detected by enzyme-immunoassay. The IgE-binding components found in the extracts have molecular weights that range from 25 to 74 KD. Three different IgE-binding components were detected in each extract of yeast and Epicoccum. Also, two allergenic components were found in Candida and Bispora. Both Candida and yeast share common bands 46 and 63 KD. These results showed much diversity of the IgE-antibody response to the variety of allergenic components found in the crude extracts of these genera.

Acknowledgments

We thank Dr. Walter H. Pattillo, Jr. for critically reviewing this manuscript. This research was supported by National Institutes of Health Grant R5 A128042A.

REFERENCES

Axelsson, I.G.K, Johansson, S.G.O., Larsson, P.H. and Zetterstrom, O. (1990). Characterization of allergenic components in sap extract from the weeping fig (Ficus benjamina). Int. Arch. Allergy Appl. Immunol., 91, 130-135.

Bradford, M.M. (1976). A rapid and sensitive method for the quantitation of microgram quantities of protein utilizing the principle of protein dye binding. Anal. Biochem., 72, 248-254.

Budd, T.W. (1986). Allergens of Alternaria. Grana, 25, 147-154.

Burge, H.A. (1986). Some comments on the aerobiology of fungus. Grana, 25, 143-146.

Burge, H.A. (1989). Airborne allergenic fungi: Classification, nomenclature and distribution. Immunol. Allergy Clin. N. Amer., 9, 307-319.

Ford, S.A., Tovey, E.R. and Baldo, B.A. (1985). Identification of orchard grass (Dactylis glomerata) pollen allergens following electrophoretic transfer to nitrocellulose. Int. Arch. Allergy Appl. Immunol., 78, 1521.

Hjelmross, P.M. (1988). The significance of mould spores in connection with allergic symptoms in children. A preliminary survey. Ann. Sci. Natur., 13e, 95-98.

Hoffman, D.R., Kozak, P.P., Gillman, S.A., Cummins, L.H. and Hallup, J. (1981). Isolation of spore specific allergens from Alternaria. Ann. Allergy, 46, 310-316.

Hoffman, D.R. (1984). Mould allergens. In: Mould Allergy. pp. 104-116. (Y. Al-Doory and J. Donson, eds.). Lea and Febiger, Philadelphia.

Lacey, J. (1981). The aerobiology of conidial fungi. In: The Biology of Conidial Fungi. pp. 373-416. (G.T. Cole and W.B. Kendrick, eds.). Academic Press, New York.

Laemmli, U.K. (1970). Cleavage of structural proteins during assembly of the head of the bacteriophage T4. Nature, 227, 680-685.

Paris, S., Fitting, C., Ramirez, E., Latge, J.P. and David, B. (1990). Comparison of different extraction methods of alternaria allergens. J. AllergyClin. Immunol., 85, 941-948.

Potter, P.C., Juritz, J., Little, F., McCaldin, M. and Dowdle, E.B. (1991). Clustering of fungal allergen-specific IgE antibody responses in allergic subjects. Ann. Allergy, 66, 149-153.

Ruffin, J., Chiao, C. and Banerjee, S. (1988). Detection of some airborne pollen grain and fungal spore allergens: Frequency of IgE-binding following electroblotting and enzyme immunoassay. Grana, 27, 14.

Savolainen, J., Viander, M., Einarsson, R. and Koivikko, A. (1989). Allergenic variability of different strains of Candida albicans. Int. Arch. Allergy Appl. Immunol., 90, 61 -66.

Tovey, E.R. and Baldo, B.A. (1984). Standardization of allergens: Quantitative definition of house dust mite extracts following electroblotting and detection of components with antibody and lectin probes. Int. Arch. Allergy Appl. Immunol., 75, 322-329.

Tovey, E.R. and Baldo, B.A. (1985). Detection of house dust mite allergens and frequency of IgE-binding following electroblotting and immunoenzyme assay. Int. Arch. Allergy Appl. Immunol., 76, 82-85.

Towbin, H., Staehelin, T. and Gordon, J. (1979). Electrophoretic transfer of proteins from polyacrylamide gels to nitrocellulose sheets: Procedure and some applications. Proc. Nat. Acad. Sci., 76, 4350-4354.

Gauci, M., Stone, B.F. and Thong, Y.H. (1990). An immunoblot technique for identification of allergens of the Australian paralysis tick Ixodes holocyclus. J. Immunol. Method., 126, 51-55.

Yunginger, J.W. (1984). Allergen standardization. J. Allergy Clin. Immunol., 73, 245-267.

STIMULATION OF ZEARALENONE PRODUCTION IN RICE CULTURES OF *FUSARIUM SEMITECTUM* BY ADDITION OF ZEARALENONE AND α-AND β-ZEARALENOL

K.E. RICHARDSON, Department of Poultry Science, North Carolina State University, Box 7608, Raleigh, NC 27695, USA

W.M. HAGLER, JR.*, Mycotoxin Laboratory, Department of Poultry Science, North Carolina State University, Box 7636, Raleigh, NC 27695, USA

P.B. HAMILTON, Department of Poultry Science, North Carolina State University, Box 7608, Raleigh, NC 27695, USA

INTRODUCTION

Zearalenone is an estrogen produced by several species of Fusarium on corn and other cereal grains (Caldwell and Tuite, 1974; Gross and Robb, 1975; Bottalico, 1979). Its presence in animal feeds and subsequent consumption have been associated with hyperestrogenism and other reproductive disorders in swine (Miller et al., 1973), poultry (Meronuck et al., 1970; Allen et al., 1981a,b) and cattle (Mirocha et al., 1968). Zearalenol, which is produced in its two epimeric forms as a metabolic precursor of zearalenone by certain species of Fusarium, exhibits estrogenic activity (Peters, 1972; Mirocha et al., 1978; Bottalico et al., 1985; Richardson et al., 1985). α-Zearalenol has three to four times as much estrogenic activity as zearalenone while β-zearalenol has about the same or slightly less estrogenicity as zearalenone (Peters, 1972; Mirocha et al., 1978; Hagler et al., 1979). Zearalenone and zearalenol also have some anabolic activity that is useful in animal nutrition (Hidy et al., 1977).

The widespread occurrence of these fungal metabolites and their possession of both useful and detrimental properties has prompted several

Biodeterioration Research 4, Edited by G.C. Llewellyn *et al.*, Plenum Press, New York, 1994

investigations into factors affecting their biosynthesis and production. Eugenio et al. (1970) reported that F. roseum would produce high levels of zearalenone in rice and corn when the moisture content was high (45-65%). The cereal grains serving as substrates themselves influence zearalenone production. When an isolate of F. roseum was cultured on several cereal grains, good production of zearalenone occurred on rice, corn, and wheat; poor production occurred on oats and barley; and no production occurred on soybeans (Eugenio et al., 1970). The effects of nutrients on toxin production in stirred and stagnant fermentation cultures of other species of Fusarium were reported (Hidy et al., 1977; Bacon et al., 1977). While studying the influence of substrate on zearalenone production, Hidy et al. (1977) discovered that the addition of a catalytic amount of zearalenone (20-30 mg/liter of liquid medium) at the time of inoculation increased zearalenone production after 14 days which implied that zearalenone stimulated its own biosynthesis. The effects of other zearalenone derivatives on zearalenone production were not examined.

The present communication reports stimulation of zearalenone production in rice cultures of F. semitectum by the addition of catalytic amounts of α-zearalenol and β-zearalenol, precursors of zearalenone and by methanol which served as carrier solvent.

MATERIALS AND METHODS

Chemicals

Solvents and chemicals were reagent-grade, except the solvents for high pressure liquid chromatography (HPLC) which were HPLC-grade (Fisher Scientific Co., Raleigh, NC). Authentic zearalenone, α-zearalenol, and β-zearalenol used for quantitative standards were provided by IMC, Inc., Terre Haute, IN.

Culture Maintenance

A strain of F. semitectum which produces zearalenone, α-zearalenol, and β-zearalenol (Hagler et al., 1980; Richardson et al., 1984) was maintained on moist, autoclaved soil at 10C. This isolate had been referred to as F. roseum 'Gibbosum' (Hagler et al., 1980) but was later identified by Marasas et al. (1984) as F. semitectum Berk. & Rav. The soil cultures were sprinkled on potato-dextrose agar plates and incubated at 24C for 5 days to provide inoculum for the experiment.

Production of Zearalenone and Preparation of Zearalenol

Pure zearalenone was prepared by growing F. semitectum in rice culture, extracting, purifying by chromatographing on a Florisil column and crystallizing as described by Hagler and Mirocha (1979). A mixture of the α- and β-epimers of zearalenol was prepared by reducing 500 mg of the zearalenone with $NaBH_4$ until 95% of zearalenone was converted to zearalenol (Shipchandler, 1975; Sarudi, 1976; Richardson et al., 1984). Zearalenol was separated from residual zearalenone by normal-phase HPLC before the epimers were separated and purified by reverse phase HPLC (Richardson et al., 1984). Identities of zearalenone, α-zearalenol, and β-zearalenol were confirmed by mass spectrophotometry (Richardson et al., 1985).

Culture Conditions and Toxin Quantitation

Sterile, moist rice (30 g rice plus 20 ml of deionized water in 250-ml flasks) was seeded with blocks of agar (0.5 cm^2) from the inoculum plates and incubated at 24C for 7 days. The cultures were then treated with 4 mg of zearalenone, 4 mg of α-zearalenol, or 4 mg of β-zearalenol, each dissolved in 0.5 ml of methanol. Other flasks received 0.5 ml of methanol (solvent control) or no treatment (control). Triplicate cultures of each treatment were extracted by the method of Richardson et al. (1984) at 0, 7, 14, 21, and 28 days after the addition of zearalenone, its derivatives, or solvent. The extracts were purified by column chromatography, and the concentrations of zearalenone, α-zearalenol, and β-zearalenol were determined by normal phase HPLC (Richardson et al., 1984). The concentrations were corrected for the amount added, for the efficiency of recovery, and the bioconversion of these compounds (Richardson et al., 1984). The data were subjected to analysis of variance by the general linear models procedure of the Statistical Analysis System (1982). Statements of significance are based on $P<.05$ unless otherwise indicated.

RESULTS

In the normal, unamended rice cultures, zearalenone, α-zearalenol, and β-zearalenol were produced by F. semitectum (Table 1) as predicted by the literature. The amount of zearalenone at all time intervals was much greater than that of α- and β-zearalenol. At the time of maximum production, zearalenone exceeded the sum of α- and β-zearalenol by a factor of about 20. Zearalenone and α-zearalenol reached their maximum concentration after 21 days incubation and plateaued there. β-Zearalenol reached its maximum after 7 days and remained there until 28 days when there was a significant ($P<.05$) decrease.

Table 1. Zearalenone, α-Zearalenol and β-Zearalenol Production by Fusarium semitectum in Normal Rice Culture.

Time (days)	Zearalenone	α-Zearalenol	β-Zearalenol
	(mg/g rice)		
0	0.085 ± 0.004^{a}	0.001 ± 0.000^{a}	0.032 ± 0.005^{a}
7	0.533 ± 0.020^{a}	0.037 ± 0.003^{b}	0.150 ± 0.023^{b}
14	2.023 ± 0.197^{b}	0.134 ± 0.001^{c}	0.162 ± 0.006^{b}
21	7.727 ± 0.710^{c}	0.182 ± 0.011^{d}	0.150 ± 0.036^{b}
28	6.218 ± 1.365^{c}	0.196 ± 0.011^{d}	0.123 ± 0.024^{c}

[a-d] Values in a column with different superscripts differ significantly ($P<.05$). Values are the mean ± SEM of triplicate cultures.

Methanol, which was used as the solvent for the addition of zearalenone and α- and β-zearalenol to the cultures at zero time, was not without effect itself. It stimulated the cultures to produce zearalenone earlier (Table 2). After 7 days incubation, the cultures receiving methanol had about a five-fold greater concentration of zearalenone than the methanol-free cultures (Table 1). This stimulatory effect, which was still observable at 14 days, disappeared by 21 days when there was no significant ($P<.05$) difference between treated and untreated cultures.

Addition of the catalytic amounts of zearalenone sparked the production of zearalenone after 7 days even more than methanol did (Table 2). However, there was no significant ($P<.05$) difference at the later time intervals. α-Zearalenol, like zearalenone, caused an early stimulation and, in addition, caused the cultures to reach maximum production at 14 days instead of the 21 days required for cultures treated with methanol or zearalenone (Table 2) and for untreated cultures (Table 1). There was no significant ($P<.05$) difference in the maximum concentration of zearalenone produced among the foregoing amendments.

The addition of β-zearalenol to the cultures sparked an early (7 days) production of zearalenone not significantly ($P<.05$) different from that sparked by zearalenone and α-zearalenol (Table 2). β-Zearalenol like α-zearalenol caused the cultures to attain their maximum concentration of zearalenone after 14 days.

Uniquely, β-zearalenol increased significantly ($P<.05$) the maximum concentrations of zearalenone attained. The increase after 14 days was about 70% over the concentration attained with α-zearalenol and with methanol and zearalenone after 21 days. The concentration of zearalenone sparked by β-zearalenol after 14 days was about three-fold greater than those sparked by methanol and zearalenone after 14 days and about seven-fold greater than the concentration found in unamended cultures (Table 1).

Table 2. Zearalenone Production by Fusarium semitectum in Amended Rice Cultures.

Time (days)	Amendment			
	Methanol	Zearalenone[a]	α-Zearalenol[a]	β-Zearalenol[a]
0	0.101 ± 0.015[b]	0.101 ± 0.022[b]	0.109 ± 0.022[b]	0.110 ± 0.014[b]
7	3.128 ± 0.127[c]	5.507 ± 0.857[d]	4.545 ± 0.132[d]	5.201 ± 0.549[d]
14	5.074 ± 0.219[d]	4.838 ± 0.152[d]	8.598 ± 0.803[e]	14.459 ± 0.762[f]
21	7.601 ± 0.795[e]	7.928 ± 0.657[e]	8.966 ± 0.474[e]	13.595 ± 0.557[fg]
28	8.114 ± 0.917[e]	8.612 ± 1.021[e]	8.484 ± 0.995[e]	12.376 ± 0.694[g]

[a]These compounds were dissolved in methanol before they were added to the cultures.
[b-g]Values with different superscripts differ significantly ($P<.05$).
Values are the means (mg zearalenone/g rice) ± SEM of 3 replicate flasks.

DISCUSSION

The ability of a carrier solvent to influence zearalenone production has been known since Richardson et al. (1984) found almost complete inhibition by ethanol. After comparing several solvents, they selected methanol for its causing only slight inhibition. The current finding of stimulation of zearalenone production by methanol (Table 2) can be reconciled with the prior report on the basis of the stimulation being an early event (7-14 day cultures) while the slight inhibition was a later (21 days) event (Richardson et al., 1984; Table 2). Another difference was the addition of methanol at the time of inoculation in the present study compared to the addition after incubation for 7 days in the prior study.

Zearalenone and α- and β-zearalenol also caused an early stimulation of zearalenone production, but their stimulatory effects (7 days) were significantly ($P<.05$) stronger than the effect of methanol (Table 2). After 14 days incubation, the rank order was β-zearalenol > α-zearalenol > zearalenone = methanol. At 21 and 28 days incubation the rank order was β-zearalenol > α-zearalenol = zearalenone = methanol. The failure of zearalenone addition to alter the maximum concentration of zearalenone attained in the cultures is contrary to Hidy et al. (1977) who reported an increase in yield of zearalenone when sparking amounts were added to fermentation cultures. The cause of the discrepancy probably lies in the fact that Hidy et al. (1977) measured zearalenone production at only one time interval when toxin production may not have been maximal, i.e., early on when zearalenone addition is stimulatory (Table 2).

The reason for the early sparking effect of the amendments tried in the present study cannot be deduced with certainty from the present data; however, there are some reasonable implications. The observation that β-zearalenol reached its maximum concentration in the unamended cultures at 7 days while zearalenone and α-zearalenol reached their maxima at 21 days implies production of β-zearalenol prior to and possibly as a precursor to zearalenone. This implication agrees with the conclusion of Richardson et al. (1984) that β-zearalenol is a biosynthetic precursor of zearalenone based on the rapid and complete incorporation of β-[^{14}C] zearalenol into zearalenone and on the complete failure of [^{14}C] zearalenone to be incorporated into β-zearalenol. α-[^{14}C] Zearalenol was slowly and incompletely converted to zearalenone without any interconversion to β-zearalenol and without conversion of [^{14}C] zearalenone to α-zearalenol. These findings and the present findings imply that there might be separate mechanisms for the conversion of α- and β-zearalenol into zearalenone. These facts and suppositions are supported by the conclusion of Steele et al. (1976) that the catabolism of zearalenone does not include zearalenol formation. Because small, essentially catalytic amounts of the amendments provoked the formation of much larger amounts of zearalenone, a likely explanation is that the regulatory mechanisms were affected. Presumably, each of the amendments affected different regulatory sites, a presumption based on methanol being so different chemically, on each amendment provoking a different temporal response and on the unique ability of β-zearalenol to increase the maximum yield. It seems possible that the normal biosynthesis of zearalenone might require the stepwise induction of several enzymatic steps and that the amendments are exerting their affects on the inductive processes.

The practical implications of the present findings are more obvious. Significant production of zearalenone can be obtained earlier by the simple addition of inexpensive methanol. While zearalenone and α-zearalenol induce a similar or even earlier production, their expense and scarcity probably rule out their use. β-Zearalenol, while equally difficult to obtain, has the virtue of almost tripling in 14 days the yield obtained with methanol.

SUMMARY

Cultures of Fusarium semitectum which produced zearalenone, α-zearalenol and β-zearalenol on rice were treated with small amounts of zearalenone, α-zearalenol, β-zearalenol and methanol at time of inocubation to determine their effect on subsequent zearalenone biosynthesis. Cultures were extracted at several time intervals and the extracts were purified by column chromatography before determining the concentration of zearalenone in treated cultures and zearalenone, α-zearalenol and β-zearalenol in untreated cultures by high pressure liquid chromatography. The addition of a catalytic amount of methanol (0.5 ml), zearalenone (4 mg) or α-zearalenol (4 mg) to cultures increased zearalenone biosynthesis after 7 or 14 days incubation, but did not affect the maximum toxin production (about 8 mg/g of rice) attained after 21 days incubation. The addition of β-zearalenol (2 mg) to cultures resulted in higher levels of zearalenone (about 14 mg/g of rice) being produced in a shorter time interval (14 days) than the other treatments.

Acknowledgements

The use of tradenames in this publication does not imply endorsement by the North Carolina Agricultural Research Service nor criticism of similar ones not mentioned. The assistance of Andrea Free-Kwiatkowski in the preparation of this manuscript is greatly appreciated.

REFERENCES

Allen, N.K., Mirocha, C.J., Aakhus-Allen, S., Bitgood, J.J., Weaver, G., and Bates, F. (1981). Effect of dietary zearalenone on reproduction in chickens. Poultry Sci., 60, 1165-1174.

Allen, N.K., Mirocha, C.J., Weaver, G., Aakhus-Allen, S., and Bates, F. (1981). Effects of dietary zearalenone on finishing broiler chickens and young turkey poults. Poultry Sci., 60, 124-131.

Bacon, C.W., Robbins, J.D., and Barter, J.K. (1977). Media for identification of Gibberella zeae and production of F-2 (zearalenone). Appl. Environ. Microbiol., 33, 445-449.

Bottalalico, A. (1979). On the occurrence of zearalenone in Italy. Mycopathologia, 67, 119-121.

Bottalico, A., Visconti, A., Logrieco, A., Solfrizzo, M., and Mirocha, C.J. (1985). Occurrence of zearalenols (diastereomeric mixture) in corn stalk rot and their production by associated Fusarium species. Appl. Environ. Microbiol., 49, 547-551.

Caldwell, R.W. and Tuite, J. (1974). Zearalenone in freshly harvested corn. Phytopathology, 64, 752-753.

Eugenio, C.P., Christensen, C.M., and Mirocha, C.J. (1970). Factors affecting production of the mycotoxin F-2 by Fusarium roseum. Phytopathology, 60, 1055-1057.

Gross, V.J. and Robb, J. (1975). Zearalenone production in barley. Ann. Appl. Biol., 80, 211-216.

Hagler, W.M., Jr., Mirocha, C.J., Pathre, S.V., and Behrens, J.C. (1979). Identification of the naturally occurring isomer of zearalenol produced by Fusarium roseum 'Gibbosum' in rice culture. Appl. Environ. Microbiol., 37, 848-853.

Hagler, W.M., Jr. and Mirocha, C.J. (1980). Biosynthesis of [^{14}C] zearalenone from [1-^{14}C]-acetate by Fusarium roseum 'Gibbosum'. Appl. Environ. Microbiol., 39, 668-670.

Hidy, P.H., Baldwin, R.S., Gresham, R.L., Keith, C.L., and McMullen, J.R. (1977). Zearalenone and some derivatives: production and biological activities. Adv. Appl. Microbiol., 22, 59-82.

Marasas, W.F.O., Nelson, P.E., and Toussoun, T.A. (1984). Toxigenic FUSARIUM Species: Identity and Mycotoxicology. pp. 99-110, The Pennsylvania State University Press, College Park, PA.

Meronuck, R.A., Garren, K.H., Christensen, C.M., Nelson, G.H., and Bates, F. (1970). Effects on turkey poults and chicks of rations containing corn invaded by Penicillium and Fusarium species. Am. J. Vet. Res., 31, 551-555.

Miller, J.K., Hacking, A., and Gross, V.J. (1973). Stillbirths, neonatal mortality and small litters in pigs associated with the ingestion of Fusarium toxins by pregnant sows. Vet. Rec., 93, 555-559.

Mirocha, C.J., Harrison, J., Nichols, A.A., and McClintock, M. (1968). Detection of a fungal estrogen (F-2) in hay associated with infertility in dairy cattle. Appl. Microbiol., 16, 797-798.

Mirocha, C.J., Pathre, S.V., Behrens, J., and Schauerhamer, B. (1978). Uterotropic activity of cis and trans isomers of zearalenone and zearalenol. Appl. Environ. Microbiol., 35, 986-987.

Peters, C.A. (1972). Photochemistry of zearalenone and its derivatives. J. Med. Chem., 15, 867-868.

Richardson, K.E., Hagler, Jr., W.M., and Hamilton, P.B. (1984). Bioconversion of α-[^{14}C] zearalenol and β-[^{14}C] zearalenol into [^{14}C] zearalenone by Fusarium roseum 'Gibbosum'. Appl. Environ. Microbiol., 47, 1206-1209.

Richardson, K.E., Hagler, Jr., W.M., and Mirocha, C.J. (1985). Production of zearalenone, α- and β-zearalenol by Fusarium spp. in rice culture. J. Agric. Food Chem., 33, 862-866.

Sarudi, I. (1976). Hydrogenation used in the detection of zearalenone (F-2 toxin). Acta Aliment., 5, 37-40.

SAS. (1982). SAS Users Guide. Statistics. Statistical Analysis System Institute, Inc., Cary, NC.

Shipchandler, M.T. (1975). Chemistry of zearalenone and some of its derivatives. Heterocycles, 3, 471-520.

Steele, J.A., Mirocha, C.J., and Pathre, S.V. (1976). Metabolism of zearalenone by Fusarium roseum 'Graminearum'. Mycopathologia, 57, 77-78.

PATHOGENICITY OF *ASPERGILLUS FLAVUS* GROUP ISOLATES IN INOCULATED MAIZE EARS IN MISSISSIPPI

N. ZUMMO* **and G.E. SCOTT**, Crop Science Research Laboratory, U.S. Department of Agriculture, Agricultural Research Service, P.O. Drawer PG, Mississippi State University, Mississippi State, MS 39762, USA

INTRODUCTION

Kernel infection and aflatoxin contamination in maize (Zea mays L.) by Aspergillus flavus Link ex Fries and A. parasiticus Speare continue to be a serious problem in the southeastern United States. Consumption of aflatoxin contaminated grain by livestock or poultry can result in weight loss or death (Wogan, 1965; Allcroft, 1969; Semeniuk et al., 1971; Anderson et al., 1975; Smith et al., 1976). Corn grain containing in excess of 20 ng/g of aflatoxin can not be sold under current laws regulating interstate commerce.

The variability of the A. flavus group on maize is well documented (Cotty, 1990; Davis and Diener, 1983; Diener, 1989; Diener and Davis, 1966; Dorner et al., 1984; and Hesseltine et al., 1970). Numerous virulent and avirulent and toxigenic and nontoxigenic isolates of the fungus have been identified (Cotty, 1990; Semeniuk et al., 1971; Wicklow et al., 1981). Wilson et al. (1986) inoculated maize ears with several color mutants of A. flavus and one of A. parasiticus and found that the isolate of A. parasiticus was more aggressive and survived better than the A. flavus isolates. Cotty (1990) reported that simultaneous inoculation of wounded cotton bolls with toxigenic and nontoxigenic strains of A. flavus resulted in reduced aflatoxin contamination in cottonseed. He did not attempt to recover and identify strains of the fungus from infected bolls.

Biodeterioration Research 4, Edited by G.C. Llewellyn *et al.*, Plenum Press, New York, 1994

Zummo and Scott (1989; 1990) developed a method for inoculating maize ears with A. flavus that results in no apparent injury to kernels and may be used to select for resistance to kernel infection or to compare aggressiveness of Aspergillus flavus group isolates on uninjured maize kernels. Using this technique, Zummo (1991) showed that individual maize kernels could become infected simultaneously with a white and a green isolate of A. flavus. Inoculation of ears with the white isolate prior to inoculation with the green isolate reduced infection with the green isolate. However, infection by the green isolate was not affected by later inoculation with the white isolate. Sequence of inoculation with the green isolate did not influence kernel infection by the white isolate. Percentage of kernel infection was higher following inoculation with both isolates than after inoculation with a single isolate.

With the identification of visually distinctive color variants of the A. flavus group that can infect maize kernels and produce aflatoxin (Wicklow et al., 1981; Zummo and Scott, 1990), it is now possible to compare Aspergillus isolates commonly occurring on maize with isolates from other sources for pathogenicity and aggressiveness.

The objectives of this study were to compare visually distinctive A. flavus group isolates known to infect maize kernels with visually distinctive isolates from other sources for ability to 1) infect maize kernels and 2) inhibit infection of kernels by other A. flavus group isolates.

MATERIALS AND METHODS

Field Plots, Harvesting, and Sampling Methods

Maize for this study was grown in replicated single-row plots at the Plant Science Research Center, Mississippi State University, Starkville, MS. Plots were 5 m long and spaced 1 m apart and were overseeded and later thinned to 20 plants spaced about 25 cm apart. The top ear of each plant was hand-harvested at 60 days after midsilk. Ears with obvious severe bird or insect damage were discarded. Immediately after harvest, ears were dried at 42 C for 7 days in a forced-air dryer. Ears were shelled and kernels from each plot were stored in paper bags at 6 C and 45% relative humidity until assayed.

Inoculum sources and production and inoculation techniques

Visually distinct isolates of Aspergillus flavus, NRRL 3357, NRRL 3251, and NRRL 20521 and A. parasiticus isolate 2999, which were able to infect maize kernels in the field, and A. flavus isolates Soybean 1 and Soybean 2 isolated from soybeans at Starkville, MS were used in this study. The six isolates were identified visually in culture using the criteria listed in Table 1. Earlier studies (Zummo and Scott, 1989; 1990) indicated that NRRL 2999 and NRRL 3357 were similar in aggressiveness in infecting maize kernels and more aggressive than NRRL 20521. NRRL 2999 and NRRL 3251 were considered producers of high amounts of aflatoxin, whereas NRRL 3357 and NRRL 20521 were intermediate aflatoxin producers (Rodriguez et al., 1983; Zummo, 1991).

Table 1. Colony color, growth habit, conidial head size, and sclerotial production on Czapek solution agar (CSA-S) amended with 7.5% NaCl for six Aspergillus flavus group isolates after 7 days incubation at 28 C.

Isolate	Colony color on CSA-S[a]	Colony growth	Conidial head size	Sclerotia in culture
NRRL 3357	olive yellow	medium low	medium/large	no
NRRL 2999[b]	yellowish oil green	medium low	medium[b]	no
NRRL 20521	ivory yellow	medium	large	no
NRRL 3251	grape green	low	small	abundant small
Soybean 1	pale olive buff	medium fine	small	no
Soybean 2	white	medium	small	scattered large

[a] Ridgeway, R. (1912). Color Standards and Color Nomenclature, 43 pp., published by the author, Washington, DC.
[b] When seeded on CSA-S, this isolate produced a conspicuous white mycelial mat within 3 days at 28 C.

Fungal cultures were grown for 14 days in 500-ml Erlenmeyer flasks containing 50 g of corncob grits (Grit-O-Cobs, Maumee, OH) and 100 ml of H_2O. Conidia were washed from the surface of the grits with 100 ml sterile distilled water containing two drops of Tween 20. All inocula were prepared at one time and stored at 4 C as an aqueous suspension of 1×10^8 conidia per milliliter. Inoculum was diluted to 9×10^6 conidia per milliliter on each day of inoculation and kept on ice in the field prior to use.

Inoculations were made using modifications of the needle inoculation techniques described by Zummo and Scott (1989). A tree-marking gun fitted with a 14-gauge hypodermic needle, 35 mm long, with the tip opening plugged and 1-mm holes drilled 6, 8, and 10 mm from the tip was used in all tests. The inoculator delivered 1.7 ml of inoculum containing 9 X 10^6 conidia per milliliter.

Assay

Only undamaged kernels were assayed for incidence of isolates on Czapek solution agar amended with 7.5% NaCl (CSA-S), which supports growth of A. flavus/parasiticus but inhibits growth of bacteria and most other fungi. To eliminate surface microbes, kernels were dipped momentarily in 70% ethanol, submersed in 1.25% NaOCl for 3 min, and rinsed in sterile distilled water. Kernels were plated on CSA-S in 100-mm Petri plates (13 kernels per plate) and incubated for 7 days at 28 C. Percentage of kernels infected with the inoculated Aspergillus isolates was determined. In addition, recovered A. flavus group isolates that differed from the inoculated isolates were recorded.

Field Experiments

In 1989, ears of Pioneer 3369A were inoculated at 6 days after midsilk with a conidial suspension of each of the six fungal isolates. Ears in uninoculated plots were retained as a control. The experiment was conducted with six replications of each treatment and was repeated in a crop planted 30 days after the initial planting.

Analysis of Data

Plot means of percentage of kernels infected by individual A. flavus group isolates were subjected to an analysis of variance. Means were separated using LSD.

RESULTS AND DISCUSSION

Infection data for the different isolates were similar for both planting dates (Table 2). In the first planting, there was more infection with isolates from maize than with the isolates from other sources. A similar but not significant trend was found in the second planting. On the other hand, there was less kernel infection with naturally occurring (wild type) isolates in ears inoculated with the maize isolates than with Aspergillus isolates from other sources (Table 3). There was significantly more kernel infection in ears inoculated with isolates NRRL 3357, NRRL 2999, and NRRL 20521 than in ears inoculated with isolates Soybean 1 and NRRL 3251 (Table 2). There was no significant difference in kernel infection in ears inoculated with isolate Soybean 2 than in kernel infection in ears inoculated with the isolates from maize. There was no significant difference in percentage kernel infection in ears inoculated with isolates NRRL 3251, Soybean 1, and Soybean 2 and percentage in kernels from uninoculated ears.

Corn ears inoculated with isolates NRRL 3357 and NRRL 2999 had significantly fewer kernels infected with wild type isolates than ears inoculated with NRRL 3251 and Soybean 1 (Table 3). The ability of an isolate to produce high or low levels of aflatoxin in maize kernels apparently did not influence its ability to infect kernels or prevent kernel infection by wild type isolates. There

Table 2. Percentage of Pioneer Brand 3369A kernels infected with distinct isolates of *Aspergillus flavus* group after inoculation in the field at two planting dates at Starkville, MS in 1989.

	Planting date		
Inoculum isolate	1	2	Mean
	%	%	%
NRRL 3357	4.8[a]	3.8	4.4
NRRL 2999	4.6	2.8	3.7
NRRL 20521	6.5	2.4	4.4
Soybean 2	2.1	2.0	2.1
NRRL 3251	1.4	0.8	1.1
Soybean 1	1.1	0.2	0.6
Control	0.4	0.7	0.5
LSD (0.05)	2.0	2.7	2.5

[a]Each value is the average percentage of kernels infected with individual isolates of *Aspergillus flavus* or *A. parasiticus* in six replications of 390 kernels following surface sterilization and 7 days of incubation at 28 C on Czapek solution agar amended with 7.5% NaCl.

was significantly more kernel infection with wild type isolates from natural infection in ears inoculated with NRRL 3251 than in ears inoculated with any of the other isolates or in uninoculated ears.

These data indicate the occurrence of physiological specialization among *A. flavus/parasiticus* isolates in regard to kernel infection in maize. This is in agreement with results of other workers (Cotty, 1990; Wicklow et al., 1981; and Wilson et al., 1986). Some isolates, such as NRRL 3357 and NRRL 2999, not only were more aggressive and better able to infect maize kernels than other *Aspergillus* isolates but are also apparently able to reduce kernel infection by other isolates. This may be due to the ability of these isolates to compete for infection sites on kernels which could prevent secondary infection by wild type isolates. When breeding for resistance to kernel infection by *A. flavus*/

Table 3. Percentage of kernels naturally infected with wild type isolates of *Aspergillus flavus* group following inoculation of maize ears in plots of Pioneer 3369A with six distinct isolates of *Aspergillus flavus* or *A. parasiticus* at two planting dates in the field at Starkville, Mississippi in 1989.

Inoculum isolate	Planting date 1	Planting date 2	Mean
	%	%	%
NRRL 3357	0.1[a]	0.1	0.1
NRRL 2999	0.1	0.1	0.1
NRRL 20521	0.4	0.3	0.4
Soybean 2	0.4	0.5	0.4
NRRL 3251	1.0	1.4	1.2
Soybean 1	0.7	0.5	0.6
Control	0.4	0.7	0.5
LSD (0.05)	0.5	0.7	0.4

[a]Each value is the average percentage of kernels infected with wild type *A. flavus* group isolates in six replications of 390 kernels for each of six distinct *Aspergillus* isolates following surface sterilization and 7 days incubation at 28 C on Czapek solution agar amended with 7.5% NaCl.

parasiticus it may be necessary to select for resistance to the prevailing *A. flavus* group races in a given area.

SUMMARY

Kernel infection and aflatoxin contamination in maize by *Aspergillus flavus/parasitcus* is a serious problem in the southeastern United States. Consumption of aflatoxin contaminated grain by livestock or poultry can result in weight loss or death. Grain containing in excess of 20 ng/g aflatoxin cannot be sold under current laws regulating interstate commerce in the United States. Isolates of the *A. flavus* group isolated from maize were significantly more agressive on field inoculated maize ears than isolates from soybeans. There was significantly less kernel infection from natural infection by wild type isolates in maize ears inoculated with isolates from corn than with isolates from soybeans. Some isolates, such as NRRL 3357 and NRRL 2999, not only were more agressive and better able to infect maize kernels than other *Aspergillus* isolates, but are apparently able to reduce kernel infection by other isolates. This may be due to the ability of these isolates to compete for infection sites on kernels which could prevent infection by wild type isolates. These findings indicate the occurrence of physiological specialization among *A. parasiticus* isolates in regard to kernel infection in maize.

REFERENCES

Allcroft, R. (1969). Aflatoxicosis in farm animals. In: Aflatoxin, pp. 237-264 (L.A. Goldblatt, ed.), Academic Press, New York.

Anderson, H.W., Nehring, E.W., and Wicher, W.R. (1975). Aflatoxin contamination of corn in the field. J. Agric. Food Chem., 23, 775-782.

Bennett, J.W., Horowitz, P.C., and Lee, L.S. (1979). Production of sclerotia by aflatoxigenic and nonaflatoxigenic strains of Aspergillus flavus and Aspergillus parasiticus. Mycologia, 71, 415-422.

Cotty, P.J. (1990). Effect of atoxigenic strains of Aspergillus flavus on aflatoxin contamination of developing cottonseed. Plant Dis., 74, 233-235.

Davis, N.D., and Diener, U.L. (1983). Some characteristics of toxigenic and nontoxigenic isolates of Aspergillus flavus and Aspergillus parasiticus, In: Aflatoxin and Aspergillus in Corn, pp. 1-5, South. Coop. Ser. Bull. 279.

Diener, U.L. (1989). Preharvest aflatoxin contamination of peanuts, corn, and cottonseed: A review. In: Biodeterioration Res. 2, pp. 217-244. (G.C. Llewellyn and C.E. O'Rear, eds.) Plenum Publishing Co., NY.

Diener, U.L., and Davis, N.D. (1966). Aflatoxin production by isolates of Aspergillus flavus. Phytopathology, 56, 1390-1393.

Dorner, J.W., Cole, R.J., and Diener, U.L. (1984). The relationship of Aspergillus flavus and Aspergillus parasiticus with reference to production of aflatoxin and cyclopiazonic acid. Mycopathologia, 87, 13-15.

Hesseltine, C.W., Shotwell, O.L., Smith, M., Ellis, J.J., Vandegraft, E., and Shannon, G. (1970). Production of various aflatoxins by strains of the Aspergillus flavus series. In: Toxic Microorganisms. pp. 202-210 (M. Herzberg, ed.), UJNR and Dept. of Interior, U.S. Government Printing Office, Washington, DC.

Ridgeway, R. (1912). Color Standards and Color Nomenclature. 43 pp. published by the author, Washington, DC.

Rodriguez, J.C., C.G. Patterson, M.F. Potts, C.G. Poneleit, and R.L. Beine. (1983). Role of selected anthropods in the contamination of corn by Aspergillus flavus as measured by aflatoxin production. In: Aflatoxin and Aspergillus flavus in Corn pp. 23-26, South. Coop. Ser. Bull. 279.

Semeniuk, G., Harshfield, G.S., Carlson, C.W., Hesseltine, C.W., and Kwolek, W.F. (1971). Mycotoxins in Aspergillus. Mycopathol. Mycol. Appl., 43, 137-152.

Smith, R.B., Jr., Griffin, J.M., and Hamilton, P.B. (1976). Survey of aflatoxicosis in farm animals. Appl. Environ. Microbiol., 31, 385-388.

Wicklow, D.T., Shotwell, O.L., and Adams, G.L. (1981). Use of aflatoxin-producing ability medium to distinguish aflatoxin-producing strains of Aspergillus flavus. Appl. Environ. Microbiol., 41, 697-699.

Wilson, D.M., McMillian, W.W., and Widstrom, N.W. (1986). Use of Aspergillus flavus and A. parasiticus color mutants to study aflatoxin contamination of corn. In: Biodeterioration VI. pp. 55-66 (S. Barry, G.C Llewellyn, and C.E. O'Rear, eds.), International Biodeterioration Society, London.

Wogan, G.N. (1965). Mycotoxins in Foodstuffs, pp. 1-291, Massachusetts Institute of Technology Press, Cambridge, MA.

Zummo, N., and Scott, G.E. (1989). Evaluation of field inoculation techniques for screening maize genotypes against kernel infection by Aspergillus flavus in Mississippi. Plant Dis., 73, 313-316.

Zummo, N., and Scott, G.E. (1990). Relative aggressiveness of Aspergillus flavus and A. parasiticus. Plant Dis., 74, 978-981.

Zummo, N. (1991). Concurrent infection of individual corn kernels with white and green isolates of Aspergillus flavus. Plant Dis. 75, 910-913.

SECTION II

PROCESSES OF WOOD DECAY AND DETERIORATION

PROCESSES OF WOOD DECAY AND DETERIORATION

INTRODUCTION

Environmental safety has emerged as the number one issue of the 1990's. Unfortunately, the major wood preservatives - creosote, penta-chlorophenol, and inorganic arsenicals - are broad-spectrum poisons that pose a serious threat to the environment. These chemicals have already been banned or restricted in some countries. At present, no harmless substitutes for toxic chemicals are available. Thus, we urgently need new, novel approaches for preventing wood deterioration. Consumer concern about environmental damage may lead to untreated wood being used in applications where it is subjected to biodeterioration.

The unsophisticated approach of using metabolic poisons can only be phased out if new and more selective alternatives are developed. This can best be accomplished if we identify and take advantage of unique features of degrading organisms or their biochemical pathways. This necessitates a thorough knowledge and understanding of the fundamentals of wood decay, including mechanisms and methods of deterioration, as well as conditions conducive to fungal proliferation and growth. If researchers understand the biochemical agents in fungal attack that are responsible for degradation, they will be able to "sabotage" the ability of these agents to decay wood. Focusing research on physiological differences between the wood-attacking fungi and nontarget organisms contributes to environmental safety; we can specifically interfere with fungal activity. The research reported in this chapter discusses several different aspects of fungal deterioration of wood, which should aid in the development of safer wood preservation methods.

Brown-rot fungi cause the most destructive for of wood decay, but the mechanism by which brown-rot fungi degrade wood is not well understood.

These fungi are unique in that they are the only known microbes that can degrade cellulose in wood without first removing the lignin. Furthermore, they degrade cellulose in an unusual manner, causing rapid decrease in depolymerization at low weight loss. A review paper addressed chemical changes in wood components and cotton cellulose as a result of brown-rot degradation. the literature indicates that chemical changes in brown-rotted lignin are oxidative in nature and result primarily in hydroxylation and demethylation of the lignin polymer. Results from studies of the chemical changes in cellulose as a result of brown-rot attack do not unequivocally demonstrate oxidative degradation. The Fenton's reagent (H_2O_2/Fe^{++}) has been hypothesized as a model by which brown-rot fungi oxidatively depolymerize cellulose. However, a review of the effects of Fenton's reagent on wood components and model compounds suggests that the effects of brown-rot decay are not duplicated by this reagent in many respects and, therefore, it does not represent and adequate model for brown-rot decay.

The extracellular enzymes produced by wood-decay fungi are important in digestion of wood cell wall constituents. Knowledge of these enzymes is beneficial for (a) inactivating them to prevent decay, (b) early detection of decay by producing antibodies to degradative enzymes, and (c) industrial and agricultural bioconversion processes. However, extracellular wood-decay enzymes are difficult to study because they are produced in such small amounts. One paper addressed an attempt to overproduce and enhance the secretion of lignocellulolytic enzymes from the white-rot fungus Trametes versicolor. In the development of the immunodiagnostic methodologies and inhibitor studies for decay prevention, using enzymes that have the same properties as those produced in wood during decomposition is particularly important. To determine if the culture medium affects enzyme characteristics, another paper compared the antigenic properties of the extracellular enzymes from the brown-rot fungus Postia placenta grown in liquid culture, malt agar, or wood. The authors concluded that the antigenic characteristics of extracellular polysaccharide-degrading enzymes of P. placenta and substrate dependent and that degradative enzymes produced in vitro differ from those produced in wood.

Earlier studies of decay by white-rot fungi suggested that the faster decay of hardwoods than that of softwoods may be partially explained by the effect of the different amounts and different types of lignin in these woods. The lignin of most softwoods is composed almost entirely of quaiacylpropyl (G) units. Lignin in hardwoods, in addition to G units, often has numerous syringylpropyl (S) units. A study addresses the influence of lignin type on decay by the white-rot fungus Trametes versicolor by using a number of woods with a wide range of S:G ratios and different lignin type distributions. The results of the study are consistent with the general premise that lignin concentration and lignin type affect the decay resistance of wood. However, not enough experimental evidence was obtained to prove that lignin character is the limiting determinant of decay. In fact, the syringyl content of the lignin had no general correlation with the rate of decay.

Using biological controls to protect wood against fungal attack is an environmentally safer alternative to chemical biocides. Biological control of wood decay exploits ecological relationships among organisms by using natural antagonists against wood-attacking fungi. these papers addressed the use of biological controls to prevent decay, mold and stain attack in wood.

One study determined the ability of a commercial biological formulation containing Tricoderma propagules to colonize and prevent decay in Douglas-fir and Southern Pine timbers exposed above ground. Tricoderma colonized Southern Pine timbers but did not prevent decay. Douglas-fir timbers were poorly colonized by Tricoderma. The author concluded that the use of biological control agents to control wood-attacking fungi should emphasize short-term protection, such as protection of lumber and logs during drying. For a biological control organism to provide long-term protection against decay in wood products would be very difficult. The biological control organism would have to spread throughout the wood and deposit a residual fungistatic material that remains after the organisms death or alter the substance to make it nutritionally undesirable to the decay fungus.

A second biological control paper investigated the suitability of using the metabolites from the bacterium Streptomyces rimosus, for controlling mold and stain fungi on green wood during drying. The results of this study showed that St. rimosus produced very potent antifungal metabolites that effectively prevented mold and stain attack on wood in laboratory tests and on green pine log sections in simulated field trials. St. rimosus has been reported to produce the antifungal antibiotic, rimocidin, but the authors have not yet identified the antifungal metabolites responsible for mold and stain prevention. If antifungal metabolites from St. rimosus are to be considered for commercialization, identifying them as soon as possible is essential.

The third paper on biological control reported the effects of the bacterial metabolite isatin (2, 3-indolinedione) on white- and brown-rot decay in vitro and discusses the relevance of the effects of fungal metabolism. Isatin concentrations above 5mM inhibited the growth of 12 wood-decay fungi. The effect of lower isatin concentrations on wood-decay and parasitic fungi implicated that isatin intervened in metabolic pathways responsible for fungal growth and development. The fact that isatin had a physiological effect on wood decay fungi and that toxic levels were identified for application is relevant. Isatin should be included in the search for microbial allelochemicals.

Compounds derived from natural products may be safer for the environment that synthetic chemical compounds and would be more acceptable to consumers. Another paper determined the fungitoxicity of extratives from cones from four different Pinus species against mold, sapstain, and wood-decay fungi. The CO_2 fraction of diethyl ether extracts of four species of pine cones inhibited the growth of representative mold, sapstain, and wood-decay fungi. However, none of the CO_2 fractions inhibited fungal growth entirely. A large portion of the fungitoxicity was probable due to the presence of abietane resin acids. Abietane resin acids have an isopropyl group attached to C-13 that may have been responsible for the fungitoxicity of these compounds.

Wood protection in the future will depend less on the use of broad-spectrum metabolic poisons. With more sophisticated tools and techniques to aid researchers in studying problems of wood degradation and protection, our understanding of decay processes should advance rapidly in the coming years. Biochemical and physiological studies of wood decay will reveal new and exciting information with essential biochemical processes and develop highly active and sharply targeted preservatives with enhanced preservative performance and safety.

Terry L. Highley
USDA Forest Service
Forest Products Laboratory
Madison, WI 53705-2398, USA

EXTRACELLULAR POLYSACCHARIDE-DEGRADING ENZYMES OF *POSTIA PLACENTA* ISOLATED FROM WOOD OR ARTIFICIAL MEDIA

C.A. CLAUSEN,* F. GREEN III, T.L. HIGHLEY, USDA Forest Service, Forest Products Laboratory, Madison, WI 53705–2398, USA

INTRODUCTION

A variety of extracellular polysaccharidases produced by brown-rot fungi are thought to play an important role in the wood-rotting process. There is evidence of early removal of hemicellulose, a principal cell wall constituent, by hemicellulases (xylanases, mannanases), which attack and shorten hemicellulose chains sufficiently for further hydrolysis by glycosidases into simple sugars, thus facilitating colonization and penetration of wood (Kirk and Highley, 1973; Kirk and Cowling, 1984; Highley, 1987). The enzyme b–1,4–xylanase, of the brown-rot fungus Postia placenta (Fr.) M. Lars. et Lomb. MAD 698, has been partially purified (Green et al., 1989b) and used as antigen to produce monoclonal antibodies. This glycoprotein typically co-migrates with carboxymethyl- cellulase and has been comprised of as much as 50 percent carbohydrate when extracted from wood.

Highley et al. (1981) qualitatively and quantitatively compared the polysaccharide-degradingenzymes produced by Postia placenta in decayed wood and liquid culture. Quantitative differences were observed in enzyme activities from decayed wood as opposed to liquid media. However, despite these quantitative differences, the authors concluded that the extracellular polysaccharide complex from liquid culture must be structurally similar to that of decayed wood because of similar isoelectric points, electrophoretic properties, and molecular sieving properties.

Culturing brown-rot fungi in artificial liquid medium is a common laboratory practice. The ability to produce large volumes of culture filtrate is an attractive means of procuring enzyme for purification. However, the availability of low molecular weight sugars in artificial media results in excess glucan production, which complicates enzyme purification (C. Clausen, personal observation). We postulated that the increased glucan production noted in liquid cultures of P. placenta MAD 698 (Micales et al., 1990) may alter the type and amount of enzyme produced or disguise antigenic epitopes, thereby altering antibody (Ab) recognition of the antigen, i.e., enzyme. An atypical isolate ofP. placenta, (ME–20), was also included in this study. As reported by Micales et al. (1990) and Green et al. (1992a), this isolate fails to produce extracellular glucan in liquid culture or to cause weight loss during colonization of wood.

The extracellular polysaccharidases and the role of the glucan sheath were the focus of this study. Our objectives were (1) to analyze differences in enzyme production, glucan composition, and antibody recognition and (2) to determine the differences in brown-rot decay organisms grown on wood as opposed to artificial growth medium.

MATERIALS AND METHODS

Culture Conditions

Postia placenta (Fr.) M. Lars., et Lomb. [= Poria placenta (Fr.) Cke.] MAD 698 and ME–20 were grown either in 2 percent malt agar (Difco, Detroit, MI), in basal salts solution (Highley, 1973) with 1 percent cellobiose (Sigma, St. Louis, MO), or on southern pine (Pinus sp.) in a standard ASTM soil-block method (ASTM, 1971).

Solubilization and Extraction Methods

A 0.1 percent solution of Triton X–100 (Sigma) in glass-distilled water was used to solubilize the mycelial mat of the liquid culture, the culture filtrate, or the decayed wood block. Gentle rocking for 2 hr at 25C sufficiently solubilized the hyphae and extracted the extracellular enzymes.

Enzyme and Protein Assays

The activities of endo–b–1,4–glucanase, endo-b–1,4–xylanase, endo–b–1,4–mannanase, and endo–b–1,3–glucanase were assayed by the reducing sugar microassay (Green et al., 1989a) using laminarin (Sigma), larch xylan (United States Biochemical Co., Cleveland, OH), guar gum (Nutritional

Biochemicals Corp., Cleveland, OH), and carboxymethylcellulose (Hercules, Wilmington, DE), respectively. One unit of enzyme activity was defined as the amount needed to liberate reducing power equivalent to 1 mg glucose per 24 hr at 40C using D-glucose (Fischer Scientific Co., Fair Lawn, NJ) as standard.

The b-D-galactosidase and a-D-galactosidase activities were assayed by determining the liberation of p-nitrophenol from respective p-nitrophenol substrates (Agrawal and Bahl, 1968). A unit of enzyme activity was defined as the amount needed to liberate 1 mM of p-nitrophenol per hour.

Total protein was determined by the BCA microassay (Pierce, Rockford, IL) using bovine serum albumin (Sigma) as a standard.

Carbohydrate Determination

Carbohydrate content of individual extracts was determined by HPLC with an Aminex HPX-87P (Biorad, Rockville Centre, NY) (Pettersen and Schwandt, 1991).

Gel Filtration

Molecular weight estimates were determined on a 90- by 26-cm Fractogel TSK HW-55(F) (EM Science, Gibbstown, NJ) column equilibrated with 0.1 M acetate buffer, pH 4.0. Gel filtration molecular weight markers were cytochrome C, carbonic anhydrase, and bovine albumin (Sigma).

Monoclonal Antibody

BALB/c mice were immunized with the xylanase fraction of *P. placenta*-decayed wood that had been partially purified by column chromatography (Green et al., 1989b). Antigen was mixed 1:1 with Freund's complete adjuvant and mice were boosted at 4-week intervals (Stahli et al., 1983). Hybridomas were produced by conventional HAT selection techniques (Galfre and Milstein 1981). Polyethylene glycol fusion of NS-1 myeloma cells to mouse spleen cells yielded hybridomas that were cloned by dilution cloning and characterized for xylanase specificity (Clausen et al., 1992).

Enzyme-Linked Immunosorbent Assay (ELISA)

Polystyrene 96-well ELISA plates were coated overnight at 25C with antigen (protein extracts) in 0.1 M Tris buffer with 0.15 M NaCl, pH 9.0, rinsed five times with coating buffer, and blocked with PBS-Tween 20 (Sigma) for 15 min. A 1:1000 dilution of monoclonal antibody (F1) to *P. placenta* xylanase was added to each well and incubated for 2 hr. The plate was washed with PBS-Tween and incubated with secondary antibody (horseradish peroxidase labeled goat anti-mouse IgG) (Hyclone, Logan, UT) for 2 hr. Plates were thoroughly washed with water, developed with o-phenylene diamine, and read at optical density (OD_{490}).

Western Blot

Protein extracts and prestained molecular weight standards (Biorad) were separated by SDS–PAGE using precast 12-percent gels (Biorad) and transferred to nitrocellulose paper using a Biorad Transblot apparatus. The transfer was conducted in 25 mM Tris–HC1, 192 mM glycine, 20 percent methanol, pH 8.3 at 100 V for 2 hr. The blot was then blocked for 1 hr in Tris buffered saline (TBS) (0.02 M Tris, 0.5 M NaC1, pH 7.5) with 3 percent gelatin, rinsed briefly with distilled water, and incubated with a 1:1000 dilution of monoclonal antibody (MAb) for 2 hr at 25C. The blot was washed with TBS (with 0.05 percent Tween 20 (TBS–Tween)) and incubated with horseradish peroxidase conjugated goat anti-mouse IgG in TBS with 1 percent gelatin for 2 hr. The was again rinsed with TBS–Tween and developed in 60 mg 4–chloro–1–naphthol (Sigma) in 20 ml methanol and 60 ml hydrogen peroxide in 100 ml TBS (Towbin et al., 1979).

Affinity Chromatography

Monoclonal antibody specific for xylanase of P. placenta was coupled to CNBr-activated Sepharose 4B (Pharmacia, Piscataway, NJ) per Pharmacia Principles and Methods. Gel swelled in 1 mM HC1 was washed and coupled to the MAb at 4C for 18 hr. Extracts were applied to the column in 50 mM phosphate buffer, 0.15 M NaC1, and 0.1 percent Tween 20. Unbound enzyme was washed from the column, and bound enzyme was eluted in 100 mM glycine–HC1, pH 2.6, 0.15 M NaC1, and 0.01 percent Tween 20, and neutralized with 1 M phosphate buffer, pH 7.5. The eluted protein peak was assayed for polysaccharidases as previously described (Green et al. 1989a).

Deglycosylation

Liquid culture supernatant of MAD 698 was treated with N–Glycosidase F (PGNase F, Boehringer–Mannheim, Indianapolis, IN) to determine the affect of deglycosylation on antigen binding to the MAb affinity column. Twenty units of PGNase were incubated for 17 h at 37C in 100 mM sodium acetate buffer, pH 5.0, before binding to the affinity column. An untreated sample served as a control. ELISA as previously described was also performed on treated and untreated samples.

RESULTS

Enzyme extracts were prepared from P. placenta MAD 698 and ME–20 grown on wood, on agar, or in liquid culture. For isolate MAD 698, the wood extract exhibited the highest protein to carbohydrate ratio (0.116) compared with culture supernatant, which had the lowest ratio (0.029). The extracted mycelial mat from the MAD 698 liquid culture had the highest galactosidase and hemicellulase activities. The water extract from ME–20-decayed wood contained galactosidase activities similar to that of the water extract from MAD 698-decayed wood. These extracts also contained similar carboxymethylcellulase (CMCase) activity and high laminarinase activity. Mannanase was low or not detectable in all the extracts.

ELISA results using a MAb to xylanase of P. placenta MAD 698 demonstrated high readings (OD_{490}) for all MAD 698 extracts regardless of growth substrate (0.72–1.60) (Table 1). The ME–20 extract, however, was not recognized by the MAb in ELISA (OD_{490} = 0.05–0.51). Likewise, all extracts of MAD 698 bound to two bands on a western blot at 32 and 36 kDa molecular weight (Clausen et al., 1992). Gel filtration chromatography estimates the molecular weight of MAD 698 xylanase at 32 kDa.

Total carbohydrate was highest in the supernatant of MAD 698 liquid culture (21.54 mg/ml) compared to the wood extract (9.40 mg/ml) (Table 2). Sugar composition varied for each extract. Sugars from the ME–20 water extract were predictably negligible. Sugar content was also low in the MAD 698 water extract, but glucose and mannose were measurable. Samples solubilized with Tritox X–100 had markedly higher carbohydrate content. The glucose, galactose, xylose, and mannose content of the MAD 698 culture was twice that of the MAD 698 wood extract. The MAD 698 agar extract contained high amounts of galactose and a negligible amount of mannose.

The results of affinity chromatography showed that the MAD 698 wood extract exhibited 10.7 times more xylanase and 2.1 times more CMCase binding specificity than did the liquid culture filtrate (Table 3). Laminarinase activity was detected in the affinity eluant. However, laminarinase binding appeared to be nonspecific. Only 10 percent of the total laminarinase activity was detected in the eluant, but 94 percent of the total xylanase activity per milliliter was bound to the affinity column. There was no relationship between laminarinase activity and ELISA binding for MAD 698 chromatography fractions (Fig. 1). The

Table 1. Comparison of Enzyme Preparations From Cultures of *P. placenta* Grown on Various Substances

Treatment	Protein (mg/ml)	Total CHO (mg/ml)	Glycosidase[a] (mM p-nitro/ ml)			Polysaccharidase[b] (mg glucose/ ml)		
			a-GAL	b-GAL	$ELISAOD_{490}$	Xylanase	CMCase	LAMase
0.1% Triton X-100								
MAD 698-wood	14.42	9.40	0.18	0.13	0.72	571	440	3,215
ME-20-wood	1.01	ND	0.72	0.60	0.05	566	500	1,479
MAD 698-agar	12.25	14.90	1.46	2.95	1.40	2,382	1,201	2,978
MAD 698-mat[c]	0.95	12.47	4.13	4.14	1.60	3,511	983	4,837
MAD 698-liquid[d]	0.95	21.54	3.77	3.55	1.40	307	178	663
Distilled water								
MAD 698-wood	13.86	0.50	0.16	0.20	1.33	615	100	1,676
ME-20-wood	1.12	0.16	0.37	0.48	0.51	292	276	1,663

[a] Micromoles of p-nitrophenol per milliliter. GAL is galactosidase
[b] CMCase is carboxymethylcellulase, LAMase is laminarinase, and MANase in mannanase.
[c] Mycelial mat
[d] Culture filtrate.

estimated molecular weight for laminarinase (55 kDa) was also not detected on western blot (i.e., antibodies did not bind to laminarinase). Affinity chromatography showed a total lack of recognition of ME–20 by the Mab. ME–20 xylanase has a molecular weight of 22 kDa as estimated by gel filtration chromatography.

Table 2. Carbohydrate Content of Extracellular Enzyme Extracts From P.placenta Grown on Various Substrates

Treatment and substrate	Carbohydrate (mg/ml)					
	Glucose	Galactose	Arabinose	Xylose	Mannose	Total
0.1% Triton X–100						
MAD 698–wood	2.45	3.17	0.12	1.00	2.66	9.40
MAD 698–agar	2.45	10.65	0.28	0.94	0.08	14.90
MAD 698–mat	2.12	5.78	0.10	0.99	3.48	12.47
MAD 698–liquid	5.87	6.80	0.17	2.20	6.80	21.54
Distilled water						
MAD 698–wood	0.23	0.04	0.03	0.06	0.13	0.50
ME–20–wood	0.04	0.04	0.04	0.02	0.02	0.16

Figure 1 shows polysaccharidase activities of fractions eluted from liquid chromatography and compares them to ELISA results. The highest ELISA readings correspond to the peak polysaccharidase activity for xylanase, with a drop to background readings around 305-ml elution volume.

Treatment of the MAD 698 liquid culture filtrate with the deglycosylating enzyme, PGNase F, eliminated the slight binding of the antigen in affinity chromatography. Similar results were demonstrated in ELISA. Untreated antigen had an OD_{490} of 0.742, and treated antigen decreased antibody recognition to an OD_{490} of 0.019.

DISCUSSION

Extracellular enzymes of the brown-rot fungus P. placenta MAD 698 and ME- 20 grown on artificial or wood substrates were analyzed for differences in enzyme activities, protein and carbohydrate content, and antigen recognition by a monoclonal antibody (Mab). Artificial medium enhanced enzyme activity and availability but also increased the glucan production twofold compared to that of the wood substrate. Despite increased enzyme activities, a Mab to b-1,4-xylanase preferentially bound to wood-derived xylanase and CMCase, suggesting that xylanase-CMCase from liquid culture is either conformationally different from xylanase-CMCase in wood or is disguised by the excess glucan

Table 3. Enzyme Activity of P. placenta MAD 698 Wood Extracts and Culture Filtrate by Affinity Chromatography

	Enzyme activity (ug glucose/ml)	
Enzyme	Wood extract	Culture filtrate
Xylanase	576	54
CMCase	493	234
Mannanase	83	239
Laminarinase	324	269

produced in liquid culture. Because the carbohydrate component of the liquid culture enzyme complex was composed of the same sugars as the enzyme complex from wood, it seems unlikely that the Mab would be recognizing a carbohydrate from wood not present in liquid culture.

To determine if glucan interference was causing problems in antibody recognition, liquid culture filtrate was treated with PGNase F to enzymatically remove mannose and glucose residues from the enzyme. We successfully used PGNase F in enzymatic digestion studies for characterizing the Mab (Clausen et al., 1992). In the previous study, antigen treated with PGNase F was analyzed by ELISA to determine if the MAb would recognize the protein portion of the glycoprotein once the mannose and glucose residues were removed. The ability of the MAb to bind to treated antigen was not altered, and Con A, which binds to a-linked mannose and glucose residues and is used to assess the completeness of the digestion, showed a 77-percent reduction in binding for the

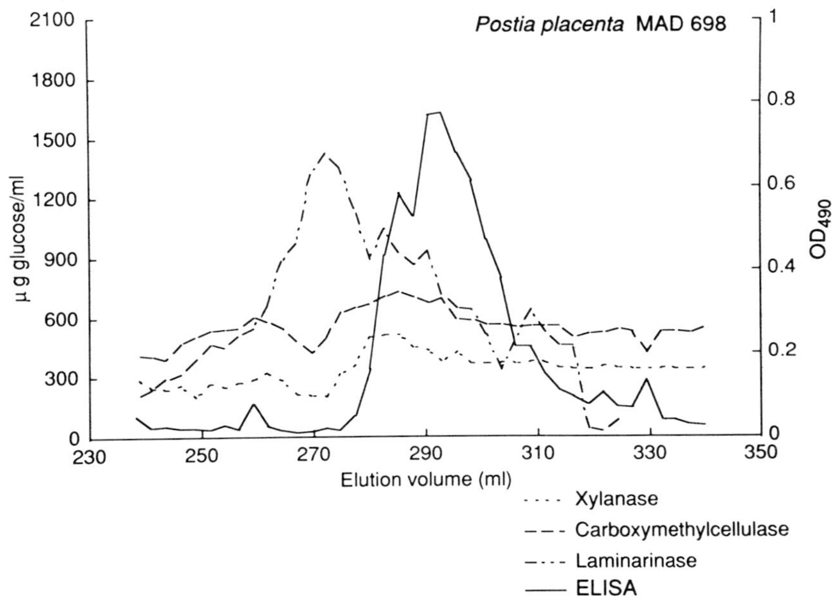

Figure 1. Polysaccharidase activity of P. placenta MAD 698 extract in gel chromatography fractions compared to ELISA readings.

control. In this study, however, untreated culture filtrate bound weakly to the affinity column whereas culture filtrate treated with PGNase F demonstrated no binding. Removing sugar residues from the antigen did not increase recognition by the MAb in affinity chromatography, again strongly suggesting the MAb is not specific for the carbohydrate moiety of the glycoprotein. In ELISA, treatment with PGNase F also reduced binding by the MAb to an undetectable amount. This suggests that although the MAb is protein specific for −1,4−xylanase, there seems to be a second smaller protein that is recognized by the MAb or a repeating epitope in extracellular enzymes from liquid culture that is altered by enzymatic digestion.

Enzyme activities of the ME–20 and MAD 698 extracts were equivalent, but the MAb did not recognize ME–20 xylanase in ELISA, western blot, or affinity chromatography eluant. The estimated molecular weight of ME–20 xylanase was 10 kDa less than the estimated molecular weight of MAD 698 xylanase. The lack of glucan production by ME–20 may account for the differences in molecular weight and antibody recognition of this enzyme. The comparative information between these two isolates of *P. placenta* is valuable for our continuing efforts to discern the inability of ME–20 to degrade wood.

We conclude that antigenic characteristics of extracellular polysaccharide-degrading enzymes of *P. placenta* are substrate dependent. The differences between enzymes from liquid culture and those from wood may not be readily apparent in the absence of immunological comparisons. This is an important consideration when developing immunodiagnostic methodologies. We feel that it is necessary to develop detection methods compatible with naturally occurring substrate conditions. Such methods are equally important for *in situ* immunolabeling experiments to optimize antibody specificity (Green et al., 1992b).

SUMMARY

Most studies on the degradative enzymes of wood-decay fungi utilize enzymes produced in liquid culture and assume that the enzymes have the same properties as those produced by the fungus in wood. To determine if the culture medium affects enzyme characteristics, we compared the antigenic properties of extracellular enzymes from the brown-rot fungus *Postia placenta* grown in liquid culture, malt agar, or wood. Availability of low molecular weight sugars in artificial media resulted in excess production of glucan. Enzyme activities were higher in artificial media, and enzyme distribution and antigenic affinity were altered, as determined by affinity chromatography using monoclonal antibodies specific for xylanase derived from wood. This study demonstrates that enzyme characteristics of *P. placenta* are substrate dependent and that degradative enzymes produced *in vitro* differ from those produced in wood. These differences are critical considerations in the design of immunological methods for early detection of brown-rot fungi.

REFERENCES

Agrawal, K.M.L. and Bahl, O.P. (1968). Glycosidases of Phaseolus vulgaris.II. Isolation and general properties. J. Biol. Chem., 243, 103–111.

ASTM. (1971). Standard method for accelerated laboratory test of natural decay resistance of woods. ASTM D 2017. American Society for Testing and Materials, Philadelphia, PA.

Clausen, C.A., Green, F. III., and Highley, T.L. (1992). Characterization of monoclonal antibodies to wood derived b-1,4-xylanase Postia placenta and their application to detection of incipient decay. Wood Science and Technology. (In press.)

Galfre, C. and Milstein, C. (1981). Preparation of monoclonal antibodies: Strategies and procedures. Methods Enzymol., 73, 3–46.

Green, F. III., Clausen, C.A., and Highley, T.L. (1989a). Adaptation of the Nelson–Somogyi reducing sugar assay to a microassay using microtiter plates. Anal. Biochem., 182, 197–199.

Green, F. III., Clausen, C.A., Micales, J.A., Highley, T.L., and Wolter, K.E. (1989b). Carbohydrate-degrading complex of the brown-rot fungus Postia placenta Purification of b-1,4-xylanase. Holzforschung, 43(1), 25–31.

Green, F. III., Clausen, C.A., Larsen, M.J., and Highley, T.L. (1992a). Immuno-scanning electron microscopic localization of wood-degrading enzymes within the fibrillar sheath of the brown-rot fungus Postia placenta. Can. J. Microbiol. (In press.)

Green, F. III, Larsen, M.J., Winandy, J.E., and Highley, T.L. (1992b). The role of oxalic acid in incipient brown-rot decay. Mat. u. Org. (In press.)

Highley, T.L. (1973). Influence of carbon source on cellulase activity of white-rot and brown-rot fungi. Wood Fiber, 5, 50–58.

Highley, T.L. (1987). Biochemical aspects of white-rot and brown-rot decay. International Research Group on Wood Preservation. Doc. No. IRG/WP/1319. 22 p.

Highley, T.L., Wolter, K.E., and Evans, F.J. (1981). Polysaccharide-degrading complex produced in wood and in liquid culture by the brown-rot fungus Poria placenta. Wood Fiber, 13(4), 265–274.

Kirk, T.K., and Cowling, E.B. (1984). Biological decomposition of solid wood, In: Chemistry of Solid Wood, Advances in Chemistry Series 207, pp. 455–487 (R.M. Rowell, ed.), American Chemical Society Press, Washington, DC.

Kirk, T.K., and Highley, T.L. (1973). Quantitative changes in structural components of conifer woods during decay by white- and brown-rot fungi. Phytopathology, 63, 1138–1342.

Micales, J.A., Richter, A.L., and Highley, T.L. (1990). Extracellular glucan production by Postia (= Poria) placenta. Mat. u. Org., 24(4), 259–269. Pettersen, R.C., and Schwandt, V.H. (1991). Improved method for wood sugar analysis. J. Wood Chem. Technol. (In press.)

Stahli, C., Staehelin, T., and Miggiano, V. (1983). Spleen cell analysis and optimal immunization for high frequency production of specific hybridomas. Methods in Enzymol., 92, 26–36.

Towbin, H., Staehelin, T., and Gordon, J. (1979). Electrophoretic transfer of proteins from polyacrylamide gels to nitrocellulose sheets: Procedure and some applications. Proceedings from the National Academy of Science. USA, 76, 4350–4354.

Note: The use of trade or firm names in this publication is for reader information and does not imply endorsement by the U.S. Department of Agriculture of any product or service.

BIOLOGICAL CONTROL OF SAPWOOD-INHABITING FUNGI BY METABOLITES FROM *STREPTOMYCES RIMOSUS*

SUKI C. CROAN* and **TERRY L. HIGHLEY**, USDA Forest Service, Forest Products Laboratory, Madison, Wisconsin 53705-2398 USA

INTRODUCTION

Benko (1988a,b) studied the suitability of several bacteria as biological control agents for preventing discoloration and deterioration of wood by sapwood-inhabiting fungi. In a study by Highley et al. (1991), metabolites produced by Streptomyces rimosus were effective in controlling sapwood-inhabiting fungi. The unconcentrated metabolites inhibited spore germination of all the sapwood-inhabiting fungi tested and prevented discoloration of pine blocks. However, Penicillium and Aureobasidium pullulans were able to attack sweetgum blocks treated with the metabolites.

Many researchers have tried to clone genes of St. rimosus for mass production of antibiotics such as oxytetracycline, chlortetracycline, tetracycline, and anhydrotetracycline. The characteristics and identification of antibiotics, including the antifungal antibiotic rimocidin, have been reported (Butler and Gedge, 1989; Cope et al., 1965; Falkowski et al., 1978; Friend et al., 1978; O(,R)ehacek, 1990; Rhodes et al., 1981). However, to the best of our knowledge, no one has studied the suitability of the antifungal metabolites produced by St. rimosus for controlling sapwood-inhabiting fungi.

The purpose of this study was to enhance and accelerate the production of antifungal metabolites from St. rimosus to evaluate the effect of the concentrated metabolites on spore germination of sapwood-inhabiting fungi on wood.

MATERIALS AND METHODS

Fungi and Bacterium

The following sapwood-inhabiting fungi were selected for the study: sapstain—Ceratocystis coerulescens (Munch) Bakeshi [C–262], C. minor (Hedgcock) Hunt [C–188], C. pilifera (Fries) Moreau [RWD 9472B], Aureobasidium pullulans (deBary) Arnand [MDX 18]; mold fungi—Aspergillus niger van Trieghens, Penicillium spp., and Trichoderma spp. The bacterium used in this study was St. rimosus Sobin et al. It was obtained from Dr. D. P. Lubeda, USDA Agricultural Research Service, Midwest Area Northern Regional Research Center, Peoria, Illinois.

Bacterial Culture

Stock cultures of St. rimosus were maintained on nutrient agar (NA) (Difco, Detroit, MI) slants supplemented with 1 percent peptone (NAP) (Difco). Cultures were grown in vegetative growth medium, 1x sporulation broth (SB), for activation and adaptation of inoculum (Highley et al., 1991). For vegetative growth, mycelia of St. rimosus from cultures grown on a NAP slant were scraped from the surface of the slant and cultured in 50 ml of SB medium in 125-ml Erlenmeyer flasks and incubated at 27C with vigorous shaking on platform shaker at 200 r/min for 2 days. The settled pellet was transferred into 100 ml of SB medium in 250-ml flasks and allowed to grow for an additional 40 hr. The settled pellet–inoculum (100 percent) was then transferred into 50 ml of SB medium at various concentrations (1x, 1/2x, 1/3x, and 1/4x) supplemented with 0.05 percent Tween 20 in 125-ml Erlenmeyer flasks to find the production medium. Flasks were incubated at 30C at 200 r/min.

Antifungal activity was determined over time by measuring the clear-zone area around the wells with 25 μl of metabolites in Aspergillus (A. niger) plate bioassay (see next section on plate bioassay). Crude metabolites were harvested after 20 to 24 hr with 1/3x medium or after 3 to 4 days with 1/2x medium. Some crude preparations were concentrated by lyophilization.

Tests of Antifungal Activity

Petri Plate Assay. The modified dual-culture techniques were similar to that described by Croan and Highley (1991). Streptomyces rimosus was tested for its antagonistic activities against sapwood-inhabiting fungi on NA or SB–agar medium. Agar plates were inoculated at the center and spread over the 20-mm-diameter surface of the plate with a 50-μl mycelial suspension of the test bacterium. (A 10-ml suspension was prepared from each NAP slant.) Twenty-five microliters of conidial suspension (200,000 conidia in 0.05 percent Tween 20) of sapwood-inhabiting fungi were streaked along the perimeters of the plate. Four sapstain and three mold fungi were inoculated per plate. The plates were incubated for 20 days in the dark at 27C and 70 percent relative humidity. The radial growth of the colony of sapwood-inhabiting fungi was measured. Three replications were examined for each experiment. Three plates inoculated with each test fungi without St. rimosus were used as controls.

Plate Bioassay. The plate bioassay was used to test the activity of inhibitory substances released by St. rimosus. A conidial suspension of sapwood-inhabiting fungi (200,000 conidia per 200 μl in 0.05 percent Tween 20) was spread-plated onto 2-percent malt extract agar plates. Four to six 5-mm-diameter wells were cut with a cork cutter, and 5 to 50 μl of culture fluid was pipetted into the wells. The clear-zone diameter or clear-zone area around the wells was measured after incubation at 27C for 2 to 5 days.

Wood-Block Tests. Blocks of two wood species, Pinus sp. (southern yellow pine) and Liquidambar styraciflua L. (sweetgum), were used. The size of the wood blocks was 3 by 28 by 42 mm in tangential, radial, and longitudinal directions, respectively. The wood blocks were soaked overnight to duplicate the moisture content of green wood. The blocks were steam sterilized for 30 min at normal atmospheric pressure for 3 consecutive days and then dipped for 60 sec into the bacterial metabolites. The blocks were treated with 1x and 10x metabolites; control blocks were dipped in water; 0.1 percent sodium pentachlorophenate (penta) was used for a positive control. All blocks were placed into sterile 140-mm-diameter petri plates lined with three layers of white paper towel, which was lined with thread and saturated with sterile water. Plates were inoculated with a 200-μl condial suspension (400,000 spores per 200 μl) of sapwood-inhabiting fungus. Plates were then placed in plastic bags and incubated for 8 weeks at 27C and 70 percent humidity.

Simulated Field Trials. Green southern yellow pine (Pinus sp.) sapwood log sections, 170 mm diameter by 12 mm long, were cut from a freshly felled tree and immediately frozen. The thawed sections were dipped for 60 sec into 1x or 10x antifungal metabolites. Control log sections were treated with water;

positive control sections were dipped into 1 percent penta. The sections were then hung on a rack and placed in an air-permeable polypropylene bag (Bel–Art Products) containing a tray with soil (40 percent moisture content) from the Valley View test site near Madison, Wisconsin. The bag was incubated for 8 weeks at 27C and 70 percent humidity.

RESULTS AND DISCUSSION

Effect of Medium Concentration

All cultures grew well in 1x SB medium, producing small (0.2–3 mm) pellets with average diameter of 0.5 mm. Approximately 2,000 to 3,000 small pellets were formed in 50 ml of growth medium in a 125-ml Erlenmeyer flask.

The result of different concentrations of SB medium was determined by measuring the development of antifungal activity in plate bioassay against about 200,000 spores from A. niger over time. The time course of antifungal activity and maximum activity reached varied with medium concentration. At all concentrations, cultures developed maximum activity on days 1 (19 hr) to 4; activity decreased sharply at the lowest concentration (1/4x) but decreased much more slowly at the highest concentration (Table 1; Fig. 1). The best result was obtained with the 1/3x concentration in the clear-zone area around the wells after 19 hr; antifungal activity was 962 $mm^2/25\ \mu l$. The next largest clear-zone area occurred with the 1/2x concentration after day 3; 775 $mm^2/25\ \mu l$. The volume of the culture and flask size did not influence the development of antifungal activity in plate bioassay when they were varied proportionately. The activity was the same in 125-ml Erlenmeyer flasks with 50 ml medium and in 1,000-ml Erlenmeyer flasks with 400 ml of the same medium. The substantial variation in activity among replicated cultures was caused by asynchrony in culture development. Crude metabolites were harvested every day with 1/3x medium or every 3 to 4 days with 1/2x medium. Fermentation was scaled up and accelerated by promoting secondary metabolism and inhibiting biomass production using large (100 percent) and tiny pellet–inoculum and a new production medium, and by varying culture volume and shaking speed. The cessation of growth signalled secondary metabolism. The pellets were recycled continuously for several months. The pellets can be recycled as long as they remain uncontaminated.

Table 1. Effect of Medium Concentration on Antifungal Activity of Metabolites From St. rimosus in Aspergillus Plate Bioassay.

	Inhibition zone[a]		
SB medium concentration[b]	Diameter (mm/25 μl)	Area (mm^2/25 μl)	Maximum activity
1x	28.5 (0.7)	638 (.4)	4 days
1/2x	31.0 (1.4)	775 (1.5)	3–4 days
1/3x	34.5 (2.1)	962 (3.5)	19 hr
1/4x	23.5 (0.7)	434 (.4)	2 days

[a]Standard deviation in parentheses.
[b]SB is sporulation broth.

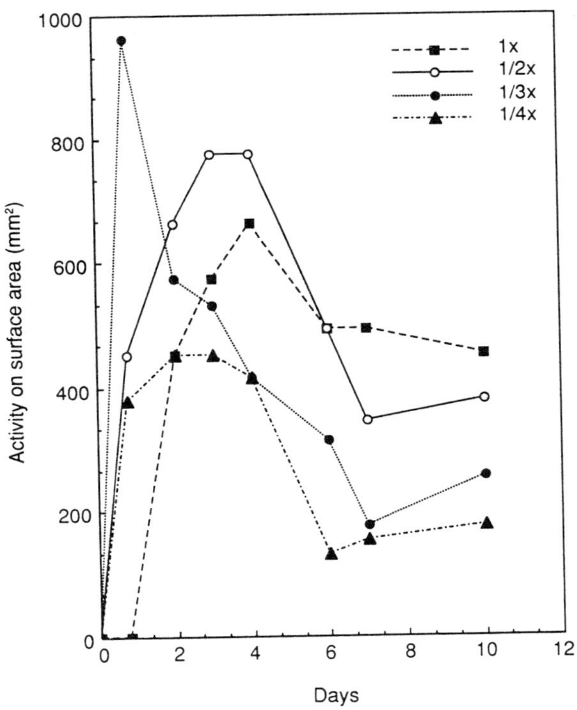

Figure 1. Effect of SB Medium Concentration on Antifungal Activity of Metabolites From St. rimosus in Aspergillus Plate Bioassay Over Time.

Antifungal Activity

Petri Plate Assay. The antagonistic effect of St. rimosus against various sapwood-inhabiting fungi tested in the petri plate assay is shown in Table 2 and Figure 2. The growth of all the sapwood-inhabiting fungi tested was inhibited at a considerable distance, leaving a clear zone of inhibition between the fungi and St. rimosus on NA plates after 20 days of incubation. Streptomyces rimosus inhibited 81 to 86 percent of the radial growth of the fungi in the petri plates (Table 2, Fig. 2). When these fungi were inoculated on the typical bacterial medium, NA or SB–agar plates, the growth of sapwood-inhabiting fungi—Aureobasidium pullulans, C. minor, C. pilifera, and Penicillium spp.—appeared to be restricted on NA plates and showed almost no growth on SB–agar plates, compared to that on 2 percent malt extract agar plates. Benko and Highley (1991) demonstrated that both wood-attacking fungi and St. rimosus grew well on 2 percent malt extract agar but failed to produce a good inhibition zone between fungi and St. rimosus in dual cultures. The composition of the medium evidently plays an important role with respect to the production of antifungal secondary metabolites. Therefore, in the current study, the SB medium (American Type Culture Collection, 1989) was used for vegetative growth of St. rimosus and a production medium (Highley et al., 1991) was used for the production of secondary metabolites to achieve maximum production. The antifungal activities against sapwood-inhabiting fungi were then tested on 2-percent malt extract agar plates using the plate bioassay method.

Plate Bioassay. The results of the interaction between sapwood-inhabiting fungi and metabolites in plate bioassay on 2-percent malt extract agar plates are given in Table 3. The conidial germination of all sapwood-inhabiting fungi was inhibited by 25 μl of 1-day-old crude metabolites from 1/3x production medium or 3- to 4-day-old metabolites from 1/2x production medium introduced in each well. The conidial germination of sapwood-inhabiting fungi was completely inhibited, leaving a clear zone around the wells. The diameter of the inhibition or clear-zone area varied from 19.3 to 38 mm per 25 μl of crude metabolites.

Wood Block Tests. The results of the interaction between sapwood-inhabiting fungi and lyophilized metabolites (10x) on wood blocks werecompared with that between fungi and 1x metabolites or 0.1 percent penta (Table 4, Figures 3 and 4). The concentrated metabolites from St. rimosus were

Table 2. Inhibition of Radial Growth of Sapwood-Inhabiting Fungi by St. rimosus in Petri Plate Assay.

Fungus	Inhibition of radial growth[a] (percent)	
Sapstain		
Aureobasidium pullulans	86.0	(2.9)
Ceratocystis coerulescens	84.6	(2.4)
C. minor	84.6	(0.9)
C. pilifera	81.3	(1.8)
Mold fungi		
Aspergillus niger	84.5	(1.6)
Penicillium spp.	85.6	(11.0)
Trichoderma spp.	81.2	(1.6)

[a]Standard deviation in parentheses.

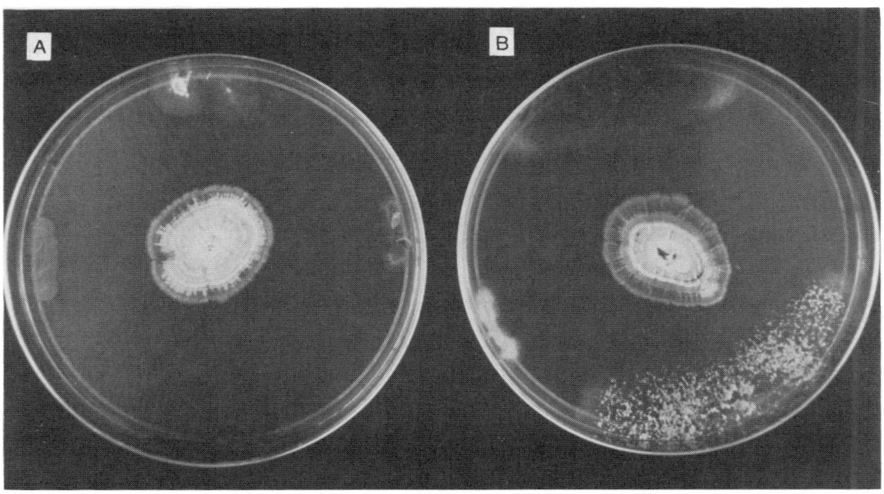

Figure 2. Inhibition of Radial Growth of Sapwood-Inhabiting Fungi; Plate A–C. coerulescens, C. minor, Aureobasidium pullulans, and C. pilifera; Plate B–Trichoderma spp., Penicillium spp., and Aspergillus niger. (M91 0224-4A).

the most effective biotic agent against all fungi tested; the metabolites protected the blocks against spore germination and discoloration. The only fungal attack observed on wood blocks treated with 1x metabolites was in those blocks exposed to Penicillium spp. Penta (0.1 percent) did not prevent attack by Trichoderma spp. and C. minor on pine, and attack by Aureobasidium pullulans and C. pilifera on both pine and sweetgum.

Table 3. Inhibition of Spore Germination of Sapwood-Inhabiting Fungi by Metabolites From St. rimosus in Plate Bioassay on 2Percent Malt Extract Agar.

	Inhibition zone[a]	
Fungus	Diameter(mm/25 μl)	
Sapstain		
Aureobasidium pullulans	30.0	(2.0)
Geratocystis coerulescens	38.0	(2.0)
C. minor	32.7	(1.2)
C. pilifera	31.3	(1.2)
Mold fungi		
Aspergillus niger	31.3	(1.2)
Penicillium spp.	19.3	(1.2)
Trichoderma spp.	28.0	(1.2)

[a]Standard deviation in parentheses.

Simulated Field Trials. The simulated field trials (Fig. 5) were conducted in an air-permeable polypropylene bag, which was confined to a limited area; therefore, micro-organisms in the soil were restricted to the bag. The entire bag was incubated for 8 weeks at 27C and 70 percent humidity, optimum conditions for sapwood-inhabiting fungi.

Table 4. Inhibition of Spore Germination of Sapwood-inhibiting Fungi on Wood Block by St. rimosus.

	Extent of Discoloration[a]													
	Aspergillus niger		Penicillium spp.		Trichoderma spp.		Aureobasidium pullulans		Ceratocystis coerulescens		C. minor		C. pilifera	
Treatment	Gum	Pine	Gum	Pine	Gum	Pine	Gum	Pine	Gum	Pine	Gum	Pine	Gum	Pine
Control	4	4	4	4	4	4	4	4	4	4	4	4	4	4
0.1% penta	0	0	0	0	0	1-2	2	4	0	0	0	4	3	4
1x metabolites	0	0	3	2	0	0	0	0	0	0	0	0	0	0
10x metabolites	0	0	0	0	0	0	0	0	0	0	0	0	0	0

[a] Blocks were rated for fungal growth or stain after 8 weeks of exposure at 27C using the following scale:

0 Clean, not stain or mold
1 Minor growth, stain or mold over <5 percent of upper surface
2 Light growth, stain or mold over <5-20 percent of upper surface
3 Moderate growth, stain or mold over 20-40 percent of upper surface
4 Heave growth, stain or mold over >40 percent or upper surface

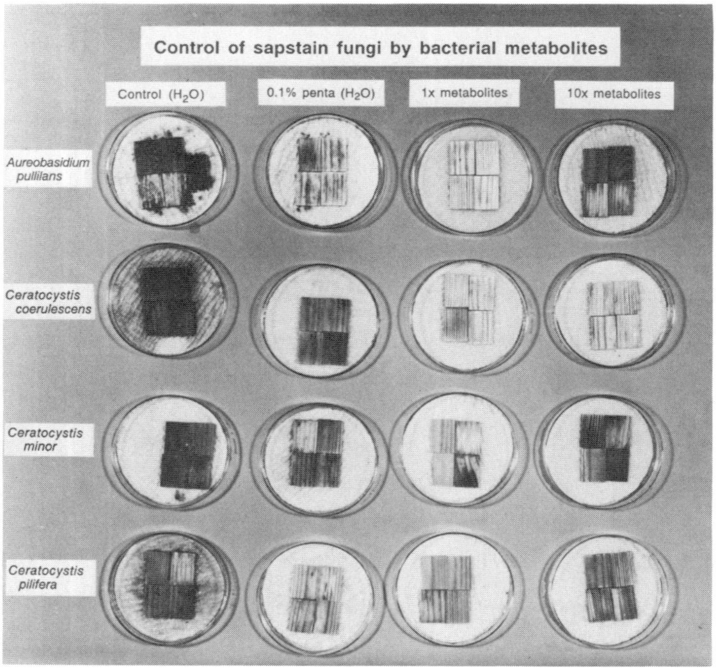

Figure 3. Inhibition of Spore Germination of Sapstain on Wood Blocks by Metabolites From St. rimosus. (M91 00877)

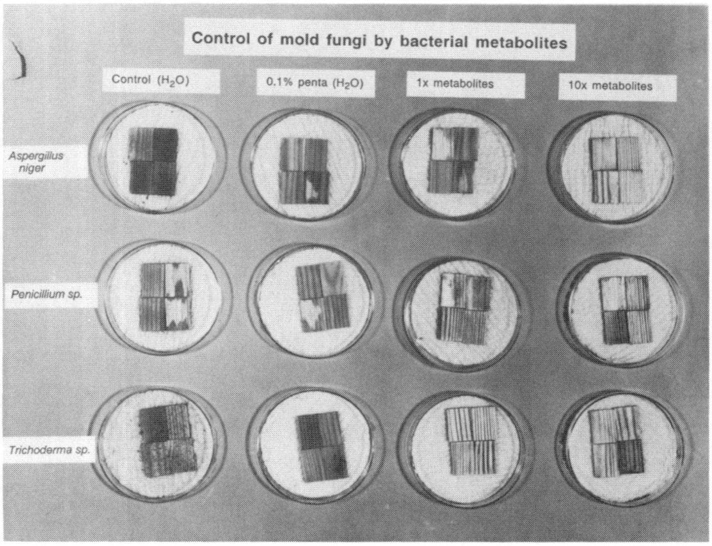

Figure 4. Inhibition of Spore Germination of Mold Fungi on Wood Blocks by Metabolites From St. rimosus. (M91 0082)

Figure 5. Green Pine Log Sections in Simulated Field Trial.(M91 0091-1)

The sapwood-inhabiting fungi were much more difficult to control in this exposure (Table 5, Fig. 6). Not even sections treated with 1 percent penta were free of attack by sapstain and decay fungi. The 10x concentrated metabolites controlled fungal attack better than did the 1x metabolites (Table 5, Figure 6).

Table 5. Control of Sapwood-Inhabiting Fungi on Green Pine Log Sections by Metabolites From S. rimosus in Simulated Field Trial[a]

Treatment	Extent of discoloration[b]
Control	4
1 percent penta	1
1x metabolites	2
10x metabolites	1–2

[a]See Figure 5 for experimental set-up.[b]Rating scale is described in Table 4.

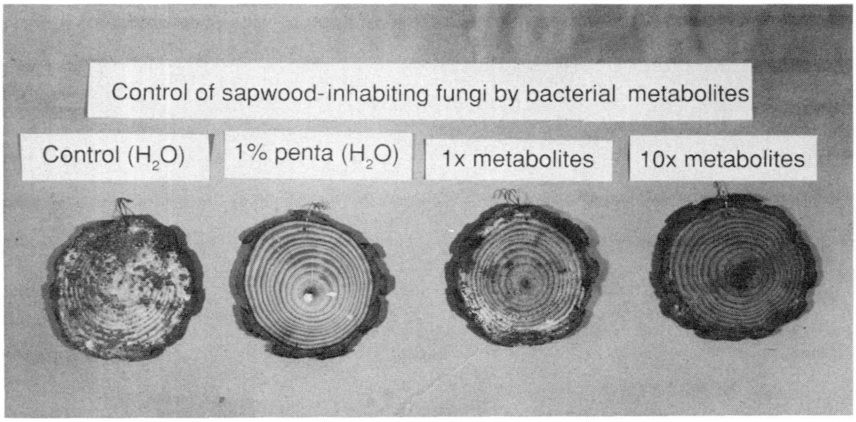

Figure 6. Control of Wood-Attacking Fungi on Green Pine Log Sections by Metabolites From St. rimosus in Simulated Field Trial. (M91 0095)

SUMMARY

Our results show that concentrated antifungal metabolites are very effective biotic agents against sapwood-inhabiting fungi. The concentrated metabolites completely protect the wood blocks from fungal attack or discoloration. Further experiments are underway to (a) identify antifungal metabolites, (b) further optimize their production by limiting or shortening the primary metabolic phase, promoting the secondary metabolic phase, and recycling pellets, (c) evaluate the synergistic effect of boron with unconcentrated metabolites, and (d) use non-wood-decay living bacterial cells as a bioprotectant (Dawson–Andoh and Morrell, 1990; Falkowski et al., 1978; Newbill and Morrell, 1991) against fungal attack and discoloration for long-term wood preservation.

REFERENCES

American Type Culture Collection. (1989). Catalogue of Bacteria and Phages. 7th ed. Rockville, MD.

Benko, R. (1988a). Bacteria as Possible Organisms for Biological Control of Blue Stain. The International Research Group on Wood Preservation. Doc. No: IRG/WP/1339.

Benko, R. (1988b). Biological Control of Blue Stain on Wood With Pseudomonas cepacia 6252, Laboratory and Field Test. The

International Research Group on Wood Preservation. Doc. No: IRG/WP/1380.

Benko, R. and Highley, T.L. (1991). Selection of Media for Screening Interaction of Wood-Attacking Fungi and Antagonistic Bacteria. 1. Interaction on Agar. Mater. Org. 25(3), 161–171.

Butler, M.J. and Gedge, B.N. (1989). Purification of Anhydrotetracycline Oxygenase From Streptomyces rimosus Using Fast Liquid Chromatography. Biotech. Tech. 3, 235–238.

Cope, A.C., Burrows, E.P., Derieg, M.E., and Sung Moon, Wirth W-D. (1965). Rimocidin. 1. Carbon Skeleton, Partial Structure, and Absolute Configuration at C-27. J. Am. Chem. Soc. 87(23), 5452–5460.

Croan, S.C. and Highley, T.L. (1991). Biological Control of the Blue Stain Fungus Ceratocystis coerulescens With Fungal Antagonists. Mater. Org. 25(4), 255–266.

Dawson-Andoh, B. and Morrell, J.J. (1990). Effects of Chemical Pretreatment of Douglas-Fir Heartwood on Efficacy of Potential Bioprotection Agents. The International Research Group on Wood Preservation. Doc. No: IRG/WP/1440.

Falkowski, L., Zielinski, J., Golik, J., and Bylec, E. (1978). The Structure of Rimocidin: Mass Spectrometric Analysis of Derivatives of the Antibiotic. J. Antibiot. (Tokyo) 31(8), 742–749.

Friend, E.J, Warren, M., and Hopwood, D.A. (1978). Genetic Evidence for a Plasmid Controlling Fertility in an Industrial Strain of Streptomyces rimosus. J. Gen. Microbiol. 106, 201–206.

Highley, T.L., Benko, R., and Croan, S.C. (1991). Laboratory Studies on Control of Sapstain and Mold on Unseasoned Wood by Bacteria. The International Research Group on Wood Preservation. Doc. No: IRG/WP/1493.

Newbill, M.A. and Morrell, J.J. (1991). Effect of Elevated Temperatures on Survival of Basidiomycetes That Colonize Untreated Douglas-Fir Poles. Forest Prod. J. 41(6), 31–33.

Rehacek, Z. (1990). Research Into Biogenesis of Biologically Active Microbial Metabolites in Czechoslovakia. Process Biochem. Int. 25(6), 197–209.

Rhodes, P.M., Winskill, N., Friend, E.J., and Warren, M..(1981). Biochemical and Genetic Characterization of Streptomyces rimosus Mutants Impaired in Oxytetracycline Biosynthesis. J. Gen. Microbiol. 124, 329–338.

Note: The use of trade or firm names in this publication is for reader information and does not imply endorsement by the U.S. Department of Agriculture of any product or service.

CHEMICAL CHANGES IN WOOD COMPONENTS AND COTTON CELLULOSE AS A RESULT OF BROWN ROT: IS FENTON CHEMISTRY INVOLVED?

DOUGLAS S. FLOURNOY, USDA Forest Service, Forest Products Laboratory, Madison, WI 53705–2398 USA

INTRODUCTION

The most destructive form of wood decay, brown rot, is caused by a relatively small number of species of Hymenomycetous basidiomycetes. These basidiomycetes are unique among cellulose destroyers because they are the only known microbes that can degrade the cellulose in wood without first removing the lignin (Cowling, 1961; Liese, 1970). Brown-rot fungi leave a brown residue—hence their name—that has been partially o-demethylated (Kirk, 1975). Furthermore, brown-rot fungi degrade cellulose in an unusual manner that differs from that of other cellulolytic organisms. Hyphae of these ubiquitous fungi invade wood cells and bring about a rapid depolymerization of the cellulose with low losses in total wood substance (Cowling, 1961; Kayama, 1962b). The average number of glucosyl residues per cellulose molecule (degree of polymerization, DP) is thereby reduced from about 10^4 (Goring and Timell, 1962) to about 200 (Cowling, 1961). The resulting fragments correspond to the size of the cellulose "crystallites." This effect is thought to be brought about by cleavages within the amorphous regions of the cellulose that separate the crystallites (Cowling, 1961). Similar depolymerization of cellulose to the "limit DP" (to the crystallites) is effected by acid hydrolysis (Battista, 1950) and by chemical oxidants (Koenigs, 1972a, 1974a,b, 1975; Highley, 1977; Kirk et al., 1991). As a result of the initial attack by brown-rot fungi and the depolymerization of the cellulose, wood strength collapses. How this rapid depolymerization occurs is a perplexing biochemical question: as Cowling and Brown (1969) recognized over two decades ago, even the smallest cellulases (approximate diameter 25 _, length 140 _) are too large to penetrate the pores of wood (median pore diameter approximately 10 _; maximum 35–100 _). Also, cellulases do not mimic the action of brown-rot fungi in generating cellulose crystallites (Chang et al., 1981; Phillip et al., 1981). Our own examination of the change in pore structure of wood as it is decayed by a brown-rot fungus suggests that the depolymerizing agent is between 12 and 38 _ in diameter (Flournoy, 1991; Flournoy et al., 1991).

Thus, brown-rot fungi seem to utilize an entirely different mechanism to degrade cellulose than do other cellulolytic organisms. The biochemistry and physiology of the degradative system employed by brown-rot fungi to accomplish this feat are far from clear. The prevailing hypothesis implicates low molecular weight transition metal (usually iron) chelates, which act in two possible ways: (1) through biochemical conversion into potent oxidizing species, which diffuse into the wood pores and oxidatively cleave the cellulose, or (2) through participation in the biochemical generation of an activated oxygen species (hydroxyl radical or equally potent metallo-oxygen species), in close juxtaposition to the cellulose, which oxidizes the cellulose leaving the lignin substantially unchanged. This review is a critical examination of this hypothesis. We begin by examining what is known about the chemical changes in wood components and cotton cellulose as a result of brown-rot decay. The literature concerning the production of reduced oxygen species by brown-rot species is then reviewed. We conclude by considering the action of Fenton's reagent on wood, lignin, cotton cellulose, and related systems.

BACKGROUND

In discussing the problem of brown-rot depolymerization of cellulose, Cowling and Brown (1969) noted that Halliwell (1965) had described the degradation of cotton cellulose by Fenton's reagent (H_2O_2/Fe^{2+}), which generates hydroxyl radical or a similar oxidant (Halliwell and Gutteridge, 1988). Based on these observations, Halliwell was the first to propose the possibility of the existence of a nonenzymatic cellulolytic system involving peroxide and iron. Subsequently, Koenigs (1972a,b; 1974a,b, 1975) demonstrated that cellulose in wood can be depolymerized by Fenton's reagent, that brown-rot fungi produce extracellular hydrogen peroxide, and that wood contains enough iron to make Halliwell's hypothesis reasonable. In commenting on this hypothesis, Koenigs (1974a) stated that "the hypothesis is reasonable pending proof that brown-rot fungi oxidize cellulose." Support for an oxidative system was later provided by Highley (1977), who obtained evidence that brown-rotted cellulose contains carbonyl and carboxyl groups. Later, Schmidt and co-workers (Schmidt, 1980; Schmidt et al., 1981) suggested that oxalic acid, which is secreted by brown-rot fungi in liquid cultures (Takao, 1965), can reduce Fe^{3+} to Fe^{2+} under certain conditions, and proposed that oxalic acid acts to generate Fe^{2+} for the Fenton system. Some authors have suggested a more direct role for oxalic acid in depolymerization of cellulose. Shimada and co-workers (Shimada et al. 1991; Akamatsu et al., 1991) reported that oxalic acid decreased the viscosity of kraft pulp and proposed that oxalic acid may depolymerize cellulose during brown-rot decay.

In an attempt to gain direct evidence for a nonenzymatic decay mechanism, Cobb (1981) grew liquid cultures of Gloeophyllum trabeum in an apparatus in which uniformly labeled [^{14}C]cellulose was separated from the fungus by an ultrafiltration membrane. The membrane was reported to prevent the passage of enzymes between the two chambers. Under these conditions, [^{14}C]CO_2 was detected and radioactive water-soluble products accumulated in the growth chamber. These observations have been taken as proof of a nonenzymatic cellulose decay mechanism being employed by Gl. trabeum (Cobb, 1981; Eriksson et al., 1990).

The existence of a low molecular weight enzyme responsible for depolymerization, a cellulase, cannot yet be ruled out. Little is known about the cellulase system of brown-rot fungi. The most striking feature is the apparent lack of C_1 (β-1,4-glucan cellobiohydrolase) activity (Nilsson, 1974). A few cellulases of brown-rot fungi have been isolated and characterized, but little is known about their chemical and physical properties (Highley, 1975). An endo-1,4-β-glucanase of Lenzites trabea has been purified and partially characterized (Herr et al., 1978). It has a M_r of ~29,000, shows maximal activity at pH 4.4 and 70C, and appears to be a typical endocellulase. Wolter and co-workers (1980) isolated a polysaccharide-degrading complex from Postia placenta with an apparent M_r of 185,000, which showed activity against eight substrates. Similar results were obtained from solid substrate cultures of P.

placenta (Highley et al., 1981; Highley and Wolter, 1982). None of these isolated proteins showed activity against highly ordered substrates such as microcrystalline cellulose (avicel) or was able to depolymerize cotton cellulose.

Enoki et al. (1989) reported the ability of brown-rot fungi to oxidize 2-keto-4-thiomethylbutyric acid (KTBA) to ethylene; KTBA is converted to ethylene by one-electron oxidants such as hydroxyl radical. Ethylene production was correlated with weight loss but not cellulose depolymerization. Recently, Enoki et al. (1990) reported the isolation of an extracellular protein from cultures of Gl. trabeum requiring H_2O_2 and capable of KTBA oxidation. These authors partially purified the protein and reported it to be an iron-containing glycoprotein of M_r ~ 1600-2000. Based on their work, Enoki and co-workers have suggested the existence of "a unique wood-component degrading system that participates directly or indirectly in the fragmentation of cellulose as well as of lignin in wood and oxidizes KTBA to give ethylene." However, it is yet to be established whether the H_2O_2-dependent KTBA-oxidizing ability of this protein is related to cellulose depolymerization. Similar glycoproteins were isolated from white-rot and soft-rot fungi (Enoki et al., 1991), raising questions about their role in wood decay.

If the brown-rot fungi do employ a Fenton-type system, it seems likely that the Fe (or other metal) is complexed and some control exerted over its action. This control may be expressed through chelation of the metal ion involved. Chandhoke and co-workers (1991) purified siderophores from the brown-rot fungus Gl. trabeum. They demonstrated that purified siderophores were able to carry out the one-electron oxidation of KTBA and the Fenton reaction. Reaction rate was influenced by siderophore, iron, manganese, and oxalate concentration and pH. Cleavage of cellulose azure in the presence of Fe was demonstrated to be directly correlated with siderophore concentration.

Rapid progress is being made in the isolation and characterization of these extracellular proteins, siderophores, and chelators (Fekete et al., 1989; Enoki et al., 1991; Jellison et al., 1990, 1991). Yet, no one has demonstrated a cell-free system capable of depolymerizing cotton cellulose in the manner characteristic of brown-rot fungi. A solid substrate system for brown-rot fungal depolymerization of isolated cellulose has been described (Highley, 1977). Using this system, some progress has been made in understanding the chemical nature of cellulose depolymerization by brown-rot fungi (Highley, 1977; Highley et al., 1988, 1989; Kirk et al., 1989, 1991).

CHANGES IN CHEMICAL PROPERTIES OF WOOD COMPONENTS AND COTTON CELLULOSE

Lignin Properties

Cowling's (1961) comparative study of wood decay by white- and brown-rot fungi showed that the lignin content of brown-rotted wood remains relatively constant when expressed on the basis of original sound wood. Thus, brown-rot fungi do not utilize lignin to an appreciable extent. However, several lines of evidence indicate that the lignin remaining after consumption of the carbohydrate fraction of the wood is altered. Early work (Cowling, 1961, and references therein) suggested that the methoxyl content of brown-rotted lignin is decreased compared to that of sound wood. In addition, the solubility properties of the lignin have been altered (Cowling, 1961; Brown et al., 1968; Kirk, 1975). Lignin from decayed samples possesses greater solubility in water and 1% sodium hydroxide than does lignin from sound wood (Cowling, 1961). These results indicate that brown-rot fungi at least modify the properties of lignin during wood decay. No major structural changes have been demonstrated to occur in the lignin, such as depolymerization, and several points of evidence argue against depolymerization (Brown et al., 1968): (1) acid-soluble lignin does not increase with carbohydrate removal, (2) the molecular weight of formamide-soluble lignin increases rather than decreases with increasing extent of carbohydrate removal, and (3) structural integrity of the wood is maintained even at high weight loss.

Wood that has been extensively degraded by brown-rot fungi consists almost entirely of lignin residue. Several structural studies have been made of this residue. In the earliest of these, Kirk and Adler (1969) detected the presence of catechol moieties in brown-rotted lignin. Using L. trabea (= Gl. trabeum), "enzymatically liberated lignin" was prepared by extensively degrading the sapwood of sweetgum. This lignin residue was ethylated with diethyl sulfate and then oxidized with permanganate and hydrogen peroxide. Gas chromatography of the methyl esters of the resulting acids revealed that four compounds were prominent. These compounds were identified as the methyl esters of 3-methoxy-4-ethoxybenzoic acid (I), 3,4-diethyoxybenzoic acid (II), 3,5-dimethoxy-4-ethoxybenzoic acid (III), and 3,4-diethoxy-5-methoxybenzoic acid (IV).

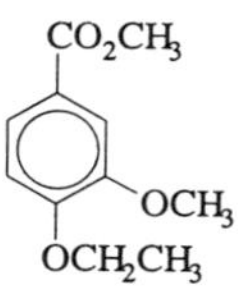

I

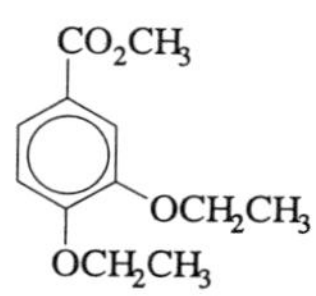

II

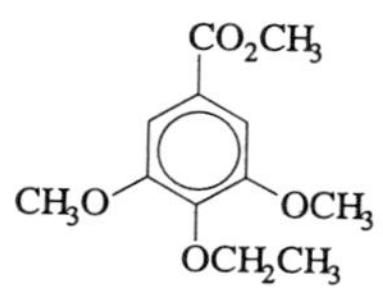

III

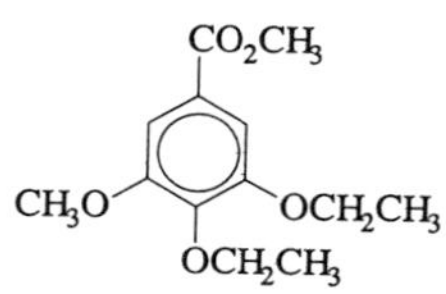

IV

Figure 1. 3-methoxy-4-ethoxybenaoic acid
Figure 2. 3, 4-diethyoxybenzoic acid
Figure 3. 3, 5-dinethooxy-4-ethoxybenzoic acid
Figure 4. 3, 4-diethoxy-5-methoxybenzoic acid

Compounds I and III are formed from phenolic guaiacyl- and syringyl-type substructures of sound lignin, as expected. But the presence of compounds II and IV indicates that catechol groups are formed in the degraded lignin by action of the brown-rot fungus *L. trabea*, (=*Gl. trabeum*). Kirk and Adler found that nearly 30% of the phenolic guaiacyl and syringyl unit had undergone net demethylation. Similar results were found with *Poria monticola* (= *Postia placenta*).

Further studies (Kirk and Adler, 1970) showed that demethylation was not limited to phenolic substructures and that a hydroxyl group adjacent (*ortho*) to

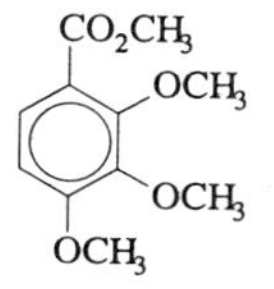

V

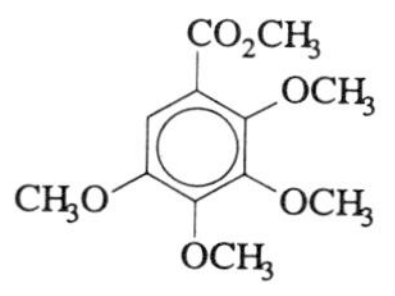

VI

Figure 5. Methyl esters of 2, 3, 4-trimethoxybenzoic acid

Figure 6. 2, 3, 4, 5-tetramethoxybenzoic acid

the methoxyl group was not required for demethylation. The extent of demethylation of phenolic units of both guaiacyl and syringyl types was found to be greater than that in corresponding nonphenolic units. Also phenolic and nonphenolic syringyl-type subunits were more extensively demethylated than were guaiacyl subunits.

Kirk et al. (1970) characterized lignin isolated from sweetgum wood decayed by L. trabea (= Gl. trabeum) and compared it to milled wood lignin of sound sweetgum wood. The lignins were methylated, oxidized with permanganate and hydrogen peroxide, and finally esterified. The resulting mixtures of methyl esters were examined by gas chromatography mass spectroscopy (GC–MS). Two compounds were produced from lignin of brown-rotted wood that were not produced from lignin recovered from sound wood. These compounds were identified by GC-MS as the methy esters of 2,3,4-trimethoxybenzoic acid (V) and 2,3,4,5-tetramethoxybenzoic acid (VI). Kirk and co-workers concluded that the lignin substructures had been hydroxylated as a result of fungal action and that this hydroxylation had occurred ortho to the side chains of phenylpropane units of lignin.

Brown-rotted lignin has similar but not the same solubility properties as that of milled wood lignin (Kirk, 1975). Also, the sulfuric acid lignin value of brown-rotted lignin is lower than that of sound lignins; carbohydrate content is similar to that of sound lignin. Elemental analysis of brown-rotted lignin indicated that the lignin was slightly enriched in oxygen and somewhat depleted in carbon and hydrogen compared to sound lignin. On a C_9-unit basis, the brown-rotted lignin was about one oxygen atom rich in nonmethoxyl oxygen. Functional group analysis indicated the carboxyl content of brown-rotted lignin was twice that of milled wood lignin. In addition, the total hydroxyl content was somewhat higher.

This increase in hydroxyl content is associated with the aromatic residues rather than the side chains. Thus, the phenolic hydroxyl content of brown-rotted lignin was more than twice that of milled wood lignin. Aliphatic hydroxyl content was depressed. This result is consistent with the observed increase in conjugated carbonyl content of brown-rotted lignin and with earlier findings that brown-rot fungi hydroxylate and demethylate lignin.

The ability of brown-rot fungi to mineralize synthetic [^{14}C]lignin was first reported by Kirk and co-workers (Kirk et al., 1975). These fungi mineralize predominately methoxyl carbon but also some side-chain and aromatic-ring carbon. These results are in accord with results of demethylation studies (Kirk and Adler, 1969; Kirk and Adler, 1970). Haider and Trojanowski (1980) examined the ability of several brown-rot fungi to mineralize synthetic [^{14}C]lignin (DHP). In liquid culture medium, all the carbons were mineralized to varying extents but the dominant reaction was demethylation. Mineralization of the methoxyl carbon was repressed by cellulose and hemicellulose, and more so by glucose, cellobiose, and xylose. It is difficult to say with certainty what these results indicate because the molecular weight distribution of the DHP was not reported and mineralization may have resulted from intracellular oxidation of a low molecular weight component of the [^{14}C]DHP. Haider and Trojanowski also reported the extensive mineralization of phenolcarboxylic and cinnamic acids and coniferyl alcohol by brown-rot fungi. Ring cleavage was observed in these latter compounds, and it is likely that these reactions occur intracellularly.

Jin and co-workers (1990b) isolated brown-rotted lignin and methylated it with $^{14}CH_3I$, producing [3, 4-O-$^{14}CH_3$]lignin. Incubation of this lignin with brown-rot fungi resulted in generation of $^{14}CO_2$ when added to solid substrate cultures. Mineralization was stimulated in an atmosphere of oxygen. Cultures supplemented with nutrient nitrogen suppressed degradation of methoxyl carbon, and supplemental carbon (glucose or glycerol) had little or no effect in the presence of nutrient nitrogen. However, addition of only these carbon sources to the cultures sharply suppressed mineralization of the methoxyl carbon.

Extraction and molecular weight analysis of undegraded [^{14}C]methoxyl-lignin from the active cultures revealed that on incubation with <u>Gl. trabeum</u>, the molecular weight of the [^{14}C]lignin actually increased during incubation. Because no [^{14}C]-labeled low molecular weight compounds accumulated in the cultures, Jin and co-workers (1990b) concluded that the demethylating activity was extracellular.

Ander et al. (1984) and other authors (Enoki et al., 1985, 1988) showed that some brown-rot species are capable of degrading lignin model compounds when grown on agar plates and somewhat less capable in liquid cultures. The nature and meaning of this degradation are unclear because the products have not been characterized. Mineralization was not demonstrated. Since these are low molecular weight uncharged molecules, it is unclear whether this degradation is intra- or extra-cellular.

Although a great deal has been learned about the chemical modification of the lignin polymer by one brown-rot fungus, _Gl. trabeum_, caution should be excercised in extending these results to brown-rot fungi in general. Brown-rot fungi, like white-rot fungi, show great variability in their behavior, which can depend on culture conditions and the nature of the substrate. Further studies concerning the action of other brown-rot fungi on lignin are needed before any broad generalizations can be made.

Cotton Cellulose Properties

To our knowledge, only two reports have addressed the chemical properties of degraded cellulose from brown-rotted wood. Bray and Andrews (1924) observed a decrease in both the quantity and quality of wood cellulose as a function of weight loss. These conclusions were based on the observation of decreasing percentage of a-cellulose content and increasing percentage of alkali solubility with increasing weight loss. Bray and Andrews also observed that reducing power increased as cellulose content decreased. Kayama (1961, 1962a) also found an increase in the copper number of pulp recovered from wood decayed by _Poria vaporaria_ relative to the copper number of pulp from sound wood.

Although little information is available on the chemical properties of cellulose decayed by brown-rot fungus, substantial information is available on chemical characterization of brown-rotted cotton cellulose. Highley (1977) compared the oxidative properties of brown-rotted cellulose and cellulose treated with oxalic acid or $H_2O_2/FeSO_4$. In conducting these studies, Highley compared the copper number, weight loss in alkaline boil, pH in 10% NaCl, and ion exchange capacity of the various celluloses. These results are presented in Table 1.

As the data in Table 1 show, brown-rotted cellulose has increased copper number, increased weight loss on boiling in sodium hydroxide, decreased pH in 10% NaCl, and increased ion-exchange capacity relative to native cellulose.

Brown-rotted cellulose also gave a positive test with ferrous sulfate—potassium ferricyanide, indicating the presence of oxycellulose. No increase in uronic acid content of the brown-rotted cellulose was detected. Similar results were obtained from cellulose treated with 1% H_2O_2 and ferrous sulfate.

Highley found that mycelium alone (P. placenta) possessed ion-exchange capacity, but he did not quantify this capacity. Although efforts were made to remove the mycelium from the samples prior to analysis, microscopic examination revealed that not all the mycelium could be removed; therefore the values in Table 1 may be slightly high.

Infrared spectra showed that carbonyl groups had been introduced into brown-rotted cellulose. Spectra of mycelium also showed the presence of carbonyl absorbance, but they did not correspond to spectra of brown-rotted cellulose. This indicated that the carbonyl groups present in the mycelium did not contribute to the carbonyl absorbance observed in the brown-rotted cellulose. The spectra of brown-rotted cellulose were similar to the spectra of cellulose treated with $H_2O_2/FeSO_4$.

In following up this work, Highley and co-workers (1988, 1989) produced infrared spectra (obtained by diffuse reflectance) of undegraded cotton cellulose that showed a low intensity broad carbonyl band centered at ~1740 cm^{-1}. This finding is consistent with the presence of a reducing-end group at the end of each cellulose molecule. In brown-rotted cellulose, this band has greatly increased intensity and is centered at a slightly lower wavenumber. Reduction of the brown-rotted cellulose sample with sodium borohydride resulted in a decrease in the intensity of the carbonyl band but did not eliminate it completely. This led the authors to conclude that part of the carbonyl absorbance was due to the presence of carboxyl groups. Complete acid hydrolysis of the brown-rotted cellulose sample led to the observation of several unidentified acids on HPLC profiles. No uronic acid was reported, and it was suggested that based on the methylene blue determination of ~0.5% mole carboxyls, there was approximately one carboxyl moiety per cellulose chain in the depolymerized sample.

Brown-rotted cellulose deploymerization products were further chemically characterized and compared with other depolymerized cellulose samples (Kirk et al., 1989, 1991). In this study, the following four depolymerized cellulose samples were prepared from pure cotton cellulose: (1) acid-hydrolyzed (HCl) to

Table 1. Properties of Untreated, Brown-Rotted, Oxalic Acid-Treatment, anc $H_2O_2/FeSO_{4-}$ Treated Cellulose[a].

Treatment and substrate	Copper number[b]	DP	Weight loss (%)	$FeSO^4$ potassium ferricyanide	pH in NaCL	Ion exchange (mM carboxyl 100g cellulose
Untreated						
Cotton	0	2,300	2	Negative	5.9	0.80
Filter paper	0.67	1,050	0	Negative	5.9	0.74
Brown-rotted						
Cotton	7.5	150	31	Positive	3.2	10.0
Filter paper	6.9	198	34	Positive	3.8	6.7
Treated Saturated Oxalic Acid						
Cotton	1.4	1,272	3	Positive	1.9	1.0
Filter paper	1.4	620	0	Positive	1.7	0.80
1% H_2O_2/Fe						
Cotton	5.6	180	32	Positive	4.2	5.0
Filter paper	2.6	390	34	Positive	3.9	5.0
0.3% H_2O_2/Fe						
Cotton	3.4	920	9	Positive	5.4	0.80
Filter paper	2.7	780	15	Positive	5.3	0.91

[a]Data from Highley, 1977.
[b]Grams reduced copper per 100 g cellulose.

the limit DP, (2) $H_2O_2/FeSO_4$-oxidized (Fenton-oxidized), (3) HIO_4/Br_2-oxidized, and (4) brown-rotted (Postia placenta). These four samples were characterized as to molecular size distribution, yield of glucose on complete acid hydrolysis, carboxyl content, uronic acid content, carbonyl content, and sugar acids released on acid hydrolysis (Table 2). Consistent with earlier results, the Fenton system, but not the other oxidation system, mimicked the brown-rot system in nearly all measured characteristics. The acid-hydrolysed sample also possessed similar characteristics. The following sugar acids were identified by GC/MS in the hydrolysates of the brown-rotted and Fenton-oxidized samples: glyceric, erythronic, arabonic, and gluconic. These results were consistent with the depolymerizing agent being related to the Fenton system, but they did not establish that the fungi employ such a system.

These studies (Highley, 1977; Kirk et al., 1991) did not establish a correlation between the oxidation of cellulose and depolymerization. We do not know, for instance, whether oxidation of the cellulose precedes depolymerization, whether oxidation is a result of post-depolymerization modification, or whether oxidation and depolymerization are coupled. The cellulose used in these studies was highly degraded. Samples from early decay were not examined; these will be the samples of most interest in understanding the depolymerization mechanism. In addition, the fungus-decayed samples in one study (Kirk et al., 1991) were washed with 0.1 N NaOH at 50C for 3 hr prior to analysis, which resulted in only 70% recovery of the original sample.

In light of the alkaline lability of oxidized cellulose (Davidson, 1934, 1936, 1940) and the high weight losses observed, we think it prudent to re-examine the carboxyl content of brown-rotted cellulose with particular attention to analysis of samples of low weight loss (early decay). Also, the cellulose should not be exposed to alkali prior to analysis. Brown-rotted cellulose should be further characterized to determine whether formation of carboxyl groups is correlated with depolymerization of cellulose. A shortcoming of techniques currently employed for carboxyl group analysis work is their lack of sensitivity. At low weight loss, the expected yield is very low ($<0.5\%$), and such a small number of carboxyls may easily escape detection. Definitive analysis will depend on the development of more sensitive techniques based on radiochemical labeling.

Table 2. Analytical Characteristics of Depolymerized Cellulose Samples.[a]

Cellulose sample	M_n[b]DP	M_wDP	Glucose onhydrolysis (%)[c]	[COOH] (per chain)	Uronic acid (per chain)	[C=O] (per chain)
Cotton cellulose	1,150	2,110	110.1	ND[d]	ND	ND
Acid-hydrolyzed	66	200	108.5	0	0	1.4
Fenton-oxidized	74	371	107.6	0.24	0	1.1
HIO_4/Br_2-oxidized	53	219	86.9	7.37	0.3	4.4
Brown-rotted	69	238	108.4	0.44	0	1.8

[a] Data from Kirk et al., 1991.
[b] M_n and M_w are number average and weight average molecular weight, respectively.
[c] Yield based on weight of sample.
[d] Not determined.

PRODUCTION OF REDUCED OXYGEN SPECIES BY BROWN-ROT FUNGI

The extracellular formation of reduced oxygen species by brown-rot fungi remains a controversial topic. This topic has attracted a great deal of research interest over the past 20 years because early models of cellulose depolymerization incorporated this reduced oxygen species. Very little information has been gained since these early studies, and the role of reduced oxygen species in cellulose depolymerization, if any, remains speculative.

Production of Hydrogen Peroxide

Koenigs (1972b) reported the first observation of extracellular hydrogen peroxide (H_2O_2) production by brown-rot basidiomycetes. In these experiments, extracellular peroxide was inferred from the green discoloration of sheep's blood medium or bovine hemoglobin agar under and around a colony. After 13 days of incubation, all brown-rot fungi tested produced green discoloration. Koenigs concluded that H_2O_2 had diffused away from the hyphae in these plates. In liquid cultures containing buffered hemoglobin, all brown-rot fungi examined caused a decrease in the absorbance of the methemoglobin peak at 630 nm and of the Soret band at 405 nm, consistent with H_2O_2 formation. Based on these kinds of experiments, Koenigs concluded that generation of extracellular H_2O_2 was a general characteristic of wood-rotting fungi. Later, Koenigs (1974a,b) reported H_2O_2 production by brown-rot fungi grown on wood and correlated its production with cellulose depolymerization. Unfortunately, the assay for H_2O_2 production employed in these studies (catalase-aminotriazole) was later established by Highley (1981) to be an invalid method for measuring H_2O_2 by wood decay fungi. Therefore, the results of Koenigs' later studies (1974a,b) must be discounted.

Highley (1982) found only one brown-rot fungus (*L. olivascens*) capable of H_2O_2 production in liquid culture even though he used many of the same species as did Koenigs (1972b). Highley (1982) also failed to detect H_2O_2 in sawdust or cellulose decayed by several brown-rot fungi (*Gl. trabeum*, *P. placenta*, *L. lepideus*). An important control of Highley's was the incubation of known concentrations of H_2O_2 with culture filtrates. He failed to observe lowering of H_2O_2 concentrations even after 24 hr of incubation, ruling out the presence of H_2O_2-degrading components.

In these studies, Highley (1982) utilized an *o*-dianisidine and a titanium tetrachloride reagent for H_2O_2 detection. He examined white-rot fungal cultures and found H_2O_2 when assayed with o-dianisidine, but the same cultures were negative when assayed with titanium tetrachloride. Efforts to detect H_2O_2 with the titanium reagent had also failed in Koenigs' hands (Koenigs, 1972). The explanation for this discrepancy is unknown. One possibility is that the technique employing o-dianisidine is not selective for H_2O_2 and that some other metabolite is responsible for the positive results attained with this method.

In his thesis, Cobb (1981) correlated growth and H_2O_2 accumulation in stationary liquid cultures of Gl. trabeum. Using the peroxytitanic acid method for peroxide quantitation, Cobb examined stationary liquid cultures using a basal salts medium supplemented with ballmilled southern yellow pine, holocellulose extracted from ballmilled southern yellow pine, a-cellulose, carboxymethylcellulose (CMC), or a hemicellulose monosaccharide mixture. Flasks showing poor or no growth (a-cellulose, hollocellulose, ballmilled southern yellow pine, and CMC) showed no peroxide accumulation. Those cultures showing good growth (hemicellulose monosaccharide mixture) gave good accumulation of peroxide. Cobb concluded that Gl. trabeum does not produce peroxide except from hemicelluloses. Growth and production of peroxide on a strictly glucose or cellobiose supplemented medium was not examined.

In cytochemical studies (Highley and Murmanis, 1985a,b), P. placenta was grown on hemlock sawdust for 6 weeks. Samples were examined for localization of H_2O_2 production by staining with 3,3'-diaminobenzidine tetrahydrochloride (DAB). On oxidation, DAB binds osmium tetroxide and can subsequently be visualized by electron microscopy. In control experiments, Coriolus versicolor, a white-rot fungus, produced large amounts of dark deposits, presumably oxidized DAB. Hyphae of P. placenta failed to produce a similar deposition. Extracts from sonicated mycelium of both fungi were positive for catalase activity. Electron micrographs of the white-rot fungus treated with aminotriazole, a catalase inhibitor, failed to produce oxidized DAB deposits. Therefore, the electron-dense regions observed were ascribed to peroxidative oxidation of DAB by catalase. An important control, omitted in this study, was the treatment of the brown-rot fungus with H_2O_2 and DAB to ensure the catalytic competence of the sample. In any case, it is apparent that H_2O_2 does not accumulate in the mycelium of brown-rot fungi as it apparently does in white-rot fungi.

White-rot fungi have been found to produce H_2O_2 and enzymes that produce and utilize hydrogen peroxide under conditions of carbon and/or nitrogen starvation. Highley (1987) proposed that this might be the case for brown-rot fungi as well; i.e., failure to detect H_2O_2 might be due to excess nitrogen or carbohydrate in the growth medium. Therefore, he examined cultures of white- and brown-rot fungi grown on agar under varying conditions of nitrogen and carbohydrate supply. H_2O_2 was detected as coloration of the medium in the presence of either o-dianisidine or 2,2'-azino-di-(3-ethylbenzthiazole-6-sulphonic acid) (ABTS) and horseradish peroxidase, which were included in the medium. Highley found that under conditions of high carbohydrate and high nitrogen, 2 of 13 brown-rot fungi and 5 of 7 white-rot fungi gave positive results with ABTS. All brown-rot fungi were negative in plates containing o-dianisidine, whereas the white-rot fungi gave similar results with both reagents.

By varying the nitrogen source, Highley (1987) was able to detect H_2O_2 production in 6 of 13 brown-rot fungi tested; 7 fungi failed to give a positive reaction regardless of the conditions employed. Of those brown-rot fungi that gave positive results, H_2O_2 production was stimulated somewhat at low nitrogen concentrations. All seven white-rot fungi employed gave positive results. By varying the carbohydate source and concentration (0.02% NH_4NO_3), Highley detected H_2O_2 production in all but 2 of the 13 brown-rot fungi tested. Peroxide production was stimulated by low concentrations of carbohydrate and supported by cellobiose, glucose, mannose, xylose, and malt extract.

Highley and co-workers (1988) were unable to detect H_2O_2 in liquid cultures of P. placenta under various oxygen concentrations (0-1.0 atm O_2) using the titanium reagent or ABTS linked with peroxidase. Similar negative results were obtained with C. puteana (Highley et al., 1989).

Illman et al. (1989) were able to detect H_2O_2 production with ABTS and horseradish peroxidase in stationary liquid cultures of Postia placenta supported by several sugars under conditions of nitrogen limitation. The authors did not report how much time was required for these cultures to develop green discoloration, which was taken as a positive indicator of peroxide, nor was the temporal nature of color development explored. Therefore, we do not know whether the H_2O_2 was produced transiently as the result of the presence of a carbohydrate oxidase for instance or whether H_2O_2 was exported by the mycelium and accumulated in the liquid medium. Hydrogen peroxide production was not strictly dependent on nitrogen limitation.

Using the horseradish peroxidase and ABTS method for detecting peroxide, Micales and Highley (1989) found that H_2O_2 was produced by degradative and nondegradative isolates of P. placenta under a variety of carbon and nitrogen conditions. No correlation was found between H_2O_2 production and the ability of the isolates to degrade wood. A nondegradative isolate (ME-20) was reported to produce large quantities of H_2O_2 but the production was not quantified.

Recently, Veness and Evans (1989) examined liquid cultures of both white-rot and brown-rot basidiomycetes, grown under a variety of conditions, for H_2O_2 production by using an oxygen electrode and catalase. They detected no H_2O_2 in culture fluids of 14- and 28-day-old carboxymethylcellulose and 7-day-old glucose cultures. Similar results were obtained with the titanium(IV) assay. However, Ritschoff et al. (1990) detected H_2O_2 in two of four brown-rot fungi when ABTS-peroxidase was added to a solid, wood-based culture medium.

The first attempt to quantify H_2O_2 production by brown-rot fungi was recently made by Ritschkoff and Viikari (1991a,b). These workers grew liquid cultures of Serpula lacrymans (Sl 1), P. placenta (FPRL 208), and Gl. trabeum (BAM Ebw 109) in liquid culture with 1% crystalline cellulose (Avicel) or amorphous cellulose (prepared from Avicel by phosphoric acid treatment). Using ABTS and horseradish peroxidase, in vivo, the authors measured the intensity of the formed color at 560 nm during the cultivation. They found that when grown on crystalline cellulose as the sole carbon source, S. lacrymans and P. placenta produced peroxide beginning at 4 and 8 days incubation, respectively. Gl. trabeum did not produce significant amounts of peroxide; the slight color that did develop in the cultures after 8 days soon disappeared. In the flasks containing amorphous cellulose, no peroxide was detectable.

The results of these various studies do not lead us to conclude that production of extracellular H_2O_2 is a general characteristic of brown-rot fungi. Several problems persist, the most serious of which is the lack of a selective and reliable assay for H_2O_2. The most serious attack of this problem was made by Veness and Evans (1989), who examined liquid cultures of basidiomycetes with an oxygen electrode and catalase. Previous studies have employed a chromagen such as ABTS or o-dianisidine in combination with horseradish peroxidase, and as noted previously, this system may not be selective for H_2O_2.

This approach allows qualitative assessments, but results have been mixed. White-rot fungi, which are generally conceded to be producers of extracellular H_2O_2, do not always give uniform results (Highley, 1982, 1987). Thus, it is difficult to interpret data from brown-rot fungi in experiments in which only one-half of the fungi tested gives a positive result. A critical examination of Koenigs' data (1972b) was suggested by Veness and Evans (1989). The correlations between three tests for H_2O_2 made by Koenigs (1972b) with sheep's blood agar, hemoglobin agar, and hemoglobin liquid medium are very poor. We agree with Veness and Evans (1989) that it is difficult to assign a meaning to these data.

It has also been noted that some fungi, two white-rot fungi in this instance, may lose the ability to produce H_2O_2 (Highley, 1987) when cultured for long periods. Similarly, loss of phenol oxidase production by the white-rot fungus C. versicolor was found to occur when the fungus was kept in culture for long periods (Szklarz et al., 1989).

It may be that brown-rot fungi produce H_2O_2 transiently or in an environment in which it is sequestered or rapidly utilized, and thus it has escaped detection in cell-free extracts. Only one research group (Ritschkoff and Viikari, 1991a,b) has carefully examined H_2O_2 production as a function of culture age. Most of the results, reported herein, are those obtained at 7 and 14 to 28 days growth (Veness and Evans, 1989) after 6 weeks growth on sawdust (Highley, 1987) or at 7, 14, and 21 days growth in stationary liquid cultures (Highley 1982) or 7 days still-culture (Koenigs, 1972b). Transient production of H_2O_2 may not be detected in such studies.

Expression of the cellulose-depolymerizing activity of brown-rot fungi has not yet been accomplished in a defined liquid medium. Production of H_2O_2 may be tightly coupled to expression of this activity and may not be observed in the absence of cellulose depolymerization. At this time, whether brown-rot basidiomycetes produce extracellular H_2O_2 is not definitively established. The answer to this question awaits the isolation and characterization of enzymes that produce or utilize H_2O_2.

Production of Radical Oxygen Species

Highley (1982) found that filtrates of 21-day-old cultures of brown-rot fungi bleached the spectrum of p-nitrosodimethylaniline when cultured on glucose, suggesting the presence of hydroxyl radical. However, this assay is not a very selective assay for hydroxyl radical. Organic radicals and easily auto-oxidizable compounds can produce the same results.

In soil-block cultures, wood blocks or cellulose impregnated with a number of different chelators and radical and peroxide quenching agents had little effect on weight loss produced by P. placenta, Gl. trabeum or L. olivascens (Highley, 1982; Highley and Murmanis, 1985a). Even blocks impregnated with chloroamphenicols had no effect on wood decay. This result indicates that brown-rot fungi may very well circumvent any toxin that may be present in wood on colonization. Two notable exceptions to this result were thiourea (1.0 M) (<2% weight loss compared to 64% for control) and fluorouracil (0.001 M) (~5% weight loss) (Highley, 1982).

Illman et al. (1989) observed the electron spin resonance signal of the dimethyl-1-pyrroline N-oxide (DMPO) derivative of the hydroxyl radical in liquid cultures of P. placenta supported by cellobiose or glucose. The appearance of this signal was time dependent and did not appear until the tenth day of growth. This signal was also detected in wood slivers of Douglas-fir and white fir. The exact chemical source of the hydroxyl radical was not determined in this study.

To our knowledge, no report of the involvement of peroxy-radical (•OOH) in cellulose depolymerization exists in the literature. This radical is easily generated by protonation of the superoxide anion $(O_2^{\bullet})^-$ whose pK_a is ~4.8. The superoxide anion has been shown to be an efficient degrader of cotton cellulose (Thompson and Corbett, 1985). The DP of 1 g of cotton cellulose in 100 mL of dimethyl sulfoxide (DMSO) containing 15 mg KO_2 is reduced from 6,700 to 313 in 5 hr. The degraded cellulose was not characterized, and how its properties relate to those of brown-rotted cellulose are unknown.

In a general test for the production of free radicals, Veness and Evans (1989) grew white- and brown-rot fungi in liquid culture in flasks whose interior surface had been silvered. Free radicals were detected by the desilvering of these mirrors in the culture flasks. Several white-rot fungi were able to completely desilver the flask below the level of the growth medium completely. Several brown-rot fungi showed little deslivering (Fibroporia vaillantii) or no deslivering (Lentinus lepideus, Serpula lacrymans). Lundborg (1988) suggests that hydroxyl radicals are formed by brown-rot fungi on the basis of depression formation in cellulose agar plates by Fomitopsis pinicola. Depression formation was inhibited by radical scavengers.

As most of these reports concerning free radicals are phenomenological in nature, it is difficult to assign any meaning to them and little can be concluded regarding their significance to brown-rot decay.

Detection of Peroxidases and Oxidases in Brown-Rot Fungi

Koenigs (1972b) reported the detection of peroxidases using o-dianisidine as a substrate in two species of brown-rot fungi, Coniophora puteana and L. lepideus, but the activities reported were very low. Koenigs also reported that production of H_2O_2 by 2- to 3-week-old mycelial mats was supported by glucose. He used the C-AT-G system to detect peroxide, a method that has since been discredited (Highley, 1981). Koenigs suggested that brown-rot fungi support peroxide production from a number of other sugars (xylose, mannose, galatose, etc.), but no report has appeared in the literature detailing these experiments. Koenigs' paper (1972b) summarizes the relevant literature concerning wood-rotting fungal carbohydrate oxidases up through the late 1960s.

Ferm and Cowling (1972) examined three brown-rot fungi for intracellular and extracellular activities of laccase, tyrosinase, and peroxidase. All these activities were detected intracellularly at some point in liquid culture growth. Poria monticola (= Postia placenta) was negative for all three extracellular activities. Very low activities of laccase and tyrosinase activities were detected in extracellular filtrate of Poria cocos (= Wolfiporia cocos) after 3-12 and 3-25 days of growth; peroxidase was detected at day 25. With L. trabea (= Gl. trabeum), no extracellular laccase was detected; low levels of tyrosinase were detected after 17-25 days of growth and peroxidase after 14-25 days.

Highley (1981, 1982) reported that L. olivascens filtrates produced peroxide from glucose. However, he could not detect peroxidase activity in that organism (Highley, 1982) or in other brown-rot fungi. Likewise, Szklarz et al. (1989) did not find any extracellular oxidase or peroxidase in cultures of L. trabea (= Gl. trabeum). Again, these studies were not concerned with a detailed temporal approach, so it can only be concluded that peroxidase activity does not accumulate. Highley and Murmanis (1985b) were unable to find alcohol oxidase activity in sonicated mycelium extracts of P. placenta recovered from liquid culture.

Enoki et al. (1989) reported the ability of brown-rot fungi to oxidize KTBA; KTBA is converted to ethylene by one-electron oxidants such as hydroxyl radical or ferric ammonium sulfate. It is also oxidized by horseradish peroxidase or lignin peroxidase in the presence of H_2O_2. Of 10 brown-rot fungi tested, ethylene production from KTBA was correlated with weight loss from Japanese cedar sapwood and Japanese beech sapwood. Some cultures reached peaks of ethylene production, which subsequently declined; most showed an increase in ethylene production in the first 15 to 30 days, which then reached a plateau. The authors did not correlate ethylene production with cellulose depolymerization.

These workers (Enoki et al., 1989) also showed that degradation of lignin model compounds correlated with ethylene production from KTBA, albeit 8 of the 10 brown-rot fungi showed no degradative activity against lignin model compounds. Cellulose (filter paper) degradation was also correlated with ethylene production. Again, most of the brown-rot fungi showed little activity against cellulose. Based on these results, Enoki and co-workers (1989, 1990) have suggested the existence of "a unique wood-component-degrading system that participates directly or indirectly in the fragmentation of cellulose as well as of lignin in wood and oxidizes KTBA to give ethylene."

Enoki and co-workers (1989) isolated the extracellular proteins from wood-containing cultures of brown-rot fungi and examined KTBA oxidation by these proteins. Under anaerobic conditions, these extracellular proteins produced ethylene in the presence of KTBA and H_2O_2. In the presence of O_2, a reductant such as nicotinamide adenine dinucleotide (NADH) (reduced form) was needed to stimulate ethylene production. This reaction was inhibited by catalase. Enoki and co-workers proposed the existence of an extracellular H_2O_2-requiring protein capable of oxidizing KTBA and of oxidizing NADH in the presence of O_2 to generate H_2O_2. These extracellular proteins were incapable of polymerizing phenols to colored products, in the presence of H_2O_2, ruling out phenoloxidase activity.

Recently, the work of these authors (Enoki et al., 1990) with Gl. trabeum and Tyromyces palustris has led to a partially purified compound reported to be an iron-containing glycoprotein of Mr ~ 1600-2000. It is yet to be established whether the H_2O_2-dependent KTBA-oxidizing ability of these compounds is related to cellulose depolymerization.

REACTION OF FENTON'S REAGENT WITH LIGNIN AND CELLULOSE

A number of studies have concerned themselves with the reaction of wood compounds with the Fenton reagent. It would be of interest to examine the results of these studies and compare these with the known chemistry of brown-rotted lignin and cellulose. Such a comparison will allow us to critically evaluate whether a role for the Fenton reagent is plausible in brown-rot decay. One of the difficulties with interpreting these experiments is the fact that, for the most part, the reactions have been conducted under conditions not typically found in brown-rot. With this caveat in mind, we believe it still instructive to examine the findings.

Reaction With Lignin

Gold and co-workers (1983) synthesized [^{14}C]-ring, [^{14}C]-side-chain, and [^{14}C]-methoxyl labeled dehydropolymerizates (DHPs), model lignin polymers. Exposure of these [^{14}C]DHPs to molar H_2O_2 and 10 mM $FeSO_4$ in dimethyl formamide (DMF) for 20 hr led to extensive depolymerization of all three lignins as demonstrated by gel filtration chromatography on LH-20. This result suggests that free hydroxyl radical is probably not operative in brown-rot fungi since only minor modification of the lignin is observed even after extensive decay by these fungi (Cowling, 1961; Kayama, 1961). Although this result does not rule out a role for the Fenton system, it requires us to conclude that the fungus will need to devise a mechanism that can deliver such a highly reactive species selectively to the cleavage sites within the cellulose molecule and that will leave the lignin polymer virtually intact.

The only other study of Fenton's reagent with lignin that this author is aware of is that of Tatsumi and co-workers (1980). These authors examined the oxidation of specifically [^{14}C]labeled pine kraft lignin with hydrogen peroxide and ferrous salts under alkaline conditions. When small amounts of H_2O_2 (0.5 M) were employed, condensation of monomer to give dimer and oligmer products was observed. When a large amount of hydrogen peroxide (1.32 M) was employed, high molecular weight lignin was considerably degraded to give low molecular weight products. Ring rupture was also observed. Analysis of gaseous products released indicated that a large portion of methoxyl carbon was converted to CO_2 and CH_4. This study suggests that demethylation of lignin by Fenton's reagent is feasible.

Several studies have addressed the action of Fenton's reagent on whole wood. Koenigs (1974a) found that $H_2O_{2/}FeSO_4$ treated sweetgum wood meal resulted in a lignin that possessed markedly increased solubility during acid hydrolysis. Up to 74% of the lignin in sweetgum could be rendered soluble. Interestingly, lignin from loblolly pine was not depolymerized. These results indicate that the substrate may have a profound affect on susceptibility to attack under otherwise identical conditions. On the other hand, Kohdzuma et al.(1991) treated sugi (Cryptomeria japoica D. Don) heartwood and tochinoki (Aesculus turbinata Blume) in acetic acid buffer solution and Fenton's reagent and found that both holocellulose and Klason lignin were extensively degraded. In addition, the change in the density of wood substances as a function of weight loss for brown-rot decay did not match the relationship obtained from wood treated with Fenton's reagent.

Reaction With Model Lignin and Aromatic Compounds

The finding of hydroxylation products in brown-rotted lignin (Kirk et al., 1970) is consistent with a role of hydroxyl radical in lignin modification. Ample evidence exists that the Fenton system under anaerobic conditions and at low pH will hydroxylate aromatic ring systems ortho- and para- to phenolic moieties (Maskos et al., 1990, Walling, 1975). Walling and Johnson (1975) concluded that most, if not all, hydroxyl radical attack on aromatics occurs via addition to yield hydroxycyclohexadienyl radicals and that in the presence of adequate oxidant (Fe^{3+}), phenolics can be made the major products. These authors further concluded that direct side-chain attack cannot be more than a minor reaction path. Halliwell (1978) found that hydroxylation of aromatic compounds by Fenton's reagent was enhanced by the presence of an iron chelator such as EDTA. Reactions of nonphenolic lignin model compounds (Kirk et al., 1985) with Fenton's reagent result predominately in demethylation of aromatic methoxy groups and hydroxylation of the aromatic nucelus.

Reaction With Carbohydrates and Cellulose

Studies have addressed the action of Fenton's reagent on carbohydrates (Moody, 1963, 1964). Although much work has been done on the chemistry of Fenton's reagent with D-glucose, the action of H_2O_2 and ferrous salts on cellulose has received relatively little attention (Moody, 1963, 1964).

In an acidic medium and in the presence of Fe^{2+}, H_2O_2 will oxidize a primary hydroxyl in monohydric alcohols to an aldehyde or carboxyl group (Ivanov et al., 1953). In polyhydric alcohols, primary hydroxyl groups will react first, followed by secondary hydroxyl groups. Presumably, if this mechanism were operative in brown-rot decay, attack would occur in the amorphous region of the cellulose and the reactivity of the primary hydroxyl group would not be diminished by hydrogen bonding, as would be expected of the primary hydroxyl groups in the crystalline region. Thus, uronic acid residues would be expected in the brown-rotted cellulose, but none has been reported (Highkey, 1977; Kirk et al., 1991). In their examination of the reaction of cotton with H_2O_2 and $FeSO_4$ at pH 4.6, Inanov and co-workers concluded that oxidation proceeds intensively in acid medium (pH < 4), that primary hydroxyl are oxidized to aldehyde or carboxyl groups and that secondary hydroxyls groups are oxidized to ketone groups with eventual rupture of the ring to form two aldehyde groups, which may be further oxidized to carboxyl groups.

In a study of the oxidation of methyl β-D-glucospyranoside, probably the simplest model for cellulose, by Fenton's reagent, de Belder and co-workers (1963) identified as products the four carbonyl compounds resulting from oxidation at C-2, C-3, C-4, and C-6, although the predominate reaction is apparently demethylation to form D-glucose. Numerous studies with cellobiose have also been conducted. Reactions of cellobiose with Fenton's reagent (Uchida and Kawakishi, 1988) led to a much more complex product distribution. However, (1Æ6)-linked disaccharides are apparently more reactive with •OH than (1Æ4)-linked disaccharides and α-linkages are more reactive than b-linkages.

The earliest report on cellulose as such simply monitored changes in its viscosity as it is treated with peroxide and various iron salts (Yamafuji and Urakami, 1950). These workers observed a decrease in cellulose viscosity and concluded that the cellulose was depolymerized but nothing else.

Halliwell (1965) examined cotton cellulose degradation by peroxide and ferrous salts under acidic conditions. He found that the cellulose was first broken down to very short fibers within a few days, with the accumulation of only a small amount of soluble products. Solubilization was maximum at pH 4.2-4.3 and greatly retarded at higher (≥5) and lower (≤3) pH values.

Koenigs (1974a) reported that the holocellulose of sweetgum and loblolly pine meal was depolymerized by the action of $H_2O_2/FeSO_4$. Cotton cellulose (Koenigs, 1972a, 1975) was also rapidly depolymerized by Fenton's reagent. High peroxide levels were required to effect efficient depolymerization (1% or 0.29 M), which was accompanied by high weight loss (DP ~180 at 32.7% weight loss). Kohdzuma et al. (1991) also found the holocellulose of wood extensively degraded by Fenton's reagent. Unfortunately, the chemical properties of these holocelluloses were not determined in these studies.

DISCUSSION

Although much work has been done on the cellulose depolymerization mechanism employed by brown-rot fungi, any role for the employment by brown-rot fungi of the H_2O_2/Fe^{2+} system in their depolymerization of wood cellulose remains speculative. No direct evidence points to a role for iron or any other transition metal and reduced oxygen species in the cellulose depolymerization mechanism employed by brown-rot fungi.

Some indirect evidence is consistent with a role for the H_2O_2/Fe^{2+} system. For instance, the finding of demethylation (Kirk and Adler, 1969, 1970) and hydroxylation (Kirk et al., 1970) of brown-rotted lignin is consistent with the presence of hydroxyl radical. Other oxygen species can be responsible for these findings as well (Hall, 1980).

Much data also argue against the presence of the H_2O_2/Fe^{2+} system in brown-rot decay. Probably the strongest argument against its presence in brown-rot fungi is the lack of ability of many researchers to demonstrate convincingly the production of H_2O_2 by cultures of brown-rot fungi in liquid culture; such production has never been demonstrated in solid substrate cultures (although, as pointed out by Koenigs (1972a), the absolute concentration of H_2O_2 at any one time may not be important since effects of H_2O_2 are expected to be cumulative). It is possible that constitutive low levels of H_2O_2 have escaped detection.

Although much work has been performed on the oxidation of cellulose over the past 70 years, nearly all has dealt with reaction in the alkaline environment (Nevell, 1985). Little experimental work has been done under acidic conditions (Daruwalla et al., 1960), and there are gaps in the literature concerning this area of cellulose oxidation. In general, two extreme types of cellulose are generated on oxidation, depending on the reaction conditions (Davidson, 1940). The first and most commonly studied is that generated by oxidants under alkaline conditions. This produces an acidic type of oxycellulose that is characterized by low copper number, high affinity for methylene blue, and increased lability toward acid hydrolysis. The second kind of oxycellulose is generated by oxidation under acidic conditions, which produces a reducing type of oxycellulose characterized by high copper number, excessive weight loss on alkaline boiling, low methylene blue absorption, and resistance to acid hydrolysis.

The action of the H_2O_2/Fe^{2+} system on cellulose is expected to be oxidative in nature, and similar changes have yet to be demonstrated in wood or cotton cellulose. Oxidative changes in brown-rotted lignin are firmly established. As Koenigs has observed (1974b), "it seems necessary to settle the point definitively for cellulose." Indirect evidence for oxidation of cellulose by brown-rot fungi include observation of high copper numbers in depolymerized cellulose (Kayama, 1961; Highley, 1977) and increased alkaline solubility of the depolymerization product (Cowling, 1961; Kayama, 1961,1962a; Highley, 1977). Based on these observations, brown-rot fungi apparently produce a reducing type of oxycellulose. Direct evidence includes detection of carboxyl groups (Highley, 1977) and identification of sugar acids (Kirk et al., 1991) in the depolymerization products. Again, all of this work was performed on highly degraded samples. An increase in copper number does not necessarily mean that the cellulose has been oxidized because acid hydrolysis of cotton cellulose also leads to an increase in copper number. Moreover, increased alkaline solubility only establishes the presence of low molecular weight cleavage components in the sample and may not result from alkaline-sensitive linkages of the kind expected in certain kinds of oxidized cellulose. Thus, unequivocal demonstration of oxidation of cellulose by brown-rot fungi has not been accomplished.

Hydroxyl radical reacts with carbohydrates at nearly diffusion-controlled rates. With cellulose, presumably radicals will be generated that will either eliminate H_2O, react with molecular oxygen to form peroxy radicals, or undergo hydrolysis (von Sonntag, 1980). Elimination of water will generate an oxycellulose that is expected to be labile in alkali and stable in acid, i.e., a reducing type of oxycellulose. Reaction with molecular oxygen will produce peroxycellulose radicals that are expected to readily eliminate $HO_2^{\bullet}$ and give the corresponding carbonyl compound, aldoses and ketoses. Hydrolysis leading to scission of the glycosidic linkage is expected to be fast and will ultimately lead to oxidation products as well.

A number of investigators have noted that the course and specificity of the Fenton reaction may be altered by the presence of substrate, pH, and buffer ions (Smidsrød, et al., 1965; Larson et al., 1967; Zbaida et al., 1988; Bottu, 1991). It has been suggested that in some iron/peroxide systems, in the presence of the EDTA or DTPA, hyroxyl radical is not even produced (Rush and Koppenol, 1986; Rahhal and Richter, 1988; Yamazaki and Piette, 1990) and the reactive species may be an iron-oxo species such as the ferryl ion. Larsen and Smidsrød (1967) determined that an iron/peroxide/substrate complex may be operative in some of their oxidations of polysaccharides by Fenton's reagent. Zbaida and co-workers (1988) found that the presence of various iron chelators can promote either seletive hydroxylation, N-demethylation, or sulfoxidation of cimetidine. Iron chelators with high redox potential, i.e., ethylenediamine, inhibited both hydroxylation and N-demethylation, whereas iron chelators with low redox potential, i.e., EDTA, promoted both hydroxylation and N-demethylation of cimetidine. This suggests that it is at least possible that brown-rot fungi can selectively O-demethylate and hydroxylate lignin with the Fenton system if the proper chelator were produced. Another factor that must be considered is the pH of the medium in which the reaction is conducted, as this will affect the redox potential of Fe(II)/Fe(III), the chelation constants, and the redox potential of H_2O_2. Work on biomimetic systems in an effort to model ligninases lent insight into these questions (Habe et al., 1985; Dolphin et al., 1987). Paszcynski and co-workers (1988) successfully demonstrated that treatment of wood chips and wood pulps with natural hemes and synthetic porphyrins in the presence of tert-butyl hydroperoxide resulted in extensive delignification with very little loss of cellulose. It may be possible, given the appropriate set of conditions, that preferential depolymerization of cellulose can be achieved with a Fenton-type system that results in little or only minor modification of the lignin.

SUMMARY

The mechanism by which brown-rot fungi degrade wood is not well-understood. The prevailing hypothesis implicates low molecular weight transition metal (usually iron) chelates, which act through biochemical conversion into potent oxidizing species, which diffuse into the wood pores and oxidatively cleave the cellulose, or through participation in the biochemical generation of an activated oxygen species (hydroxyl radical or equally potent metallo-oxygen species), in close juxtaposition to the cellulose, which oxidizes the cellulose leaving the lignin substantially unchanged.

This review has addressed chemical changes in wood components and cotton cellulose as a result of brown-rot fungal degradation. The literature indicates that chemical changes in brown-rotted lignin are oxidative in nature and result primarily in hydroxylation and demethylation of the lignin polymer. Results from studies on the chemical changes in cellulose are less definitive, and oxidative changes in this substrate as a result of brown-rot decay have not been unequivocally demonstrated.

A review of the effects of Fenton's reagent on wood components and model compounds suggests that the effects of brown-rot decay are not duplicated by this reagent in many respects and therefore, it does not represent an adequate model for brown-rot decay.

REFERENCES

Akamatsu, Y., Ohta, A., Takahashi, M., and Shimada, M. (1991). Enzymatic formation of oxalate from oxaloacetate with cell-free extracts of the brown-rot fungus Tyromyces palustris in relation to the biodegradation of cellulose. Mokuzai Gakkaishi, 37(6), 575–577.

Ander, P., Eriksson, K.E., and Yu, H.S. (1984). Metabolism of lignin-derived aromatic acids by wood-rotting fungi. J. Gen. Microbiol., 130, 63–68.

Battista, O.A. (1950). Hydrolysis and crystallization of cellulose. Ind. Eng. Chem., 42, 502–507.

Bottu, G. (1991). The effect of buffers and chelators on the reaction of liminol with Fenton's reagent near neutral pH. J. Biolumin. Chemilumin., 6, 147–151.

Bray, M.W. and Andrews, T.M. (1924). Chemical changes of groundwood during decay. Ind. Eng. Chem., 16, 137–139.

Brown, W., Cowling, E.B. and Falkehag, S.I. (1968). Molecular size distributions of lignins liberated enzymatically from wood. Svensk Papperstidn., 71, 811–821.

Chandhoke, V., Jellison, J., Goodell, B., and Fekete, F. (1991). Action of siderophores from Gloeophyllum trabeum on 2-keto-4-thiomethylbutyric acid and cellulose azure substrates. Phytopathology, 81(1), 121.

Chang, M.M., Chou, T.Y.C., and Tsao, G.T. (1981). Structure, pretreatment and hydrolysis of cellulose. pp. 15–42, In: Advances in Biochemical Engineering.(A. Ziechter, ed.), Springer-Verlag, Berlin.

Cobb, C.E. (1981). The nonenzymatic decomposition of cellulose by the brown-rot fungus Gloeophyllum trabeum. M.S. Thesis, Michigan Technological University.

Cowling, E.B. (1961). Comparative biochemistry of the decay of sweetgum sapwood by white-rot and brown-rot fungi. Tech. Bull. No. 1258, U.S. Department of Agriculture, Washington, D.C. 79 pp.

Cowling, E.B. and Brown, W. (1969). Structural features of cellulosic materials in relation to enzymatic hydrolysis. In Cellulases and Their Applications. pp. 152-187. (R.F. Gould, ed.), American Chemical Society Advances in Chemistry Series 95, Washington, D.C.

Daruwalla, E.H., Kangle, P.J., and Nabar, G.M. (1960). Acid hydrolysis of chemically modified celluloses. Text. Res. J., 30, 469–471.

Davidson, G.F. (1934). The dissolution of chemically modified cotton cellulose in alkaline solutions. Part I—In solutions of sodium hydroxide, particularly at temperatures below the normal. J. Textile Inst., 25, T174–T196.

Davidson, G.F. (1936). The molecular structure of cellulose. Part I. A review of current theories. Part II. The evidence of the chemically modified cotton celluloses. J. Textile Inst., 27, P144–P168.

Davidson, G.F. (1940). The modification of cellulose by oxidants. J. Soc. Dyers and Colourists, 56, 58–63.

de Belder, A.N., Lindberg, B., and Theander, O. (1963). The oxidation of glycosides XIII. The oxidation of methyl b-D-glucopyranoside with fenton's reagent. Acta. Chem. Scand. 17, 1012–1014.

Dolphin, D., Nakano, T., Maione, T.E., Kirk, T.K., and Farrell, R. (1987). Synthetic model ligninases. pp.157-162. In: Lignin enzymatic and microbial degradation: Symposium International; 1987 April 23–24; Paris: Institut Natioanl De La Recherche Agronomique.

Enoki, A., Takahashi, M., Tanaka, H., and Fuse, G. (1985). Degradation of lignin-related compounds and wood components by white-rot and brown-rot fungi. Mokuzai Gakkaishi, 31(5), 397–408.

Enoki, A., Tana H., and Fuse, G. (1988). Degradation of lignin related compounds, pure cellulose, and wood components by white-rot and brown-rot fungi. Holzforschung, 42, 85–93.

Enoki, A., Tanaka, H., and Fuse, G. (1989). Relationship between degradation of wood and production of H_2O_2-producing one-electron oxidases by brown-rot fungi. Wood Sci. Technol., 23, 1–12.

Enoki, A., Yoshioka, S., Tanaka, H., and Fuse, G. (1990). Extracellular H_2O_2-producing and one-electron oxidation system of brown-rot fungi. The International Research Group on Wood Preservation, Document No. IRG/WP/1445, IRG Secretariat, Box 5607, S–114 86 Stockholm, Sweden.

Enoki, A., Fuse, G., and Tanaka, H. (1991). Extracellular H_2O_2-producing and H_2O_2-reducing compounds of wood decay fungi. The International Research Group on Wood Preservation, Document No. IRG/WP/1516, IRG Secretariat, Box 5607, S–114 86 Stockholm, Sweden.

Eriksson, K.-E. L., Blanchette, R.A., and Ander, P. (1990). Microbial and Enzymatic Degradation of Wood and Wood Components, Springer-Verlag, Berlin, p. 122.

Fekete, F.A., Chandhoke, V., and Jellison, J. (1989). Iron-binding compounds produced by wood-decaying basidiomycetes. Appl. Environ. Microbiol., 55(10), 2720–2722.

Ferm, R. and Cowling, E.B. (1972). A new procedure for analysis of phenol-oxidizing enzymes in wood-destroying fungi. Svensk Papperstidning, 75, 767–772.

Flournoy, D.S. (1991). Changes in pore structure and cell wall volume in wood decayed by brown- and white-rot fungi. The International Research Group on Wood Preservation IRG Document No. IRG/WP/1501, IRG Secretariat, Box 5607, S-114 86 Stockholm, Sweden.

Flournoy, D.S., Kirk, T.K., and Highley, T.L. (1991). Wood decay by brown-rot fungi: Changes in pore structure and cell wall volume. Holzforschung, 45(5), 383–388.

Gold, M.H., Kutsuki, H., and Morganm, M.A. (1983). Oxidative degradation of lignin by photochemical and chemical radical generating systems. Photochem. Photobiol., 38(6), 647–651.

Goring, D.A.I. and Timell, T.E. (1962). Molecular weight of native celluloses. TAPPI, 45, 454–460.

Habe, T., Shimada, M., and Higuchi, T. (1985). Biomimetic approach to lignin degradation. I. H_2O_2-dependent C-C bond cleavage of the lignin model compounds with a natural iron porphyrin and imidazole complex. Mokuzai Gakkaishi, 31(1), 54–55.

Haider, K. and Trojanowski, T. (1980). A comparison of the degradation of 14C-labeled DHP and corn stalk lignins by micro- and macrofungi and bacteria. In: Lignin Biodegradation: Microbiology, Chemistry and Potential Applications. pp.111–134.(Kirk, T.K., Huguchi, T, Chang, H.-M., eds.), Vol 1, Boca Raton, FL, CRC Press.

Hall, P.L. (1980). Enzymatic transformations of lignin. 2. Enzyme Microb. Technol., 2, 170–176.

Halliwell, G. (1965). Catalytic decomposition of cellulose under biological conditions. Biochem. J., 95, 35–40.

Halliwell, G. (1978). Superoxide-dependent formation of hydroxyl radicals in the presence of iron chelates. Is it a mechanism for hydroxyl radical production in biochemical systems? FEBS Lett., 92(2), 321–326.

Halliwell, B. and Gutteridge, J.M.C. (1988). Iron as a biological pro-oxidant. ISI Atlas of Sci.: Bioch., 1, 48–52.

Herr, D., Baumer, F., and Dellweg, H. (1978). Purification and properties of an extracellular endo-1,4-b-glucanase from Lenzites trabea. Arch. Microbiol., 117, 287–292.

Highley, T.L. (1975). Properties of cellulases of two brown-rot fungi and two white-rot fungi. Wood and Fib., 6(4), 275–281.

Highley, T.L. (1977). Requirements for cellulose degradation by a brown-rot fungus. Mat. und Org., 12, 25–36.

Highley, T.L. (1981). Catalase-aminotriazole assay—an invalid method for measurement of hydrogen peroxide by wood-decay fungi. Appl. Environ. Microbiol., 42, 925–927.

Highley, T.L. (1982). Is extracellular hydrogen peroxide involved in cellulose degradation by brown-rot fungi? Mater. und Org., 17(3), 205–214.

Highley, T.L. (1987). Effect of carbohydrate and nitrogen on hydrogen peroxide formation by wood decay fungi in solid medium. FEMS Microbiology Letters, 48, 373–377.

Highley, T.L. and Murmanis, L.L. (1985a). Involvement of hydrogen peroxide in wood decay by brown-rot and white-rot fungi. The International Research Group on Wood Preservation, IRG Document No. IRG/WP/1256, IRG Secretariat, Box 5607, S–114 86 Stockholm, Sweden.

Highley, T.L. and Murmanis, L.L. (1985b). Determination of hydrogen peroxide production in Coriolus versicolor and Poria placenta during wood degradation. Mater. und Org., 20(4), 241–252.

Highley, T.L. and Wolter, K.E. (1982). Properties of a carbohydrate-degrading enzyme complex from the brown-rot fungus Poria placenta. Mater. und Org., 17(2), 127–134.

Highley, T.L., Wolter, K.E., and Evans, F.J. (1981). Polysaccharide-degrading complex produced in wood and in liquid media by the brown-rot fungus Poria placenta. Wood and Fib., 13(4), 265–274.

Highley, T.L., Ibach, R., and Kirk, T.K. (1988). Properties of cellulose degraded by the brown-rot fungus, Postia placenta. The International Research Group on Wood Preservation, IRG Document No. IRG/WP/1350, IRG Secretariat, Box 5607, S–114 86 Stockholm, Sweden.

Highley, T.L., Kirk, T.K., and Ibach, R. (1989). Effects of brown-rot fungi on cellulose. In: Biodeterioration Research II. pp. 511-525 (C.E. O'Rear and G.C. Llewellyn, eds.), General biodeterioration, degradation, mycotoxins, biotoxins, and wood decay. Proceedings, 2d meeting, Pan American Biodeterioration Society: 1989 July 28–31; Washington D.C., Plenum Press, N.Y.

Illman, B.L., Meinholtz, D.C., and Highley, T.L. (1989). Oxygen free radical detection in wood colonized by the brown-rot fungus, Postia placenta.In: Biodeterioration Research II. General biodeterioration, degradation, mycotoxins, biotoxins, and wood decay., pp. 497–509,.(C.E. O'Rear and G.C. Llewellyn, eds.), Proc., 2d meeting, Pan American Biodeterioration Society: 1989 July 28–31; Washington, DC, Plenum Press, N.Y.

Ivanov, V.I., Kaverzneva, E.D., and Kuznetsova, Z.I. (1953). Chemical changes produced in the cellulose macromolecule by oxidizing agents. Communication 8. Chemical changes in cellulose produced by oxidation with hydrogen peroxide. Div. Chem. Sci. Acad. Sci. USSR Bull. (Engl. Transl.), 2, 341–350.

Jellison, J., Goodell, B., Fekete, F., and Chandhoke, V. (1990). Fungal siderophores and their role in wood biodegradation. The International Research Group on Wood Preservation, Document No. IRG/WP/1442, IRG Secretariat, Box 5607, S–114 86 Stockholm, Sweden.

Jellison, J., Chandhoke, V., Goodell, B., and Fekete, F. (1991). The action of siderophores isolated from Gloeophyllum trabeum on the structure and crystallinity of cellulose compounds. The International Research Group on Wood Preservation, Document No. IRG/WP/1479, IRG Secretariat, Box 5607, S–114 86 Stockholm, Sweden.

Jin, L., Schultz T.P., and Nicholas, D.D. (1990a). Structural characterization of brown-rotted lignin. Holzforschung, 44(2), 133–138.

Jin, L., Nicholas, D.D., and Kirk, T.K. (1990b). Mineralization of the methoxyl carbon of isolated lignin by brown-rot fungi under solid substate conditions. Wood Sci. Techol., 24(3), 277–288.

Kayama, T. (1961). Chemical studies on decayed wood. I. The chemical composition of decayed wood and some properties of resultant pulps. Mokuzai Gakkaishi, 7(3), 161–166.

Kayama, T. (1962a). Chemical studies on decayed wood as a raw material for pulp. IV. The behavior of decayed wood cellulose during pulping process (sulfite cooking and bleaching). Mokuzai Gakkaishi, 8(2), 87–90.

Kayama, T. (1962b). Chemical studies on decayed wood as a raw material for pulp. V. Progressive changes in degree of polymerization of decayed wood cellulose and effect of decay on degree of polymerization and crystalline region of pulp from decayed wood. Mokuzai Gakkaishi, 8(5), 197–203.

Kirk, T.K. (1975). Effects of brown-rot fungus, Lenzites trabea, on lignin spruce wood. Holzforschung, 29(3), 99–107.

Kirk, T.K. and Adler, E. (1969). Catechol moieties in enzymatically liberated lignin. Acta Chem. Scand., 23, 705–707.

Kirk, T.K. and Adler, E. (1970). Methoxyl-deficient structural elements in lignin of sweetgum decayed by a brown-rot fungus. Acta Chem. Scand., 24, 3379–3390.

Kirk, T.K, Larsson, S., and Miksceal, G.E.. (1970). Aromatic hydroxylation resulting from attack of lignin by a brown-rot fungus. Acta Chem. Scand,. 24, 1470–1472.

Kirk, T.K., Connors, W.J., Bleamm R.D., Hackett, W.F., and Zeikus, J.G. (1975). Preparation and microbial decomposition of synthetic [^{14}C] lignins. Proc. Natl. Acad. Sci., USA 72(7), 2515–2519.

Kirk, T.K., Mozuch, M.D., and Tien, M. (1985). Free hydroxyl radical is not involved in an important reaction of lignin degradation by Phanerochaete chrysosporium Burds. Biochem. J., 226, 455–460.

Kirk, T.K., Highley, T.L., Ibach, R., and Mozuch, M.D. (1989). Identification of terminal structures in cellulose degraded by the brown-rot fungus Postia placenta. The International Research Group on Wood Preservation, Document No. IRG/WP/1389, IRG Secretariat, Box 5607, S–114 86 Stockholm, Sweden.

Kirk, T.K., Ibach R.E., Mozuch, M.D., Conner, A.H., and Highley, T.L. (1991). Characterization of cotton cellulose depolymerized by a brown-rot fungus, by acid, or by chemical oxidants. Holzforschung, 45(4), 239–244.

Koenigs, J.W. (1972a). Effects of hydrogen peroxide on cellulose and on its susceptibility to cellulase. Mat. und. Org., 7, 133–147.

Koenigs, J.W. (1972b). Production of extracellular hydrogen peroxide by wood-rotting fungi. Phytopathology, 62, 100–110.

Koenigs, J.W. (1974a). Hydrogen peroxide and iron: A proposed system for decomposition of wood by brown-rot basidiomycetes. Wood Fib., 6(1), 66–80.

Koenigs, J.W. (1974b). Production of hydrogen peroxide by wood-rotting fungi in wood and its correlation with weight loss, depolymerization, and pH changes. Arch. Microbiol., 99, 129–145.

Koenigs, J.W. (1975). Hydrogen peroxide and iron: A microbial cellulolytic system? Biotechnol. Bioeng. Symp. No. 5, 151–159.

Kohdzuma, Y., Minato, K, Katayama, Y., Okamura, K. (1991). Preparation of artificial waterlogged wood. II. Comparison of some characteristics of wood degraded by Fenton's reagent with those of waterlogged wood. Mokuzai Gakkaishi, 37(5), 473–480.

Larsen, B. and Smidsrød, O. (1967). The effect of pH and buffer ions on the degradation of carbohydrates by Fenton's reagent. Acta Chem. Scand., 21(2), 552–564.

Liese, W. (1970). Ultrastructural aspects of woody tissue disintegration. Ann. Rev. Phytopathol., 8, 231–258.

Lundborg, A. (1988). Deformation of agar by wood decaying fungi--a possible indication of the occurrence of radicals. Mater. und Org., 23(4), 259–269.

Maskos, Z., Rush, J.D., and Koppeno, W.H. (1990). The hydroxylation of the salicylate anion by a Fenton reaction and G-radiolysis: A consideration of the respective mechanisms. J. Free Radical Biol. Med., 8, 153–162.

Micales, J.A. and Highley, T.L. (1989). Physiological characteristics of a non-degradative isolate of Postia (=Poria) placenta. Mycologia, 81(2), 205-215.

Moody, G.J. (1963). The action of Fenton's reagent on carbohydrates. Tetrahedron, 19, 1705–1710.

Moody, G.J. (1964). The action of hydrogen peroxide on carbohydrates and related compounds. Advances in Carbohydrate Chem., 19, 149–179.

Nevell, T.P. (1985). Oxidation of cellulose. In: Cellulose Chemistry and its Applications. pp. 243–265 (T.P. Nevell and S.H. Zeronian, eds.), Ellis Horwood Limited, Chichester, England.

Nilsson, T. (1974). Comparative study on the cellulolytic activity of white-rot and brown-rot fungi. Mater. und Org., 9, 173–198.

Paszcynski, A., Crawford, R.L., and Blanchette, R.A. (1988). Delignification of wood chips and pulps by using natural and synthetic porphyrins: Models of fungal decay. Appl. Environ. Microb., 54(1), 62–68.

Phillip, B., Dan D.C., and Fink, H.P. (1981). Acid and enzymatic hydrolysis of cellulose in relation to its physical structure. Proc. Inter. Symp. on Wood and Pulping Chem., (Stockholm) 4, 79–83.

Rahhal, S. and Richter, H.W. (1988). Reduction of hydrogen peroxide by the ferrous iron chelate of diethylenetriamine-N,N,N',N'',N''-pentaacetate. J. Am. Chem. Soc., 110, 3126–3133.

Ritschkoff, A.-C., Paajanen L., and Viikari, L. (1990). The production of extracellular hydrogen peroxide by some brown-rot fungi. The International Research Group on Wood Preservation, IRG Document No. IRG/WP/1446, IRG Secretariat, Box 5607, S–114 86 Stockholm, Sweden.

Ritschkoff, A.-C. and Viikari, L. (1991a). The production of extracellular hydrogen peroxide production by brown-rot fungi. Mat. und Org., 26(2), 157–167.

Ritschkoff, A.-C. and Viikari, L. (1991b). The influence of crystalline and amorphous cellulose on extracellular hydrogen peroxide production by brown-rot fungi. The International Research Group on Wood Preservation, IRG Document No. IRG/WP/1482, IRG Secretariat, Box 5607, S–114 86 Stockholm, Sweden.

Rush, J.D. and Koppenol, W.H. (1986). Oxidizing intermediates in the reaction of ferrous EDTA with hydrogen peroxide. J. Biol. Chem., 261(15), 6730–6733.

Schmidt, C.J. (1980). The role of oxalic acid in the nonenzymatic decomposition of cellulose. M.S. Thesis, Michigan Technological University.

Schmidt, C.J., Whitten B.K., and Nicholas, D.D. (1981). A proposed role for oxalic acid in non-enzymatic wood decay by brown-rot fungi. Proc. Am. Wood Preserv. Assoc., 77, 157–164.

Shimada, M., Akamatsu, Y., Ohta, A., and Takahashi, M. (1991). Biochemical relationships between biodegradation of cellulose and formation of oxalic acid in brown-rot wood decay. The International Research Group on Wood Preservation, IRG Document No. IRG/WP/1472, IRG Secretariat, Box 5607, S–114 86 Stockholm, Sweden.

Smidsrød, O., Haug, A., and Larsen, B. (1965). Kinetic studies on the degradation of alginic acid by hydrogen peroxide in the presence of iron salts. Acta Chem. Scand., 19, 143–152.

Szklarz, G.D., Antibus, R.F., Sinsabaugh, R.L., and Linkins, A.E. (1989). Production of phenol oxidases and peroxidases by wood-rotting fungi. Mycologia, 81(2), 234–240.

Takao, S. (1965). Organic acid production by basidiomycetes. Applied Microbiology, 13, 732–737.

Tatsumi, K., Araki, H., and Terashima, N. (1980). Oxidative degradation of lignin. II. Degradation of specifically ^{14}C-labeled kraft lignin with hydrogen peroxide and ferrous salts. Mokuzai Gakkaishi, 26(5), 327–333.

Thompson, N.S. and Corbett, H.M. (1985). The effect of potassium superoxide on cellulose. TAPPI, 68(12), 68–71.

Uchida, K. and Kawakishi, S. (1988). Interaction of (1Æ4)- and (1Æ6)-linked disaccharides with the Fenton reagent under physiological conditions. Carbohydr. Res,. 173, 89–99.

Veness, R.G. and Evans, C.S. (1989). The role of hydrogen peroxide in the degradation of crystalline cellulose by basidiomycete fungi. J. Gen. Microbiol., 135, 2799–2806.

von Sonntag, C. (1980). Free-radical reactions of carbohydrates as studied by radiation techniques. Advances in Carbohydrate Chem. and Biochem., 37, 7–77.

Walling, C. (1975). Fenton's reagent revisited. Acc. Chem. Res., 8, 125–131.

Walling, C. and Johnson, R.A. (1975). Fenton's reagent. V. Hydroxylation and side-chain cleavage of aromatics. J. Chem. Soc., 97(2), 363–367.

Wolter, K.E., Highley, T.L., and Evans, F.J. (1980). A unique polysaccharide- and glucoside-degrading enzyme complex from the wood-decay fungus Poria placenta. Biochem. Biophys. Res. Commun., 97(4), 1499–1504.

Yamafuji, K. and Urakami, M. (1950). The action of iron and hydrogen peroxide on polysaccharides. J. Fac. Agr. (Kyushu Univ.), 9, 333–339. Chem Abstr., 46, 11117.

Yamazaki, I. and Piette, L.H. (1990). ESR spin-trapping studies on the reaction of Fe^{2+} ions with H_2O_2-reactive species in oxygen toxicity in biology. J. Biol. Chem., 265(23), 13589–13594.

Zbaida, S., Kariv, R., and Fischer, P. (1988). The role of iron chelates on the selectivity of Fenton reagent in hydroxylation, N-demethylation, and sulfoxidation of cimetidine: A novel biomimetic model for the regioselectivity of cytochrome P-450. Arch. Biochem. Biophys., 261, 12–15.

COLONIZATION AND CONTROL OF DECAY BY *TRICHODERMA* IN DOUGLAS-FIR AND SOUTHERN PINE EXPOSED ABOVE GROUND

TERRY L. HIGHLEY, USDA Forest Service, Forest Products Laboratory, One Gifford Pinchot Drive, Madison, WI 53705-2398, USA

INTRODUCTION

Using biological controls to protect wood against fungal attack is an attractive alternative to chemical biocides because of their environmental effects. Much research on biological control of wood-attacking fungi has involved Trichoderma spp. and, to a lesser extent, Scytalidium spp. (Freitag et al., 1991). Laboratory studies of antagonism between Trichoderma, Scytalidium. and wood decay fungi have been promising, but field studies have met with mixed results.

Glaser et al. (1969) reported that the presence of living T. viride prevented decay of pine logs by Stereum sanguinolentum. Lindgren and Harvey (1952) noticed reduced decay in Southern Pine pulpwood infected with Trichoderma. The application of T. viride to the ends of freshly cut birch bolts protected the bolts from decay (Shields and Atwell, 1963; Hume and Shields, 1975).

Most subsequent field tests have employed Binab-T powders or pellets, which are formulated in Sweden by Bio-Innovation AB (Binab). The Binab-T products have been approved by the U.S. Environmental Protection Agency for non-agricultural uses. The powders contain propagules of Trichoderma spp. (ATCC 20475 and 20476, American Type Culture Collection). By combining the species, the temperature at which Trichoderma will suppress casual organisms is broadened to a range of 4C to 27C. Most 2 species stop growing below 10C to 15C. The powder is mixed with water and sprayed or brushed on the wood.

Biodeterioration Research 4, Edited by G.C. Llewellyn *et al.*, Plenum Press, New York, 1994

Pellets are inserted into the wood. Bruce and King (1986a) destructively sampled 166 poled inoculated with Binab-T and found Trichoderma established in 94 percent of the poles. Trichoderma could be isolated for up to 4 years following inoculation. Decay was reduced by Trichoderma but establishment of biocontrol agent was somewhat inhibited by other fungi (Bruce and King 1986b). Wood removed from the interior of a Trichoderma-treated pole 6 years after inoculation was found to be resistant to decay by Neolentinus lepideus and Antrodia carbonica. However, resistance was lost when the wood was sterilized (Bruce et al., 1991).

Morris et al. (1984) found Binab-T ineffective as a remedial treatment for N. lepideus in poles. Trichoderma effectively colonized the poles, but N. lepideus was detected in as many poles as the controls. These authors conclude that the Binab system might have some application as a protective treatment but is ineffective as a remedial treatment.

The objective of our study was to determine the ability of Binab-T pellets and wettable powder to colonize and prevent decay in Douglas-fir and Southern Pine above ground..

MATERIALS AND METHODS

Test units were inoculated with Binab-T wettable powder or pellets (Bio-Innovation AB (Binab), Algaras, Sweden). The pellets contained propagules of Trichoderma polysporum ATCC 20475 and T. harzianum ATCC 29476.

Squared Timbers

Untreated Douglas-fir and Southern Pine timbers, 15.2cm by 4.26m, were inoculated with Binab-T pellets placed into holes drilled on the upper surface. The holes were plugged with a wooden plug. The timbers were exposed on racks elevated above ground at the Valley View Exposure Site near Madison, Wisconsin, or at the Harrison National Exposure Site near Gulfport, Mississippi. Only Southern Pine timbers were exposed in Mississippi.

Treated (ammoniacal-copper-arsenic and creosote) Douglas-fir curbs on a pier near Seattle, Washington, were inoculated with Binab-T pellets in the same manner as the timbers exposed at the other sites.

Piling

Creosoted Douglas-fir piles were cut into 0.6m sections with a chainsaw and the sections randomized into different groups. The pile sections were placed upright on the ground. Binab-T wettable powder was applied by brush to the upper surface of the pile. The wettable powder was also applied combined with 5 percent ammonium bifluoride, 10 percent glucose, or both. The piles were exposed at the Gulfport site.

L-Joints

L-joint test units were constructed of Southern Pine according to Carey et al. (1981). the design of the L-joint favors penetration of rainwater into the wood end grain. Ammonium bifluoride (5 percent) was brushed on the surface of all test units. In some units Trichoderma was applied to the joint areas as follows: pellets alone, pellets plus wettable powder, and wettable powder alone. Units were exposed at the Gulfport site for 4 years.

Decay Resistance

Areas from Southern Pine timbers exposed at the Madison test site that were colonized by Trichoderma were removed for subsequent exposure to decay fungi. Blocks removed from the timbers were evaluated for decay resistance using the standard American Society for Testing and Materials (ASTM) soil-block test (1971). Both unsterilized and steam-sterilized blocks were tested. Postia placenta M. Lars. et Lomb (MAD-698), Neolentinus (=Lentinus) lepideus (Fr.:Fr.) Redhead and Ginns (MAD-534), and Trametes (=Coriolus), versicolor (L:Fr.) Pilate (MAD-697) were the test fungi. After incubation at 27C and 70 percent relative humidity for 12 weeks, the blocks were conditioned and weighed, and weight losses were calculated to measure the extent of decay. Five replications were used for each treatment evaluated.

Treatment of Log Ends

Green Southern Pine log sections, 2.54cm thick, were flooded with a blend of Trichoderma (P74H) in 1 percent malt extract plus 0.2 percent yeast extract. The sections were exposed at the Madison site for 5 months and evaluated for sapstain discolorations in the interior wood.

RESULTS

Squared Timbers

The Southern Pine timbers treated with Binab-T pellets were well-colonized by Trichoderma (Table 1) at both the Madison and Gulfport sites. Recovery of Trichoderma reached a maximum at 20 months, but fell below 50 percent at 3 years.

Table 1. Percentage of Moisture Content and Trichoderma Recovery in Southern Pine and Douglas-fir Squared Timbers Exposed Above Ground.

		Southern Pine		Douglas-fir	
Exposure Site	Time (months)	Recovery	Moisture content	Recovery	Moisture content
Madison	0	-	19	-	17
	8	64	20	8	17
	12	72	-	65	-
	20	96	30+	39	19
	36	46	30+	37	17
Gulfport	0	-	30+	-	-
	6	32	30+	-	-
	20	76	30+	-	-
	36	26	30+	-	-
Seattle	0	-	-	-	21-30+
	12	-	-	37	21-30+
	24	-	-	0	21-30+

Trichoderma colonization was not as prolific in the Douglas-fir timbers as in the Southern Pine (Table 1). Maximum recovery was at 12 months in timbers exposed in Madison, but isolation of Trichoderma dropped considerably in subsequent samplings. Colonization of the Douglas-fir timbers in Seattle was much poorer than that of timbers exposed in Madison, despite very high moisture content in the Seattle timbers. the Seattle timbers were much older than the Madison timbers, and competition from other wood-inhabiting fungi might have been a factor in the poor colonization.

The spread of Trichoderma from the point of pellet insertion was determined in Douglas-fir and Southern Pine timbers exposed above ground in Madison (Table 2). As in the other Southern Pine timbers, maximum recovery occurred at 20 months. The Southern Pine was heavily colonized with Trichoderma at every distance from pellet installation. Recovery of Trichoderma dropped at every distance at 36 months. In Douglas-fir, maximum recovery of Trichoderma occurred at 8 months and dropped substantially by 36 months (Table 2). Trichoderma colonized Douglas-fir 2.54cm from pellet installation, but recovery at greater distances was very poor.

Although timbers were not tested for the presence of the Trichoderma beyond 36 months, they were left in place for observation of decay. After 5 years, decay was not present in any of the Trichoderma-treated or untreated timbers of Douglas-fir. However, visual decay developed in both Trichoderma-treated and untreated controls of Southern Pine by 5 years. Gloeophyllum trabeum fruiting bodies were observed on the pine timbers exposed in Madison; this was the only fungus isolated from the timbers.

Table 2. Recovery of Trichoderma From Exposed Southern Pine and Douglas-fir at Various Distances From Pellet Installation.

	Recovery of Trichoderma (percent) at various distances from pellet installation					
Southern Pine						
Time (months	2.54cm	0.3m	0.6m	0.9m	1.8m	Moisture content (percent)
0	-	-	-	-	-	19
8	16	42	50	8	8	25-40
20	92	100	83	75	75	28-30+
36	67	75	66	50	42	28-30+
Douglas-fir[a]						
0	-	-	-	-	-	17-20
8	75	17	0	8	0	18-22
20	69	8	25	8	8	23-26
36	42	0	17	17	8	24-30+

[a]Douglas-fir timbers exposed in ground contact.

Piling

The heartwood interior of the Douglas-fir piling was poorly colonized by Trichoderma (Table 3). Glucose and ammonium bifluoride supplements were of no benefit in establishing . Trichoderma was not present in any of the Douglas-fir piling at 5 years. Decay was present in many piles by this time.

Table 3. Recovery of Trichoderma and Presence of Decay in Piling Tops Treated With Binab-T Wettable Powder.

Treatment	Total Number of Piles	Number of Piles with Trichoderma[a] (at 16 months)	Number of Piles with decay (at 5 years)
Ammonium bifluoride + Trichoderma	11	2	8
Ammonium bifluoride	5	0	0
Ammonium bifluoride +glucose +Trichoderma	5	1	3
Trichoderma	9	4	1
Trichoderma + glucose	6	0	3
Control	5	1	3

[a]Trichoderma was not isolated from the piles at 5 years of exposure.

L-Joints

Ammonium biofluoride (5 percent) was applied to the joint area of the L-joints because fluoride is reported to stimulate Trichoderma (Lindgren and Harvey, 1952). The effect of ammonium biofluoride on stimulation of Trichoderma could not be determined because ammonium biofluoride alone protected the L-joints during the 4 year exposure.

Trichoderma was isolated from all units throughout the 4 year exposure at the Gulfport site (Table 4). However, decay was present by 2 years in some units that were not treated with ammonium biofluoride.

Table 4. Decay of Southern Pine L-Joints Treated with Binab-T Pellets. Binab-T Wettable Powder, and Ammonium Biofluoride.

Treatment		Decay Rating[a]			
Ammonium biofluoride[b]	Binab-T	1 year	2 years	3 years	4 years
Yes	No	0	0	0	0
No	No	0	0	0	0-20
No	Pellets	0	0-40	0-40	0-80
Yes	Pellets	0	0	0	0
No	Powder	0	0	0	0-40
Yes	Powder	0	0	0	0
No	Pellets and Powder	0	0	0	40
Yes	Pellets and Powder	0	0	0	0
Control[c]	-	0	20	60	60

[a]O-no evidence of decay, 20-decay suspected, 40-decay definite, but limited, 60-general decay, 80-decay warrants replacement of unit, 100-failure.

[b]All test units received surface treatment with 5 percent ammonium biofluoride. Some units were treated in the joint area as well.

[c]No surface or joint treatment with ammonium biofluoride or Binab-T.

Decay Resistance

Both unsterilized and sterilized wood removed from Trichoderma-colonized Southern Pine timbers after 3 years was not resistant to decay by Postia placenta or Trametes versicolor in soil-block tests (Table 5). Weight loss in unsterilized Trichoderma-treated pine caused by Neolentinus lepideus was considerably reduced compared to that of untreated controls. However, decay resistance was lost when the wood was sterilized.

Table 5. Decay Resistance of Trichoderma-Treated Southern Pine Timbers.

Decay fungus	Weight loss (percent)		
	Unsterilized treated timbers	Steam-sterilized treated timbers	Untreated controls
Neolentinus lepideus	11	33	39
Postia placenta	54	57	54
Trametes versicolor	6	13	7

DISCUSSION

In this study, we examined the ability of a commercial preparation of Trichoderma to colonize and survive in Douglas-fir and Southern Pine timbers exposed above ground in various climatic areas. Under field conditions Trichoderma readily colonized Southern Pine (mostly sapwood) squared timbers and L-joints, but the percentage of recovery in the squared timbers dropped considerably in the third year of exposure. Trichoderma was isolated up to 1.8m from the point of pellet insertion in the Southern Pine timbers. Colonization of the Douglas-fir heartwood by Trichoderma is squared timbers and piling exposed above ground was much poorer than that in Southern Pine. Douglas-fir timbers exposed in ground contact in Madison were better colonized, but recovery was very poor beyond 2.54cm from pellet insertion.

The Binab-treated Southern Pine timbers in this study were better colonized than the interior of creosoted Scotch pine distribution poles treated with Binab-T pellets (Bruce and King, 1986a,b; Bruce et al., 1990). Recovery of Trichoderma from the heartwood center of the distribution poles was only 25.6 percent after 18 months and 23.6 percent after 6 years. Trichoderma did not appear to develop and colonize throughout the poles. Bruce et al. (1990) attributed the lack of distribution to resident fungi, but another factor may have been the high natural durability of heartwood and lack of nutrients for favorable growth.

Although Trichoderma was able to develop and spread through the Southern Pine timbers , it did not stop decay. The timbers were eventually decayed by Gloeophyllum trabeum. In previous laboratory tests (Highley and Ricard, 1988), we found that Trichoderma did not control G. trabeum decat.

Control of G. trabeum is important because this fungus is a prevalent cause of decay of pine on the exterior of buildings and decay of other pine products off the ground and subject to rain wetting.

On the other hand, Neolentinus lepideus is easily controlled by Trichoderma in laboratory tests, but control in field tests is rather disappointing (Bruce et al., 1990). Bruce and King, (1986b) reported that inoculation of poles with Binab pellets either before or after infection with N. lepideus reduced the incidence of decay approximately 50 percent. However, in another study, Bruce et al. (1990) found that prior inoculation reduced decay only 17 percent. If N. lepideus was established prior to pole infection, no reduction in decay occurred. The authors concluded that for successful control of N. lepideus the fungus must be eradicated in the early stages of colonization. Failure of Trichoderma to prevent decay to a greater extent was probably due to poor colonization of the heartwood interior.

A biocontrol treatment would have had a better chance for success as a groundline treatment for poles than as a treatment for aboveground wood because of the continual moisture. It is difficult to envisage active control of decay by antagonists in aboveground structures, such as window frames and waterfront curbing, when the moisture content fluctuates substantially. Any biological control agent would have to maintain itself during the times when the timber remained dry, only becoming active when wetting occurred. On the other hand, if the biocontrol agent left behind a fungistatic residue, resistance to decay might be improved. Trichoderma has been reported to produce fungistatic substances (Bruce and King, 1983; Bruce and Highley, 1991). However, this evidently did not occur in the field exposure in our tests because wood removed from Trichoderma-infected areas of pine timbers and sterilized was not decay-resistant. (Bruce et al., 1991) found similar results with sterilized wood removed from Trichoderma-treated poles after 7 years. However, the authors found that unsterilized wood was resistant to N. lepideus but not to Trametes versicolor. Likewise, in the present study, unsterilized Trichoderma-treated wood removed from pine timbers reduced decay by N. lepideus, but not T. versicolor or Postia placenta.

Although Trichoderma did not successfully control decay in Douglas-fir and Southern Pine exposed above ground in our study, there are many examples of successful control of wood-inhibiting fungi by Trichoderma. In our study, control of stain in green pine log sections was successful. Hulme and Shields (1975) found that Trichoderma protected freshly-cut birch bolts from decay. Lindgren and Harvey (1952) prevented decay by Peniphora by spraying green bolts with ammonium biofluoride and Trichoderma. Pottle et al. (1977) showed that be pruning wounds of a red maple could be protected for more than 2 years by inoculating the wounds with a suspension of Trichoderma. In a comprehensive study that compared various chemical commercial compounds with Trichoderma preparations, Mercer and Kirk (1984) found that Trichoderma out-performed chemical compounds, both in acceleration of callus formation by sapwood and in prevention of basidiomycete infection of heartwood. The biggest success story for Trichoderma as a bioprotectant is its use on fruit trees to protect curl silver leaf by basidiomycete Chondrostereum purpureum (Dobos and Ricard, 1974). Chondrostereum purpureum spores infect the tree via contact with unprotected wood, most often pruning wounds. The preventative effect of Trichoderma in this disease was clearly demonstrated (Groslande, 1970).

The success stories cited here deal with relatively short-term protection of wood against attacking fungi. Thus, initial research on the use of biocontrol agents to control wood-attacking fungi should emphasize short-term protection, such as protection of lumber and logs during drying. Control of decay in wood products has to be very long-term, many years, if not decades. For a biocontrol organism to accomplish this would be very difficult unless the organism spreads throughout the wood and deposits as residual fungistatic material that remains after the death of the organism or alters the substrate to make it nutritionally undesirable to the decay fungus.

SUMMARY

In this study, we determined the ability of a biological control product of Trichoderma to colonize and prevent decay in Douglas-fir and Southern Pine timbers exposed above ground. Trichoderma colonized Southern Pine timbers but did not prevent decay from Gloeophyllum trabeum. Douglas-fir timbers were poorly colonized by Trichoderma. Recovery of Trichoderma from timbers decreased after 3 years. Material removed from timbers colonized by Trichoderma resisted decay by Neolentinus lepideus, but not Postia placenta or Trametes versicolor. Any decay prevention was lost, however, when the wood was sterilized prior to exposure to the decay fungi. Treatment of green Southern Pine log ends with Trichoderma protected the wood from sapstain fungi.

Acknowledgements

The donation on Binab-T pellets and wettable powder and technical advice from Dr. Jacques Ricard, Bio-Innovation AB, Algaras, Sweden, is sincerely appreciated. The author also thanks Les Ferge of Forest Products Laboratory for technical assistance.

Note: The use of trade or firm names in this publication is for reader information and does not imply endorsement by the U.S. Department of Agriculture of any product or service.

REFERENCES

American Society for Testing Materials. (1971). Standard method for accelerated laboratory test of natural decay resistance of woods, ASTM D 2017. Philadelphia, PA.

Bruce, A. and Highley, T.L. (1991) Control of growth of wood decay basidionycetes by soluble metabolites from Trichoderma spp. and other potentially antagonistic fungi. Forest Products J., 4(2), 63-67.

Bruce, A. and King, B. (1983). Biological control of wood decay by Lentinus lepideus (Fr.) produced by Scytalidium and Trichoderma residues. Mat. und Org., 18, 171-181.

Bruce, A. and King, B. (1986a). Biological control of decay in creosote-treated distribution poles. I. Establishment of immunizing commensal fungi in poles. Mat. und Org., 21, 1-13.

Bruce, A. and King, B. (1986b). Biological control of decay in creosote-treated distribution poles. II. Control of decay in poles by immunizing commensal fungi. Mat. und Org., 21, 165-179.

Bruce, A., Fairnington, A., and King, B. (1990). Biological control of decay in creosote-treated distribution poles. III. Control of decay in poles by immunizing commensal fungi after extended incubation period. Mat. und Org., 25, 15-28.

Bruce A., King,B., and Highley, T.C. (1991). Decay resistance of wood removed from poles biologically treated with Trichoderma. Holzforschung, 45(4), 307-311.

Carey, J.K., Purslow, D.F., and Savory, J.G. (1981). Proposed method for out-of-ground contact trails of exterior joinery protection system. The International Research group on Wood Preservation: IRG/WP/2157.

Dubos, B. and Ricard, J. (1974). Curative treatment of peach trees against silver leaf disease (Stereum purpureum) with Trichoderma viride preparations. Plant Disease Reporter, 58, 147-150.

Freitag, M., Morrell, J.J., and Bruce, A. (1991). Biological protection of wood: Status and prospects. Biodeterio. Abst., 5(1), 1-13.

Glaser, T.E., Tarocinski, E., and Banza, J. (1969). Study of the interaction of Discula brunneo-tingens and the most important fungi accompanying it in the coffee-brown discoloration of pine sawlogs. Pr. Inst. Technol. Drewna, 6, 45-58.

Groslande, C. (1970). Premiers essais de protection biologique des blessures de taille vis a vis due Stereum purpureum. Ann. Phytopathal., 2, 507-516.

Highley. T.L. and Ricard, J. (1988). Antagonism of Trichoderma spp. and Gliocladium virens against wood decay fungi. Mat. und Org., 23, 157-169.

Hulme, M.A. and Shields, J.K. (1975). Antagonistic and synergistic effects for biological control of decay. In: Biological Transformation of Wood by Microorganisms. pp. 52-63 (W. Liese, ed.) Springer Verlag, Berlin, German Federal Republic.

Lindgren, R.M. and Harvey, G.H. (1952). Decay control and increased permeability in Southern Pine sprayed with fluoride solutions. J. Forest Prod. Res. Soc., 5, 250-256.

Mercer, P.c. and Kirk, S.A. (1984). Biological treatments for the control of decay in tree wounds. II. Field Tests. Annals Appl. Biol., 104, 221-229.

Morris, P.I., Dickenson, D.J., and Levy, J.F. (1984). The nature and control of decay in creosoted electricity poles. In: Records of the British Wood Preserving Association Annual Convention. pp. 42-53.

Pottle, H.W., Shigo, A.L., and Blanchard, R.O. (1977). Biological control of wounds by hymenomycetes by Trichoderma harzianum. Plant Dis. Rep., 61, 687-690.

Shields, J.K. and Atwell, E.A. (1963). Effect of mold, Trichoderma viride, on decay of birch by four storage-rot fungi. Forest Prod. J., 13,(7), 262-265.

IN VITRO EFFECT OF ISATIN ON WOOD DECAY FUNGI

BARBARA L. ILLMAN, USDA Forest Service, Forest Products Laboratory, One Gifford Pinchot Drive, Madison, WI 53705–2398 USA

INTRODUCTION

Control of biodegradation is an ongoing problem in ecologically sensitive areas, requiring new approaches to deter degrading organisms. Biological control is an approach that takes advantage of the ecological relationship among organisms by using natural predators against degrading organisms. This approach has led to an interest in using predator metabolites as pesticides.

Biological control methods offer promise as a deterrent to fungal degradation of wood by using bacteria that are natural predators of fungi (Benko, 1988a; Benko, 1988b; Benko and Highley, 1990; Croan and Highley, 1991; Dawson–Andoh and Morrell, 1990; Highley et al., 1991). Several metabolite-producing bacteria are antagonists of important wood decay fungi (Benko and Highley, 1990; Croan and Highley, 1991). Metabolites from Streptomyces rimosus have been used to inhibit spore germination of molds and sapstain fungi, thereby preventing discoloration and deterioration of wood (Croan and Highley, 1991). Little is known about the fungitoxic properties of most bacterial metabolites.

The bacterial metabolite 2,3-indolinedione (Figure 1.) has been shown to have antifungal properties (Gil–Turnes et al., 1989). The commercial name for 2,3-indolinedione is isatin (Aldrich, Milwaukee, WI); it is produced in a cost-effective method as an intermediate in the synthesis of indigo dyes and known to have pharmacological properties (Joshi and Chand, 1982).

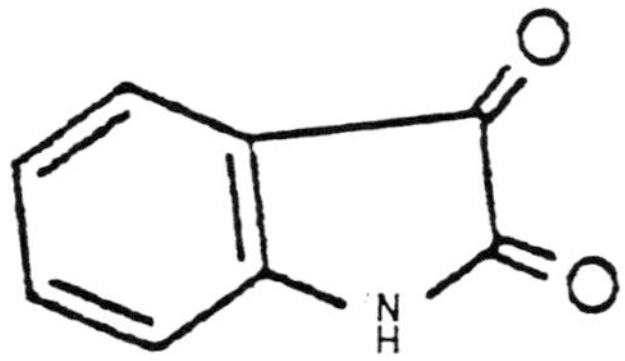

Figure 1. Molecular Structure of 2,3-Indolinedione.

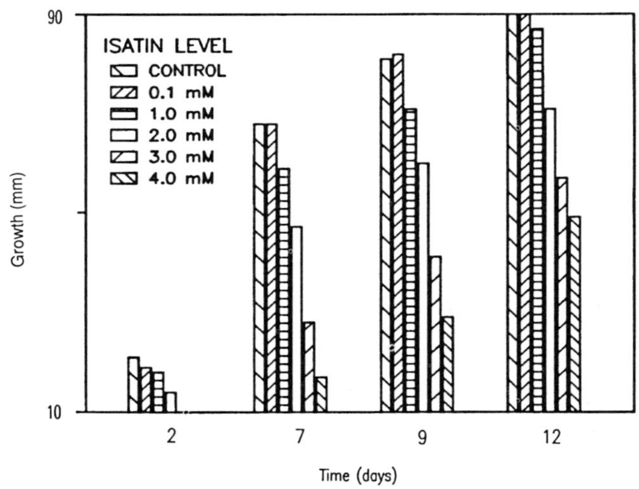

Figure 2. Effect of Isatin on Growth Rate of Postia placenta.

This paper reports the effects of isatin on white- and brown-rot wood decay fungi *in vitro* and discusses the relevance of the effects to fungal metabolism.

MATERIALS AND METHODS

Brown-rot fungi were *Postia placenta* (Fr.) M.Lars. et Lomb. (MAD–698), *Coniophora puteana* (Schum.:Fr.) Karst.(MAD–515), *Gloeophyllum trabeum* (Pers. ex. Fr.) Murr. (MAD–617), *Lentinus lepideus* Fr.(MAD–534), *Antrodia carbonica* Fr. (HHB–5104), and *Fomitopsis meliae* (Underw.) Gilbn. (FP–10002–R). White-rot fungi were *Coriolus versicolor* (Mad–697), *Phanerochaete chrysosporium* (ME–461), *Phlebia brevispora* (HHB–7030), *Ganoderma applanatum* (HHB–7823–s), *Irpex lacteus* (HHB–7328), and *Bjerkandera adusta* (L–15359).

Cultures were maintained on slants of malt extract agar (MEA) at 4C. Mycelia from slant cultures were transferred to stock culture plates containing a chemically defined, nitrogen-limited medium, mimicking nitrogen concentration in wood. The stock medium contained a previously described basal medium (Illman et al., 1989) with the addition of 1% cellobiose and 0.1 mM ammonium tartrate. Stock cultures were incubated in the dark at 27C and 70% relative humidity.

Isatin-amended medium was prepared by adding a filter-sterilized aqueous solution of isatin (Aldrich) to autoclaved, cool basal medium containing 1% cellobiose and 0.1 mM ammonium tartrate and dispensed into 100 (90 i.d.) x 20 mm sterile petri plates. Each experimental plate was inoculated by inverting a 10-mm mycelial disc from the margin of 7-day-old colonies grown on stock culture plates. Five replicate plates were prepared for each species–concentration combination. Experiments were repeated twice. Plates were incubated in the dark at 70% RH and 27C.

Diameters of colonies were measured after inoculation on days 1, 2, and 3 for the white-rot fungi *B. adusta*, *I. lacteus*, and *P. chrysosporium*; on days 1, 2, 3, and 6 for the white-rot fungi *C. versicolor*, *G. applanatum*, and *P. brevispora*; and on days 3, 6, 8, and 10 for brown-rot fungi (slower growing). Three measurements were taken of each colony and the averages compared.

To determine the concentrations of isatin to use on wood decay fungi, the effects of 0.1 to 50 mM concentrations of the chemical were tested on the growth rate of *P. placenta* (Figure 2). The experiment was repeated five times.

No growth was obtained at concentrations above 10 mM. The concentrations selected for subsequent experiments were 0, 0.1, 0.5, 1.0, and 5.0 mM.

RESULTS AND DISCUSSION

Isatin inhibited the growth rate of P. placenta at concentrations of 0.1 to 4 mM (Figure 2). The inhibition increased with increasing concentration. Growth was totally inhibited at 10 to 50 mM concentration.

The effects of isatin on growth of brown- and white-rot fungi was species-dependent with no general effect on a fungal group (Figs. 3 and 4). At the 0.1-mM concentration, isatin did not inhibit B. adusta, I. lacteus, and F. meliae during the experiments. At this concentration, isatin had no effect on G. trabeum, G. applanatum, and L. lepideus at day 3, and P. placenta after day 7; isatin had no effect on C. puteana at day 3 and stimulated growth from days 6 to 10. At concentrations >0.5 mM, isatin inhibited growth of these fungi.

Isatin stimulated the growth of the white-rot fungus P. brevispora at 0.1 to 1.0 mM, but inhibited growth at 5 mM (Fig. 4). The bacterial metabolite inhibited growth of C. versicolor, P. chrysosporium, and A. carbonica at all concentrations.

In general, the concentration of isatin had to be above 0.1 mM to inhibit growth on all fungi except P. brevispora. In future applications, a concentration between 4 and 5 mM (0.07%) and >5 mM would have to be used to be inhibitory (Figure 2). The inhibitory concentration range is similar to the 1.4-mM concentration of isatin used to inhibit growth of Lagenidium callinetes (Gil–Turnes et al., 1989).

The variable response of wood decay fungi to isatin (Figures 2, 3, and 4) is not unusual. These fungi have variable responses to several treatments, including a chitin synthetase inhibitor (Johnson, 1980; Johnson et al., 1991;) and antagonistic bacteria (Benko and Highley, 1990).

The isatin mechanism of action on wood decay fungi is not known. By using low concentrations more closely resembling physiological concentration (as opposed to toxic) to test decay fungi, we observed the dynamics of isatin effects (Figures 2, 3, and 4). At 0.1 mM, this indole was stimulatory to some wood decay fungi. Effects of 0.1, 0.5, and 1.0 mM isatin concentrations implicate a physiological role for the metabolite in growth.

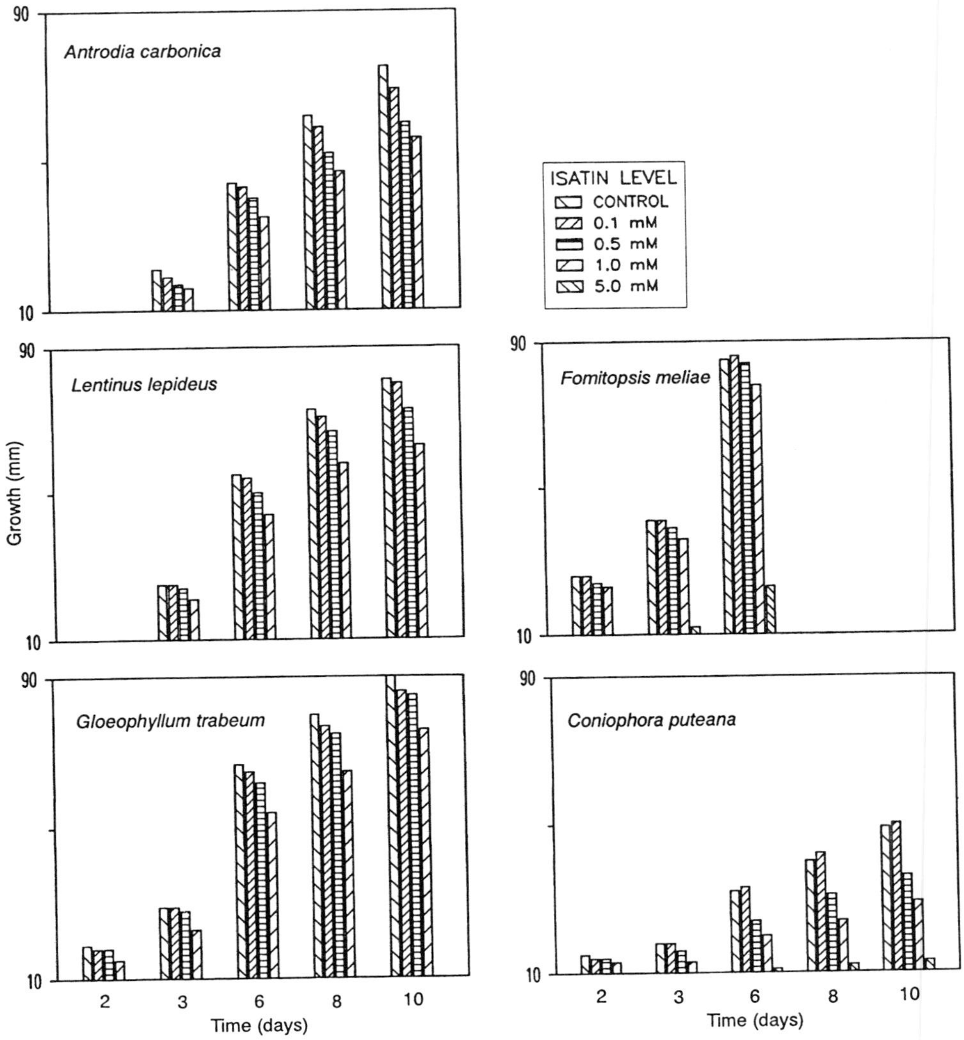

Figure 3. Effect of Isatin on Growth of Brown-rot Fungi.

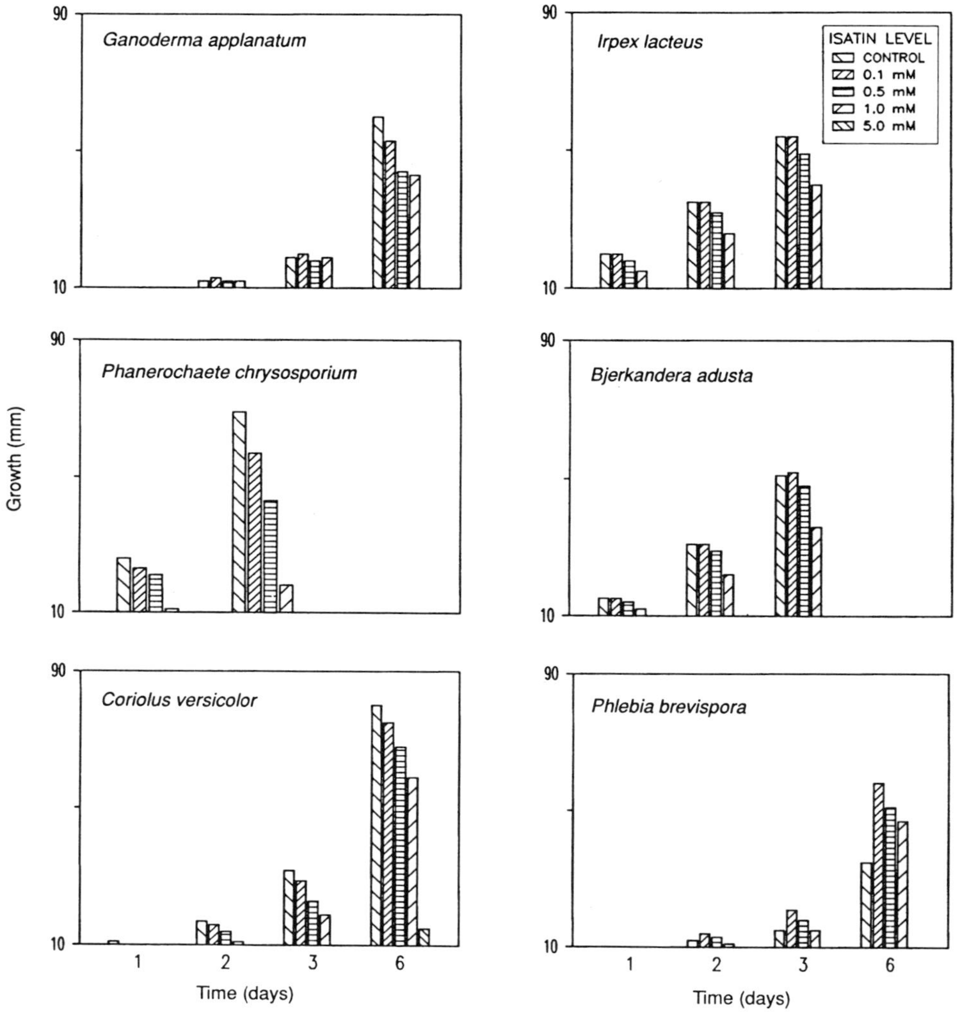

Figure 4. Effect of Isatin on Growth of White-rot Fungi.

Recent discoveries in separate scientific fields have shown that isatin has importance in ecological interactions of organisms and in mammalian physiology. Isatin has ecological significance because this indole is produced by bacteria as an antifungal metabolite (Gil–Turnes et al., 1989).

In a sequence of experiments, Gil–Turnes et al. (1989) investigated a parasitic mediation between three different organisms in an ecological interaction. A bacterial metabolite produced by Alteromonas species was shown to be toxic to the fungus L. callinetes. The bacteria were symbiontic to embryos of the shrimp Palaemon macrodactylus. The antifungal metabolite was secreted by the bacteria on the surface of the embryo, preventing colonization of the fungus and thereby protecting the susceptible embryo from a lethal bacterial infection. The bacterial metabolite was isolated and identified as 2,3-indolinedione. The mechanism of toxicity to the fungus was not studied.

Isatin was recently isolated from humans and rats (Glover et al., 1988), establishing the presence of this indole in mammalian tissues. Glover et al. identified isatin as the metabolite previously known as tribulin, a monoamine oxidase (MAO: EC.1.4.3.4) inhibitor. As an MAO inhibitor, isatin prevents oxidation of several amine substrates such as tryptamine, tyramine, and benzylamine (Zeidan, 1990). Isatin has a potent inhibitory action on MAO B, a form of the oxidase that may be an isozyme or an active form of the enzyme only under special solvent conditions. The function of isatin in mammalian physiology is not clear.

The effect of isatin on wood decay and parasitic fungi implicates its intervention in metabolic pathways responsible for fungal growth and development. The mechanism of action of isatin in mammals raises questions about fungal metabolism. Is isatin an amine oxidase inhibitor in fungi? What role would such an oxidase have on growth? A putative role for amine oxidases on growth and development would involve nitrogen availability and assimilation. To our knowledge, the effect of isatin on other amine oxidases has not been tested.

Many micro-organisms produce secondary metabolites. These allelochemicals are toxic to competitors and pathogens, protect symbiontic organisms, and thereby have an important role in ecology (Price et al., 1986). Isatin secretion by bacteria has such a role as an antibiotic.

Given the discovery that isatin is secreted by bacteria, that is is present in several mammals, and that it has at least one key function in mammalian tissue, a search should be made for the presence and mode of action of isatin in other organisms. Isatin may well be a ubiquitous compound in organisms, serving as a modulator of amine oxidases with a role in nitrogen metabolism, availability, and cycling.

SUMMARY

The in vitro effect of the bacterial metabolite 2,3-indolinedione (isatin) was studied in 12 wood decay fungi. Results indicate that growth of all the fungi can be inhibited at concentrations above 5 mM. The effect of lower isatin concentrations on growth rate was species-dependent. At 0.5 and 1 mM, isatin inhibited growth of 11 fungi. In the white-rot fungus Phlebia brevispora, growth was stimulated at 0.5 and 1 mM. Effects of isatin implicate a physiological role for the metabolite in fungal growth.

In studies on the physiological activity of metabolites in two unrelated areas, the chemical structure of a bacterial metabolite (unnamed) that has fungitoxic properties and that of a mammalian metabolite (tribulin) that inhibits monoamine oxidase were identified as 2,3-indolinedione. Isatin may be present in many more organisms. It is therefore relevant that isatin has been shown to have a physiological effect on wood decay fungi and that toxic levels have been identified for application. Perhaps isatin should be included in the search for microbial allelochemicals.

REFERENCES

Benko, R. (1988a). Bacteria as possible organisms for biological control of blue stain. The Inter. Res. Group on Wood Preserv. Doc. No: IRG/WP/1339.

Benko, R. (1988b). Biological control of blue stain on wood with Pseudomonas cepacia 6252, Laboratory and field test. The Inter. Res. Group on Wood Preserv. Doc. No: IRG/WP/1380.

Benko, R. and Highley, T.L. (1990). Biological Control of Wood-Attacking Fungi Using Bacteria. Biodeterioration Research 3, pp. 327–332 (G. C. Llewellyn and C.E. O'Rear, eds.), Plenum Press, New York.

Croan, S.C. and Highley T.L. (1991). Antifungal activity in metabolites from Streptomyces rimosus. Inter. Res. Group on Wood Preser. Doc. No: IRG/WP/1495.

Dawson–Andoh, B. and Morrell, J.J. (1990). Effects of chemical pretreatment of Douglas-fir heartwood on efficacy of potential bioprotection agents. The Inter. Res. Group on Wood Preserv. Doc. No: IRG/WP/1440.

Gil–Turnes, M.F.; Hay, M.E.; and Fenical, W. (1989). Symbiotic marine bacteria chemically defend crustacean embryos from a pathogenic fungus. Sci., 246, 116–118.

Glover, V.; Halket, J.M.; Watkins, P.J.; Clow, A.; Goodwin, B.L.; and Sandler, M. (1988). Isatin: Identity with the purified endogenous monoamine oxidase inhibitor tribulin. J. Neurochem., 51(2), 656–659.

Highley, T.L.; Benko, R.; and Croan, S.C. (1991). Laboratory studies on control of sapstain and mold on unseasoned wood by bacteria. The Inter. Res. Group on Wood Preserv. Doc. No: IRG/WP 1493.

Illman, B.L.; Meinholtz, D.C.; and Highley, T.L. (1989). Oxygen free radical detection in wood colonized by the brown-rot fungus, Postia placenta. In: Biodeterioration Research 2, pp. 479–509. (C.E. O'Rear and G.C. Llewellyn, eds.), Plenum Press, NY.

Johnson, B.R. (1980). Responses of wood decay fungi to polyoxin D, an inhibitor of chitin synthesis. Mat. und Org. 15(1), 9–24.

Johnson, B.R.; Croan, S.; and Illman, B. (1991). Chitin synthetase activity in cellular fractions of wood-decay fungi. The Intern. Res. Group on Wood Preserv. Document IRG/WP 1497.

Joshi, K.C. and Chand, P. (1982). Biologically active indole derivatives. Pharmazie 37, 1–12.

Price, P.W.; Westoby, W.; Rice, B.; Atsatt, P.R.; Fritz, R.S.; Thompson, J.N.; and Mobley, K. (1986). Parasite mediation in ecological interactions. Ann. Rev. Ecol. Syst. 17, 487–505.

Zeidan, H.M. (1990). The active site of monoamine oxidases. "The key enzyme of biogenic amines degradation" as seen by magnetic relaxation spectroscopy. In: Biodeterioration Research 3, Biodeterioration Research 2, pp. 555–564 (C.E. O'Rear and G.C. Llewellyn, eds.), Plenum Press, NY.

CHEMICAL COMPOSITION AND FUNGITOXIC ACTIVITIES OF PINE CONE EXTRACTIVES

J.A. MICALES*, J.S. HAN, USDA Forest Service, Forest Products Laboratory, Madison, WI, USA
J.L. DAVIS, and R.A. YOUNG, Department of Forestry, University of Wisconsin, Madison, WI, USA

INTRODUCTION

Pine cones, a renewable resource, are not used to their potential. Large quantities of cones are produced annually throughout the world, especially in pine plantations grown for the pulp and paper industry. They are collected, dried to facilitate seed release, and generally discarded. Some nurseries grind the cones into mulch or they may be sold for crafts, but consumer demand for cones is small compared to by-products from other industries. New uses for pine cones could provide additional income for forest landowners.

A better understanding of the chemical composition of pine cones may lead to the discovery of desirable products. Hot water extracts of Pinus parviflora Sieb et Zucc. cones are a popular folk medicine in Japan, especially for the treatment of gastric cancer (Nagata et al., 1990). Medical researchers have found tannins and other lignin-related compounds that appear to have antibacterial, antiviral, and antitumor properties in extracts from pine cones (Fukuchi et al., 1989a, Harada et al., 1991; Kurakata et al., 1989; Nagata et al., 1990; O–Hara et al., 1990; Sakagami et al., 1991). Pine cones are also known to contain monoterpenes (Mattson and Strauss, 1986; Kossuth and Biggs, 1981) and diterpenoid resin acids (Kossuth and Biggs, 1981).

Pine cones accumulating on the forest floor for many years appear to resist decay. Whether this resistance is due to fungitoxic extractives in the cones, or some other mechanism, such as lignin content or lack of contact with soil moisture, is unknown. The possibility of unidentified fungitoxic extractives should be explored. Currently, new wood preservatives are needed to replace traditional compounds that have been banned or restricted due to their general toxicity. Compounds derived from natural products may have less impact on the environment than synthetic agents and would be more acceptable to consumers. This project was designed to study the chemical composition of pine cones and to determine whether cones from four different Pinus species contain any extractives active against the fungi that cause mold, sapstain, and wood decay.

MATERIALS AND METHODS

Extraction and Analysis

Pine cones from lodgepole pine (Pinus contorta Dougl.), Jeffrey pine (Pinus jeffreyi Grev. & Balf.), and sugar pine (Pinus lambertiana Dougl.) were collected from Toiyabe National Forest near Carson City, Nevada. They were air dried and ground to pass through a 6-mm screen and then stored at -4C until needed. Fresh cones off the tree and 1-year-old oxidized cones from the ground were collected from Scots pine (Pinus sylvestris L.) in the University of Wisconsin Arboretum, Madison, Wisconsin, and were processed immediately upon collection.

The ground samples were analyzed according to Pettersen and Schwandt's method (Petterson and Schwandt, 1991) for sugars and the TAPPI standard method (TAPPI, 1989) for lignin. Resin acids were removed by Soxhlet extraction using diethyl ether for 24 hr. The samples were then extracted with acetone and methanol.

Resin Acid Analysis

The diethyl ether extracts from each species were further separated using DEAE-Sephadex column chromatography (Zinkel, 1983; Zinkel and Rowe, 1964). The extracts were eluted sequentially to obtain a neutral fraction (by eluting with a 89% diethyl ether, 10% methanol, 1% water), a CO_2 fraction (by eluting with a 89% diethyl ether, 10% methanol and 1% water saturated with carbon dioxide), and an acetic acid fraction (by eluting with 88% diethyl ether, 10% methanol, 1% water and 1% acetic acid). The fractions were methylated with diazomethane to form methyl esters prior to gas chromatography. Fractions were analyzed by DB-1 (Han and Zinkel, 1990) and BDS (Han and Zinkel, 1991) fused silica capillary columns. Compounds were identified by comparing the retention times of their methyl esters with those of standard samples of softwood extractives.

Culture Maintenance

The following fungal species and isolates were used in this study: Ceratocystis coerulescens (Muench) Bakshi, C256; Aureobasidium pullulans (deBary) G. Arnaud, MDX18; Trichoderma harzianum Rifai, ATCC20476; Postia placenta (Fr.) M. Lars. & Lomb., MAD698; Neolentinus lepideus (Fr.:Fr.) Redhead and Ginns, MAD534; Gloeophyllum saepiarium (Wulfen:Fr.) P. Karst., FD18; Gloeophyllum trabeum (Pers. ex Fr.) Murr., MAD617; Trametes versicolor (L.:Fr.) Pilat, MAD697; and Schizophyllum commune Fr., MAD566S. The isolates were maintained on 2% malt-extract agar (MEA) slants at 4C for the duration of the study.

Fungal Growth Inhibition

The neutral and CO_2 fractions of the diethyl ether extracts were evaluated for their abilities to inhibit mold, sapstain, and wood-decay fungi. The fractions were dissolved in absolute ethanol to final concentrations of 50 or 100 μg/ml (depending on the amount of sample available). These stock solutions were further diluted to final concentrations of 5, 10, 50, and 100 μg/ml by adding 0.5 ml of stock solution to 250 ml of autoclaved, molten 2% MEA. Control cultures contained 0.5 ml absolute ethanol in 250 ml media. The media were inoculated with mycelial plugs from the margins of 14-day-old colonies of the fungi grown on 2% MEA. Two replicate plates were made for each concentration and each fungal species. The cultures were incubated at 27C. Colony diameters were measured at 4 and 7 days, depending on the growth rate of the fungus. The percentage growth inhibition was calculated based on the diameter growth of control colonies.

The ability of nine known resin acids to inhibit fungal growth was also determined. Pimaric, isopimaric, sandaracopimaric, levopimaric, dehydroabietic, abietic, neoabietic, palustric, and lambertianic acids were obtained from the resin acid collection of the Forest Products Laboratory. (Some of the resin acids were purchased, some were donated, and others were by-products from specific experiments. All the resin acids were purified using appropriate combinations of silver nitrate-alumina column and gas liquid chromatography. All the samples were analyzed by DB–1 and BDS capillary columns prior to the actual experiments.) Stock solutions were prepared by dissolving the resin acids in absolute ethanol. They were then incorporated into 2% MEA for a final concentration of 200 μg/ml. All the resin acids, with the exceptions of pimaric and isopimaric, were further diluted to final concentrations of 5, 10, and 50 μg/ml in the medium. Control media contained 0.5 ml absolute ethanol per milliliter. The media were inoculated with plugs from the margins of 14-day-old colonies of fungi growing on MEA. Two replicate plates were made for each concentration and each fungal species. The cultures were incubated at 27C. Colony diameters were measured after 3, 6, 8, or 10 days, depending on the growth rate of the fungus. The percentage growth inhibition was calculated by comparing these growth rates with the radial growth rate of colonies on control media.

Table 1. Chemical Analysis of Scot Pine Cone.

Chemicals	Content[a] (%)
Glucose	46.2
Mannose	24.6
Galactose	4.0
Xylose	3.5
Arabino	0.0
Klason lignin	23.8
Acid soluble lignin	0.7
Ash	0.4
Ethanol/toluene extractives	6.4

[a]Sugars and lignin given as a percentage of extractive-free wood.

RESULTS

Chemical Composition of Pine Cones

Table 1 shows the chemical composition of Scots pine cones. The cones contained large amounts of glucose (46%) and mannose (25%), and minor quantities of galactose and xylose. The cones also contained significant levels of Klason lignin (24%) but only barely detectable quantities of acid-soluble lignin (0.7%). Ethanol/toluene extractives made up 6% of the sample.

Identification of Extractives

Analysis of the neutral fractions of diethyl ether extracts of four pine cones by gas chromatography (using a DB-1 column) revealed the presence of about six primary compounds that have not been identified. The acetic acid fractions of diethyl ether extracts of the same four samples contained approximately 15 unidentified compounds. The CO_2 fractions of diethyl ether extracts contained nearly 24 resin acids. The 16 that were identified are presented in Table 2. The different species of pine contained different distributions of resin acids, but there were common patterns. All four species contained >18% of dehydroabietic acid.

Lodgepole, Jeffrey, and sugar pine also exhibited significant quantities (>23%) of isopimaric acid. Pimaric acid made up 15% of the extractives of Scots pine but <2% in the other species. Lodgepole pine contained much higher levels of abietic acid than the other species (14.3% compared with 3% to 6%). Similarly, sugar pine was the only species that possessed a large quantity (25%) of lambertianic acid. Most other resin acids detected were present at very low levels in all four species.

The resin acid compositions of fresh and dried cones from Scots pine are shown in Table 3. The concentrations of pimaric, isopimaric, and sandaracopimaric acids were relatively constant. Upon drying, quantities of levopimaric, palustric, abietic, and neoabietic acids decreased, while dehydroabietic acid substantially increased.

Fungal Growth Inhibition

The neutral fraction of diethyl ether extracts from Scots pine cones exhibited little fungitoxic activity (Table 4). Growth reductions were observed for C. coerulescens, a sapstain fungus, and Trametes versicolor, a white-rot fungus, but growth of the other sapstain, mold, and wood-decay fungi was not affected. Higher levels of fungitoxic activity were present in the CO_2 fraction of diethyl ether extracts derived from Scots pine cones (Table 5). Aureobasidium pullulans, a sapstain fungus, and P. placenta, a brown-rot fungus, displayed no sensitivity to a 50-μg/ml preparation, but all other species tested were inhibited by 14% to 53%.

The fungitoxic properties of the neutral and CO_2 fractions of diethyl ether extracts from cones of sugar, lodgepole, and Jeffrey pines are shown in Table 6. The neutral fractions from these species displayed little fungitoxic activity. At

50 μg/ml, the growth of two white-rot fungi was slightly inhibited (2% to 17%) by neutral fractions of cones from all three species. The neutral fractions from sugar and lodgepole cones inhibited the radial growth of the brown-rot fungus G. trabeum, but only by 5% to 9% at 50 μg/ml. The neutral fraction from Jeffrey pine cones was inactive against G. trabeum but inhibited the growth of G. saepiarium by 17%. Similar to Scots pine, the CO_2 fractions from sugar, lodgepole, and Jeffrey pine cones exhibited greater fungitoxic activity than did the neutral fraction. The CO_2 fraction from sugar pine inhibited the growth of G. trabeum, G. saepiarium, Trametes versicolor, and S. commune by 28% to 44% at 100 μg/ml. The CO_2 fractions from Jeffrey and lodgepole pines displayed similar inhibitory properties against G. trabeum, Trametes versicolor, and S. commune, but were not active against G. saepiarium.

The fungitoxicity of nine known resin acids is shown in Table 7. Pimaric and isopimaric acid had little or no effect on the growth of C. coerulescens, Trichoderma harzianum, G. trabeum, or Trametes versicolor at 200 μg/ml. Sandracopimaric, levopimaric, dehydroabietic, abietic, neoabietic, palustric, and lambertianic acids all displayed some fungitoxic activity. The growth of Trichoderma harzianum was resistant to some of these compounds. The fungitoxicity of these active compounds was determined for other mold, sapstain, and decay fungi (Table 8). Abietic, dehydroabietic, and levopimaric acids exhibited the greatest fungitoxicity; in most cases, dehydroabietic acid was the most effective compound. Dehydroabietic acid was particularly inhibitory to S. commune, reducing growth 75% at 50 μg/ml. This was greater than the inhibition caused by any of the CO_2 fractions. Lambertianic, palustric, and sandracopimaric acids were the least effective compounds, although all the resin acids had some activity against G. trabeum. The presence of lambertianic, levopimaric, palustric, and sandaroacopimaric acids in the media selected for a nonpigmented colony type of C. coerulescens. This nonpigmented phenotype was more resistant to these compounds than was the commonly expressed, pigmented colony type.

The fungitoxicity of the CO_2 fractions is not due solely to the presence of known resin acids. The portion of fungitoxicity of the CO_2 fractions that could be derived from abietic, dehydroabietic, and sandaracopimaric acids is calculated in Table 9. The greatest amount of fungitoxicity is probably due to dehydroabietic acid, but in most cases additional compounds, or interactions among compounds, must also be involved. The toxicity of isocupressaic and acetylisocupressaic acids, which were also identified in the CO_2 fraction of Scots pine (Table 2), could not be determined since the purified compounds were not available for comparison. The presence of these and other unidentified extractives may account for the remaining fungitoxic effect.

Table 2. Resin Acids in Pine Cones.

Resin acid	Content (%)			
	Lodgepole pine	Jeffrey pine	Sugar pine	Scots pine
Myrcecommunic acid	0.7	0.0	0.3	0.0
Secodehydroabietic acid	0.2	1.9	0.9	0.5
Pimaric acid	1.5	1.3	0.3	14.6
Sandaracopimaric acid	2.3	2.5	3.0	7.0
Isopimaric acid	34.8	31.0	22.5	5.9
Levopimaric acid	3.6	1.0	0.4	0.9
Palustric acid	2.7	0.4	0.5	5.0
Lambertianic acid	0.4	0.3	25.3	0.0
Dehydroabietic acid	19.5	41.1	18.3	34.0
Imbricataloaic acid	0.5	2.6	0.3	0.0
Abietic acid	14.3	3.9	3.7	6.2
Neoabietic acid	3.7	0.0	0.0	3.2
Imbricatoloaic acid	0.0	2.1	0.0	0.0
Isocupressaic acid	0.9	0.7	0.6	1.4
Acetylimbricatoloaic acid	0.0	1.6	3.0	3.4
Acetoxyisocupressaic acid	1.4	2.1	0.0	0.0

Table 3. Resin Acids in Fresh and Oxidized Scots Pine Cones.

Resin acid	Content (%)	
	Fresh	Oxidized
Pimaric acid	16.7	18.3
Sandaracopimaric acid	6.5	8.7
Levopimaric Acid	3.6	0.9
Palustric acid	17.0	5.2
Isopimaric acid	7.9	7.2
Abietic Acid	13.4	8.6
Dehydroabietic acid	11.8	47.0
Neoabietic acid	23.0	4.1

Table 4. Inhibition of Colony Diameter Growth of Wood-Rotting and Sapstain/Mold Fungi by Neutral Fraction of Diethyl Ether Extracts From Scots Pine Cones.

		Inhibition[a] (%)		
Fungus	Decay/mold	10 μg/ml	50 μg/ml	100 μg/ml
Ceratocystis coerulescens	Sapstain	0.0	18.7	28.4
Aureobasidium pullulans	Sapstain	0.0	0.0	0.0
Trichoderma harzianum	Mold	0.0	0.0	0.0
Postia placenta	Brown rot	0.0	0.0	0.0
Neolentinus lepidius	Brown rot	0.0	0.0	0.0
Gloeophyllum trabeum	Brown rot	0.0	0.0	0.0
Trametes versicolor	White rot	27.9	13.0	16.9
Schizophyllum commune	White rot	0.0	0.0	0.0

[a]Each reading represents the average of two replicate measurements each of two replicate plates.

Table 5. Inhibition of Colony Diameter Growth of Wood-Rotting and Sapstain/Mold Fungi by CO_2 Fraction of Diethyl Ether Extracts From Scots Pine Cones.

		Inhibition[a,b] (%)		
Fungus	Decay/mold	5 μg/ml	10 μg/ml	50 μg/ml
Ceratocystis coerulescens	Sapstain	0.0	11.3	36.7
Aureobasidium pullulans	Sapstain	0.0	0.0	0.0
Trichoderma harzianum	Mold	0.0	0.0	38.0
Postia placenta	Brown rot	0.0	0.0	0.0
Lentinus lepidius	Brown rot	6.2	8.2	19.2
Gloeophyllum trabeum	Brown rot	0.0	0.0	16.1
Gloeophyllum saepiarium	Brown rot	0.0	0.0	14.4
Trametes versicolor	White rot	0.0	0.0	21.7
Schizophyllum commune	White rot	0.0	0.0	53.1

[a]Sufficient compound was not available to test 100 μg/ml.

[b]Each reading represents the average of two replicate measurements each of two replicate plates.

Table 6. Inhibition of Colony Diameter Growth of Wood-Rotting Fungi by Neutraland CO_2 Fractions of Diethyl Ether Extracts of Pine Cones.[a]

Fungus	Concentration μg/ml	Inhibition (%) Neutral Sugar pine	Jeffrey pine	Lodge pole pine	CO_2 Sugar pine	Jeffrey pine	Lodge pole pine
Brown-rotters							
Gloeophyllum trabeum	100	ND[b]	ND	ND	36.3	46.4	42.9
	50	4.8	0.0	9.0	21.9	39.3	33
	10	0.0	0.0	0.0	2.4	17.9	8.9
	5	0.0	0.0	2.3	0.0	7.1	5.4
Gloeophyllum saepiarium	100	ND	ND	ND	44.2	1.8	0.0
	50	0.0	17.4	0.0	48.8	1.8	0.0
	10	0.0	17.4	0.0	0.0	0.0	0.0
	5	0.0	10.9	0.0	ND	0.0	0.0
White-rotters							
Trametes versicolor	100	ND	ND	ND	27.1	30.5	25.9
	50	8.5	16.7	6.4	20.8	23.7	20.7
	10	0.0	0.0	0.0	10.4	15.2	10.3
	5	0.0	0.0	6.4	8.3	8.5	6.9
Schizophyllum commune	100	ND	ND	ND	27.7	40.7	25.0
	50	4.2	2.1	6.5	20.8	23.7	20.7
	10	0.0	0.0	0.0	0.0	20.4	19.2
	5	0.0	0.0	0.0	6.4	22.2	7.6

[a] Each meqasurement based on the diamater growth of two replicate colonies. Colonies incubated at 27C for 4 to 7 days, depending on the growth rate of the organism.
[b]ND, not determined. Sufficient neutral fraction was not available to test 100 μg/ml.

Table 7. Inhibition of Radial Growth of White-Rot, Brown-Rot, Mold, and Sapstain Fungi (5 Day Growth) by Resin Acids (200 μg/ml).

Resin acid	Inhibition (%) Ceratocystis coerulescens (sapstain)	Trichoderma harzianum (mold)	Gloeophyllum trabeum (brown-rot)	Trametes versicolor (white-rot)
Pimaric acid	0	0	0	0
Isopimaric acid	0	10	0	0
Sandaracopimaric acid	26	0	15	11
Levopimaric acid	25	66	64	37
Dehydroabietic acid	17	49	17	37
Abietic acid	42	15	42	26
Neoabietic acid	41	35	41	33
Palustric acid	19	4	16	19
Lambertianic acid	26	2	12	14

DISCUSSION

The chemical composition of Scots pine cones generally resembled that of solid pine wood (Pettersen, 1984). The cones contained large quantities of glucose, probably derived from cellulose, and smaller quantities of mannose, galactose, and xylose, probably derived from hemicellulose. They also contained significant amounts of lignin and ethanol/toluene extractives. The yields of extractives increased as the polarity of the solvent increased. This suggests an abundance of polyphenols and tannins in the cones and a smaller amount of terpenes, fats, and waxes. The CO_2 fractions obtained by column chromatography contained resin acids; 16 were identified by gas chromatography.

Table 8. Fungitoxic Effect of CO_2 Fraction From Scots Pine Attributable to Known Resin Acids.[a]

	Fungitoxic effect (%)		
Organism	Abietic acid	Dehydroabietic acid	Sandaracopimaric acid
C. coerulescens	8.4	50.9	0.0
T. harzianum	3.9	44.7	2.0
N. lepidius	4.8	33.8	1.8
G. trabeum	12.7	88.8	10.0
G. saepiarium	7.7	42.4	1.0
T. versicolor	9.1	65.9	1.6
S. commune	4.4	48.0	0.0

[a]Percentage of fungitoxic effect calculated by (expected inhibition) ÷ (actual inhibition (Table 5)) X 100, where expected inhibition = (inhibition by pure resin acid (Table 8)) X (concentration of resin acid in CO_2 fraction (Table 2)). Pimaric and isopimaric acids are not included since they were not fungitoxic at 50 μg/ml.

The toxicity of diterpenes, of which resin acids are one group, is well-known from water pollution studies associated with paper mills. The fungitoxicity of individual resin acids has not been widely studied. The LD_{50}s of abietic and dehydroabietic acids to mice have been reported as 2.5 g/kg and 3 to 4 g/kg, respectively. The median lethal concentrations to rainbow trout, measured after 96 hr bioassays, were as follows: dehydroabietic acid – 1.1 g/kg; pimaric acid - 0.8 g/kg; abietic acid - 0.7 g/kg; palustric acid - 0.5 g/kg; and isopimaric acid - 0.4 g/kg (Zinkel and Russell, 1989). Apparently these compounds are relatively nontoxic to vertebrates.

Certain resin acids detected in the pine cone extracts were fungitoxic to representative mold, sapstain, and wood-decay fungi. Levopimaric, dehydroabietic, abietic, and neoabietic acids inhibited the mycelial growth of this assortment of fungi from 15% to 64% at 200 μg/ml. Pimaric, isopimaric, and sandaracopimaric acids displayed only limited activity. This difference in toxicity can be explained by the chemistry of these compounds. Pimaric, isopimaric, and sandaracopimaric acids belong to the pimarane group of resin acids. In these compounds, C-15, C-16, and C-17 are attached to C-13 as vinyl and methyl groups (Fig. 1). The other five resin acids with the greatest fungitoxicity belong to the abietane group of compounds. These acids have isopropyl groups from C-15, C-16, and C-17 attached at C-13 (Fig. 2). Apparently this isopropyl group is required for fungitoxicity.

A previous study (Walter et al., 1989) showed that some resin acids are more stable than others. Levopimaric acid tends to be somewhat unstable, while dehydroabietic acid is usually less reactive. This reactivity can be explained by the chemical structure of the compounds. Levopimaric acid can be oxidized into dehydroabietic acid, either directly or with abietic acid as an intermediary step. Neoabietic, abietic, and palustric acids, which are isomers, can also be oxidized into dehydroabietic acid (Fig. 2). These differences in stability are reflected in theresin acid composition of the pine cones and in the changes observed in resin acid composition of Scots pine cones upon drying. All four species contained significant quantities of dehydroabietic acid and only minor quantities of levopimaric acid. The substantial increase in dehydroabietic acid and simultaneous decrease in levopimaric, palustric, abietic, and neoabietic acids upon cone drying supports the explanation based on oxidation sequences. However, dried cones from lodgepole pine contained substantial levels of abietic acid; reduced forms of resin acids can also exist in nature.

In most cases, the amount of fungitoxicity exhibited by the CO_2 fractions was higher than could be explained based on the content of the known resin acids alone. The response of the fungi to the preparations from different species of Pinus was variable; some fungi were more sensitive than others to extracts from the different pines. These two observations support the idea that other compounds besides the known resin acids may be fungitoxic. The identity of these additional fungitoxic agents may vary among pines. For example, the CO_2 fraction from sugar pine was the only preparation with significant activity against the brown-rot fungus G. saepiarium. Sugar pine was the only species that produced measurable quantities of lambertianic acid, but this compound was not active against G. saepiarium; some other agent must also be involved. The combination of two or more resin acids with even low levels of fungitoxicity could greatly increase the efficacy of the preparation and the range of target organisms it is active against, perhaps due to synergistic effects. None of the CO_2 fractions from different Pinus species inhibited fungal growth entirely, but it is possible that some fungitoxic agents are present in very small quantities and would be more toxic at higher concentrations. Purifying these compounds further would be necessary to produce a more effective wood preservative.

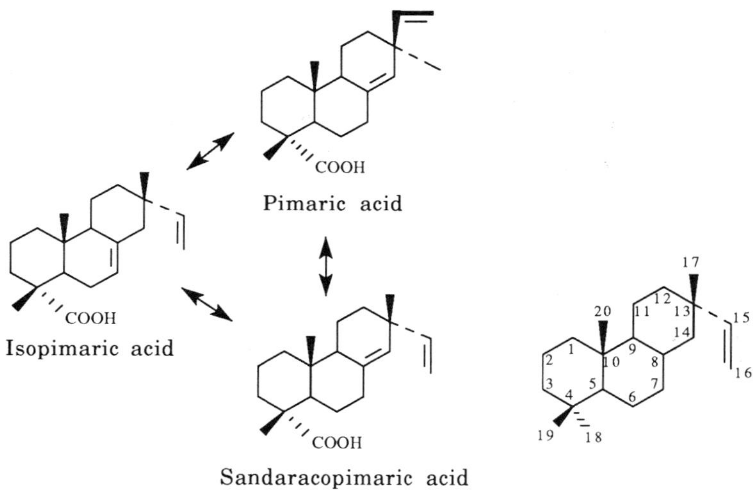

Figure 1. Isomerization of Pimarane Group Resin Acids.

Incorporating a compound into an agar medium may not be the best means of evaluating fungitoxicity (Hart and Hillis, 1974; Scheffer and Cowling, 1966). Agar growth media provide an extremely artificial testing environment. Compounds that successfully prevent the growth of fungi in agar may be ineffective in wood due to problems of penetration, inactivation, or differences in fungal physiology. More research is needed to properly evaluate the preservative potential of these preparations. Formulation research would also be needed to maximize the fungitoxic effect of any compound. For example, resin acid soaps have been shown to have an LD_{50} of 2 mg/l to Daphnia pulex; free acid forms did not cause mortality at 6 mg/l (Zinkel and Russell, 1989). This demonstrates how different formulations of a compound or group of compounds can affect fungitoxicity.

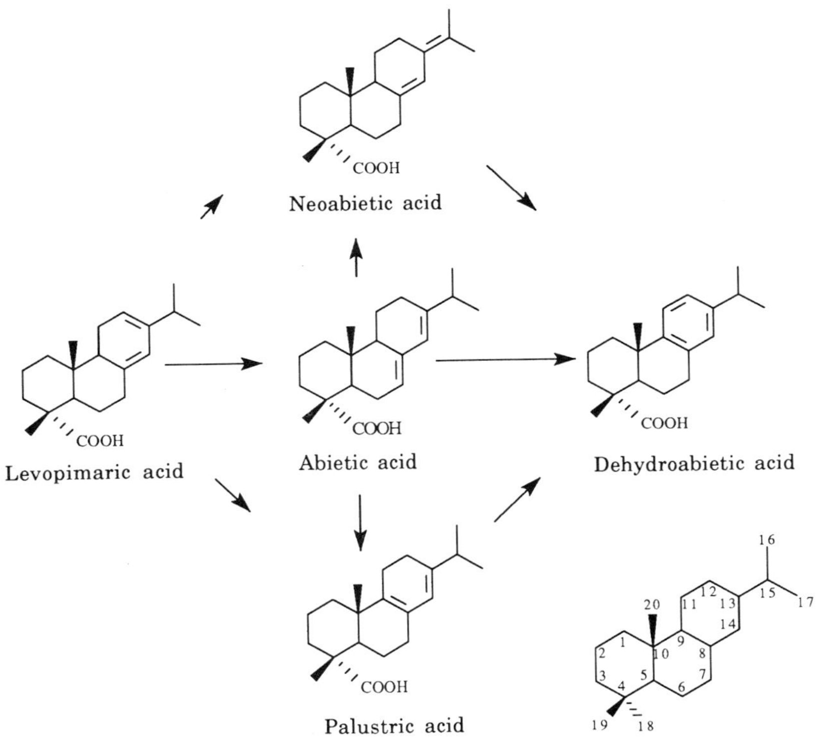

Figure 2. Isomerization and Oxidation of Abietane Group Resin Acids.

Further research is needed to identify the remaining unknown extractives present in pine cones and to evaluate the true potential of resin acids as wood preservatives. Associating the isopropyl group with fungitoxicity is an important first step towards understanding the mode of action of resin acids and for predicting the fungitoxic activity of individual compounds.

SUMMARY

The CO_2 fraction of diethyl ether extracts of four species of pine cones inhibited the growth of representative mold, sapstain, and wood-decay fungi. The fractions contained more than 24 resin acids; 16 were identified by gas chromatography as diterpenoid resin acids. A large portion of the fungitoxicity of the CO_2 fractions is probably due to the presence of abietane resin acids. The abietane resin acids have an isopropyl group attached to C-13; this moiety may be responsible for the fungitoxicity of these compounds. Pimaric, isopimaric, and sandaracopimaric acids, which have methyl and vinyl groups attached to C-13, displayed little biological activity. The fungitoxicity of the CO_2 fractions was higher than could be explained by the presence of known resin acids alone; additional resin acids and other extractives found in low concentrations may add to the fungitoxicity of the preparation.

Acknowledgments

A portion of this research was supported by the Grainger Foundation. The authors also thank Adrian Richter (Forest Products Laboratory, retired) and Barry Stern (Toiyabe National Forest) for collecting the pine cones.

Note: The use of trade or firm names in this publication is for reader information and does not imply endorsement by the U.S. Department of Agriculture of any product or service.

REFERENCES

Fukuchi, K., Sakagami, H., Ikeda, M., Kawazoe, Y., Oh-Hara, T., Konno, K., Ichikawa, S., Hata, N., Kondo, H. and Nonoymama, M. (1989a). Inhibition of Herpes simplex virus infection by pine cone antitumor substances. Anticancer Res. 9, 313-318.

Fukuchi, K., Sakagami, H., Okuda, T., Hatano, T., Tanuma, S.I., Kitajima, K., Inoue, Y., Inoue, S., Ichikawa, S (1989b). Inhibition of Herpes simplex virus infection by tannins and related compounds. Antiviral Res. 11, 285-298.

Han, J.S. and Zinkel, D.F. (1990). Gas chromatography of resin acids with a methyl silicon fused-silica capillary column. Naval Stores Rev. 100, 11-15.

Han, J.S. and Zinkel, D.F. (1991). Gas chromatography of resin acids with a BDS fused-silica capillary column. Naval Stores Rev. 101, 13-16.

Harada, H., Sakagami, H., Nagata, K., Oh-Hara, T., Ishihama, A., Hata, N.and Misawa, Y. (1991). Possible involvement of lignin structure in anti-influenza virus activity. Antiviral Res. 15, 41-50.

Hart, J.A. and Hillis, W.E. (1974). Inhibition of wood-rotting fungi by stilbenes and other polyphenols in Eucalyptus sideroxylon. Phytopathology, 64, 939–948.

Kossuth, S.V. and Biggs, R.H. (1981). Role of apophysis and outer scale tissue in pine cone opening. Forest Sci. 27, 828-836.

Kurakata, Y., Sakagami, H., Takeda, M., Konno, K., Kitajima, K., Ichikawa, S.,Hata, N., and Sato, T. (1989). Mitogenic activity of pine cone extracts against cultures splenocytes from normal and tumor-bearing animals. Anticancer Res. 9, 961-966.

Mattson, W.J. and Strauss, S.H. (1986). Are cone volatiles involved in cone finding by the red pine cone beetle, Conophthorus resinosae (Coleoptera: Scolytidae)? In: Proceedings of the 2nd Conference of the Cone and Seed Insects Working Party S2.07-01. pp. 185–204, September 3-5, 1986., Braincon, France.

Nagata, K., Sakagami, H., Harada, H., Nonoyama, M., Ishihama, A., Konno, K. (1990). Inhibition of influenza virus infection by pine cone antitumor substances. Antiviral Res. 13, 11-21.

Oh–Hara, T., Sakagami, H., Kawazoe, Y., Kaiya, T., Komatsu, N., Ohsawa, N., Fujimaki, M., Tanuma, S.I., and Konno, K. (1990) Antimicrobial spectrum of lignin-related pine cone extracts of Pinus parviflora Sieb. et Zucc. In Vivo 4, 7-12.

Petterson, R.C. (1984). The chemical composition of wood, In: The Chemistry of Solid Wood, R. Rowell (Ed.), pp. 57–126. American Chemical Society, Washington, D.C.

Petterson, R. C., and Schwandt, V. H. (1991). Wood sugar analysis by anion chromatography. J. Wood Chem. Technol. 11, 495-502.

Pullianen, E. (1972). Summer nutrition of crossbills (Loxia pytyopsittacus, L. curvirostra, and L. leucoptera) in northeastern Lapland in 197. Annales Zoologici Fennici 9, 28-31.

Sakagami, H., Kawazoe, Y., Komatsu, N., Simpson, A., Nonoyama, M., Konno,

K.,Yoshida, T., Kuroiwa, Y., and Tanuma, S.I. (1991). Antitumor, anitviral and immunopotentiating activities of pine cone extracts: potential medicinal efficacy of natural and synthetic lignin-related materials. Anticancer Res. 11, 881-888.

Sakagami., H., Oh-Hara, T., Kaiya, T., Kawazoe, Y., Nonoyama, M., Konno, K. (1989). Molecular species of the antitumor and antiviral fraction from pine cone extract. Anitcancer Res. 9, 1593-1598.

Schaffer, T.C. and Cowling, E.B. (1966). Natural resistance of wood to microbial deterioration. Ann. Rev. Phytopathol. 4, 147–170.

TAPPI (1989). Official Test Method T–222 OM–88. Acid-insoluble lignin wood and pulp. Technical Association of Pulp and Paper Industry, Atlanta, GA. TAPPI Test Methods, Vol. 1.

Walter, J., Han, J.S., and Zinkel, D.F. (1989). Fate of resin acids in Kraft pulping. Naval Stores Rev. 99, 17-19.

Young, R.D. and Davis, J.L. (1990). Utilization of pine cones. Report submitted to the Grainger Foundation. 23 p.

Zinkel, D.F. 1983. Quantitative separation of ether-soluble acidic and neutral materials. J. Wood Chem. Technol. 3, 131-143.

Zinkel, D.F. and Rowe, J.W. (1964). A rapid method for the quantitative separation without alteration of ether-soluble acidic and neutral materials. Analytical Chem. 36, 11601161.

Zinkel, D.F. and Russell, J. (eds.) (1989). Naval Stores Production, Chemistry, and Utilization. New York, Pulp Chemicals Association. 1,060 p.

DISTRIBUTION OF POLYPHENOL OXIDASE IN ORGANELLES OF HYPHAE OF THE WOOD-DETERIORATING FUNGUS, *CORIOLUS VERSICOLOR*

NINA L. MOORE * and **L.A. BRAKO**, Department of Biology, Clark Atlanta University, Atlanta, GA 30314, USA

C. A. CLAUSEN, USDA Forest Service, Forest Products Laboratory, Madison, WI 53705-2398. USA

B.R. JONES, Department of Biology, Morris Brown College, Atlanta, GA, 30314, USA

W.V. DASHEK, Department of Biology, Clark Atlanta University, Atlanta, GA 30314, USA

INTRODUCTION

Coriolus versicolor, a white-rot Basidiomycete, secretes cellulolytic and ligninolytic enzymes (Higuchi, 1953; Rosenberg, 1979; Harvey et al., 1987) as well as polyphenol oxidase (Taylor et al., 1987; Moore et al., 1989). Whereas the former degrade certain wood polymers (Higuchi, 1985), the latter can convert o-diphenols to o-diquinones (Fric, 1976) and oligomerize syringic acid (Liu et al., 1981), a lignin derivative. In this connection, certain phenolic compounds can serve as disease-resistance factors possibly regulating the proliferation of wood-decay fungi within host tissues (Shigo, 1984; Dashek et al., 1990). Because ligno-cellulolytic enzymes are of commercial value to the paper-pulp industry (Zadrazil and Reiniger, 1988) and the agricultural community (Van der Meer et al., 1987), an inexpensive and "readily-available" supply of these enzymes is highly desirable. As C. versicolor can be "batch-cultured", over-production and enhanced secretion of these enzymes are feasible (Fahraeus and Reinhammer, 1967). While over-production could possibly be achieved by substrate induction and/or recombinant DNA technologies (Maniatis et al., 1982), secretion might be enhanced by controlling cellular energy gradients and/or possible glycosylation of PPO.

Previously, we reported the time-dependent appearances of intracellular and extracellular polyphenol oxidase (PPO) as well as its substrate specificity, de novo synthesis and partial purification (Taylor et al. 1987; Moore et al. 1989). Here are reported results from a combined biochemical and electron microscopical approach to delineate PPO's subcellular route of secretion culminating with regulating its appearance in the external milieu. Concomitant with these investigations are studies employing substrate induction (Taylor et al., 1987) and recombinant DNA technologies (Williams et al., 1990, 1991) to yield PPO over-production.

MATERIALS AND METHODS

Growth Conditions

Mycelia of a Coriolus versicolor (l. ex fr.) Quel isolate, which was provided by the USDA-Northeastern Forest Experiment Station (Durham, NH) and maintained at 4°C upon agar slants, were aseptically transferred to a defined culture medium (Kirk and Kelman, 1965) containing 2% Difco agar. Following mycelial growth in the dark at 25 ± 2°C upon an agar plate's surface to a disc 30 mm in diameter, discs were aseptically excised with a sterile scalpel and then homogenized for 30s into 25ml autoclaved distilled H_2O with a sterile Wareing blender cup. One ml aliquots of the mycelial homogenates were transferred with a sterile manostat to autoclaved 250 ml Erlenmeyer flasks containing 25 ml of the above sterile medium lacking agar. The cultures were maintained at 25 ± 2°C and 150 rpm within a model 3529 Lab-line gyrotory shaker (Melrose Park, IL) for various times depending upon the described experiments.

Cytochemical Distribution of Hyphal Macromolecules

Hyphae were cultured for 5 days upon a 2% agar-solidified defined medium. The mycelial mat was removed from the surface of the medium and fixed for 15 min in Carnoy's solution (ethanol:glacial acetic acid, 75:25) or 10% formalin. Following fixation, the hyphae were dehydrated through a graded ethanol series (30, 50, 70, 95 and 100%, 2X each % for 10 min each time except for 100% which was 3X with the third wash being 30 min). In the case of Carnoy-fixed material, dehydration began with 75% ethanol. Following dehydration, the hyphae were washed twice with two changes of xylene (2 min each time) and then progressively embedded in xylene/tissue prep (3:1, 1:1 and 1:3 one h each) prior to final embeddment in tissue prep (Fisher Scientific, Norcross, GA). Sections were "cut" with an AO Spencer microtome and mounted onto pre-cleaned slides with Mayer's adhesive. Next, sections were deparaffinized with 2 changes of xylene, and then hydrated through 2 changes each of 100, 90, 70, 50 and 30% ethanol for PAS staining (Sheehan and Hrapchak, 1980) or 100, 90 and 70% ethanol for safranin/fast green staining.

Electron Microscopy of Hyphae

Scanning Electron Microscopy: Hyphae from 26 ml, 3-day old liquid cultures of Coriolus versicolor were "pre-fixed" 30 min in 3.0% glutaraldehyde buffered with pH 7.4, 0.05M cacodylate/cacodylic acid buffer, washed twice (5 min each time) with cacodylate buffer and then "post-fixed" for 2 h in 2.0% OsO_4 buffered with cacodylate (Sabatini et al., 1965). Following post-fixation, hyphae were washed (5 ml each time) with cacodylate buffer and dehydrated through a graded acetone series consisting of 30, 50, 70, 95 and 100% "EM grade (Polysciences, Warrington, PA) acetone. Then, the hyphae were critically point dried, mounted onto aluminum stubs and sputter-coated with gold/palladium. Some hyphae were subjected to immunoelectron microscopy utilizing colloidal gold tagged antibody to PPO (see below). The hyphae were then viewed with a scanning electron microscope.

Antigen Purity

To obtain antibody, the purity of commercial mushroom PPO (Sigma, St. Louis, MO) was assessed. The purity was determined by ion exchange chromatography on DEAE-50, gel filtration on G-150 and Laemmli SDS-PAGE. For ion exchange chromatography, 10 mg PPO dissolved in 100 μl 25mM Na_2HPO_4-NaH_2PO_4, pH 6.5 were layered onto a 0.7 x 15.0 cm Econo column packed with DEAE. The column was eluted with buffer lacking 1 M NaCl and then the buffer containing the salt. Since PPO was the only protein on the column, column eluates were assayed at 280nm. In addition, 10 mg PPO in 100 μl acetate buffer, pH 5.0 were layered onto a 0.7 x 15.0 cm Sephadex G-150 column which was eluted with 100 mM, pH 5.0 acetate buffer. Column eluates were assayed at 280 nm. As an additional means to assess the purity of PPO, 10 mg PPO were subjected to Laemmli (1970) SDS-PAGE tube gel electrophoresis. The tube gels were calibrated with Bio-Rad high molecular weight marker proteins.

Immunization

Two New Zealand white rabbits were bled from the control ear artery to derive control serum. Then, they were inoculated intramuscularly (500 μl hind muscle^{-1}) with an emulsion consisting of phosphate-buffered saline (PBS-tyrosinase, Sigma Chemical Co., St. Louis, MO) and Freund's complete adjuvant. To obtain the emulsion 150 μl tyrosinase (10 mg ml^{-1}) were diluted with 450 μl PBS. Then, the 600 μl were mixed with 600 μl of the adjuvant. Nearly three weeks later, the rabbits were re-inoculated with 500 μl emulsion per hind muscle. This emulsion was prepared by diluting 200 μl tyrosinase (10 mg ml^{-1}) with 500 μl PBS. The 700 μl of diluted tyrosinase were mixed with 700 μl of Freund's complete adjuvant. Approximately, one month later the rabbits were again re-inoculated with 500 μl emulsion hind muscle^{-1}. The emulsion was prepared by mixing 500 μl tyrosinase (10 mg ml^{-1}) and 650 μl PBS. Then, 700 μl of diluted tyrosinase were mixed with 700 μl Freund's incomplete adjuvant. Fifteen days later 8 ml and 25 ml of sera were obtained from rabbits, 1 and 2, respectively. The sera were rapidly frozen and maintained at -20°C. Two months later, the rabbits were inoculated as per the third inoculation. Four days later, the rabbits were bled to yield 27 ml and 28 ml of sera from rabbits 1 and 2, respectively. Two weeks later, the rabbits were inoculated as per the third inoculation. Eleven days later, they were bled to yield 22 and 13 ml of sera from rabbits 1 and 2, respectively.

Quantification of Antibody Titer

To determine whether the serum contained an antibody to PPO, an ELISA assay (Voller et al., 1976) was performed. Antigen, (PPO), at 1 mg ml^{-1} PBS was diluted 1:10, 1:100, 1:200, 1:400, 1:800, 1:1600 and 1:3200 with "coating" buffer (carbonate-bicarbonate, pH 9.6). Then, 100 μl aliquots of each dilution were added to the wells of Falcon 3192 test III microtiter plates (Beckton-Dickinson, Oxnard, CA). Following an overnight incubation at 4°C, the antigen was washed 3x with PBS/Tween 20 prior to adding 200 μl of 10 mg BSA ml^{-1} to each well. After 1 h at 25°C, the antigen was washed 3x with PBS/Tween. Next, 100 μl aliquots of antiserum at 1:10, 1:100, 1:200, 1:400, 1:800, 1:1600 and 1:3200 were added to the wells (Table 1). The control wells consisted of: 1) no antiserum, 2) diluted normal rabbit serum, 3) no secondary antibody or 4) no antigen subsequent to either a 2 h incubation at room temperature or 1 h at 37°C, each well was washed and 10 μl of secondary antibody plus 10 ml PBS/Tween (1:100 dilution) added. The secondary antibody consisted of goat anti-rabbit IgG conjugated to alkaline phosphatase (ELISA Pierce starter kit/rabbit anti-goat IgG H & L, AP). The possible occurrence of antibody in the antiserum was quantified by employing paranitrophenylphosphate (PNP) dissolved in 5 mls of DE buffer kit (diluted) at room temperature. One hundred μl aliquots were added to each well and allowed to sit for 10 minutes. Then, 25 μl of 3M NaOH were added to each well to stop the reaction. The ELISA plate was read at 405nm immediately after preparation by an ELISA microplate reader (Titertek Multiskam MCC/340).

Transmission Electron Microscopy

Mycelia grown upon a defined growth medium containing agar as well as mycelia grown for 3 days in a defined liquid growth medium were harvested and pre-fixed for 30 min in 0.1M, pH 7.2 K_2HPO_4/KH_2PO_4 buffered 2.5% glutaraldehyde (Sabatini et al., 1963), washed twice with buffer (5 min each time) and the sample divided into two. One half was post-fixed with 2% aqueous $KMnO_4$ for 60 min and the other half with 2% phosphate- buffered OsO_4 for 120 min (Taylor et al., 1988). Following post-fixation, hyphae were washed twice (5 min each time) with dH_2O ($KMnO_4$ post-fixed) or buffer (OsO_4 post-fixed) and then dehydrated. Dehydration involved 2 changes each 30%, 50%, 70%, 90% and 100% EM grade acetone (Polysciences, Warrington, PA). Dehydrated hyphae were washed a third time with 100% acetone for 30 min. and then suspended in a 1:1 mixture of acetone/Spurr's reagent for 18 h. Next, the mixture was removed and pure Spurr's (1969) reagent was added for final embeddment in "beam capsules" and polymerization at 70°C for 8 h.

Table 1. Summary of procedure for ELISA ASSAY.

antigen (top) / antiserum (bottom)

	A	B	C	D	E	F	G	H
1	1/10 1/10	1/100 1/100	1/200 1/10	1/400 1/10	1/800 1/10	1/1600 1/10	1/3200 1/10	
2	1/10 1/100	1/100 1/100	1/200 1/100	1/400 1/100	1/800 1/100	1/1600 1/100	1/3200 1/100	
3	1/10 1/200	1/100 1/200	1/200 1/200	1/400 1/200	1/800 1/200	1/1600 1/200	1/3200 1/200	
4	1/10 1/400	1/100 1/400	1/200 1/400	1/400 1/400	1/800 1/400	1/1600 1/400	1/3200 1/400	
5	1/10 1/800	1/100 1/800	1/200 1/800	1/400 1/800	1/800 1/800	1/1600 1/800	1/3200 1/800	
6	1/10 1/1600	1/100 1/1600	1/200 1/1600	1/400 1/1600	1/800 1/1600	1/1600 1/1600	1/3200 1/1600	

	A	B	C	D	E	F	G	H
7	1/10 1/3200	1/100 1/3200	1/200 1/3200	1/400 1/3200	1/800 1/3200	1/1600 1/3200	1/3200 1/3200	
8	1/10 1/NAS	1/100 1/NAS	1/200 1/NAS	1/400 1/NAS	1/800 1/NAS	1/1600 1/NAS	1/3200 1/NAS	
9	1/10 1/NRS	1/100 1/NRS	1/200 1/NRS	1/400 1/NRS	1/800 1/NRS	1/1600 1/NRS	1/3200 1/NRS	
10	1/10 1/NSA	1/100 1/NSA	1/200 1/NSA	1/400 1/NSA	1/800 1/NSA	1/1600 1/NSA	1/3200 1/NSA	
11	1/10	NA 1/10	NA 1/100	NA 1/200	NA 1/400	NA 1/800	NA 1/1600	
12	1/10							

Pipet antigen 100 μl/well of each dilution
Pipet antiserum 100 μl/well

NAS = No antiserum
NRS = Normal Rabbit serum
NSA = No secondary antibody
NA = No antigen

Thin sections were cut with an LKB ultramicrotome and collected onto coated, washed 300 mesh copper grids. Some sections were then stained with lead citrate (Reynolds, 1963) and viewed with an RCA EMU4 electron microscope at an accelerating voltage of 30kV. Multiple hyphae were examined.

Substrate Localization of PPO. To assess whether PPO could be detected within hyphal organelles by a TEM substrate localization method (Czaninski and Catesson, 1974; Eppig, 1974), hyphae from 12 day-old cultures were "pre-fixed" in 2.5% glutaraldehyde in 0.1M, pH 7.4 cacodylate buffer for 30 min. After "washing-out" the glutaraldehyde with cacodylate buffer (2X, 5 min each time), hyphae were incubated with 50 mg 10 ml^{-1} dihydroxyphenylalanine (DOPA) in cacodylate buffer (control cacodylate buffer only) for 18 h on ice and subsequently for 1 h at 37°C. Then, unreacted substrate was removed by washing hyphae with buffer. Next, the hyphae were homogenized with a mortar and pestle into pH 5.0, 100 mM acetate buffer on ice. Finally, the homogenates were dialyzed 18 h against 41 acetate buffer at 4°C and aliquots assayed for PPO and 280nm-absorbing substances as below.

Because enzymatic activity utilizing dihydroxyphenylalanine (DOPA) as a substrate was detected subsequent to light glutaraldehyde pre-fixation, interposing DOPA between glutaraldehyde pre-fixation and OsO_4 post-fixation was performed for the TEM substrate localization of PPO (Czaninski and Catesson, 1974; Eppig, 1974). Hyphae were pre-fixed, post-fixed, dehydrated and embedded for sectioning as above.

Cell Fractionation

Hyphae cultured for 0-16 days were collected from 26 ml cultures as above and sonicated 15 secs on ice with a Heat Systems-Ultrasonics sonicator (model W-22OF) into 4 ml of 250 mM sucrose, 3 mM EDTA and 25 mM Tris, pH 7.2 (Walker et al., 1984). To remove cell walls, sonicates were centrifuged at 500 xg, 15 min prior to the sequential differential centrifugations depicted in Figure 1. Whereas the 1,000 and 10,000 xg centrifugations were accomplished in a Beckman J-21C preparative centrifuge, the 40,000 and 105,000 xg centrifugations were performed with a Beckman L8-70M ultracentrifuge utilizing a Beckman SW-Ti65 rotor. Following centrifugation, the organelles were resuspended in 2 ml pH 5.0, 100 mM acetate buffer with a Ten-Broeck tissue homogenizer. After resuspension, ^{A}280nm-absorbing substances and PPO spc.

acts. were quantified as in assays. Subsequent to assay, the organelles were frozen in liquid nitrogen and stored at -85°C for marker enzyme analysis (see Discussion).

<u>Assay of Total Protein and PPO Specific Activity</u>

Total Protein. The residues from lyophilization of either homogenized mycelia or media were each reconstituted in pH 5.0, 100 mM acetate buffer and subsequently divided into two equal portions. Whereas one of these was dialyzed (Spectrofluor tubing, MW "cut-off" = 14,000) with stirring for 18 h at 4°C against 2-41 acetate buffer, the other fraction was maintained for 18 h at 4°C. Then, non-dialyzed and dialyzed fractions were assayed for both 280nm-absorbing substances and PPO activity as below.

To quantify total protein, 60-140 μl aliquots of both dialyzed and non-dialyzed media were diluted to 2 ml and assayed at 280 nm using quartz cuvettes and a stock of 140 μl bovine serum albumin (Sigma, St. Louis, MO) 100 μl^{-1} acetate buffer for standard curve construction. The quantified 280nm-absorbing substances within aliquots were corrected to mg total protein total fraction volume^{-1}.

Polyphenol oxidase for dialyzed homogenates was assayed by pippetting 10 μl of the reconstituted fractions (usually 2 ml acetate buffer) into an one ml reaction mixture consisting of 100 mM catechol (Fisher Scientific, Norcross, GA) dissolved in "pre-warmed", pH 5.0 100 mM acetate buffer and then maintaining the mixture for 1 min at 30°C with shaking. At 2, 4, 6, 8 and 10 minutes (occasionally for longer time periods), PPO activity was measured by determining absorbances at 440nm. To generate a 0 time "reading", 10 μl aliquots of the fractions were added to "ice-cold" reaction mixtures and their absorbances recorded immediately. The 0 time was subtracted from the 2, 4, 6, 8 and 10 minute readings and the resultant values divided by time to yield A min^{-1} to derive units of PPO activity. The value was multiplied by 1,000, corrected to total volume of the reconstituted fractions and divided by the appropriate mg protein. Thus, PPO activity was expressed as Units where:

$$1 \text{ Unit} = \frac{\text{A min}^{-1} \times 1000}{\text{mg protein}}$$

rather than umol product formed min^{-1} mg protein^{-1} (Lehninger, 1982) because of the lack of certitude regarding the product's identity (Worthington, 1982).

Culture Mycelia on Defined Medium (Kirk and Kelman, 1965)
Containing Agar for 5 Days
Sub-culture Into Defined Liquid Medium
For 6, 9, 12 and 15 Days

Harvest Mycelia By Centrifugation, 500 xg, 5 minutes

Supernatant (Medium) | Pellet Sonicate into 25mM Tris buffer, pH 7.2 containing 250 mM Sucrose and 3 mM EDTA

Centrifuge 500 xg, 15 minutes

Supernatant | Pellet

Centrifuge 1,000 xg, 10 minutes

Supernatant | Pellet

Centrifuge 10,000 xg, 30 minutes

Supernatant | Pellet

Centrifuge 40,000xg, 30 minutes

Supernatant | Pellet

Centrifuge 80,000 xg, 30 minutes

Supernatant | Pellet

Centrifuge 105,000 xg, 60 minutes

Supernatant | Pellet

Adapted from Danley *et al* (1981) and Walker *et al* (1984).

Resuspend all pellets in 2 ml pH 5.0, 100 mM acetete buffer with a Ten-Broeck tissur homogenizer and assay 10 ul aliquots for PPO activity according to Evans and Palmer (1983)

Figure 1. Summary of the Method Employment for the Isolation of Subcellular Organelles from *Coriolus versicolor*. Adapted from Dashek (1970) Walker *et al*., (1984).

Statistical Analyses

Each experiment was routinely replicated and, therefore, data are means and standard deviations (Snedecor and Cochran, 1979) throughout with the number of replicates being experiment dependent as detailed in the Results and Discussion section below.

RESULTS AND DISCUSSION

Cytochemistry

Polysaccharide and total protein appeared to be distributed throughout the length of hyphae, i.e., preferential localizations were not observed in sections of hyphae.

Scanning Electron Microscopy

Scanning electron microscopy of 3 day-old hyphae revealed that the surface topography of the growing tip (pollen tubes, root hairs and fungal hyphae are tip growers) was smooth in contrast to the convoluted appearance of the surface of the non-growing, mature region of the hyphae (Figure 3A).

Immunoelectron Microscopic Localization of PPO

Because SDS-PAGE of growth medium did not yield PPO purified to homogeneity, antibody was prepared against authentic mushroom tyrosinase (see Materials and Methods) which migrated as a single band on SDS-PAGE. Table 2 displays the ELISA quantification of antibody titer in crude serum. It is apparent that PPO antibody was present in the serum and could be detected at a 1:3200 dilution of the serum. The antibody could be separated from other serum proteins by ProteinG chromatography (Figure 2). Whereas the broad peak was not ELISA positive, the "sharp" peak was "tagged" with colloidal gold (Abrahamson, 1986) and subsequently employed for the scanning immunoelectron microscopic localization of hyphal PPO. Colloidal gold was associated with the hyphal sheath (Figure 3). This suggests that before PPO is elaborated free into the growth medium, it associates with the external surface of the hyphae. To determine the mechanism by which the enzyme becomes extracellular, in progress transmission immunoelectron microscopy is serving to determine the time-dependent intracellular distribution of PPO. This approach is an alternative to the substrate localization procedure and PPO kinetic analysis of cell fractions method.

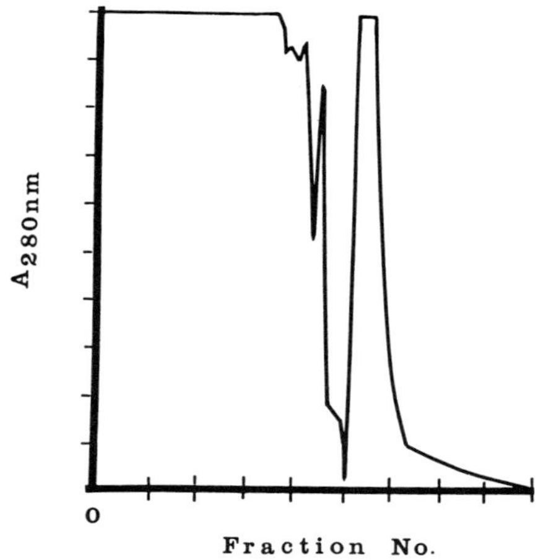

Figure 2. Chromotography of Crude Serum from Rabbits Immunized with Agaricus biosporus on a Protein G Column. One ml Collected Per Fraction.

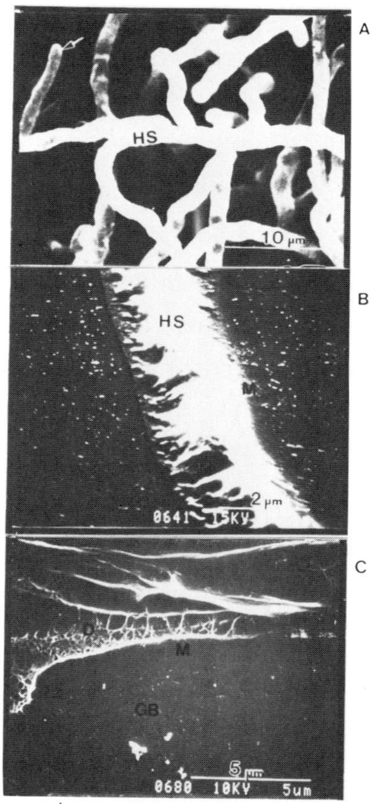

Figure 3. Scanning Electron Micrograph of Hyphae. Panel A: Before Tagged with Antibody. Panel B: Colloidal Gold Labeled with Triton X. Panel C: Colloidal Gold Labeled without Triton X.

Table 2. Quantification of PPO Antibody in Crude New Zealand Rabbit Serum by ELISA Assay[a].

	A	B	C	D	E	F	G	H
1	0.613[b]	0.530	0.515	.0484	0.459	0.396	0.333	0.073
2	0.861	0.725	0.668	0.689	0.530	0.439	0.329	0.068
3	0.864	0.764	0.656	0.606	0.516	0.404	0.260	0.072
4	0.805	0.669	0.587	0.643	0.443	0.276	0.205	0.074
5	0.691	0.553	0.549	o.491	0.376	0.266	0.151	0.069
6	0.682	0.489	0.438	0.365	0.321	0.181	0.142	0.075
7	0.438	0.378	0.357	0.296	0.222	0.151	0.123	0.068
8	0.073	0.079	-.075	0.072	0.081	0.074	0.072	0.071
9	0.109	0.099	0.099	0.084	0.099	0.091	0.090	0.087
10	0.078	0.076	0.073	0.072	0.079	0.074	0.075	0.074
11	0.084	0.188	0.088	0.090	0.089	0.087	0.081	0.075
12	0.093	0.082	0.075	0.081	0.099	0.086	0.077	0.069

[a] = Letters are well rows down, numbers are well rows across Microtiter Plate. [b] = A_{490nm}

Transmission Electron Microscopy

Because SEM demonstrated that the hyphal wall of the growing tip was morphologically different from that of the growing region, TEM was performed to assess whether this difference was reflected in the subcellular organization of the two hyphal regions. In addition, TEM was carried out to ascertain the distribution of the organelles in spatial relation to the wall at the hyphal growing tip.

To establish hyphal ultrastructure, three day old hyphae were pre-fixed in glutaraldehyde and then post-fixed in either OsO_4 or $KMnO_4$. Figure 4A, B compared the subcellular morphologies of hyphae pre-fixed in glutaraldehyde and then post-fixed in OsO_4 (A) or $KMnO_4$ (B).

The non-growing, mature region of a 3-day-old C. versicolor OsO_4 post-fixed hyphae (Figure 4A-b,c) is composed of prominent mitochondria (M) with well-developed cristae, a nucleus (N), lipid (L), electron luscent amorphous masses (EL), electron luscent vesicles (V) containing electron dense material and particles of much smaller and variable dimensions. Some of the latter are most likely non-membrane bound ribosomes. Although not clearly discernible on Figure 4A other micrographs (Figure 4A-b,c) revealed the occurrence of rough endoplasmic reticulum (Taylor et al., 1988).

In contrast to the cytoplasm of the non-growing region, the hyphal growing tip's cytoplasm (Figure 4A-a) appeared to be devoid of mitochondria,lipid and nuclear material. Instead, the cytoplasm seemed to be characterized by an abundance of the electron luscent vesicles (V) containing electron dense material. Certain of these masses appeared to occupy a position adjacent to the plasmalemma at the very tip of the hypha where there seemed to be an accumulation of electron dense material between the plasmalemma (P) and wall (W). This acumulation did not appear to be visible in the space (symplasma) (S) between the plasmalemma and the multilayered wall of the non-growing hyphal region. Like the non-growing hyphal's cytoplasm, that of the growing tip abounded with electron dense particles of very small dimensions (Figure 4A-a). Both the non-growing and growing portions of the hyphae contained what appeared to be invaginations of the plasmalemma (arrows) partially enclosing circular structures of uncertain origin.

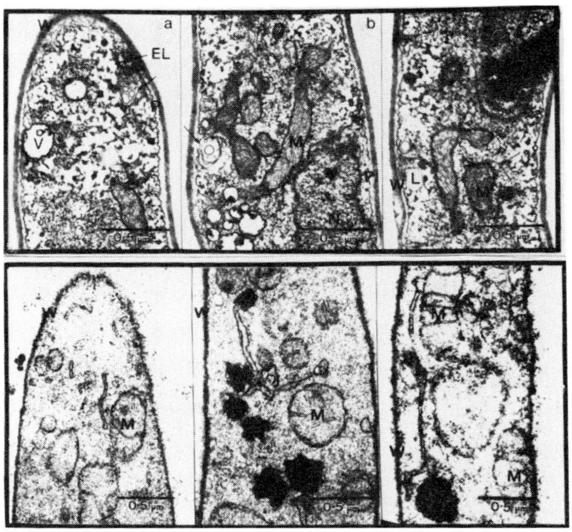

Figure 4A, B. Transmission Electron Microscopy of Hyphae Cultures for Three Days in a Defined Liquid Medium. A - Longitudinal Section through the Hyphal Tip, B and C - Longitudinal Sections of the Hyphal Non-Growing Region. (a Top Panel = Glutaraldehyde Pre-Fixation, OsO_2 Post-Fixation. b Bottom Panel = Glutataldehyde Pre_Fixation, $KmnO_4$ Post-Fixation.)

In contrast to OsO_4 post-fixed hyphae, $KMnO_4$ post-fixed hyphae exhibited leaching but well-preserved mitochondrial membranes. The organization of the growing-tip appeared to be different from that of OsO_4 post-fixed hyphae, i.e., the presence of membrane delimited electron luscent masses was not observed.

The origin, chemical composition and function of the electron luscent amorphous masses are unknown at the present time. The observation that certain masses were closely associated with the plasmalemma at the very tip may reflect their involvement in secretion of intracellularly-produced chemicals to the external milieu. Indeed, the tip-most 3-5 μm of the growing lily pollen tube contains Golgi-complex-derived vesicles whose contents are apparently deposited into the tube wall (Rosen et al., 1964). Furthermore, accumulations of vesicles have also been reported in the growing tips of Achylia hyphae (Hill and Mullins, 1980). Whether the electron luscent masses are vehicles for the transport of intracellular chemicals to the wall and/or external milieu could be established by high resolution radioautography following pulse-chase experiments with appropriate radionuclides (Dashek and Rosen, 1966) and chemical characterization of the masses following their possible isolation and purification from hyphae. If the electron luscent amorphous masses are similar to the vesicles present in Lilium pollen tubes (Dashek and Rosen, 1966) and Achylia (Hill and Mullins, 1980) hyphal tips, they may counterstain with phosphotungstic (PTA) acid which when combined with chromic acid as PTA-chromic acid-positive material cold suggest the presence of glycoproteins.

The origins of the amorphous masses, which upon examination at high magnification demonstrated that they were, at least, in part membrane-bound, are unknown. If the masses represent the coalescence of vesicles, they do not appear to have originated from Golgi complex-derived vesicles as the micrographs in Figure 4A,B do not possess distinct dictyosomes.

The function of the purported plasmalemma invaginations (arrows) remains to be established. Indeed, the possibility exists that they are not invaginations but rather membrane-encased cytoplasmically-derived structures en route to the wall and/or the external milieu.

The copious occurrence of mitochondria within the non-growing region of the hyphae suggests a high rate of metabolic activity. This would be in keeping with a tip-growing system as these systems exhibit very rapid growth rates (Rosen et al., 1964).

Lastly, it may be of value to compare the ultrastructures of hyphae grown in liquid culture with those grown in situ. In this connection, Hale and Eaton (1985) have examined the ultrastructure of soft rot-fungal hyphae found in wood cell walls.

Substrate Localization of Polyphenol Oxidase Within Hyphae

Table 3 presents the spc. acts. of PPO for hyphae cultured 12 days and subsequently either "pre-fixed" with buffered glutaraldehyde or not "pre-fixed". The substrates that were employed were catechol and dihydroxyphenylalanine (DOPA). When these were withheld, PPO activity was not present in either buffered glutaraldehyde-treated or buffer-treated hyphae. If hyphae were treated with either cacodylate buffer or cacodylate-buffered glutaraldehyde, the PPO spc. act. was greater than that for hyphae which were not treated when DOPA was employed as the substrate. In contrast, if catechol was utilized as the substrate, the PPO spc. act. was similar if the hyphae were either not treated or treated with cacodylate. In contrast, a diminution in PPO spc. act. was observed when hyphae were "pre-fixed" with glutaraldehyde.

Table 3. Comparison of Polyphenol Oxidase Specific Activities for Mycelia Cultured for 12 Days and Subsequently Not-Treated or Treated with Glutaraldehyde in Cacodylate Buffer or Cacodylate Buffer Only[a].

Mycelial Treatment Condition	Substrate	Polyphenol Oxidase Specific Activity (Units)[a]	
2.5% Glutaraldehyde in 0.1M Cacodylate Buffer pH 7.4, 30 Minutes on ice	10mM catechol	15.12±	10.66
	50mg $10ml^{-1}$ Dihydroxy-phenylalanine	146.12±	89.90
	None	0.00±	0.00
0.1M Cacodylate Buffer pH 7.4, 30 minutes on ice	100nM Catechol	41.84±	22.60
	50mg $10ml^{-1}$ Dihydroxy-phenylalanine	128.40±	86.31
	None	0.00±	0.00
None, 30 minutes on ice	100nM Catechol	43.14±	53.02
	50mg $10ml^{-1}$ Dihydroxy-	71.56±	0.00

[a]Data are xs and os where N=3 (glutaraldehyde in cacodylate) and 4 (cacodylate and no treatment) replicate flasks.

The detection of PPO in hyphal homogenates following "light" pre-fixation with glutaraldehyde supports the findings of Czaninski and Catesson (1974) and Eppig (1974) who reported PPO activity subsequent to pre-fixation with glutaraldehyde. Because PPO activity against DOPA occurred after pre-fixation and OsO_4 post-fixation was performed to obtain the TEM substrate localization of PPO (Martyn et al., 1979; Vaughan and Duke, 1981) within hyphae. This in progress effort together with the utilization of colloidal gold "tagged" antibody (Knox and Clarke, 1978) for the TEM immunocytochemical localization of PPO will be reported at another time (Moore et al., in preparation).

Cell Fractionation

Another approach to defining the organelles mediating PPO secretion centered about following alterations in PPO spc. act. in subcellular fractions prepared by differential centrifugation of hyphal homogenates (Figure 1). The time-dependent changes in PPO spc. act. in these fractions are depicted in Figure 5. Whereas the spc. act of PPO in both the total homogenate and 105,000 xg supernatant remained constant with time, marked alterations in this activity were apparent for other fractions. For example, PPO spc. act. within the 10,000 40,000 and 105,000 xg pellets increased from day 0 to day 3 of culture and thereafter declined. In contrast, PPO spc. act. for the 1,000 pellet rose to day 6 and then diminished. If mycelia were homogenized with a mortar and pestle, the pellet was rendered cytosolic. It is pre-mature to render conclusions regarding PPO synthesis and transport via the organelles constituting these fractions until an in progress combined TEM and marker enzyme analysis (Quail, 1979) is completed. Each fraction is being assayed for pH 6.0 and 9.0 K-stimulated ATPase (mitochondrial and plasmalemma marker enzymes, respectively - Hodges and Leonard, 1974), inosine diphosphatase (dictyosome marker, Dauwalder et al., 1969) and NADH cytochrome C reductase (endoplasmic reticulum marker, Hatefi and Rieske, 1973).

In summary, the current results of hyphal TEM, the projected results of TEM substrate localization and immunogold labeling localization of hyphal PPO and the present and future results of combined cell fractionation/kinetic analysis should yield the route of PPO secretion. This may possibly enable the control of secretion, e.g., are there differential energy gradients for synthesis vs. secretion? This is currently being explored through the time-dependent addition of

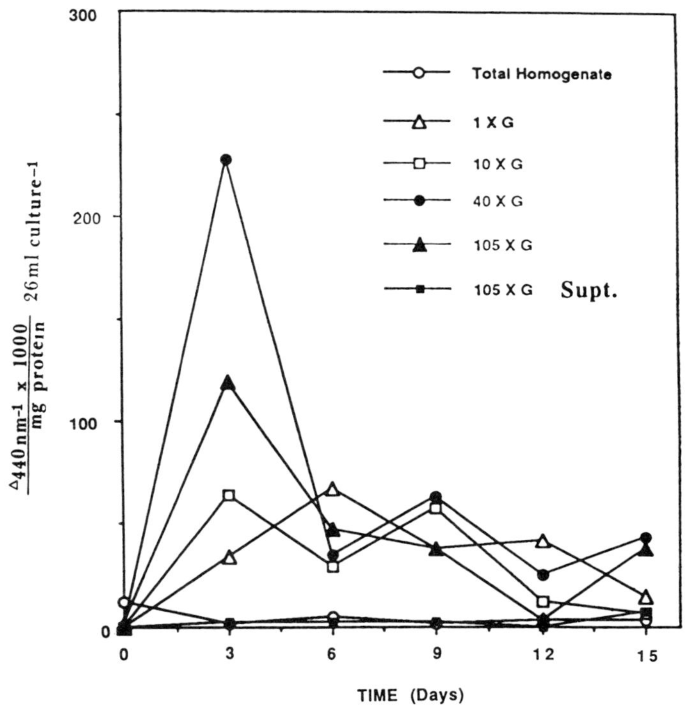

Figure 5. Time-Dependent Changes in PPO Spc. Act. in Subcellular Fractions Prepared by Differential Centrifugation of Hyphal Homogenates. The RCFs are XI,000 each. Data are Means Where N = 5-18 Replicates.

respiration inhibitors such as sodium azide and sodium fluoride (Ross, 1974) followed by the quantification of intracellular and extracellular PPOs.

Acknowledgements

We thank Ms. Doris Donnell for typing the manuscript and Mr. Subrasua Ibim for graphics. This research was supported by DOE-ECUT sub-contract #957697 and its extension 19K SG071V, a Sigma Xi Grant-in Aid to NLM, and NIH-RCMI Grant #G12RR03062. The technical assistance of Dr. Rosannah Taylor is appreciated.

Finally, experiments have been initiated to quantify the time-dependent alterations of intracellular and extracellular cellulose and ligninase because of their significances to the paper-pulp industry.

SUMMARY

Coriolus versicolor elaborates lignocellulolytic enzymes which possess industrial and agricultural applications. Thus, an attempt is in progress to overproduce and enhance the secretion of these enzymes using polyphenol oxidase (PPO) as a model. Detailed light-electron microscopies/biochemistry establish the PPO (o-diphenols → o-diquinones and oligomerizes syringic acid) secretion route permitting its regulation. TEM of glutaraldehyde pre- and OsO_4 post-fixed hyphae revealed mitochondria, ER, ribosomes, vacuoles & osmiophilic bodies in the non-growing region's cytoplasm, but electron luscent masses appressed to the plasmalemma in the growing tip's. To verify this, results from OsO_4 and $KMnO_4$ post-fixations are compared. TEM PPO localization involved interposing dihydroxyphenylalanine (DOPA) between pre- and post-fixations. As an additional approach, antibody was prepared by immunizing rabbits with purified PPO and their bleed at 81 d followed by "boosting". An ELISA indicated that crude serum could be employed at 1:3200. Chromatography of crude serum on protein G yielded a single, ELISA-positive, UV-absorbing peak. The putative antibody was "tagged" with colloidal gold and immunoelectron microscopy localized PPO over the disrupted hyphal sheath. As for PPO in centrifuged organelles, O-16 d fractions' PPO spc. acts. indicated that it was particulate rather than cytosolic. TEM and marker enzyme analysis of fractions are mentioned.

REFERENCES

Abrahansom, D.R. (1986). Post embedding gold immunolocalization of laminin to the lamina rara interna, lamina densa and lamina rara externa of renal glomerular basement membranes. J. Histochem. Cytochem, 34, 847-853.

Czaninski, Y. and Catesson, A.M. (1974). Polyphenol oxidases (plants). In: Electron Microscopy of Enzymes. (M.A. Hayat, ed.), Van Nostrand Reinhold, NY.

Dashek, W.V. and Rosen, W.G. (1966). Electron microscopical localization of chemical components in the growth zone of Lilium pollen tubes. Protoplasma, 61, 192-204.

Dashek, W.V., Moore, N.L., Williams, A.C., Williams, A.L., O'Rear, C.E., and Llewellyn, G.C. (1990). Wood-decay - A mini review. In: Biodeterioration Research III, pp. 391-404, (Llewellyn G.C. and O'Rear, C.E., eds.), Plenum Press, NY.

Dauwalder, M., Whaley, W.G., and Kephart, J.E. (1969). Phosphatase and differentiation of the Golgi apparatus. J. Cell Sci., 4, 455-497.

Eppig, J.J. (1974). Tyrosinase In: Electron Microscopy of Enzymes. pp. 79-91 (M.A. Hayat, ed.), Van Nostrand Reinhold, NY.

Evans, C.S. and Palmer, J.M. (1983). Ligninolytic activity of Coriolus versicol or. J. Gen. Microbiol., 129, 2103-2108.

Fahraeus, G. and Reinhammer, B. (1967). Large scale production and purification of laccase from culture of the fungus Polyporus versicolor and some properties of laccase. Acta Chem. Scand,. 21, 2367-2378.

Fric, F. (1976). Oxidative enzymes, Physiological Plant Pathology. pp. 617-631 In: (R. Hertefussi and Williams, P.H., eds). Springer Verlag, Heidelberg.

Green, F. III, Clausen, C.A., Larsen, M.J., and Highley, T.L. (1991). Ultrastructural characterization of the hyphal sheath of Postia placenta by selective removal and immunogold labeling. In: Biodeterioration and Biodegradation 8, pp. 530-532, (H.W. Rossmore, ed.), Elsevier Applied Science, London and NY.

Hale, M.D. and Eaton, R.A. (1985). The ultrastructure of soft rot-fungi, I. Fine hyphae in wood cell walls. Mycologia., 77, 447-463.

Harvey, P.J., Schoemaker, H.E. and Palmer, J.M. (1987). Lignin degradation by white fungi. Plant Cell and Environ., 10, 709-714.

Hatefi, Y., and Rieske, J.S. (1973). The preparation and properties of DPNH-cytochrome reductase (Complex I-II) of the respiratory chain. In: (R.W. Estabrook and M.E. Pullnam, Eds.), Methods In Enzymology, Academic Press, NY.

Higuchi, T. (1985). Biosynthesis and Biodegradaton of Wood Components. Academic Press, NY.

Hill, T.W. and Mullins, J.T. (1980). Hyphal tip growth in Achylia. II. Subcellular localization of cellulase and associated enzymes. Can. J. Microbiol., 26, 1141-1146.

Hodges, T.K and Leonard, R.T. (1974). Purification of plasma membrane bound adenosine triphosphatase from plant roots. In: Methods in Enzymology, 31B. (S. Fleisher and I. Packer, eds.), Academic Press, NY.

Kirk, K.T. and Kelman, A. (1965). Lignin degradation as relaetd to polyphenoloxidases of selected wood decaying basidiomycetes. Phytopahtol. 55, 739-745.

Knox, R.B. and Clarke, A. (1978). Localization of proteins and glycoproteins by binding to labeled antibodies and lectins. In: Electron Microscopy and Cytochemistry of Plant Cells. pp. 150-185, (J.L. Hall, ed.), Elsevier/North Holland Biomedical Pess, Amsterdam.

Laemmli, U.K. (1970). Cleavage of structural proteins during the assembly of the head of bacteriophaJean-Marc. Nature, 227, 680-685.

Liu, Shu-Yen, Minard, R.D and Bollag, Jean-Marc, (1981). Oligomerization of syringic acid, a lignin derivative, by phenoloxidase. Soil Sci. Am. J,. 45, 1100-1105.

Maniatis, T.E., Fritsch, T. and Sambrook, J. (1982). Molecular Cloning: A Laboratory Manual. Cold Spring Harbor, NY.

Martyn, R.D., Samuleson, D.A., and Freeman, T.E. (1979). Ultrastructural localization of polyphenoloxidase activity in leaves of healthy and disease water hyacinth. Phytopathol., 69, 1278-1287.

Moore, N.L., Mariam, D.H., Williams, A.L., and Dashek, W.V. (1989). Substrate specificity, de novo synthesis and partial purificaton of polyphenol oxidase derived from the wood-decay fungus, Coriolus versicolor. J. Indust. Microbiol., 4, 349-364.

Quail, P.H. (1979). Plant cell fractionation. Ann. Rev. Plant. Physiol., 30, 425-484.

Reynolds, E.S. (1963). The use of lead citrate at high pH as an electron opaque stain in electron microscopy. J. Cell Biol., 17, 208-212.

Rosen, W.G., Gawlik, S.R., Dashek, W.V. and Siegesmund, K.A. (1964). Fine structure and cytochemistry of Lilium pollen tubes. Amer. J. Bot., 51, 61-67.

Rosenberg, S.L. (1979). Physiological studies of lignocellulose by the thermotolerant mold, Chrysosporium pruninosum. Dev. Ind. Microbiol., 20, 133-142.

Ross, C.W. (1974). Plant Physiology Laboratory Manual. Wadsworth Publishing Co., Inc., Belmont, CA.

Sabatini, D., Bensch, K,. and Barnett, R.J. (1963). Cytochemistry and electron microscopy. J. Cell. Biol., 17, 19-58.

Sheehan, D. and Hrapchak, B.B. (1980). Theory and Practice of Histotechnology. The C.V. Mosby Company, St. Louis, MO.

Shigo, A.L., (1984). Compartmentalization: A conceptual framework for understanding how trees grow and defend themselves. Ann. Rev. Phytopathol., 22, 181-214.

Snedecor, G.W. and Cochran, W.C. (1979). Statistical Methods. Iowa State University Press, Ames, IA.

Spurr, A.W. (1969). A low viscosity epoxy embedding medium for electronmicroscopy. J. Ultrastruct. Res., 26, 31-43.

Taylor, R., Dashek, W.V., Williams, A.L., Llewellyn, G.C., Shortle, W.C., and Mayfield, J.E. (1988). Ultrastructure of the wood-decay fungus, Coriolus versicolor, in relation to a catechol-induced bimodal growth response. International Biodeterioration 24, 343-358.

Taylor, R., Llewellyn, G.C., Mayfield, J.E., Shortle, W.C., and Dashek, W.V. (1987). Time-dependent appearance of extracellular polyphenol oxidase in relation to the bimodal growth response of Coriolus versicolor to catechol. In: Biodeterioration Research I. pp. 63-74. (G.C. Llewellyn and C.E. O'Rear eds.), Plenum Press, NY.

Voller, A., Bidwell, D. and Ballett, A. (1976). Microplate enzyme immunoassay for the immodiagnoses of virus infections. Manual of Clinical Immunology (Rose, V. and Friedman, H., eds.).

Van der Meer, J.M., Rijkes, B.A., and Ferranti, M.P. (1987). Degradation of Lignocellulosics in Ruminants and or Industrial Processes. Elsevier Science Publishers, Amsterdam, pp. 120.

Vaughn, K.C. and Duke, S.O. (1981). Tissue localization of polyphenol oxidase in Sorghum. Protoplasma, 108, 319-327, (O'Rear eds.), Plenum Press, N.Y.

Walker, S.J., G.C. Llewellyn, E.B. Lillehoj, and W.V. Dashek. (1984). Uptake and distribution of aflatoxin B in excised, soybean roots and toxin effects on root elongation. J. Environ. Exp. Bot., 24:113-122.

Williams, A.C., Moore, N.L., Dashek, W.V., and Williams, A.L. (1990). Coriolus versicolor, a model system to investigate the biotechnology of wood-deteriorating enzymes. In: Biodeterioration Research 3, pp. 405-417, (G.C. Llewellyn and C.E. O'Rear eds.), Plenum Press, NY.

Williams, A.L., Williams, A.C., Moore, N.L. and Dashek, W.V. (1991). Biotechnology of wood deteriorating enzymes synthesized and secreted by Coriolus versicolor, a white rot basidiomycete. In: Biodeterioration and Biodegradation 8, pp. 547-549, (Rossmoore, H.W., ed.), Elsevier Applied Science, London.

Worthington Enzymes and Related Biochemicals. (1982). Worthington Diagnostic Systems, Inc., Freehold, N.J.

Zadrazil, F and Reiniger, P. (1988). Treatment of Lignocelulosics With White Rot Fungi. Elsevier Science Publishers, Amsterdam, pp. 122.

INFLUENCE OF LIGNIN TYPE ON THE DECAY OF WOODY ANGIOSPERMS BY *TRAMETES VERSICOLOR*

JOHN R. OBST*, TERRY L. HIGHLEY, and REGIS B. MILLER, USDA Forest Service, Forest Products Laboratory, Madison, WI 53705-2398, USA

INTRODUCTION

Numerous studies of the fungal decay of wood have been conducted, especially in the past few decades. The major motivation in the search to understand wood decay is the potential application of research results in essentially two areas: (1) decay protection of wood and wood products in use and (2) biological processing of wood for the production of pulp for paper manufacture, pulp bleaching, and effluent treatments.

Two types of wood decay by white-rot fungi have been described (Blanchette et al., 1990). In one type, selective removal of lignin occurs, leaving the secondary cell walls relatively intact. The second type of decay is nonselective, removing lignin and carbohydrates at about the same rate. In this latter type, fiber cell walls generally are removed preferentially, with middle lamellae and cell corners showing decay resistance (Cowling, 1961; Wilcox, 1968). This sequence of attack has often been related to the distribution of lignin and polysaccharides. The lignin content is highest in the compound middle lamella (hereafter referred to as middle lamella) and cell corners and imparts greater decay resistance to these regions. The polysaccharides are more abundant in the secondary wall, contributing to a faster degradation of the secondary wall.

The faster decay of angiosperm woods than decay of gymnosperm woods may be partly explained by the effect of the different amounts and different types of lignin in these woods (Butcher and Nilsson, 1982; Faix et al., 1985;

Biodeterioration Research 4, Edited by G.C. Llewellyn *et al.*, Plenum Press, New York, 1994

Highley, 1982; Highley and Illman, 1990; Highley and Murmanis, 1987; Kirk et al., 1975; Leightley, 1984; Nilsson et al., 1988). The observed decay resistance for woody angiosperm middle lamellae, cell corners, and vessel elements is generally attributed to two lignin effects: the lignin concentration in these regions is greater, and the type of lignin present in these regions is more like that of woody gymnosperms.

The lignin in the normal wood of most gymnosperms is composed almost entirely of guaiacylpropyl (G) units that arise from the lignin precursor coniferyl alcohol. However, lignin in woody angiosperms, in addition to G units, often has a significant syringylpropyl (S) unit composition derived from the additional lignin precursor sinapyl alcohol. The exact distribution of S and G units in angiosperm lignins has not been determined unambiguously. Futhermore, the distribution is unlikely to be the same for all woods. White birch (Betula papyrifera) has been the most studied, and it was initially determined by ultraviolet microscopy that the middle lamella between fibers had lignin with a S:G ratio of 1:1 (Fergus and Goring, 1970). A subsequent study indicated that the middle lamella lignin between fibers was "mostly guaiacyl-type" (Musha and Goring, 1975). A third investigation of white birch (Saka and Goring, 1988) indicated that the middle lamella lignin had a S:G ratio of 12:88. For the purposes of our discussion, the more recent description of white birch will be used: middle lamella, cell corner, and vessel wall lignins are predominantly G lignins; the secondary wall lignin of fibers is mainly S lignin; whereas, the lignin in walls of ray parenchyma cells is considered to be a 51:49 S:G mix (Saka and Goring, 1988). However, the lignin concentration determined by a different method for birch (Betula verrucosa) ray cell walls was considerably greater (Eriksson et al., 1988) than that found for white birch rays (Saka and Goring, 1988). Increased lignin content could affect the estimate of the syringyl content of the rays. Note that the distribution of S and G units found in white birch may not be the same in all angiosperm woods (Musha and Goring, 1975; Obst and Landucci, 1977).

The generally observed pattern of decay resistance in woody angiosperms can be correlated with the guaiacyl-rich lignin regions in white birch. However, Highley and Murmanis (1987) reported that Trametes versicolor, a nonselective white-rot fungus, severely degraded the middle lamella and cell corners of sweetgum without significant degradation of adjacent cell walls. This study reexamines this finding and extends the investigation to describe the decay of vessels and rays as well as fibers. Furthermore, we address the influence of lignin type by using additional woods that have a wide range of S:G ratios and different lignin type distributions (Musha and Goring, 1975; Obst and Landucci, 1977).

MATERIALS AND METHODS

Wood Samples and Decay Tests

Sapwood blocks 6.35 by 6.35 by 3.18 mm (the small dimension in the fiber direction) were cut from sweetgum (Liquidambar styraciflua L.), white birch (Betula papyrifera Marsh.), red oak (Quercus rubra L.), American elm (Ulmus americana L.), red maple (Acer rubrum L.), boxelder (Acer negundo L.), and pacific madrone (Arbutus menziesii Pursh). The blocks were numbered, conditioned to constant weight at 27C and 70% relative humidity, and then weighed. Blocks were decayed by Trametes (=Coriolus) versicolor(L. ex Fr.) Pilate (MAD–697), using the American Society for Testing and Materials Standard soil-block method (1971). Following incubation, the blocks were removed, reconditioned, weighed, and their weight losses calculated. Noninoculated blocks served as controls.

Light Microscopy

Degraded samples were sectioned on a sliding microtome without embedding or chemical treatment. Sections were stained with picro-aniline-blue and safranin, dehydrated with ethanol, and mounted in a toluene-based permanent mounting media.

Scanning Electron Microscopy

Degraded samples removed from bottles at 2 and 4 weeks were fixed in 3% glutaraldehyde in 0.1 phosphate buffer, pH 7.0, in an evacuation chamber for 12 h. The samples were then rinsed in four changes of distilled water for 1 h and infiltrated with polyethylene glycol (PEG) (1:1) at 60C for 12 h, followed by infiltration in 100% PEG at 60C for 12 h. The samples were embedded in 100% PEG at 60C and then placed in a refrigerator at 4C to form white solid blocks. Using a razor blade and a stereo microscope, the faces of the embedded blocks were cut smoothly and rinsed in five changes of distilled water for 1-1/2 h. The wood blocks were then dehydrated in a series of increasing concentrations of acetone (30%, 50%, 70%, 90%, 95%, and 100%) for 2 h. This was followed by critical-point drying in liquid carbon dioxide and coating with gold for 2.5 min. Specimens were observed with the aid of a Hitachi S–530 scanning electron microscope at an accelerating voltage of 20 to 25 kV.

Transmission Electron Microscopy

After 2, 4, and 6 weeks, the degraded blocks were removed from bottles and fixed in formaldehyde-glutaraldehyde mixture in 0.1 M phosphate buffer, pH 7.2, postfixed in 1.33% OsO_4 in s–collidine buffer, pH 7.2, and stained with 1.5% uranyl acetate. All samples were dehydrated in acetone series, followed by propylene oxide. Embedment was in Araldite, Epon, and dodecenyl succinic anhydride in proportions 1:1:3. Polymerization was carried out at 80C, and micrographs were taken with a Hitachi H–300 electron microscope.

Chemical Analyses

We followed the method of Effland (1977) for lignin analysis and the method of Pettersen et al., (1985) for carbohydrate analysis. For alkaline nitrobenzene oxidations, the method used was that as described by Kirk and Obst (1988). Except for boxelder, the molar syringaldehyde to vanillin ratio was divided by three to give a more accurate estimate of the syringyl:guaiacyl (S:G) ratio (Manders, 1987).

RESULTS AND DISCUSSION

Weight Loss by Decay

Weight losses were determined after 2, 4, and 6 weeks of incubation. Although all woods suffered extensive weight loss after 6 weeks, the relative rates and extents of decay differed (Table 1). Madrone was the most resistant to decay; whereas, white birch was the least resistant. Boxelder, red oak, and elm decayed somewhat more rapidly than did madrone. Decay was rapid in sweetgum and red maple and was more similar to that in birch.

The S:G ratios, as approximated by nitrobenzene oxidation analysis, of the sound woods are also given in Table 1. It has often been argued that the ease of decay of woody angiosperms, compared to that of gymnosperms, is due to the syringyl content of angiosperm lignins. This hypothesis has been extended to suggest that among angiosperms, those with high syringyl lignins will decay more readily than those with high guaiacyl lignins. From the results using seven angiosperms of varying syringyl content in this study, no simple correlation was apparent between syringyl content and the rate or extent of decay. Indeed, the woods with the highest and lowest syringyl contents (madrone and boxelder) are the most resistant to degradation by Trametes versicolor. Birch and sweetgum, which have high and similar syringyl contents, exhibit different behavior on decay. Red maple, which has a moderate syringyl content, is easily decayed. Red oak, which has a syringyl content similar to that of red maple, is more resistant to decay than red maple. Because Trametes versicolor is a very active white rot, the differences in the rate of decay might be greater if less-active fungi were used.

Table 1. Weight Loss of Wood After Incubation With Trametes versicolor.

Species	S:G[a]	Weight loss (%) after 2 weeks	4 weeks	6 weeks
Madrone	2.3	11.5	29.0	70.0
Boxelder	0.1	15.0	45.4	77.0
Red oak	0.8	16.2	44.3	78.8
American elm	0.6	16.5	49.2	79.2
Sweetgum	1.1	16.5	52.8	89.3
Red maple	0.8	22.5	61.2	91.8
White birch	1.2	30.2	55.0	94.0

[a]Syringyl:guaiacyl ratio.

Correlation of the decay behavior of various woods with the syringyl content of their lignin may be very difficult because, in most cases, many significant differences exist: the amount and effect of extractives, the pH of the wood, the composition and distribution of polysaccharides, the amount of lignin, the distribution of lignin, the distribution of S and G units in the lignin, the relative proportion of fibers, vessels, and parenchyma cells, cell wall thickness, vessel element size, the presence of tyloses, and the ratio of earlywood to latewood. With this in mind, it is interesting to compare the decay of boxelder and red maple. Of the woods used, boxelder and red maple are the most similar in floral morphology, anatomy, and chemical composition, although lignin and lignin S and G distributions have not been described. Red maple lignin has a considerably higher syringyl content than does boxelder lignin and decayed faster. Although this suggests that the difference in the syringyl contents of the lignin explains the decay difference, care must be exercised in claiming that lignin composition is the dominant factor. It is likely that a complex interplay takes place among all the physical and chemical characteristics of wood in the decay process. It must also be recognized that thousands of white-rot fungi differ markedly in decay (Eriksson and Blanchette, 1990).

Morphology of Decay Summary

The decayed woods were examined using light microscopy, scanning electron microscopy (SEM), and transmission electron microscopy (TEM). Fungal growth was assessed by visual estimation. All woods exhibited a similar pattern of decay. After 2 weeks of decay, hyphae were very evident in fiber and vessel lumens but generally did not appear in ray parenchyma lumens. Degradation of fiber cell walls was slight or nonexistent. Decay typically observed after 4 weeks is represented in Figure 1. Fiber cell walls were extensively degraded, although latewood fibers were less degraded than were earlywood fibers. In some cases, cell wall thinning was extensive, but middle lamellae and cell corners remained (Fig. 2). This general pattern of degradation of angiosperm fibers is the same as that of gymnosperm tracheids (Highley and Murmanis, 1987). Vessels were much less degraded than were the fibers, and ray cells appeared intact (Fig. 1). After 6 weeks of decay, all fiber cell walls were degraded, but middle lamellae and cell corners often persisted. Although hyphae were present in the lumens of vessels and rays, these cell types were much less degraded and appeared mainly intact.

The decay resistance of rays and vessels compared to the fibers is interesting. Vessel wall lignin is characterized as predominantly guaiacyl type; whereas, ray parenchyma wall lignin has a higher syringyl content (Fergus and Goring, 1970; Musha and Goring, 1975; Saka and Goring, 1988). If lignin type exerted the greatest influence on decay, vessels would be expected to be decay resistant and rays and fibers should decay more readily. The decay resistance of vessels was verified, but the decay resistance of rays does not correlate with lignin type.

Particularly interesting is that the decay pattern for madrone is the same as that for the other woods. Madrone was chosen for this study because its ray cell and vessel wall lignins have a high syringyl content similar to the lignin in its fiber cell walls (Musha and Goring, 1975). If syringyl content is the determinant of decay, madrone ray cell and vessel element walls should decay as rapidly as fiber walls, and this clearly was not observed.

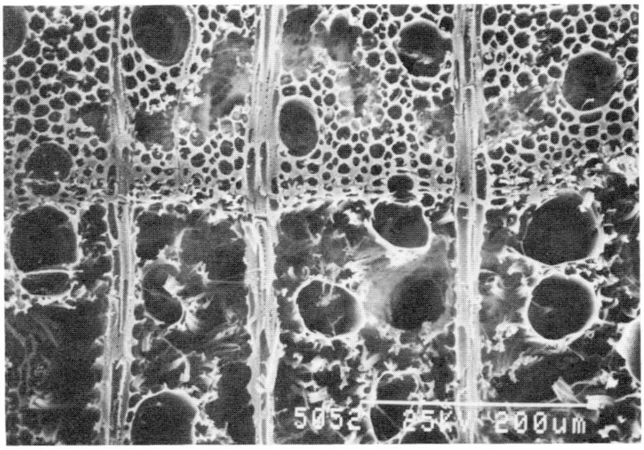

Figure 1. SEM of Red Maple, Decayed 4 Weeks. Vessels and ray cells are more resistant to decay than the fibers. Thin-walled earlywood fibers are degraded before the thicker-walled latewood fibers.

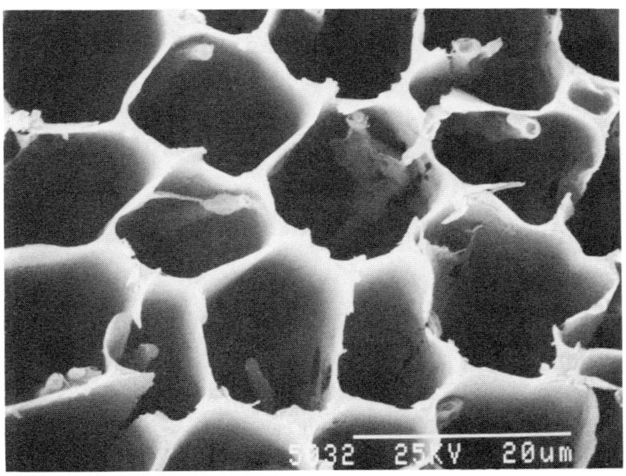

Figure 2. SEM of White Birch, Decayed 4 Weeks. Compound middle lamellae and cell corners are all that remain of the fibers in this region.

A comparison of the degradation of birch vessels to vessels of the other woods also does not support the suggestion that the syringyl content determines the rate of decay. Except for madrone, vessel wall lignin was suggested to be uniformly guaiacyl rich for those woods examined (Musha and Goring, 1975). Although the vessels of all woods were resistant to decay, birch vessels appeared to be the least degraded. Yet, it seems likely that the vessels of the other woods, and especially boxelder, would not have a higher syringyl content than that of birch.

As mentioned previously, lignin concentration may have a significant role in wood decay. Most determinations of lignin distribution in woody angiosperms have been on birch. Because angiosperms differ significantly in chemical composition and anatomy, it is likely that the lignin distributions will vary also. However, if the lignin concentrations found in Betula verrucosa are considered representative of those woods used in this study, then the decay pattern can be correlated roughly with lignin concentration. The relative lignin concentrations found for fiber, ray parenchyma, vessel walls, and fiber cell corners were 1, 1.5, 1.6, and 3, respectively (Eriksson et al., 1988). Higher lignin concentration corresponds to the observed order of decay resistance.

One goal of this study was to reinvestigate the previous finding of Highley and Murmanis who reported the preferential degradation of fiber cell corners in sweetgum (Highley and Murmanis, 1987). Although we did not find evidence to support the finding of preferential cell corner decay, we observed cell corner degradation of fibers in several instances for red oak, red maple, American elm, madrone, and sweetgum. However, we did not observe cell corner degradation in boxelder and birch. Examples of cell corner degradation are shown for red maple and red oak in Figures 3 and 4. Fungal hyphae can be seen in some of what appears to be the cell corner voids, and in some cases, erosion of the cell corner and even the cell wall, is apparent. Originally some of these areas may not have been intercellular spaces or voids. According to Daniel et al. (1991), middle lamella cell corners in many species show a nonhomogeneous structure or less dense regions, which they call electron-lucent regions. Their results show variable lignin content in these areas suggesting incomplete lignification and an elevated level of nonlignified components (presumably carbohydrates). We suggest that fungal hyphae invade intercellular spaces or electron-lucent regions through pit canals, bore holes, or minute cracks in the cell wall. We suggest that

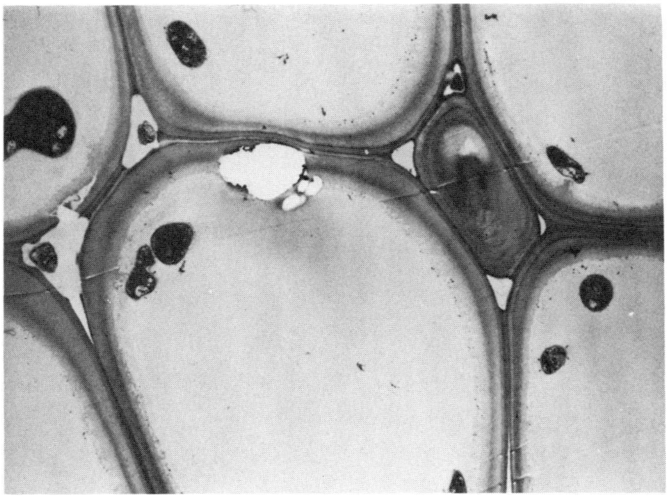

Figure 3. TEM of Red Maple, Decayed 2 Weeks. Area shows fiber cell corner intercellular spaces; some spaces contain fungal hyphae. Most voids show erosion of the cell corner; in some cases, the degradation extends into the cell wall.

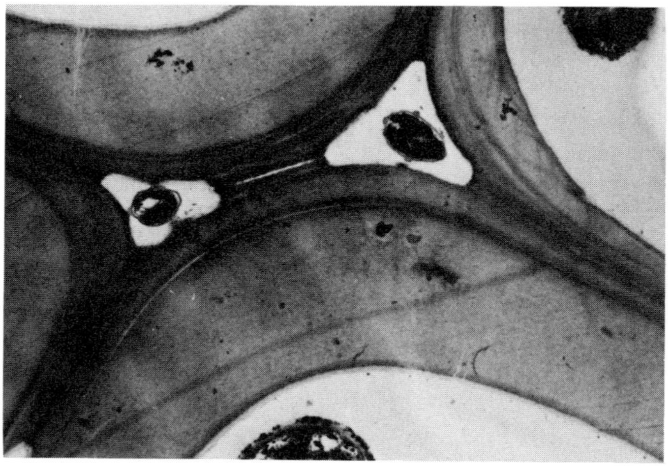

Figure 4. TEM of Red Oak, Decayed 2 Weeks; area shows hyphae in fiber cell corner intercellular spaces.

these openings are existing intercellular spaces invaded by fungal hyphae through pit canals, bore holes, or minute cracks in the cell wall. Although intercellular spaces are not common among fibers, they can occur sporadically in all species (Carlquist, 1988).

Hyphae were also found in intercellular spaces (electron-lucent regions?) between ray parenchyma cells (Fig. 5). We suggest that these openings are naturally occurring intercellular spaces that are invaded by fungal hyphae. Intercellular spaces are more common in rays than in other parts of the secondary xylem (Carlquist, 1988), which explains the abundance of intercellular spaces in Figure 5.

A final microscopic observation concerned the degradation of gelatinous fibers. Gelatinous fibers occur typically in tension wood but are present

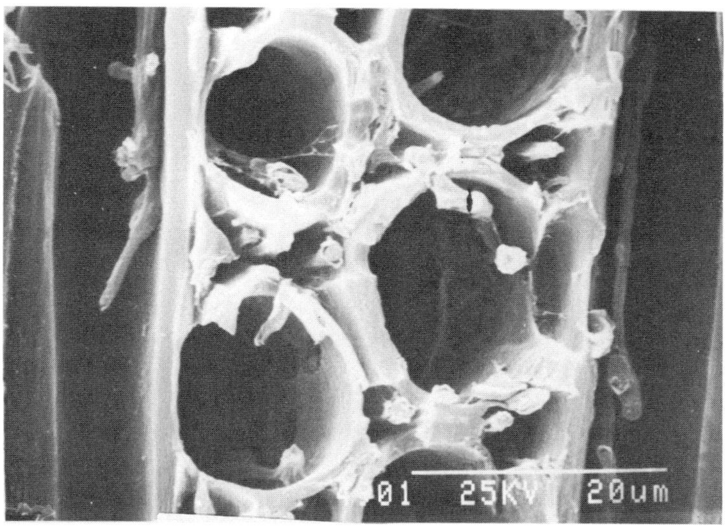

Figure 5. SEM of Sweetgum, Decayed 2 Weeks. No significant degradation had occurred, but hyphae were observed in the lumens and intercellular spaces of ray parenchyma cells.

occasionally in normal wood. Examination of the sound woods before incubation showed some gelatinous fibers. After decay, it appeared that the gelatinous layer of these fibers was resistant to degradation (Fig. 6). This observation was also made in a detailed study of the decay of reaction wood (R. Blanchette, unpublished). The factors contributing to the persistence of the gelatinous layer are not known.

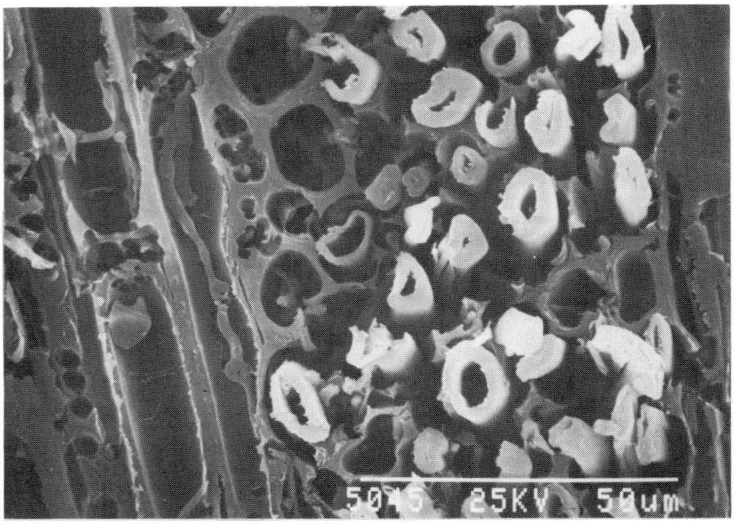

Figure 6. SEM of American Elm, Decayed 4 Weeks; area shows the resistant nature of the gelatinous layer of gelatinous fibers.

Chemical Composition of Decayed Woods

The analyses for Klason lignin, acid-soluble lignin, total sugars after acid hydrolysis, and five individual sugars in undecayed wood are presented in Table 2; the percentage of loss of these constituents during decay is presented in Table 3.

Removal of cell wall components by T. versicolor differed with the wood species (Table 3). Differences in removal of cell-wall constituents by this white-rot fungus were observed in other studies (Highley and Illman, 1990; Kirk and Highley, 1973; Kirk and Moore, 1972). The syringl content of the lignin did not correlate with the rate of lignin removal. Klason lignin was removed slowest in madrone, which has the highest S/G ratio. In contrast, lignin removal in boxelder, which has a low syringl content, was relatively fast. Lignin removal in boxelder was also preferentially removed compared to the carbohydrates. This differs from a previous study where T. versicolor preferred carbohydrate to lignin in a hardwood (Ceibo) containing primarily guaiacyl lignin (Highley, 1982).

Table 2. Analysis of Sound Wood.

Species	S:G[a]	Content (%)							
		Klason lignin	Acid-soluble lignin	Total sugars	Glucose	Xylose	Galactose	Arabinose	Monnose
Madrone	2.3	26.0	5.75	64.6	60.8	32.0	2.9	2.0	2.2
Box-elder	0.1	28.2	1.06	66.5	69.3	21.8	2.1	1.0	5.9
Red Oak	0.8	23.8	3.93	69.4	64.7	27.6	2.5	1.1	4.1
American Elm	0.6	2.50	3.30	65.2	65.3	27.7	2.8	1.2	2.9
Sweetgum	1.1	23.8	4.80	68.5	68.3	26.6	2.0	0.9	2.2
Red Maple	0.8	24.0	3.97	68.2	69.1	23.5	2.0	1.6	3.8
White Birch	1.2	20.7	4.48	72.5	61.2	32.8	1.8	1.2	2.9

[a]Syringyl:guaiacyl ratio.

Hemicelluloses were generally removed at about the same rate as cellulose or slightly faster. The glucomannan hemicellulose was removed at a faster rate than cellulose in all species but madrone, sweetgum, and white birch.

Syringyl:Guaiacyl Ratios of Residual Lignins

The S:G ratios of the residual lignins were determined by alkaline nitrobenzene oxidation (Table 3). This method was shown to provide a useful measure of composition (Obst et al., 1987) and was employed in wood-decay studies (Blanchette et al., 1988; Nilsson et al., 1989). A general trend was observed that as the weight loss by decay increased, the S:G ratio of the residual lignin decreased. The degree of the decrease was dependent on wood species.

The S:G ratio of the residual lignins of several woods (sweetgum, red maple, and white birch) decreased significantly as the extent of decay increased. This observation is consistent with preferential fiber cell wall degradation and with the fiber cell wall lignin being syringyl rich. Lacking rigorous quantitative data, it is not possible to determine whether the decrease in syringyl content after decay is always a consequence of the removal of cell wall lignin or whether there is a selective removal of syringyl lignin from all morphological regions.

Madrone can be considered an example of wood where the S:G ratio did not decrease significantly. The S:G ratios of the madrone residual lignins decreased a little at low weight loss and then remained relatively constant. The syringyl content of the fiber and vessel wall lignins in madrone were previously been found to be high (Musha and Goring; 1975), but the middle lamella lignin in our study had a lower syringyl content. Boxelder lignin S:G ratios changed little on decay. However, emphasis should not be placed on this apparent lack of selective lignin monomer-type removal, because the syringyl content of this wood was initially very low.

Although American elm fiber cell wall lignin was indicated to have a higher syringyl content than its middle lamella lignin (Musha and Goring, 1975), another study suggested that lignin monomers were more uniformly distributed for American elm and red oak (Obst and Landucci, 1986). Therefore, American elm and red oak were specifically chosen for this study. The S:G ratios of their residual lignins after decay did not appear to decrease significantly. This would be consistent with removal of fiber cell wall lignin, which had a S:G ratio similar to that of the middle lamella.

SUMMARY

Trametes versicolor decayed seven species of woody angiosperms such that fiber walls were degraded preferentially to cell corners, middle lamellae, vessels, and ray cells. However, in some instances, cell corners were degraded prior to fiber walls. In these cases, it is considered likely that fungal hyphae had easy access to existing cell corner intercellular spaces (electron-lucent regions?) through pit canals or bore holes. This suggests that for fungi such as _Trametes versicolor_ physical accessibility is an important factor in wood decay.

Ray cells and vessels were rapidly colonized by fungal hyphae but were resistant to decay compared to the fibers. In this case, physical accessibility is very good but other factors may retard degradation. The high degree of resistance exhibited by rays may be due in part to extractives because rays are often repositories for such materials. Indeed, in some cases, fibers near ray cells were less degraded than were those further away. To a lesser extent, extractives also may play a part in vessel decay resistance.

Our results are consistent with the general premise that lignin concentration and lignin type affect the decay resistance of wood. However, sufficient experimental evidence has not been obtained to state that lignin character is the limiting determinant of decay. Indeed, in our study, the effect of the syringyl content of lignin on decay was difficult to ascertain. For woods with fairly uniform syringyl content throughout (red oak, American elm, and boxelder), middle lamellae and cell corners were still resistant to decay. Madrone vessels and rays were resistant to decay, although these cell walls have high syringyl contents similar to the fiber walls that decay readily. Analysis of vessel-rich fractions from red oak indicated a significant syringyl content of the lignin in these vessels (Obst et al., 1987), yet red oak vessels were also resistant to decay.

In general, the syringyl units of the lignin were preferentially degraded. This effect was significant for some woods, but other woods showed only slight preferential removal of syringyl units. The syringyl content of the lignin had no general correlation with the rate of decay. However, red maple and boxelder, woods that are very similar in chemical composition and anatomy, exhibited decay rates that reflected the syringyl contents of their lignins. That is, boxelder, which has a low syringyl lignin content, decayed more slowly than red maple, which has a higher syringyl lignin content. However, removal of Klason lignin was similar in the two woods.

Acknowledgments

Transmission electron micrographs were obtained by the late Lidija Murmanis. Scanning electron micrographs were taken by Yongjing Li, visiting scientist, Nanjing Forestry University, Nanjing, China.

REFERENCES

Blanchette, R.A.; Abad, A.R.; Cease, K.R. [and others] (1990). Enzyme Immunocytochemistry and Ultrastructural Localization of Cell Wall Components by Enzyme–Gold Complexes, In: Biotechnology in Pulp and Paper Manufacture, Applications and Fundamental Investigations, pp. 69–81 (T.K. Kirk and H.–M. Chang, Eds.), Butterworth–Heinemann, Boston, MA.

Blanchette, R.A.; Obst, J.R.; Hedges, J.I.; and Weliky, K. (1988). Resistance of Hardwood Vessels to Degradation by White Rot Basidiomycetes. Can. J. Bot., 66, 1841–1847.

Butcher, J.A. and Nilsson, T. (1982). Influence of Variable Lignin Content Amongst Hardwoods on Soft-Rot Susceptibility and Performance of CCA Preservatives. The International Research Group on Wood Preservation: IRG/WP/1151.

Carlquist, S. (1988). Comparative Wood Anatomy. Berlin, Springer–Verlag.

Cowling, E.B. (1961). Comparative Biochemistry of the Decay of Sweetgum Sapwood by White-Rot and Brown-Rot Fungi. USDA Technical Bulletin 1258.

Daniel, G.; Nilsson, T.; and Pettersson, B. (1991). Poorly and non-lignified regions in the middle lamella cell corners of birch (Betula verrucosa) and other wood species. IAWA Bull. n.s. 12:, 70–83.

Effland, M.J. (1977). Modified Procedure to Determine Acid Insoluble Lignin in Wood and Pulp. Tappi J., 60(10), 143–144.

Eriksson, I.; Lidbrandt, O.; and Westermark; U. (1988). Lignin Distribution in Birch (Betula Verrucossa) as Determined by Mercurization With SEM and TEM–EDXA. Wood Sci. Technol., 22(3), 251–257.

Erikkson, K.–E.L.; Blanchette, R.A.; and Ander, P. (1990). Microbial and Enzymatic Degradation of Wood and Wood Components. Berlin, Springer–Verlag.

Faix, O.; Mozuch, M.D.; and Kirk, T.K. (1985). Degradation of Gymnosperm (Guaiacyl) vs. Angiosperm (Syringyl/Guaiacyl) Lignins by Phanero- chaete Chrysoporium. Holzforschung, 39(4), 203–208.

Fergus, B.J. and Goring, D.A.I. (1970). The Location of Syringyl and Guaiacyl Lignins in Birch Xylem Tissue. Holzforschung, 24(4), 113–117.

Highley, T.L. (1982). Influence of Type and Amount of Lignin on Decay by Coriolus Versicolor. Can. J. For. Res., 12(2), 435–438.

Highley, T.L. and Illman, B. (1990). Changes in Cell Wall Components of White Pine and Maple by White-Rot Fungi. Biodeterioration Res., 3, 349–360.

Highley, T.L. and Murmanis, L.L. (1987). Micromorphology of Degradation in Western Hemlock and Sweetgum by the White-Rot Fungus Coriolus versicolor. Holzforschung, 41(2), 67–71.

Kirk, T.K.; Chang, H.--C.; and Lorenz, L.F. (1975). Topochemistry of the Fungal Degradation of Lignin in Birch Wood as Related to the Distribution of Guaiacyl and Syringyl Lignins. Wood Sci. Technol., 9, 81–86.

Kirk, T.K. and Obst, J.R. (1988). Lignin Determination: In: Methods in Enzymology, Volume 161, Biomass, Part B, Lignin, Pectin, and Chitin. pp. 87–101 (W.A. Wood and S.T. Kellogg, Eds.), Academic Press, Inc., New York.

Leightley, L.E. (1984). The Use of C–13 CP/MAS NMR in the Chemical Identification of Decayed and Undecayed Tropical Timber Species. The International Research Group on Wood Preservation: IRG/WP/1224.

Manders, W.F. (1987). Solid-State C–13 NMR Determination of the Syringyl/Guaiacyl Ratio in Hardwoods. Holzforschung, 40(1), 13–18.

Musha, Y. and Goring, D.A.I. (1975). Distribution of Syringyl and Guaiacyl Moieties in Hardwoods as Indicated by Ultraviolet Microscopy. Wood Sci. Technol., 9, 45–58.

Nilsson, T.; Daniel, G.; Kirk, T.K.; and Obst, J.R. (1989). Chemistry and Microscopy of Wood Decay by Some Higher Ascomycetes. Holzforschung, 43(1), 11–18.

Nilsson, T.; Obst, J.R.; and Daniel, G. (1988). The Possible Significance of the Lignin Content and Lignin Type on the Performance of CCA-Treated Timber in Ground Contact. The International Research Group on Wood Preservation: IRG/WP/1357.

Obst, J.R. and Landucci, L.L. (1986). Quantitative C–13 NMR of Lignins---Methoxyl:Aryl Ratio. Holzforschung, 40(Suppl.), 87–92.

Obst, J.R.; Landucci, L.L.; and Manders, W.F. (1987). Characterization of Hardwood Lignin: Solution and Solid State C–13 NMR. Fourth International Symposium of Wood and Pulping Chemistry. Volume 1, 273–277, Paris.

Pettersen, R.C.; Schwandt, V.H.; and Effland, M.J. (1985). An Analysis of the Wood Sugar Assay Using HPLC: A Comparison With Paper Chromatography. J. Chromatogr. Sci., 22, 478–484.

Saka, S. and Goring, D.A.I. (1988). Distribution of Lignin in White Birch Wood as Determined by Bromination with TEM–EDXA. Holzforschung, 42(3), 149–153.

Syafii, W., Samejima, M., and Yoshimoto, T. (1988). The role of lignin on decay resistance of Ulin (Eusidearoxylon zevageri) to wood-rotting fungi. Bull. Tokyo Univ. For., 79, 119–126.

Wilcox, W.W. (1968). Changes in Wood Microstructure Through Progressive Stages of Decay. USDA, Forest Service, Research Paper FPL 70.

Note: The use of trade or firm names in this publication is for reader information and does not imply endorsement by the U.S. Department of Agriculture of any product or service.

SECTION III

BIODETERIOGENS AND PLANT STRESS INJURY

BIODETERIOGENS AND PLANT STRESS INJURY

INTRODUCTION

A variety of naturally-occurring biological and industrially-originated environmental deterioges can cause marked deterioration of highly significant agro-economic plants resulting in regional as well as global economic losses. In very recent years, there has been an unprecedented interest by the worldwide community in protecting the quality of the earth's terrestrial, aquatic and atmospheric environments. Thus, considerable basic and applied research has been performed worldwide to either prevent or ameliorate the injurious effects of certain biologically and industrially-generated pollutants upon plant growth and development. This section of the current volume contains a limited number of papers representative of some of the plant stress injury research being carried out in the Unites States as collaborative efforts between scientists located in academic institutions and those situated in federal laboratories. While two of the papers report results connected with reducing Forest species economic loss, another describes the employment of certain lower plants in a novel manner to deplete fresh-water environments of unwanted toxic metals. The reader is cautioned not to conclude that these papers cover the breadth of contemporary research investigating environmental deterrioges and plant stress injury. Instead, the reader is also directed to other sections of the volume for enlightened discussions of the deleterious effects of biodeterioges upon the health and well-being of the earth's Forests.

William V. Dashek
Department of Biology
Clark Atlanta University
J.P. Brawley at Fair St., SW
Atlanta, GA 30314 USA

PROTEIN CONTENTS OF OZONE- AND AIR-FUMIGATED *PINUS STROBUS* NEEDLES

EUNICE T. DLAMINI, ARTHUR L. WILLIAMS AND WILLIAM V. DASHEK, Department of Biology, Clark Atlanta University, Atlanta, GA 30314, USA
WAYNE T. SWANK AND JAMES M. VOSE, United States Department of Agriculture, Coweeta Hydrologic Laboratory, Otto, NC 28763, USA

INTRODUCTION

Ozone (0_3) is normally present within the earth's troposphere at 20-30ppb concentration. However, as a result of increased industrialization and urbanization, e.g., automobile emissions (Elstner, 1984), 0_3 has become an undesirable atmospheric pollutant in many densely-populated areas of the United States. However, monitoring of atmospheric 0_3 at the Coweeta Hydrologic Laboratory (Otto, NC) during 1988 revealed April through September fluctuations of 40-70 ppb (Figure 1), and mean hourly concentrations sometimes exceed 90 ppb. Thus, even rural areas experience elevated atmospheric 0_3 levels, presumably as a result of long-range transport of ozone precursors (e.g., NO_x).

Although considerable literature exists regarding the effects of elevated 0_3 concentrations on crop plant growth and yield (Taylor, 1984), a computer search of the literature yielded limited information (Wilkinson and Barnes 1972; Berry 1973; Kress and Skelly 1982; Miller and Stottle 1984; Hogsett et al., 1985; Peterson et al., 1987; Pye, 1988) regarding the effects of adverse 0_3 levels on forest species' chemistry and metabolism. What literature does exist is mainly concerned with 0_3 effects on photosynthetic rate (Barnes, 1972) and usually from a whole-plant prospective (Koslowski, 1980; Smith, 1981). In this regard, Pye (1988) reported that data from 25 experiments utilizing seedlings of 43 tree species and hybrids revealed that 0_3 can reduce growth and

Biodeterioration Research 4, Edited by G.C. Llewellyn *et al.*, Plenum Press, New York, 1994

photosynthetic rate in many areas of the U.S.A. Ozone is responsible for the growth decline of pollutant sensitive eastern white pine genotypes in the Southeastern United States and of ponderosa and jeffrey pines in the mountains of Southeastern United States and Central California (Chevone and Linzon, 1988).

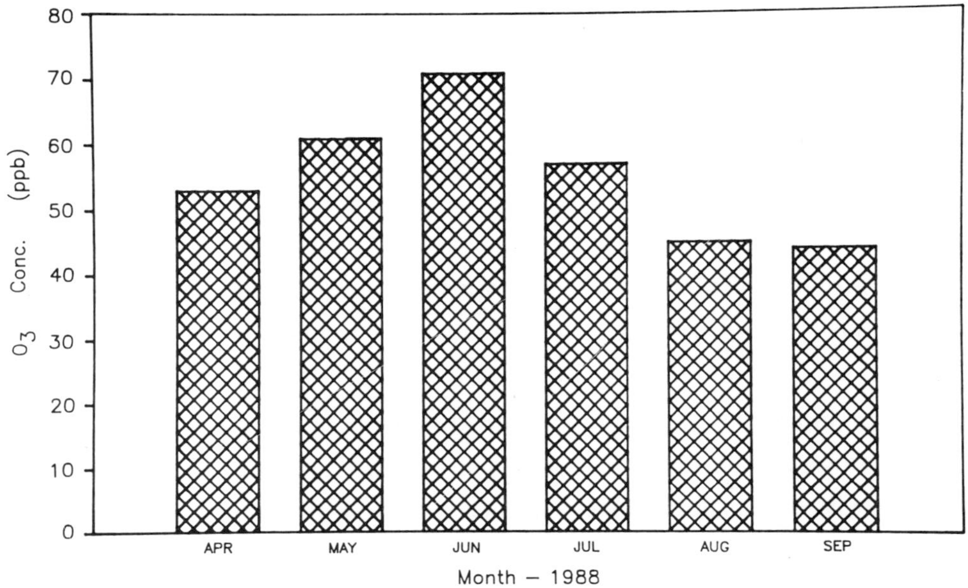

Figure 1. Ozone Concentrations As a Function of Month at the Coweeta Hydrologic Laboratory Site.

There appear to be few published reports regarding O_3 effects on amino acid (Wilkinson and Barnes, 1972) and protein contents of pine species. Defining these contents is important since proteins are the end-products of gene expression and enzymatic proteins can profoundly affect growth and development. In addition, the marked accumulation of certain amino acids, e.g., proline in leaves of certain plants is stress-related (Dashek and Erickson, 1981; Erickson and Dashek, 1982) and thus, may prove useful as an indicator of O_3 stress.

Here, we compare the ethanol-insoluble protein contents of charcoal-filtered air and O_3-fumigated new foliage derived from 30-yr old Pinus strobus trees. This investigation is an initial attempt to identify conifer proteins which may be markedly sensitive to O_3 with the intent of genetically-engineering air-pollutant resistant pine trees (Maniatis et al., 1982; Wu et al., 1983).

MATERIALS AND METHODS

Fumigations of Needles

Twigs comprised of one yr-old needles were sampled from 30 yr-old trees (Coweeta Hydrologic Laboratory, Otto, NC), grafted onto two yr-old seedlings and fumigated with 4,000 and 8,000 ppb hr^{-1} O_3 (controls were fumigated with charcoal-filtered air). The three genotypes used in this study had previously been observed to show visual O_3 damage under field conditions (Swank and Vose, in press). The ozone was generated by the UV-photolysis of pure oxygen (Ozone Research Equipment Corporation, Phoenix, AZ). Subsequent to fumigation, 0.27 to 2.32 g fr wt lots of needles were excised, added to ~ 250 ml 80% EtOH and stored at 4°C until analysis.

Extraction and Quantification of Needle Protein

Needles, which were soaked in 80% EtOH, were blotted, weighed and homogenized on ice with a mortar and pestle into 95% aqueous EtOH (Stewart, 1972). Following homogenization, the homogenates were centrifuged at 12,000 xg for 30 min at 4°C utilizing a Beckman J21C centrifuge.

Whereas the EtOH-soluble fraction (12,000xg supernatant) was evaporated to dryness and stored at -20°C over drierite for amino acid determination (subject of another communication), the EtOH-insoluble fraction (12,000xg pellet) was suspended in 3ml of 0.15M NaCl for protein quantification. Aliquots of the resuspended EtOH-insoluble protein were diluted to 1 ml for assay at 280nm utilizing bovine serum albumin (BSA) for standard curve construction. The UV-spectroscopic quantification of protein is dependent upon the presence of aromatic amino acids in protein. Therefore, a colorimetric procedure, Bradford (1976), was also employed to quantify the protein contents of air-fumigated and O_3-fumigated needles.

The procedures utilized to extract and quantify needle protein are summarized in Figure 2. In this connection, alcohol has been employed by Durzan and Stewart (1967) to derive the soluble nitrogen content of jack pine leaves.

Assessment of UV Assay Specificity

To assess whether the EtoH-insoluble fractions of new foliage needles contained compounds which could interfere with the UV quantification of protein, UV spectra (200-300nm) of EtOH-insoluble substances and BSA (Sigma, St. Louis, MO) suspended in 0.15M NaCl were performed utilizing an LKB ultraspectrophotometer. Three to six replicates were carried out for both the needle EtOH-insoluble and BSA fractions to obtain means which were entered into an IBM computer employing a Lotus 123 program to "plot" the absorption spectra.

Recovery of Needle Protein

To assess the efficiency of the EtOH extraction/UV quantification procedure, 10, 25, 50 and 100mg BSA were added to 3ml 95% EtOH and processed as in Figure 2.

Statistical Analyses

The number of replicates for needle protein quantification, protein recovery and UV spectra are presented in the legends to the figures and captions to the tables. Statistically significant differences in 280 nm contents between air - and O_3-fumigated needles were sought with an analysis of variance utilizing a 95% confidence level (Elzey, 1987).

RESULTS AND DISCUSSION

Needle Total Protein Content

The data in Figure 3 reveal that substantially elevated O_3 levels (> 200 ppb) are required to show a statistically significant reduction in total protein content of new foliage as there were no significant differences amongst the

Fumigate Needles
with 0, 100 and 200ppb O_3

Excise Needles
and Add to 80% Ethanol

Decant 80% Ethanol and
Add 95% Ethanol

Homogenize Mortar and Pestle
with Alumina 15 mins

Pull through a Single Layer
of Miracloth

Centrifuge Filtrate, 12,000x*g* 30 min

Ethanol Soluble

Evaporate to Dryness
and Store at -20°C

Ethanol Insoluble

Resuspend in 0.15M NaCl

Assay at 280nm by
UV - Spectroscopy and at
595nm by Bradford
Colorimetry

Figure 2. Summary of EtOH Procedure for the Extraction of EtOH-insoluble Protein and-Soluble Amino Acids; Adapted With Modification From Stewart (1972).

means at $p = 0.05$. Mean protein content was more than 50% lower in needles exposed to 200ppb O_3 relative to controls; however, the data were highly variable, making statistically significant differences difficult to detect. The trend of decreased total protein content in fumigated needles may still be biologically important. The apparent O_3-induced diminution in protein content raises the question as to whether the reduction occurred in each needle protein or alternatively preferentially in certain types of proteins. For example, ribulose 1,5-bisphosphate carboxylase (RUBISCO), a photosynthetic enzyme, composes a significant amount of the protein of the leaves of certain plants. Therefore, it may be of value to compare the RUBISCO activities of O_3- and air-fumigated needles. In addition, the use of starch gel electrophoresis coupled with appropriate substrate methods could serve to assess whether various classes of enzymes are affected, e.g., dehydrogenases, esterases, peroxidases, etc. In this regard, peroxidase has been suggested as an enzyme monitor of air-pollution (Dashek et al., 1990). In addition, a variety of chlorophyll synthesizing enzymes could be assayed to determine whether O_3-induced needle tip browning is a result of an impairment of chlorophyll synthesis thereby reducing needle photosynthesis. Whether selected or all proteins are decreased in amount could be established by electrophoresis of EtOH-insoluble needle proteins by SDS-PAGE (Laemmli, 1970).

The data in Table 1 suggest that the recovery of a known protein, bovine serum albumin, through the procedure in Fig. 2 was limited. This raises the possibility that some needle proteins were trapped upon the Miracloth. The extent to which this affected our analysis of the differences between treatments is unknown and indicates that an alternative method, e.g., centrifugation for collecting cell walls should be employed.

Assessment of UV Assay Specificity

Figure 4A, B presents the UV spectra for EtOH-insoluble needle substances (A) and bovine serum albumin (B). Similar spectra were observed for each suggesting a limited occurrence of 280nm interfering substances. Indeed, the EtOH-insoluble material exhibited a positive Coomassie blue colorimetric reaction (Bradford, 1976) but the quantity of detected protein was less than that detected by UV assay.

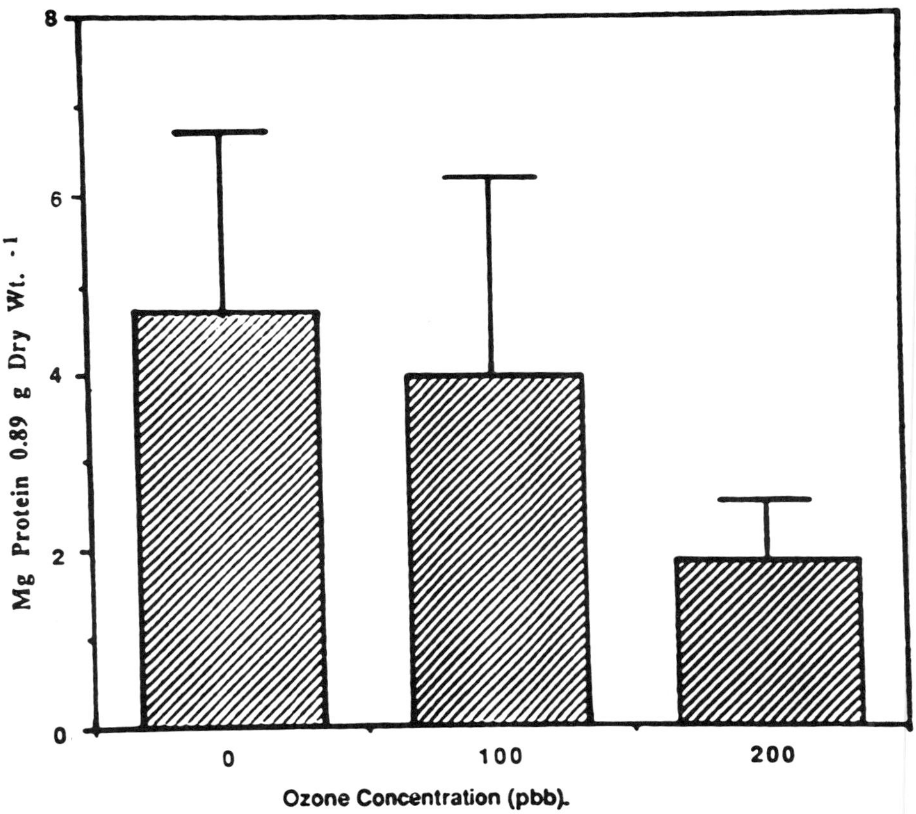

Figure 3. Histogram Showing Total Protein Reduction.

Table 1. Recovery of bovine serum albumin through the ethanol extraction procedure.

NOT HOMOGENIZED INTO ETHANOL			HOMOGENIZED INTO ETHANOL	
mg BSA Added	mg BSA Recovered	% of mg BSA Added	mg BSA Recovered	% of mg BSA Recovered for Non-homogenized BSA
10	7.81 ± 1.03	78.09	1.94 ± 1.24	24.84
25	13.82 ± 0.87	55.28	3.61 ± 0.88	26.12
50	28.20 ± 4.01	56.40	9.83 ± 3.25	34.86
100	61.92 ± 5.38	61.92	25.07 ± 1.13	40.49

Data are X's and O's where N = 3-6 replicates.

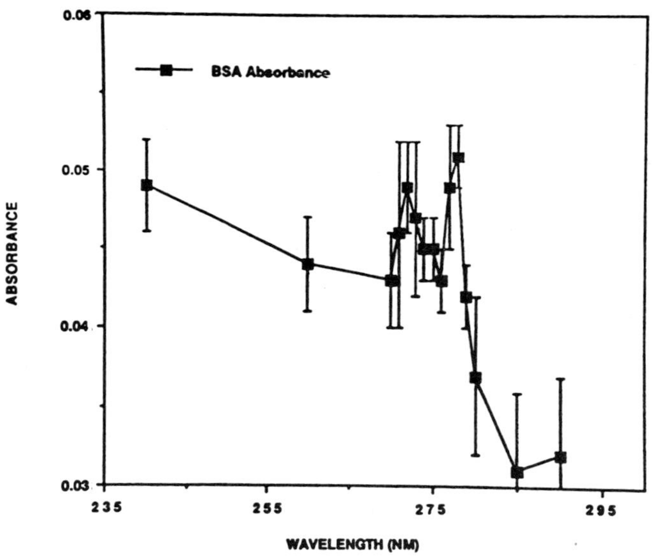

Figure 4A. Absorption Spectrum of Bovine Serum Albumin.

SUMMARY

The earth's ozone background is 20-30 ppb. Monitoring atmospheric O_3 at the Coweeta Hydrologic Laboratory (Otto, NC) revealed mean monthly concentrations of 40-70 ppb in the April - September period. A computer-based literature search yielded little regarding the effects of adverse atmospheric O_3 levels on needle chemistry and metabolism. Here are reported ethanol-insoluble, 280nm contents of charcoal-filtered air-and O_3-fumigated (100 and 200 ppb) for one yr-old needles on twigs from 30 yr-old Pinus strobus trees grafted onto two-yr-old seedlings at Coweeta. Ozone fumigations employed UV-photolysis of pure

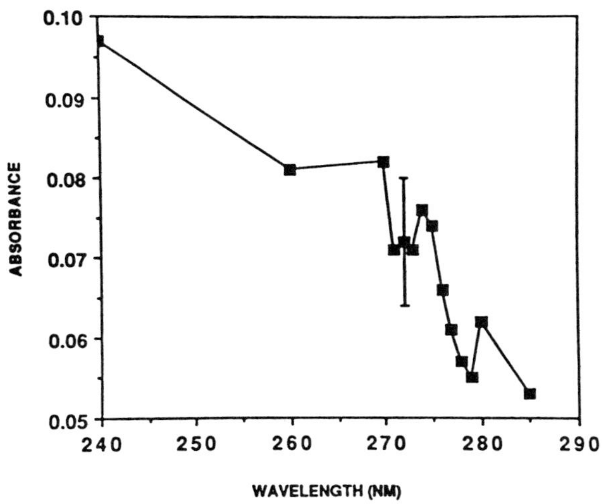

Figure 4B. Absorption Spectrum of Needle Protein.

oxygen (Ozone Research Equipment Corporation, Phoenix, AZ). Tissue protein was derived by soaking 0.27-2.32 g fr wt air- and O_3-fumigated needles in 80% ethanol, decanting the alcohol and subsequent homogenization of "pre-chopped" needles into 95% ethanol. After "pulling" through Miracloth, the homogenate was centrifuged (12,000xg, 30 min.) and the pellet suspended in 0.15M NaCl for assay. Whereas the 280nm substances content of air-fumigated needles was 4.72 $\pm$ 2.02mg 0.89g needle dry wt^{-1}, the contents of 100 and 200ppb O_3-fumigated needles were 16 and 68% less. An absorption spectrum

(200-300nm) revealed no interference with the quantification of protein which was also detected by colorimetry but less than that at 280 nm. These results suggest that greater than 200ppb is required to significantly depress the total protein content of new foliage.

Acknowledgment

We thank Mr. S. Ibim and Mr. L. Brako for graphics, Ms. Corinda Belbodha for technical assistance, Ms. Tondaleyo Conley for thoughtful time typing this manuscript and Dr. J. Stewart is acknowledged for assistance with statistics. This work was supported by USDA-Co-operative Agreement #29-345 and a Grant-in-Aid from Sigma Xi, NIH-RCMI Grant No. G12RR03062 and NSF-RIMI Grant No. RII-8912781.

REFERENCES

Barnes, R.L. (1972). Effects of chronic exposure to ozone on photosynthesis and respiration of pines. Environ. Pollut. 3, 133-138.

Berry, C.R. (1973). Age of pine seedlings with primary needles affects sensitivity to ozone and sulfur dioxide. Phytopathol. 64, 207-209.

Bradford, M.M. (1976). A rapid and sensitive method for the quantitation of microgram quantities of protein utilizing the principle of protein dye binding. Anal. Biochem. 72, 248.

Chevone, B.I. and Linzon, S.M. (1988). Tree decline in North America. Environ. Pollut. 50, 876-99.

Dashek, W.V. and Erickson, S.S. (1981). Isolation, assay, biosynthesis, metabolism, uptake and translocation and functions of proline in plant cells and tissues. Bot. Review. 47, 340-385.

Dashek, W.V., Williams A.L., and Zeidan, H., (1990). Needle peroxidase, a potential monitor for ozone-induced damage to eastern white pine seedlings. Plant Physiol. 93, 156.

Durzan, D.J. and Stewart, F.C. (1967). The nitrogen metabolism of Picea glauca (Mensch) Voss and Pinus bankstrana lamb as influenced by mineral nutrition. Can. J. Bot. 45, 695-710.

Elstner, E.F. (1984). Comparison of "Inflammation" in pine needles and humans. In: Oxygen radicals in Chemistry and Biology. (Bos W., Sara, M., and Tait, D., eds.) Walter Dehydryterate Co., Berlin.

Elzey, F.F. (1987). Introductory Statistics: A microcomputer approach. Brooks/Cole Publishing Company, Monterey, CA.

Erickson, S.S. and Dashek, W.V. (1982). Accumulation of foliar soluble proline following fumigation of Glycine max, cv. "Essex" and "Excelsior" seedlings with sulphur dioxide. Environ. Pollut. 28, 89-108.

Hogsett, W.E., Plocker, M., Wildman, V., Tingey D.T., and Bennett, J.P. (1985). Growth response of two varieties of slash pine seedlings to chronic ozone exposures. Can. J. Bot. 63, 2369-2376.

Kozlowski, T.T. (1980). Impacts of air pollution on forest ecosystems. American Institute of Biological Sciences, WI.

Kress, L.W. and Skelly, J.M. (1982). Response of several - forest tree species to chronic doses of ozone and nitrogen - Plant Dis. 66, 1149-1152.

Laemmli, V.K. (1970). Cleavage of structural proteins during the assembly of the head of bacteriophage. Nature 227, 680-685.

Maniatis, T., Fritsch, E.F., and Sambrock, J. (1982). Molecular Cloning, A laboratory Manual. Cold Spring Harbor Laboratory, Cold Spring Harbor, NY.

Miller, P.R. and Stottle, K.W. (1984). Response of forest species to O_3, SO_2 + NO_2 mixtures. Proceedings APCA Annual Meeting 77, 84-305.

Peterson, D.L., Arbaugh, M.J., Wakefield, V.A., and Miller, P.R. (1987). Evidence of growth reduction in ozone-injured Jeffrey pine (Pinus jeffreyi, Grev. and Balf.) in Sequoia and Kings Canyon National Parks. J Air Pollut. Control Assoc. 37, 906-912.

Pye, J.M. (1988). Impact of ozone on the growth and yield of trees: A review. J. Environ. Qual. 17, 347-360.

Smith, W.H. (1981). Air Pollution and Forests. Interaction Between Air Contaminants and Forest Ecosystems. Springer-Verlag, New York.

Stewart, C.R. (1972). Effects of proline and carbohydrates on the metabolism of exogenous proline by excised bean leaves in the dark. Plant Physiol. 50, 551-555.

Swank, W.T. and Vose, J.M. Watershed-scale response to some events in a Pinus strobus L. plantation. Water, Air and Soil Pollution (in press).

Taylor, O.C. (1984). Organismal responses of higher plants to atmospheric pollutants: Photochemical and Other. In: Air Pollution and Plant Life (M. Treshow, ed.). John Wiley and Sons, Ltd. pp. 215-223.

Wilkinson, T.G. and Barnes, R.L. (1972). Effects of ozone on $^{14}CO_2$ fixation patterns in pine. Can. J. Bot. 51, 1573-1578.

Wu, R., Grassman, L., and Moldave, K. (1983). Recombinant DNA, parts B and C, Academic Press, NY.

CONE ANALYSIS OF *PINUS TAEDA L.* (LOBLOLLY PINE) CLONES

ROSLYN A. MARCH, Department of Biology, Clark Atlanta University, Atlanta, GA 30314, USA

DAVID L. BRAMLETT, USDA Southeastern Forest Experiment Station, Dry Branch, GA 31020, USA

JOHN E. MAYFIELD, North Carolina Central University, Department of Biology, Durham, NC 27707, USA

WILLIAM V. DASHEK*, Department of Biology, Clark Atlanta University, Atlanta, GA 30314, USA

INTRODUCTION

Loblolly pine is the most important forest tree species in the Southeastern United States, producing the major volume of both commercial timber and pulpwood for the region. Inadequate regeneration of this species has been identified as one of the causes of lower yields of softwood in the Southeast. Part of the solution would be a readily available seed supply for forest nurseries. In the mid 1950's, an intensive program was initiated to produce genetically improved seeds from loblolly pine for reforestation of harvested stands. Scions from superior phenotypes in natural stands were grafted onto seedlings and clonal seed orchards were established (Zobel et al., 1958). Even with substantial insect and disease protection, annual seed losses are observed. Any seed loss has an economic effect because of the high value of the genetically improved seed. The objective of the present investigation was to evaluate seed production efficiency in the Georgia Forestry Commission's Arrowhead Seed Orchard, Pulaski County, Georgia. Here, we report data regarding analysis of cones collected from clones in the seed orchard. Analysis of 21 sample trees revealed that two clones, 558 and 593, were the most productive based upon seed efficiency data. In some ovules, the deterioration of the ovular contents began in June and resulted in the production of "empty seeds".

Biodeterioration Research 4, Edited by G.C. Llewellyn *et al.*, Plenum Press, New York, 1994

MATERIALS AND METHODS

Cone Collection

Five cones from each of 15 trees of the 1983 cone crop were collected in October, 1983 as part of an inventory-monitoring system in the orchard and held for cone analysis within the laboratory. Five cones from each of 6 trees of the 1984 cone crop were collected in September, 1984, and subjected to cone analysis. Trees produced from scions from the same parental tree have been given the same clone number. For all samples, cones were selected at random, beginning with the lower branches, proceeding to the middle branches, and ending with the top branches. Collections were made with a hydraulic lift that was rotated so that cones could be collected from various sides of each tree. Because of experiments begun on trees after October 1983; it was not possible to use the same clones in 1984 as in 1983. Two of the clones (617 and 593) were analyzed both years.

Cone Analysis

To evaluate production efficiency in the seed orchard, the number of seeds, the percentage of filled seeds, the seed potential and the seed efficiency of five cones per tree were determined (Bramlett et al., 1977). The percentage filled seeds is equivalent to the number of seeds with embryos divided by the number of seeds with mature integuments X 100. The seed potential equals 2 times the number of fertile scales and the seed efficiency equals the number of seeds with embryos/seed potential X 100. Seeds were radiographed to count the number of seeds with gametophytes and embryos.

RESULTS

Cone Analysis of the 1983 Cone Crop

The numbers of developed seeds/cone ranged from 70 to 180. The frequency distribution is shown in Figure 1A. The mean number of seeds/cone was 134.5 $\pm$ 22.6 (N=75). Sixty-five percent of the cones possessed seed numbers within the range of 129.5 to 189.5. Only 15% had seed numbers in the range of 159.5 to 189.5. The first column of Table 1 groups the means and standard deviations of the numbers of seeds/cone according to the clones. Clones 558 and 593 exhibited means within the 149.5 to 189.5 (Figure 1A) range. These clones produced the highest numbers of seeds/cone while clone 617 produced the lowest.

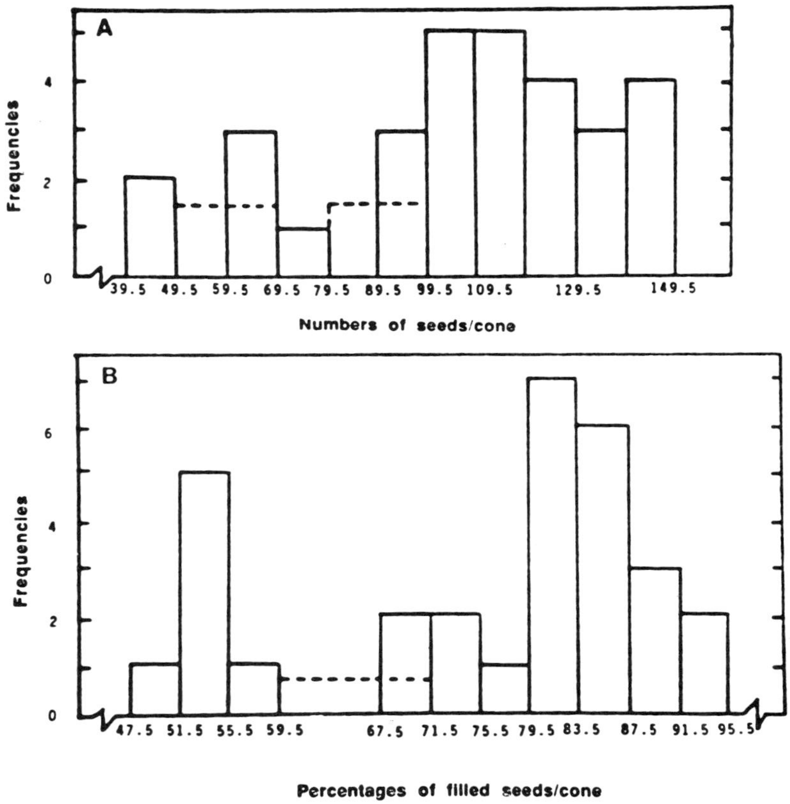

Figure 1. Frequency Distribution of Numbers of Seeds/Cone (A) and Percentages of Filled Seeds/Cone (B) in the 1983 Cone Crop (N = 75).

Table 1. Cone Analysis of the 1983 Cone Crop.

Clone No.	X and SD of # seeds in cone[a]	X and SD of % filled seeds[a]	X and SD of seed potentials[a]	X and SD of seed efficiencies[a]
617	84.8 ± 10.5	60.0 ± 17.1	159.6 ± 18.0	31.4 ± 6.6
504	115.6 ± 20.4	86.0 ± 2.2	152.8 ± 8.3	65.2 ± 11.1
502	124.4 ± 9.1	85.7 ± 7.9	151.6 ± 7.8	70.4 ± 8.7
618	130.2 ± 27.7	76.2 ± 4.8	196.4 ± 11.9	51.7 ± 9.1
599	132.0 ± 7.2	39.3 ± 10.6	152.0 ± 9.0	33.8 ± 7.9
551	133.2 ± 10.7	85.0 ± 9.1	151.2 ± 8.8	75.0 ± 10.5
578	133.6 ± 17.2	79.4 ± 13.5	166.4 ± 12.3	63.8 ± 12.7
521	136.8 ± 35.3	83.4 ± 24.2	184.0 ± 11.8	64.6 ± 26.1
574	137.4 ± 10.5	49.6 ± 8.3	172.4 ± 10.1	39.5 ± 7.4
584	139.8 ± 12.9	74.0 ± 14.5	166.4 ± 13.5	62.6 ± 15.2
550	141.8 ± 10.4	73.0 ± 19.4	171.2 ± 15.3	60.4 ± 15.6
509	143.6 ± 7.5	80.6 ± 4.4	168.4 ± 6.5	56.7 ± 28.8
594	148.4 ± 10.8	68.0 ± 5.4	176.8 ± 12.0	57.1 ± 4.6
558	157.0 ± 15.2	92.6 ± 8.9	171.2 ± 13.6	85.1 ± 10.8
593	158.4 ± 9.6	73.8 ± 9.8	179.2 ± 7.7	65.3 ± 9.4

[a]Data are means and standard deviations were N = 5.

The percentage of filled seeds/cone ranged from 27 to 98%. The frequency distribution is presented in Figure 1B. The mean and standard deviation were 73.8 ± 17.9 (N=75). Sixty-five percent of the cones fell within the range of 69.5% to 100.0%. The second column of Table 1 shows that clone 558 was the only one with a mean > 90%. Clones 504, 502, 551, 521 and 509 possessed means in the 80's. Clone 593, which had a mean number of seeds/cone similar to clone 558, gave a mean percentage filled seeds of only 73.8.

The seed potentials ranged from 140 to 208. The frequency distribution is presented in Figure 2A. The mean and standard deviation were 168.0 ± 16.5

(N = 75). Forty-seven percent of the cones possessed a seed potential of 167.5 and above while only 12% had means of 191.5 and above. The third column of Table 1 shows both the means and standard deviations of the seed potentials of the clones. There is the expected correlation between seed potential and the number of seeds produced. The first six clones with the reduced seed potentials

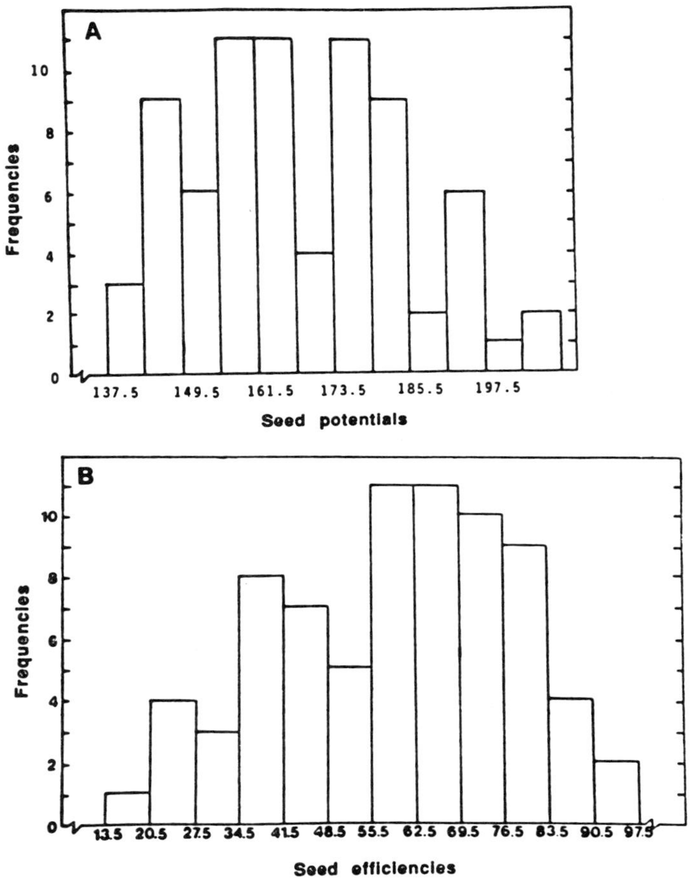

Figure 2. Frequency Distribution of Seed Potentials (A) and Seed Efficiencies (B) in the 1983 Cone Crop (N = 75).

produced the lowest number of seeds. The remaining nine clones produced the highest number of seeds. There was poor correlation between the seed potential and the number of seeds. For example, clone 618 had the highest mean for seed potential but ranked fifth in the mean number of seeds/cone.

The frequency distribution of the seed efficiencies is shown in Fig. 2B. The mean and standard deviation were 59.6 ± 18.0 (N = 75). Sixty-three percent of the cones had mean efficiencies within the range of 55.5% to 97.5%. Only 8% displayed efficiencies within the range of 83.5% to 97.5%, and only 3% had efficiencies between 90.5% and 97.5%. The means and standard deviations of the seed efficiencies according to clone number are shown in the last column of Table 1. Clone 558 had the highest seed efficiency and ranked second in the mean number of seeds/cone.

An analysis of variance ($p < 0.01$, Scheffe' test) performed on the data in Table 1 revealed that among the top 5 clones for the number of seeds per cone, there was no significant difference among the clones; nor was there a significant difference between their percentages of filled seeds and seed potentials. In the seed efficiency category, clone 558 was significantly different from clones 594, 509 and 550; but not significantly different from clone 593. Based upon its seed efficiency, and the fact that it had a relatively high number of seeds per cone, clone 558 was the better seed producer among the clones. If the two best clones was to be selected, then clone 593 would be included. In fact, clone 593 proved to be the better seed producer among the six clones studied the following year.

<u>Cone Analysis of the 1984 Cone Crop</u>

The numbers of seeds/cone ranged from 49 to 146 (Figure 3A). The mean number of seeds/cone was 107.1 ± 28.5 (N = 30). Twenty-three percent of the cones had a mean within the range of 129.5 to 149.5. Table 2 shows that clone 593 displayed the highest mean followed by clone 520 (2) between clones 520 (1), 520 (2) and 593 for the number of seeds per cone, the percentage of filled seeds and the seed potential; clone 593 had a significantly different seed efficiency that clone 520 (2). Clone 593 was not significantly different from clone 520 (1) in the area of seed efficiency. Since clone 593 not only had the highest seed efficiency, but also a relatively high number of seeds per cone; it was the better seed producer among the clones.

Clones 593 and 520 (2) exhibited the highest percentages of filled seeds (Table 2). Figure 3B is the frequency distribution of the percentages of filled seeds/cone. The overall mean was 76.0 ± 13.9 (N = 30) with a range of 47.6 to 93.7. Sixty-three percent of the cones were between 75.5% and 95.5%. Only 7% possessed a mean within the interval of 91.5% to 95.5%.

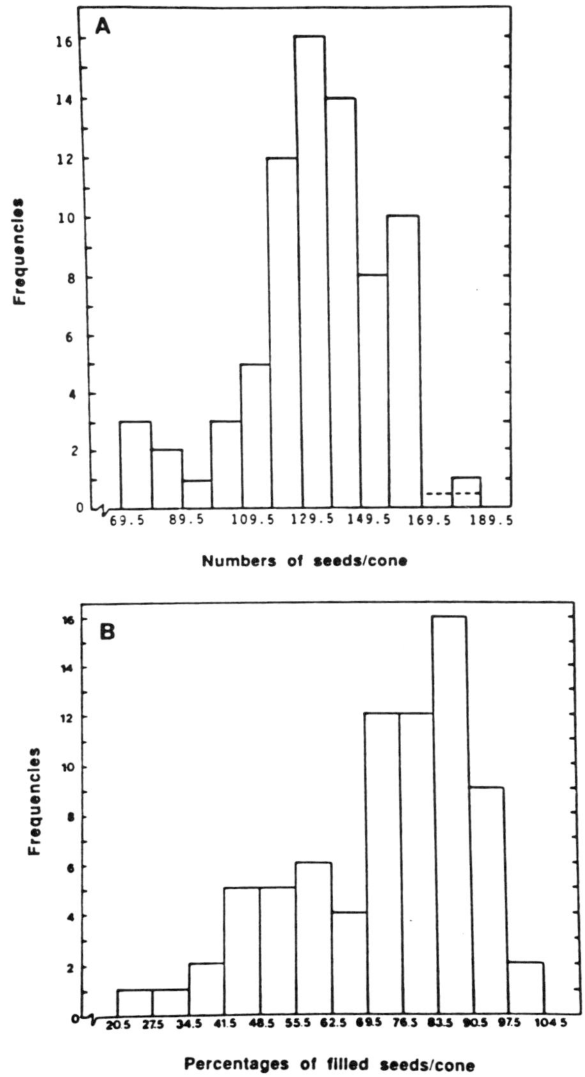

Figure 3. Frequency Distribution of Number of Seeds/Cone (A) and Percentages of Filled Seeds/Cone (B) in the 1984 Cone Crop (N = 30).

The seed potentials ranged from 148 to 210 for the individual cones. The frequency distribution is represented in Fig. 4A. The overall mean was 173.1 $\pm$ 13.2 (N = 30). Forty-three percent of the cones showed potentials within the 173.5 to 215.5 range with ten percent falling within the interval of 191.5 to 215.5. As shown in Table 2, there was not a good correlation between the mean seed potentials and the mean number of seeds within the clones. Clone 617 had the highest mean seed potential but produced the second lowest mean number of seeds. Clone 593, with the highest mean number of seeds, produced

Table 2. Cone Analysis of the 1983 Cone Crop.

Clone No.	X and SD of # seeds in cone[a]	X and SD of % filled seeds[a]	X and SD of seed potentials[a]	X and SD of seed efficiencies[a]
628	59.4 ± 12.0*	73.9 ± 11.0*	170.0 ± 7.5*	26.1 ± 7.1*
617	102.8 ± 24.0**	75.5 ± 5.7*	192.0 ± 14.0*	40.8 ± 8.0*
538	111.4 ± 17.8**	53.1 ± 4.1**	165.2 ± 11.7*	35.6 ± 4.4*
520(1)[b]	111.6 ± 17.0**	79.8 ± 15.4*	162.8 ± 6.4*	55.4 ± 15.5*,**
520(2)[b]	115.6 ± 12.0**	83.9 ± 1.4*	181.2 ± 2.3*	53.5 ± 5.9*
593	139.6 ± 9.4**	89.5 ± 4.4*	167.2 ± 5.6*	74.8 ± 7.5**

[a]Data are means and standard deviation where N = 5.
[b]Number in parenthesis refers to tree number.
*The same number of asterisks shared by values indicates that the values are not significantly different at $p < 0.01$ (Scheffe' test).

the highest mean percentage of filled seeds but showed the third lowest seed potential.

The seed efficiencies ranged from 16% to 85% for the individual cones with the overall mean being 47.7% $\pm$ 17.9 (N = 30). Only 10% of the cones displayed seed efficiencies in the interval of 71.5% to 89.5% (Fig. 4B). Clones 520 (trees 1 and 2) and 593 showed both the highest means for seed efficiencies and the highest mean number of seeds.

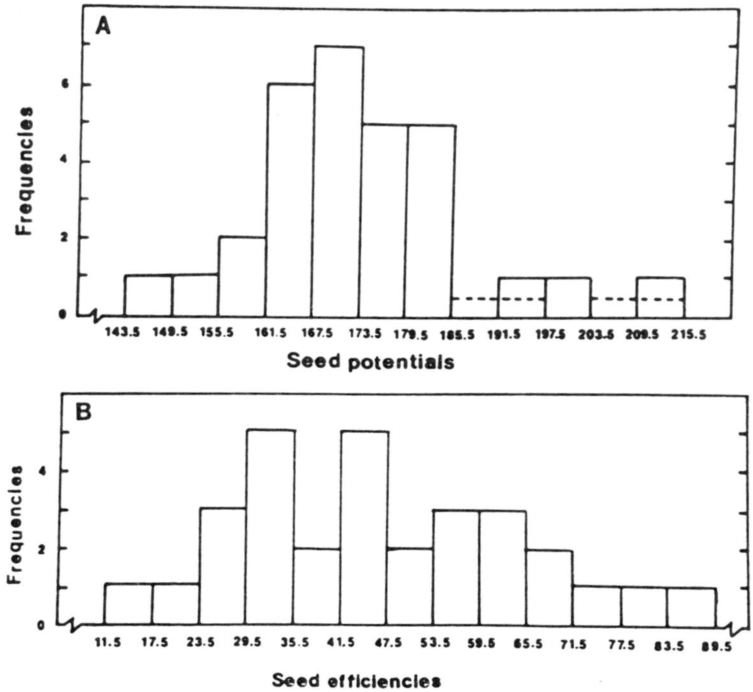

Fig. 4. Frequency distribution of (A) seed potentials and seed efficiencies (B) in the 1984 cone crop N = 30.

An analysis of variance ($p < 0.01$, Scheffe' test) performed on the data in Table 2 showed that although there was no significant difference between clones 520(1), 520(2) and 593 for the number of seeds per cone, the percentage of filled seeds and the seed potential; clone 593 had a significantly different seed efficiency than clone 520(2). Clone 593 was not significantly different from clone 520(1) in the area of seed efficiency. Since clone 593 not only had the highest seed efficiency, but also a relatively high number of seeds per cone; it was the better seed producer among the clones.

During June of year 2 development it was observed that some developing seeds contained ovules that were beginning to dry and shrivel up. In September, the ovules were completely dry. These were the seeds which appeared empty during x-ray analysis. Seeds from three cones for clones 628, 520 (1) and 538 were examined for deterioration. The means and standard deviations for the empty seeds were 14.0 ± 3.2, 35.8 ± 21.1 and 40.6 ± 4.9, respectively.

DISCUSSION

Since failure of pollen to germinate results in ovule abortion during year 1, this failure can be ruled out as an explanation for the empty seeds found during analysis of year 2 cones. These empty seeds could have resulted from fertilization not taking place because of developmental problems in the female gametophyte. Also, embryo abortion soon after fertilization would result in empty seed formation. In this connection, failure of fertilized eggs to undergo embryonic development has been suggested for reductions in seed viability in tissue culture/derived tobacco plants (Hughes, 1986). Embryo abortion may be the result of problems inherent in the egg cell, the sperm cell or the female gametophyte. An intensive investigation with samples from daily collections in June is needed to reveal such problems.

From the 1983 cone crop, clone 558 is the most desirable clone because of its high number of seeds produced and its high seed efficiency (Table 1). Clone 558 is chosen over 593 because of the difference observed for seed efficiency. Statistically however, there is no significant difference. Of the six clones studied the following year, 593 was indeed the most productive.

From the 1984 cone crop, as stated before, the most productive clone was 593. It ranked the highest in all instances except for seed potential. Clone 617 did not come close to its seed potential, resulting in a low seed efficiency. Clone 617 was the least productive clone in both years. Observations of the most productive clones and the least productive clones need to be made to determine if physical differences such as cone size and shape contribute to differences in seed potentials. In cases of similar seed potentials, the reasons for the difference in the number of seeds obtainable should be determined. The difference has to arise either from year 1 abortions or the empty seeds formed during year 2 or a combination of both.

Acknowledgements

We thank Ms. Doris Donnell for typing this manuscript and Mr. Lawrence Brako for making prints of the histograms. This work was supported by USDA Co-operative Agreement 29-075 and a Grant-in-Aid from Sigma Xi. Dr. March's present address is NIH, NIDDK, 9000 Rockville Pike, Bldg. 8, Rm. 2A-11, Bethesdsa, MD 20892.

REFERENCES

Bramelett, D.L., E. W., Belcher, G. L. DeBarr, Jr., G. D. Hertel, R. P. Karfalt, C. W. Lantz, T. Miller, K. D. Ware, and H. O. Yates, III (1977). Cone analysis of southern pines, a guidebook. U.S.D.A. For. Ser. Res. Note SE-13. Southest For. Ex. Stn.

Hughes, K. W. (1986). Tissue culture derived crossing barriers. Amer. J. Bot. 73: 323-329.

Zobel, B. J., J. Barber, C. L. Brown, and T. V. Perry (1958). Seed orchards-- their concept and management. J. For. 56: 815-825.

COUPLED REDUCTION AND REMOVAL OF HEAVY METALS BY A MIXED MICROBIAL ECOSYSTEM

VICTOR M. IBEANUSI*, Department of Biology, Spelman College, Atlanta, GA 30314, USA
ERROL ARCHIBOLD, Department of Biology, Morehouse College, Atlanta, GA 30314, USA

INTRODUCTION

Contamination of surface water by conglomerate mixtures of metals is a significant health problem in aquatic environments near mining and metal (tannery) processing operations. Metals, such as hexavalent chromium (CrVI), manganese (Mn), lead (Pb), selenium (Se), and zinc (Zn) have been shown to occur at elevated levels in many near-shore environments (Capone et al., 1983). _In situ_ bioremediation of these metals has been suggested as a mechanism to alleviate the presence of these contaminants. In this regard, our laboratory has examined the bioremediation and removal of various individual heavy metals from soil and water, in simulated ponds using a mixed microbial ecosystem. In these experiments 80-90% of the metals examined were reduced and recovered from both soil and water in the biomass formed on the pond surface (Ibeanusi et al., 1989; Archibold et al., 1989; Bender et al, 1987).

The present study was undertaken to determine the potential of using the mixed microbial ecosystem for the detoxification and removal of conglomerate metals in solution. A possible mechanism for the coupled reduction and precipitation of individual metals in the mixed microbial ecosystem is presented.

Biodeterioration Research 4, Edited by G.C. Llewellyn *et al.*, Plenum Press, New York, 1994

MATERIAL AND METHODS

A 5-ml inoculum containing a mixture of multiple-metal resistant bacterial strains (originally isolated from the laboratory's simulated ecosystem), was inoculated in simulated ponds (Fig. 1.) containing CrVI, Mn, Pb, Se, and Zn. The concentration of each metal added was 20 ppm or 60 mg/pond. During the 12-day culture period, the mixed microbes (bateria and algae) in the simulated ponds were allowed to progress through their natural successional stages until a stable

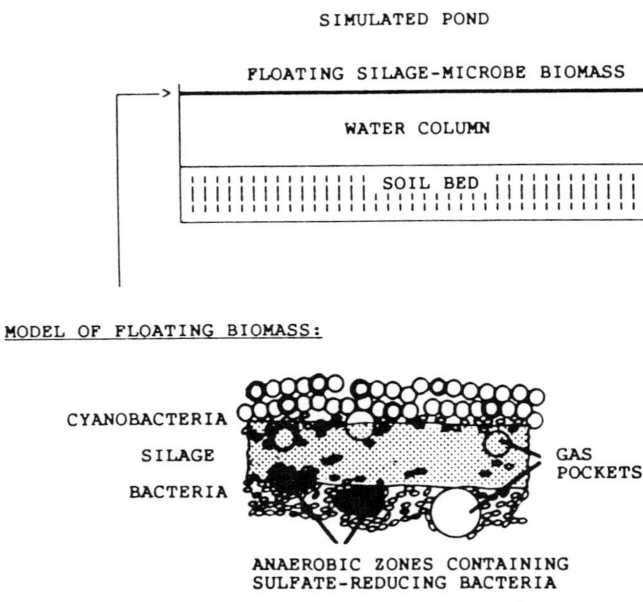

Figure 1. Model of the Silage-Microbe Ecosystem.

mixed microbial mat was formed at the pond surface. During this culture period, daily samples (5 gm, and 5 ml, respectively) of soil, and water, were completely acid hydrolyzed (APHS, 1985) and assayed for the presence of each metal using the atomic absorption unit, AA775 (Varian Instruments, Sugarland, Texas). Finally, the floating mixed microbial mat was completely harvested and also assayed for the presence of each metal added in the pond.

RESULTS

The daily uptake dynamics of each metal present in the mixed metal solution, and their equivalent uptake per gram dry weight of microbial biomass are representated in Figure 2. These results indicated that of the total amount of metals (333 mg), 240 mg were recovered in the mixed microbial biomass. The range of recovery of the metals tested was 61-87%, which represented an average of 72%. under the prevailing experimental conditions, the concentration of metals in the biomass was 28 mg total metals per gram biomass. This

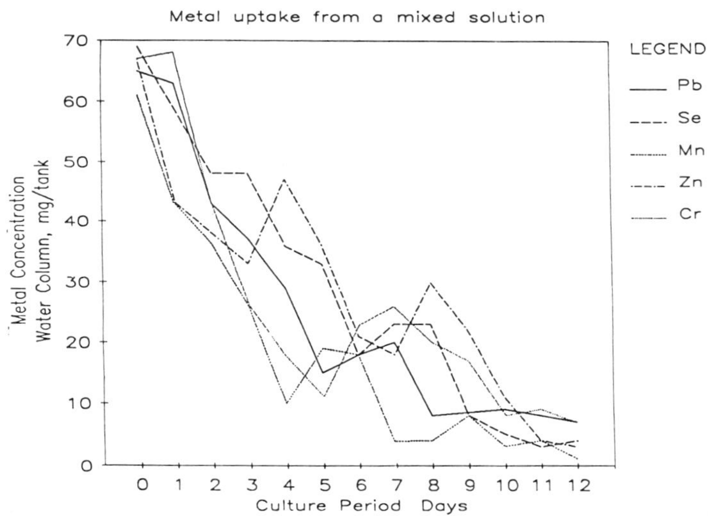

Figure 2. Metal Uptake From a Conglomerate Solution

represents the removal of an average of 8.0 mg of total metal tested per liter per day.

DISCUSSION

Metal-to metal interactions in biological systems has been observed in coastal areas (Dressler et al., 1991). In these interactions, and in the presence of appropriate reducing condition, the ecosystem may favor the precipitation of a particular metal. In this present study, a significant complexation and precipitation of selenium (Se^{+4}) as in selenite to metal selenide(s) were observed.

Studies on selenium reduction have shown that the process is facilitated by a membrane bound reductase (Doran and Alexander, 1977). This reduction process has also been comfirmed in studies with Se-adapted cultures of the mixed microbial ecosystem (Ibeanusi, 1988). As evidenced by the predominace area diagram (Fig. 3), the chemical equilibrium generated in the simulated pond favored the reduction of the selenite to elemental selenium. Under the prevailing condition, the presence of any of the fast exchange metals (such as Pb or Zn), may simply drive the reduction of the elemental selenium to selenides (Se^{-2}) through precipitation to form either PbSe or ZnSe. It was not possible to determine the specific metal which provided the energy for the co-reduction process. Perhaps, the specific stereochemistry of the complex (PbSe or ZnSe) may only be of a minor importance. On the other hand, the complexation process may simply involve one ligand reactions - which requires a close proximity of two ligands within the coordination sphere, called the insertion center (Norris and Kelly, 1970). In this regard, the precipitated elemental selenium provided the insertion center for Pb or Zn (or a combination of the two) to give the metal selenide(s).

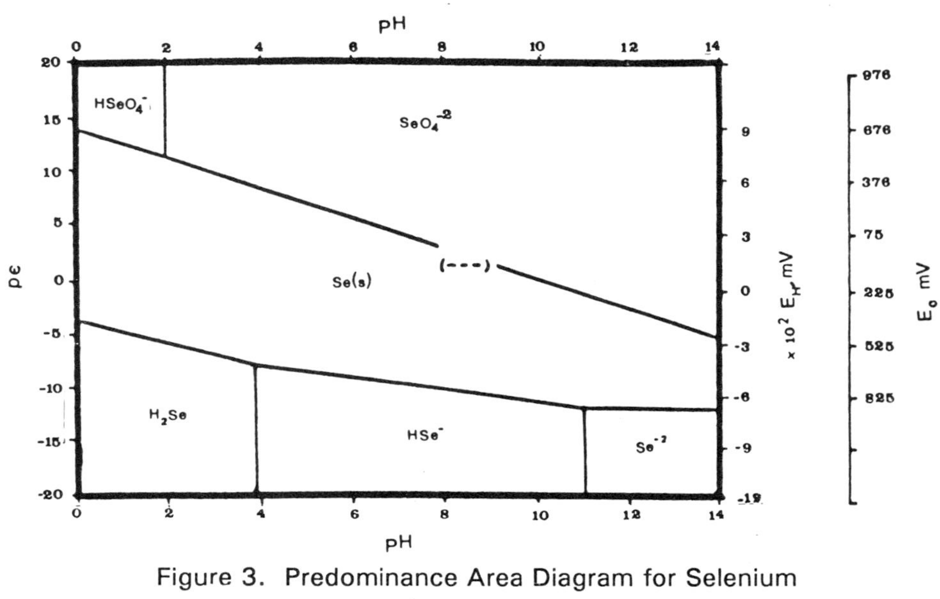

Figure 3. Predominance Area Diagram for Selenium

The observed increased complexation and precipitation of selenium may have implications in the removal of selenites and elemental selenium found in various coastal areas.

SUMMARY

A simulated mixed microbial ecosystem (composed of bacteria, cyanobacteria, and silage) have been shown to be effective in the co-reduction and removal of heavy metals from water and soil. In these experiments, the environmental conditions generated supported an increased complexation of selenium (as in selenite) to metal selenides.

Acknowledgements

The editorial suggestions of Dr. R. Jones, the cooperation of Dr. William V. Dashek, and the clerical assistance of Ms. Toni Conley are highly appreciated.

REFERENCES

Archibold, E.R., Ibeanusi, V.M., Bender, J.A., and Gould, J.P. (1989). Use of a mixed microbial system in the removal of lead from contaminated water. In: Biodeterioration Research 2, pp 161-174 (C. O'Rear and Llewellyn, ed.), Plunum Press, N.Y.

Bender, J., and Ibeanusi, V. (1987). Effects of supplements on the bioaccumulation of lead in Anabaena sp. Bull. Environ. Contam. Toxicol., 39, 209-213.

Capone, D.G., Reese, D., and Kiene, R. (1983). Effects of metals on methanogenesis, sulfate reduction, carbon dioxide evolution, and microbial biomass in anoxic salt marsh sediments. Appl. Environ. Microbiol., 45, 1586-1591.

Doran, J., and Alexander, M. (1977). Microbial transformations of selenium. Appl. Environ. Microbiol., 33, 31-37.

Hong, W., and Wood, J. (1983). Microbial resistance to heavy metals. Environ Sci. Technol., 17, 582-590.

Ibeanusi, V., and Archibold, E., Bender, J, and Gould, J. (1989). Mechanisms of mixed microbial mobilization and recovery of heavy metals in a simulated pond system. Abst. 89th Annu Meeting. Am. Soc. Microbiol., 89, 362.

SECTION IV

PARASITES AS BIODETERIOGENS AND HEALTH-RELATED ISSUES

PARACITES AS BIODETERIOGENS AND HEALTH-RELATED ISSUES

INTRODUCTION

This section will attempt to introduce infectious parasites as biodeteriogens and their role(s) as causative agents in human disease. Diseases caused by parasitic infections continue to represent a serious and major threat to the health and well-being of millions of people in the world, especially in tropical and subtropical regions. It has been estimated that more than 800 million people suffer from malaria at any time, of who 1.5 million die annually, while 800,000 of the 200 million infected with schistosomiasis die each year. More than 800 million people are believed to be infected with hookworm and 900 million with ascaris. A multidisciplinary is needed to counteract parasitic diseases utilizing health education, sanitation programs, vector control, drug treatment and more funding for basic and clinical research. The World Health Organization and the United Nations development program have a commitment over the next 20 years, to eradicate what is considered the six most important infections: Malaria, Schistosomiasis, Filariasos, African Trypanosomiasis, and Leishmaniasis. Leprosy, and infection caused by mycobacterium, is also included. these diseases affect children and adults causing significant morbidity and mortality.

This section will consist of a group of papers comprised of studies on Schistosomiasis, a trematode infection; Dracunculiasis, a nematode infection, and Myiasis, an infection caused by certain species of Diptera.

The first paper on Schistosomiasis invokes kinetic studies in the Schistosoma mansoni snail intermediate host, Biomphalaria glabrata. Specific acetylcholoneasterase and non-specific butyrylcholonesterase activity are shown in infected and non-infected snail vectors. A significant increase in enzyme activity is observed in infected snails.

The second paper is a study of urinary Schistosomiasis in children in ILE-IFE, Nigeria. The prevalence, severity and determinants of the disease among school children in a defined community in Nigeria, is described.

The third paper is concerned with the guinea worm disease, Dracunculiasis. This pilot study provides, on a small scale, information that may be used for large scale control of the disease, using surgical extraction.

The fourth paper on dermal Myiasis is a unique case study in an infant who was being treated for Malaria.

Betty Ruth Jones, Ph.D.
Department of Biology
Morris Brown College
Atlanta, GA 30314, USA

SURGICAL EXTRACTION AS A COMPLIMENTARY CONTROL OPTION FOR DRACUNCULIASIS

BABATUNDE O. AMOLE, Department of Medical Microbiology and Parasitology, Faculty of Health Science, Obafemi Awolowo University, Ile-Ife, Nigeria Laboratory of Parasitology, Department of Biology, Morris Brown College, Atlanta, GA 30314, USA
OLUSEGUN A. FADIRAN, Department of Surgery, Obafemi Awolowo University, Ile-Ife, Nigeria
EMMANUEL I. OFOEZIE, Institute of Ecology, Obafemi Awolowo University, Ile-Ife, Nigeria
TITUS A. (B) OGUNNIYI, Department of Medical Microbiology and Parasitology, Obafemi Awolowo University, Ile-Ife, Nigeria
BETTY R. JONES*, Laboratory of Parasitology, Department of Biology, Morris Brown College, Atlanta, GA 30314, USA

INTRODUCTION

Dracunculiasis is one of the oldest parasitic diseases known to man, and currently affects approximately 50 millon people in Africa, India and the middle east (Hopkins, 1991). The disease is commonly referred to as guinea worm disease. Recently, there have been more concerted efforts to control the disease; and more attention directed towards many ways of controlling its economic impact in various countries (Hopkins, 1991). While the mortality rate is low, the disease's socio-economic impact is probably greater than that of many fatal infections. The alleviation of this disease will result in a dramatic improvement in the health and agricultural productivity and general well-being of millions of affected people in endemic areas of the developing world.

One of the most important intervention points for the control of dracunculiasis is the infected individuals. An infected person with a protruding guinea worm on the body is an important link in the life cycle of the disease. This infected person supplies the embryos required to infect the cyclops vector. Thus, any control model that is capable of breaking the link may lead to mechanisms of alleviating the disease. One model which may prove to be efficient involves the extraction and proper disposal of the guinea worm as soon as it emerges.

The earliest attempts involving extraction are by gradually rolling the worm onto a stick. This method has been found to be inefficient, painful and uncertain because the worm could break at any point (Rab et al., 1991). In addition, it does not break the envisaged link between embryo and vector because the infected person often wades into water bodies carrying the worm on the stick (Imevbore et al., 1987). Drugs such as niridazole, mebendozole, thiabendazole and metronidazole have been administrated to patients with little success at instant extrusion (Eberhard et al., 1990). In addition, serious side effects have also been demonstrated with the use of these drugs (Eberhard et al., 1990; 1991).

With the ineffectiveness of manual extraction and chemotherapeutic approaches, plans for an immediate cure of patients and control of the disease through breakage of the embryo-vector link have diminished. Control methods and approaches are now directed to long-term strategies of preventing transmission by way of alternative safe water supply to endemic communities. However, this approach is very costly and does not eliminate the pain of patients already infected. In addition, patients may continue to utilize pools containing the cyclops vector. Perhaps, more teaching and training in health education could alleviate some of the socio-cultural beliefs and practices that have been known to contribute to the transmission of the disease.

Thus, an efficient, reliable and cost-effective complementary strategy is needed to minimize the suffering of patients and to break the embryo-vector link. Surgical extraction may prove to be a very attractive option. If it is successful and followed with routine post-operative dressings, the suffering of the patients may be reduced and the guinea worm will be disposed of efficiently, thus such an extraction would break the embryo-vector link. More data is needed on the efficacy and cost effectiveness of such an approach and its applications on a larger scale.

The objective of this pilot study is to provide information needed for a large scale control usage of surgical extraction approaches on patients from Akowide village.

MATERIALS AND METHODS

Patients were transported by van over a distance of 25 kilometers from Akowide to the Obafemi Awolowo University Teaching Hospitals Complex (OAUTHC), Ile-Ife, Nigeria. Surgical extraction was performed on nine patients, divided into two groups as follows:

Group A: Patients with acute suppuration, discharging sinus and with the adult guinea worm partly extruded.

Group B: Patients with acute suppuration and without either a discharging sinus or a partly extruded guinea worm.

The extraction followed the standard surgical procedure, using ketamine hydrochloride and Diazepam as intravenous anaesthesia. Wound swabs were taken for microscopy, culture and sensitivity. Patients were monitored post-operatively and fresh dressings were applied as needed.

RESULTS

The age and sex distribution of the patients is shown in Table 1. For Group A (GA) patients, worm extraction appeared successful, relatively easy, and rapid, lasting between 15 and 20 minutes. Whereas, extraction from Group B (GB) patients appeared unsuccessful, extremely difficult and time consuming. Usually, each attempt was abandoned after 40 minutes. However, in 2 out of 4 cases, a second attempt was made two weeks later when it was thought that the same worms had emerged at different sites from the previous.

At this stage, extraction appeared successful as in GA patients. In both the GA and GB patients and in the two GB patients who had a second operation, post operative ulcers healed within one week of daily dressings.

The cost of the surgical operation in United States dollars is shown in Table 2. The total cost for the entire project, which lasted 38 days was $ 202.80. Therefore, the cost of extraction per patients was $ 22.50.

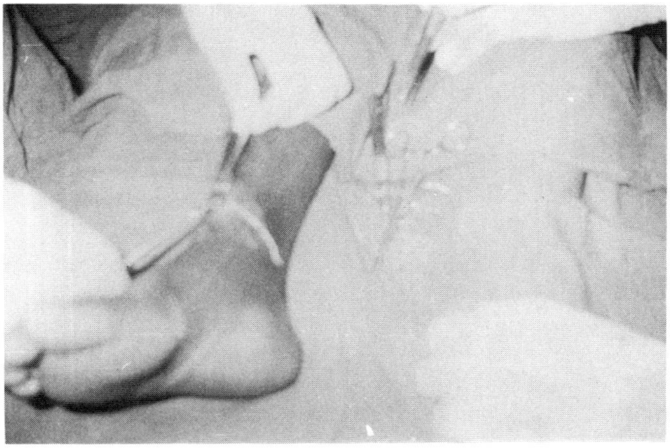

Figure 1. A Patient Undergoing Surgery to Extract a Protruding Guinea Worm.

Table 1. Age and Sex Categories of Patients in Group A and B.

	GROUP A		GROUP B		
AGE	MALE	FEMALE	MALE	FEMALE	TOTAL
1-4	-	-	-	-	-
5-9	1	2	2	-	5
10-14	-	1	-	1	2
15-19	-	-	-	1	1
20-24	-	-	-	-	-
25-29	1	-	-	-	1
30	-	-	-	-	-
TOTAL	2	3	2	2	9

The most common type of bacteria found in discharging sinuses were *Staphylococcus aureus* and *Corynebacterium diptheriae*. These bacteria were found to be sensitive to gentamicin and streptomycin.

Table 2. Cost Analysis of Surgical Extraction of Guinea Worm from Nine Patients at Obafemi Awolowo University Teaching Hospitals Complex (OAUTHC), Ile-Ife, Nigeria.

Surgical materials including Anesthesiology agents	$ 137.00
Transportation	$ 28.75
Other Hospital charges	$ 37.00
Total	$ 202.75
(i.e., cost/patient)	$ 22.50

DISCUSSION

The results from group A patients, showed clinical manifestations such as inflammation of the tissues around the extruding worm, a tract with minimal adhesions and relatively ease extraction of the worm. This was not the case with group B patients. The presence of a pointing abscess was not synonymous with the presence of the adult worm in a similar nodule. In addition, there appeared to be no evidence of the worm location based on general clinical features. Extraction was therefore difficult in this group of patients. It is our opinion that a great deal of radical dissection would appear to be necessary to make extraction feasible. This may lead to inoculation of normal tissues with virulent bacteria and thus greater morbidity. We believe that the cost of treatment of secondary bacterial infections would lead to a higher cost, thus reducing the cost effectiveness of this measure as a feasible control strategy.

As to our knowledge, this pilot study represents the first reported evidence of a strategy to control dracùnculiasis using surgical extraction. We could not find any other reported studies in the literature. Based on the preliminary results of this study we concluded the following:

(i) Age is not an important factor in extraction and that the only determining factor may be the stage of worm extrusion;

(ii) Surgical extraction is feasible and easy to perform if undertaken at the appropriate time;

(iii) The most appropriate time to perform surgical operation is within a few days after the emergence of the guinea worm;

(iv) Surgical complications and secondary infections did not occur in association with the extraction if the wounds were properly dressed post-operatively.

An important attribute of this procedure is that extracted worms have no chance of reaching water-bodies; hence, the break in embryo-vector link has a high probability of occurring. Furthermore, patients regained their full health within days of operation. Even, a completely immobilized patient who had surgery was capable of returning to the farm three days after extraction.

The significance of these findings are two fold: A 100% break in the embryo-vector link should allow patients to be monitored early and the worm extracted. Such clinical practices may lead to a more rapid alleviation of the disease from endemic communities in a relatively short period of time. The overall process of monitoring of patients could be enhanced through scheduled and structured educational approaches with the community. Another important attribute resulting from this study is exemplified through improved health of patients and a feeling of well being a few days after extraction of the worms. Prior to such treatments, patients usually experienced 2 to 3 months of serious incapacitation (Edungbola, 1984, Ofoezie and Adeniyi, 1989). The surgical extraction method appears to offer an effective control mechanism of the guinea worm because (1) it reduces the duration of the pain suffered by victims; (2) it breaks the embryo-vector link; and (3) it is cost-effective. The present pilot was carried out at a cost of $22.50 per patient. We believe this amount could be reduced by 50-65% when carried out on a larger scale. Some of the possible ways to reduce cost will include the use of a mobile medical team on the field using the facilities available in the local dispensaries.

Therefore, in view of the reported problems and limitations associated with other control measures, namely chemotherapy (Eberhard et al., 1990; 1991), provisions of portable water (Edungbola et al., 1988; Udonsi, 1987) or even no control (Edungbola and Watts, 1985; Watts, 1986), surgical extraction may offer the best cost-effective control of guinea worm. However, further field studies are needed on other selected villages.

SUMMARY

The phase of worm protrusion in dracunculiasis is vital in the life cycle of the parasite. Therefore any control model capable of breaking the host-vector link at this point will enhance mechanisms for the disease. A pilot study was carried out to provide information on the efficacy, reliability and cost effectiveness of surgical extraction as a control strategy for dracunculiasis. Surgical extraction was performed on 9 patients in two categories (A and B). While both groups showed clinical manifestations of suppuration. Only group A revealed discharging sinuses and partly extruded guinea worm. Staphylococcus aureus and Corynebacterium diptheriae both sensitive to gentamicin and streptomycin were commonly cultured from post-operative wound of patients by way of swabs. Worm extraction was relatively easy and fast in group A while it was difficult, time consuming and unsuccessful in group B patients. Overhead surgical extraction costs were $22.50. In conclusion, surgical extraction based on the results of our findings 1) reduces duration of suffering by patients; 2) breaks host-vector link; and 3) is cost-effective and may offer effective control.

Acknowledgements

This study was supported in part by a grant from the Third World Academy of Sciences, Trieste, Italy and in part by a grant from the National Science Foundation. I am indebted to the parasitology laboratory at Morris Brown College for providing me with the space and time to complete the various writing aspects of this paper. I extend a special thanks to Mr. Charles Nelson, an undergraduate parasitology research associate for all word processing associated with this manuscript and for making several revisions prior to initial review. In addition, I extend my gratitude to Dr. Betty Ruth Jones, one of the authors of this manuscript for providing her time and support towards all aspects of this project.

REFERENCES

Eberhard, M.L., Brandt, F.H., Ruiz-Tiben, E. and Hightower, A. (1990). Chemoprophylactic drug trials for treatment of dracunculiasis using the Dracunculus insignis-ferret model. J. Helminthol. 64, 79-86.

Eberhard, M.L., Brandt, F.H., Kaiser, R.L. (1991). Chlortetracycline for Dracunculiasis. Lancet 337, 500.

Edungbola, L.D. (1984). Dracunculiasis at Igbon, Oyo State, Nigeria. J. Trop. Med. Hyg., 87, 153-158.

Edungbola, L.D., Watts, S.J. (1985). Epidemiological assessment of the distribution and endemicity of guinea worm infection an Asa, Kwara State, Nigeria. Trop. Geog. Med., 37, 22-28.

Edungbola, L.D., Watts, S.J., Alabi, T.O. and Bello, A.B. (1988). The impact of a UNICEF-Assisted rural water project on the prevalence of guinea worm disease in Asa, Kwara State, Nigeria. Am. J. Trop. Med. Hyg., 39 (1), 79-85.

Hopkins, D.R. (1991). Eradication of Dracunculiasis: Update. Epidemiologic Reviews 13, 316-319.

Imevbore, A.M.A., Ofoezie, I.E. and Obot, E.A. (1987). WHO Report of the study on small and cyclopoid borne disease problems of small scale water resources development projects in Nigeria. 207 pp.

Ofoezie, I.E. and Adeniyi, I.P. (In Press) The prevalence and endemicity of dracunculiasis in Akowide Village, Oyo State, Nigeria. Ann. Trop. Med. Parasitol.

Rab, M.A., Khan, R.N., Atiq, A., and Ahmed, S.A. (1991). Dracunculiasis: An approach to hasten worm expulsion. J. Trop. Med. Hyg. 94, 325-326.

Udonsi, J.K (1987). Control of endemic dracontiasis by provision of water supply in rural communities of Imo State, Nigeria. Public Health, 101, 63-70.

Watts, S.J. (1986). The comparative study of patterns of guinea worm prevalence as a guide to control strategies. Soc. Sci. Med., 23 (10), 975-982.

URINARY SCHISTOSOMIASIS AMONG SCHOOL CHILDREN IN ILE-IFE, NIGERIA

BABATUNDE O. AMOLE, Department of Medical Microbiology and Parasitology, Faculty of Health Sciences, Obafemi Awolowo University, Ile-Ife, Nigeria
Laboratory of Parasitology, Department of Biology, Morris Brown College, Atlanta, GA 30314, USA
MOHAMED K. JINADU, Department of Community Health and Nutrition, Faculty of Health Sciences, Obafemi Awolowo University, Ile-Ife, Nigeria
BETTY R. JONES, Laboratory of Parasitology, Department of Biology, Morris Brown College, Atlanta, GA 30314, USA

INTRODUCTION

Urinary schistosomiasis, a water-born parasitic infection is a debilitating disease of great importance in most parts of developing countries (Hatz, 1990). Studies have shown that the disease is endemic in Nigeria and that its epidemiological patterns vary from one part of the country to another (Pugh and Giles, 1979; Ejeize and Ade-Seramo, 1982; Arene et al., 1966; Adekolu John and Abolarin, 1986).

In Oyo State of Nigeria, the disease is generally believed to be widespread among school-age children. Thus, there is a need to delineate prevalence of the infection in children and to initiate planning strategies which may lead to effective control measures for the infection. The purpose of this study is to determine the prevalence, severity and determinants of the disease among the school children in a defined community in the Oyo State of Nigeria.

MATERIALS AND METHODS

This study was conducted in 1990 among primary and secondary school children in Ile-Ife, a town in Oyo State, Nigeria. In the town, there were 42

Biodeterioration Research 4, Edited by G.C. Llewellyn *et al.*, Plenum Press, New York, 1994

primary and 21 secondary schools with a total population of 26,890 school children in 1990. The schools were scattered throughout the town. Using a map (Figure 1), the town was divided into four quadrants and a list of primary and secondary schools were randomly selected from each quadrant. An arm of each class from classes 1-6 and 1-5 in the selected primary and secondary schools respectively were also randomly selected for the study. Following permission from the head teacher and parents, urine samples were obtained from the school children in the selected classes between the hours of 10:00 a.m. and 12 GMT for examination. Quantification of schistosome eggs in the urine was done by the filtration method (Peters et al., 1976). Egg output of between 1-32/10 ml urine was regarded as a light infection; 33-256/10 ml as moderate, while greater than 256/10 ml was regarded as a heavy infection (Pugh and Giles, 1979). The sediment was then examined microscopically for evidence of gross hematuria. Information was also obtained about the children's age and water contact behaviors.

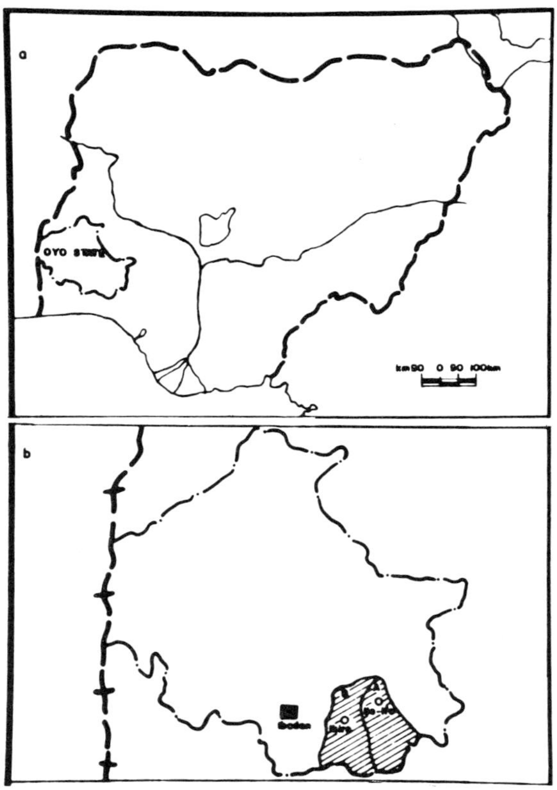

Figure 1. [a]A Map of Nigeria Showing Oyo State. [b]A map of the Oyo State of Nigeria Showing the Town of Ile-Ife.

RESULTS

Out of the 553 school children examined for urinary schistosomiasis in the town, 268 (48.5%) were infected. Table 1 shows the age distribution of the school children. Ninety children were infected between the age of 6-7 years, their first year in primary school. The percent of the infected school children increased with increasing age of the children until it reached a peak of 65.1% at age-group 12-13 years of age. It then declined slightly to 57.6% at age-group 16 years and above when the children were ready to leave high school.

Table 2 shows the prevalence of hematuria and the intensity of infection, as measured by egg outputs in the urine, among the various age-groups of the school children. The percent of school children with hematuria increased from 4.4% in the age-group 6-7 years to 55.9% in the age-group 16 years and above.

Table 1. Age Distribution of the School Children With and Without Urinary Schistosomiasis.

AGE - GROUP	NO. EXAMINED	NO. (%) WITH SCHISTOSOMIASIS (EGG POSITIVES)
6 - 7	90	21 (23.3)
8 - 9	85	25 (29.4)
10 - 11	101	52 (51.5)
12 - 13	106	69 (65.1)
14 - 15	86	52 (60.5)
16 +	85	49 (57.6)
TOTAL	553	268 (48.5)

Overall prevalence of hematuria among the school children was 26.2%, which represents about half of the infected population. The table also shows that light to moderate infection accounted for 29.8% and 14.3% respectively of the school children examined, while heavy infection accounted for 4.3%. The majority of the lightly infected were over 13 years and majority of the moderately and heavily infected were between 8-15 years and 8-13 years respectively. The intensity of the infection was cross-tabulated with the presence or absence of hematuria in table three. Among those with hematuria, 29% were lightly infected, 54.5% were moderately infected and 16.5% were heavily infected. These data show a very strong association between the intensity of infection and the presence of hematuria.

Table 2. Age Distribution of School Children According to the Presence of Hematuria and Urinary Egg Output.

AGE-GROUP	NUMBER EXAMINED	NUMBER WITH HEMATURIA	EGG OUTPUT LIGHT	MODERATE	HEAVY
6-7	90	4 (4.4)	21 (23.3)	-	-
8-9	85	8 (9.4)	17 (20.0)	4 (4.7)	4 (4.7)
10-11	101	43 (42.5)	26 (25.6)	33 (31.1)	9 (8.9)
12-13	106	59 (55.9)	25 (23.6)	33 (31.1)	11 (10.9)
14-15	86	15 (17.4)	35 (40.7)	17 (19.8)	0 -
16+	85	16 (8.4)	41 (48.2)	8 (9.4)	0 -
TOTAL	553	145 (26.2)	165 (29.8)	79 (14.3)	24 (4.3)

Light = 1-32/10 ml of urine
Moderate = 33-256/10 ml of urine
Heavy = greater than 256/10 ml of urine

Table 3. .Intensity of Infection with Urinary Schistosomiasis and the Presence of Hematuria Among the School Children.

INTENSITY OF INFECTION	NUMBER WITH HEMATURIA		NUMBER WITHOUT HEMATURIA		TOTAL	
	Number	(%)	Number	(%)	Number	(%)
Light	42	(25.5)	123	(74.5)	165	(100)
Moderate	79	(100)	0	(0)	79	(100)
Heavy	24	(100)	0	(0)	24	(100)
TOTAL	145	(54.1)	123	(45.9)	268	(100)

Bathing in streams and ponds, which were quite abundant in the township or its vicinity, was a common preoccupation of the school children. Nearly all the infected school children (93.7%) readily admitted to have bathed regularly in the streams or ponds, compared with 45% of the non-infected children.

DISCUSSION

This study shows that urinary schistosomiasis is a very common health problem among school children in Ile-Ife. About half of the school children in our study were infected with this serious and debilitating illness which compares favorably with 41.6% in Oyedirian's (1976) findings. During this investigation, discussions were held with some of the parents of the infected school children and they appeared not to appreciate the seriousness of the problem. There was either a tendency for the parents to ignore the problem or resort to the use of traditional remedies.

Our study also confirms a similar study in the country by Ejeize (1982) that shows a characteristic relationship between the prevalence and intensity of infection and age. There was a sharp increase in both the prevalence and the intensity of infection as age of the children increased as shown in tables one and two of our study.

Hematuria was observed in about 50% of the infected school children and these were mainly children with moderate and heavy infections. All children without hematuria were however, lightly infected, showing a significant relationship between hematuria and intensity of infection. In other words, the more severe the infection, as measured by egg out put in urine, the more severe the degree of hematuria in the infected persons.

In conclusion, this preliminary investigation has highlighted the severity of urinary schistosomiasis as a health problem among school children in the Ile-Ife township. There is an urgent need to institute control measures against the infection in order to reduce its impact on the health of children and the community.

Health education to discourage the children form bathing in streams or ponds may not achieve the desired impact because children simply love bathing, especially during the dry season of the year. Therefore, the main strategy recommended for the control of schistosomiasis in this area as well as in other parts of the country is regular chemical disinfection and adequate treatment of infected school children. These measures should be combined with health education on how to avoid and recognize the disease, and the importance of adequate medical treatment, as an aspect of school health programs in the country.

SUMMARY

A study of urinary schistosomiasis among 553 randomly selected primary and secondary school children in Ile-Ife township in 1990 shows that half (48.5%) of the school children were infected. There was a sharp increase in both the prevalence and intensity of the infection up to age 13 years which then declined slightly by age 14. About 50% of the infected school children had gross hematuria. There was an association between the intensity of the infection and the presence of the infection and the presence of hematuria. The main strategies recommended for the control of the infection were regular disinfection of ponds and streams in the town and adequate treatment of infected school children, combined with school health education programs.

Acknowledgements

We are indebted to the parasitology laboratory at Morris Brown College for providing me with the space and time to complete the various writing aspects of this paper. We extend a special thanks to Mr. Charles Nelson, an undergraduate parasitology research associate for all word processing associated with this manuscript and for making several revision prior to initial review.

REFERENCES

Aedkolu-John, E.O., and Abolarin, M.O. (1986). Status of Human Schistosomiasis on the Eastern Side of Kainji Lake Area of Nigeria. East Afri. Med. J., 63 (7), 463-470.

Arene, F.O., Powell, C.B. and Okwodu, N.E. (1986). Prevalence of Schistosoma haematobium among Inhabitants of Ahoada District. Public Health, 100 (5), 302-308.

Ejeize, G.C. and Ade-Seramo, M.A. (1982). A Survey of some Parasitic Infections of in Epe Villages of Lagos State, Nigeria. Nig. Med. J., 12, 81-87.

Hatz, C. (1990). Measurement of Schistosomiasis-related morbidity at community level in areas of different endemicity. Bull. W. H. O., 68 (6), 777-787.

Oyediran, A.B. (1976). Renal Function in Vesical Schistosomiasis. An M.D. Thesis Submitted to University of Ibadan, pp. 77-82.

Peters, P.A., Mahmoud, A.A.F., Warren, K.S. et. al., (1976). Field Studies of a Rapid, Accurate Means of Quantifying Schistosoma haematobium Eggs in Urine Samples. Bull. W.H.O., 54, 159-162.

Pugh, R.N.H. and Giles, H.M. (1976). Malumfashi Endemic Diseases Research Project, X, *Schistosoma haematobium* and Bacteriuria in the Malumfashi Area. *Ann. Trop. Med. Parasitol.*, 73, 349-354.

ACETYLCHOLINESTERASE LEVELS IN BRAINS OF RABBITS INFECTED WITH *Trypanosoma brucei brucei*: A PRELIMINARY STUDY

BABATUNDE O. AMOLE, Department of Medical Microbiology and Parasitology and Chemical Pathology, Faculty of Health Sciences, Obafemi Awolowo University, Ile-Ife, Nigeria ; Laboratory of Parasitology, Department of Biology, Morris Brown College, Atlanta, GA 30314, USA

KASHOPE D. THOMAS, Department of Medical Microbiology and Parasitology and Chemical Pathology, Faculty of Health Sciences, Obafemi Awolowo University, Ile-Ife, Nigeria

BETTY R. JONES*, Laboratory of Parasitology, Department of Biology, Morris Brown College, Atlanta, GA 30314, USA

CHARLES A. NELSON, Laboratory of Parasitology, Department of Biology, Morris Brown College, Atlanta, GA 30314, USA; Department of Biology, Morehouse College, Atlanta, GA 30314, USA

INTRODUCTION

Humans infected with African trypanosomes develop severe neurological and neuropsychiatric disturbances during the terminal stages of the disease (Lambo, 1966; WHO report, 1985) and the mechanisms causing these disturbances are not known. Currently, much attention is focused on the possible biochemical basis for these disturbances (Tanowitz et al., 1981, 1983; Brennessel et al., 1985). These investigators reported that some alterations of enzyme activity in the central nervous system may play a role in the pathogenesis. Acetylcholinesterase (AChE) is an enzyme found in the cholinergic synapses and neuromuscular junctions of various mammalian tissues. We postulate that AChE levels in infected brains of rabbits with Trypanosoma brucei brucei may be directly or indirectly related to cerebral symptoms in animals and possibly humans. Thus, the objective of this investigation was to study AChE activity in the brains of rabbits infected with T. brucei brucei.

MATERIALS AND METHODS

Maintenance of Animals

The experiments were performed on a total of 14 adult male New Zealand White (NZW) rabbits, weighing between 1.6 and 2.3 kg at the beginning of each experiment. The rabbits were bred in our animal house on high protein rabbit pellets containing 21% protein. They were divided into two groups. Group A consisted of uninfected controls and Group B was comprised of animals infected with T. brucei brucei. All animals were administered 1.0 ml of penicillin-streptomycin for 3 days, and the following 3 days antibiotics were not administered.

Parasites and Infection

Parasites were maintained in the laboratory by passage of parasites from mouse to mouse. Trypanosoma brucei brucei (NSUKKA/62NITR/8 strain) was used for the infection. This strain produces a chronic infection in rabbits lasting about 5 weeks and is usually characterized by a terminal high parasitemia. Rabbits were infected by intraperitoneal (IP) injection of approximately 10×10^6 of T. brucei brucei. Infected rabbits and their controls were sacrificed after 4 weeks of infection.

Preparation of Tissues

The rabbits were sacrificed by cervical fracture, followed quickly by decapitation, removal and dissection of brains on a thick glass slide placed on ice. The olfactory lobes and the pineal tissue were not included with the rest of the brain. Approximately 4 mm of spinal cord was left attached to the medulla. Brains were carefully dissected into right and left halves by making a longitudinal incision with a scalpel and the two halves were immediately stored at -70°C and thawed on the day of assay. For the regional determination of enzyme, the brains were dissected into 5 regions (Cortex, striatum, brain stem, hippocampus and cerebellum). One gram of tissue was homogenized in 5.0 ml of distilled water using a hand homogenizer placed on ice. The homogenate was then diluted 1 in 50, with distilled water and 0.4 ml of this was assayed. Contamination with red blood cells was not observed.

Preparation of Parasite Homogenates

Pooled blood from infected mice was collected from bronchial arteries, diluted with cold phosphate-saline glucose (PSG) buffer as described by Lanham (1968) (0.10 M sodium phosphate, 0.073 M NaCl and 1.5% glucose, pH 8.0). Heparin was used as an anticoagulant. The diluted blood was then centrifuged at 1000 xg for 15 min, and the buffy coat was collected and resuspended in PSG. The enriched trypanosome fraction was further purified by passage through a diethylaminoethylcellulose column (Whatman, DE 52, Whatman, Inc., Clifton N.J. U.S.A.) prepared as described by Lanham (1968). Parasites were eluted from the column free of host cells and washed before use. They were washed three times in PSG. Approximately 1.0 X 10^6 purified parasites were suspended in 0.2 ml of PSG and homogenized with a hand homogenizer. The 0.2 ml homogenate was made up to 1.0 ml, with PSG, centrifuged at 1000 xg for 15 minutes and 0.4 ml of the supernatant taken for enzyme assay.

Enzyme Assay

The assay procedure utilized in this study was according to Ellman et al. (1961), using acetylthiocholine as a substrate. The AChE activity was expressed as micromoles of AChI hydrolysed/min/g tissue. For homogenate samples, activity was expressed as micromoles/min/ml of homogenate. The same procedure was also used with butyrylthiocholine as substrate for the measurement of the activity of the non-specific esterases in whole brains of control and infected rabbits.

Statistics

Statistical analysis was performed using the student's test, for unpaired data. Statistical significance was considered at $P<0.05$.

Histological Examination

Tissues from the cortex, brain stem, corpus striatum, hippocampus and the cerebellum were fixed in 10% formalin buffered saline for one hour. The sections were cut 6-10 microns in diameter and subsequently stained with hematoxylin and eosin. The sections were screened for the presence of parasites and inflammatory changes, using a light microscope.

RESULTS

Acetylcholinesterase activity in whole brains of T. brucei brucei infected rabbits are shown in figure 1. The mean value for infected whole brains was 46 micromoles as compared to 11 micromoles AChI hydrolysed/min/g tissue for control uninfected rabbits. This represents a 300% increase in AChE activity in brain tissue of infected rabbits as compared to uninfected controls ($P<0.01$). Regional concentrations of the enzyme are shown in table 1. The highest enzyme concentration was recorded in the striatum and the lowest in the brain stem. The enzyme concentration was significantly higher in the striatum of infected rabbits than in control rabbits ($P<0.05$). Acetylcholinesterase activity in the cortex, brain stem, hippocampus and cerebellum respectively showed no significant difference between the control and infected rabbits. There was no AChE activity detected in the homogenates of parasite suspensions. Histological examination of the cortex, brain stem, corpus striatum, hippocampus and the cerebellum appeared normal with no evidence of inflammatory changes. Parasites were not observed in the tissues.

DISCUSSION

The results in this study show high levels of AChE activity in the brains of rabbits infected with T. brucei brucei, the causative parasite for African Trypanosomiasis. Non-specific esterase activity did not contribute to this increase in enzyme activity. Inflammatory cells which could have produced non-specific esterases were not observed during histological examination of the brain tissues. Studies by WHO (1985), Poltera et al. (1982), Morrison et al. (1983) have reported extensive lymphocytic, histolytic and plasma cell infiltration of the meninges accompanying cerebral trypanosomiasis. Our findings are not in total accord with these studies. The regional increase in AChE activity in the striatum as compared with other parts of brain tissues, points to localized neuronal injury in the infected rabbits. Few parasites were observed in the brain tissues and AChE activity was not seen in the parasite homogenates. This may indicate that the changes observed in infected brains tissue could be due to alterations in level(s) of host enzyme activity.

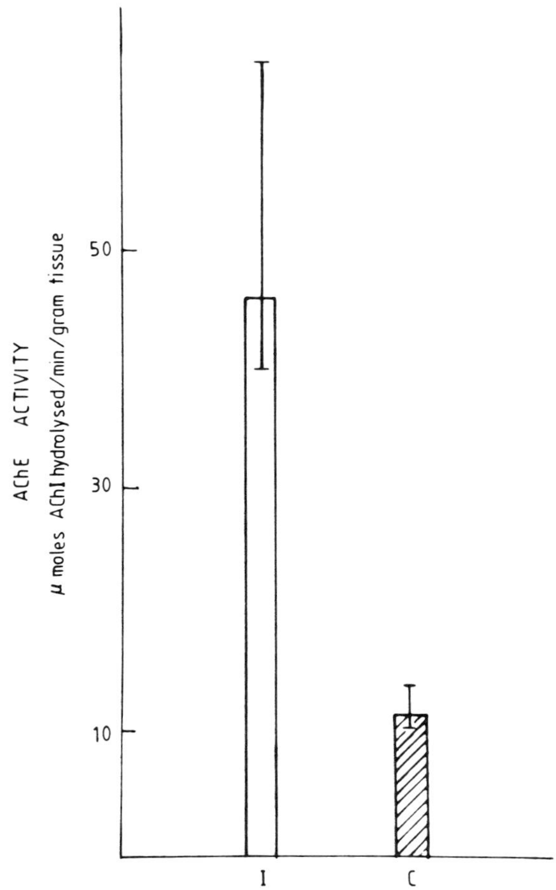

Figure 1. Mean acetylcholinesterase activity in whole brains obtained from infected rabbits. (i) (N = 9) 30 days post-infection and control (C) (n = 5).

Acetylcholinesterase is an integral part of cholinergic synapses (for example, the reticular formation in the brain) and neuromuscular junctions and it is an enzyme involved in several clinical situations including paralysis of skeletal muscle and autonomic dysfunctions (Engel et al., 1977, Elias and Appel, 1981). Our data suggest that major abnormalities associated with cerebral African trypanosomiasis such as hemiplegia, paraplegia, fine tremor, convulsions, unsteady gait and slurred speech may be affected by cholinergic functions in the brain.

Reports of neurochemical changes which underlay some of the neurological and psychiatric manifestations have only recently started to appear. Brennessel et al. (1985) reported a significant increase in AChE activity in B6 mice, 20 days post-infected with Trypanosoma cruzi, and more recently we have just reported alterations in monoamines in African trypanosomiasis, which we believe may form the basis for the neuropsychiatric abnormalities associated with this disease (Amole, et al. 1989). More work is needed to locate the source and mechanism(s) of these AChE alterations in brain the in T. brucei brucei infections.

Table 1. AChE activities in Brains Regions of Infected and Uninfected Rabbits.

	Striatum	Cortex	Brain stem	Hippocampus	Cerebellum
Uninfected [n=5]	45.75 ± 8.25	10.73 ± 0.08	6.63 ± 5.13	9.80 ± 2.20	20.40 ± 1.20
Infected [n=9]	134.25 ± 37.59*	24.75 ± 2.073	7.60 ± 3.54	11.75 ± 0.94	20.50 ± 2.12

AChE activity is measured in micromoles/min/g tissue. Each value is presented as the mean with the standard deviation.
*$P<0.05$

SUMMARY

Acetylcholinesterase activity was measured in homogenates of whole brain from uninfected and infected rabbits with Trypanosoma brucei brucei. Enzyme activity was also measured in brain regions including the cortex, brain stem, corpus striatum, hippocampus and the cerebellum respectively. There was 300% increase in activity in whole brains of rabbits 30 days post-infection as compared to controls. The highest degree of enzyme activity was found in the striatum. The striatum of infected rabbits had a significantly higher degree of activity ($p<0.05$) of AChE than their respective uninfected controls. Non-specific esterases did not contribute to this increase in enzyme activity and there was no evidence of inflammatory changes due to infection. Acetylcholinesterase activity was not observed in homogenates of parasitic suspensions. This data may suggest that there are alterations in the cholinergic neuronal system in African trypanosomiasis.

Acknowledgements

We wish to thank Mr. Charles Nelson, an undergraduate research associate working in the laboratory of Dr. Betty Jones in the area of parasitology for the word processing work associated with this paper. We extend thanks to also to Dr. William Dashek, Associate Professor of Biology at Clark Atlanta University for further editing this manuscript. In addition, I extend my gratitude to Dr. Betty Jones for providing her time and support toward all aspects of this project. This research is supported by the following agencies: National Institutes of Health/Minority Biomedical Research Support Program, grant number 1-S06-GM-45913-01 and National Science Foundation Research in Minority Institution (NSF-RIMI), grant number RII-80-6958.

REFERENCES

Amole, B.O., Sharpless, N., Wittner, M. and Tanowite, H.B. (1989). Neurochemical measurements in the brains of mice infected with Trypanosoma brucei brucei (TREU 667). Ann. Trop. Med. Parasit. 83 (3), 226-232.

Brennessel, D.J., Whittner, M., Braunstein, V. and Tanowitz, H.B. (1985). Acetyl cholinesterase levels skeletal muscle of mice infected with Trypanosoma cruzi. Am. J. Trop. Med. Hyg. 34 (3), 460-464.

Ellman, G.L., Courtney, K.D., Andres, V. and Feather-Stone, R.M. (1961). A new and rapid coloriometric determination of acetylcholinesterase activity. Biochem. Pharm. 7, 88-95.

Elias, S.B. and Appel, S.H. (1981). Biochemistry of myoneural junction and of its disorders. In: Basic Neurochemistry, pp. 513-528 (G. Siegel et al., eds.), Little Brown and Company, Boston.

Engel, A.G., Lambert, E.H. and Gomez, M. (1977). A New myasthenic syndrome with end plate cholinesterase deficiency, small nerve terminals and reduced acetylcholinesterase release. Ann. Neurology 1, 315-330.

Lambo, T.A. (1966). Neuropsychiatric syndromes associated with human trypanosomiasis in tropical Africa. Acta. Psychatr. Scand. 42, 474-484.

Lanham, S.M. (1968). Separation of trypanosomes from the blood of infected rats and mice by anion exchanger. Nature (London) 218, 1273-1274.

Long, C., King, E.J. and Sperry, W.M. (1971). Biochemist's Handbook pp. 275-277, E and F.N. Spon ltd., London, U.K.

Ikede, B.O. (1979). Genital lesions in experimental chronic Trypanosoma brucei infection in rams. Res. Vet. Sci. 26, 145 -151.

Morrison, W.I., Murray, M., Whitelaw, D. W. and Sayer, P.D. (1983). Pathology of infection with Trypanosoma brucei. Disease syndromes in dogs and cattle revisiting from se severe tissue damage. Contrib. Microbiol. Immunol. 1, 103-119.

Poltera, A.A., Hochmann, A. and Lambert, P.H. (1982). Trypanosoma brucei gambiense: Cerebral immunopathology in mice. Acta. Trop. 39, 205-218.

Tanowitz, H.B., Davies, P., Factor, S.M., Minase, T., Herskowitz, A. and Whittner, M. (1981). Trypanosoma cruzi: Choline acetyltransferase activity in tissue of susceptible and resistant mice infected with the Brazilian strain. Exp. Parasit. 51, 269 -278.

Tanowitz, H.B., Davies, P. and Whittner, M. (1983). Alterations in acetylcholine receptors in experimental Chagas Disease. J. Inf. Dis. 147, 460-466.

W.H.O. report, (1985). African trypanosomiasis, In: Tropical Disease Research (TDR), 7th programme report January 1, 1983 -December 31, 1984, UNDP/World Bank/W.H.O. Special programme for research and training in tropical diseases.

A COMPARATIVE FIXATION ANALYSIS OF *TETRAHYMENA PYRIFORMIS* ULTRASTRUCTURE

B.J. BELL*, Department of Biology, Clark Atlanta University and Spelman College, Atlanta, GA 30314, USA

L.A. BRAKO, Department of Biology, Clark Atlanta University, Atlanta, GA 30314, USA

B.R. JONES, Department of Biology, Morris Brown College, Atlanta, GA, 30314, USA

W.V. DASHEK Department of Biology, Clark Atlanta University, Atlanta, GA 30314, USA

INTRODUCTION

The ciliate, Tetrahymena pyriformis, has been of major importance to scientific inquiry since its successful growth as axenic cultures in 1922 (Hill, 1972). The pear-shaped fresh water protozoan has been referred to as the "Escherichia coli" of the non-photosynthetic eukaryotes (Elliott, 1973), much more than the classical Paramecium. Tetrahymena has been widely accepted as a research model for studies in genetics, biochemistry, endocrinology and toxicology. Cell and developmental biologists have enjoyed a lasting relationship with Tetrahymena as a result of its remarkable resemblance in structure and function to a typical cell.

Most of the research on the morphology of Tetrahymena is based on growing, probably logarithmic-phase, cells. The earlier light microscopic observations revealed the general morphology, but it was not until ultrastructural studies were possible that the subcellular details became known (Elliott, 1973). The importance of electron microscopy in biological research is evidenced by its extensive use in biology, biochemistry and medical research (Dawes, 1990).

Tetrahymena ultrastructure has been studied subsequent to treatments with DMSO (Nilsson, 1977), mercuric chloride (Tingle et al., 1973), phosphonoacetic acid (Smith et al., 1982) and many other cytotoxic compounds with varying results. Low concentrations of chlorpromazine (0.01 mM) inhibit growth and nucleic acid synthesis in the ciliate, Tetrahymena pyriformis (Blum, 1980).

A pilot study of psychotropic (chlordiazepoxide) drug interactions has been performed using Tetrahymena. Unicellular organisms, particularly Tetrahymena, prove to be model systems to demonstrate certain basic effects of these drugs at cellular levels. The use of these cells added the parameters of motility and readily recognizable survival which lend themselves to direct visual and photographic observations (Forrest et al., 1964).

Valium and Librium (Chlordiazepoxide and Diazepam) are termed anti-anxiety drugs or minor tranquilizers, and they appear to be prescribed primarily for the relief of anxiety. Unfortunately, the rapid growth in the use of these prescription drugs has not been associated with any striking improvements in overall health (Waldron, 1977). Since the introduction of these compounds, reports have been published concerning their use during pregnancy (Gill et al., 1988). Preliminary results in the pilot study focused on dose-dependent ciliary alterations and ciliary loss. If impairments of the protozoan cilia can be considered a model for tranquilizing effect, then this would indeed reflect the tranquilizing effect of these drugs in humans (Forrest et al., 1964). With time, chlordiazepoxide has a profound effect on the overall growth of Tetrahymena cells (Bell et al., 1987).

The present investigation assesses the ultrastructure of Tetrahymena pyriformis after the utilization of two fixation methods. The glutaraldehyde-osmium tetroxide method is compared to the glutaraldehyde-permanganate fixation procedure. Both log and stationary phase Tetrahymena cells were treated in this manner. The fixatives and their reactions with macromolecules and organelles will be compared. The findings of ultrastructural effects of chlordiazepoxide on Tetrahymena cells will be the subject of a separate report.

MATERIALS AND METHODS

Preparation of Tetrahymena pyriformis Liquid Shake Cultures

Tetrahymena pyriformis strain E were grown in axenic cultures using American Type Culture Collection (ATTC) 357 media. Cultures were incubated in the dark at 28°C using the Freas Control Temperature Incubator. Liquid shake cultures were maintained by serial transfer of 1 ml of exponential phase cells to 100 ml of fresh ATTC 357 media contained in 500 ml Belco Erlenmeyer flasks. Stock liquid shake cell lines were subcultured every 48 hr under constant agitation in a New Brunswick Reciprocal Shaker operating at 100 strokes/min.

Quantification of Growth

After 48 hr of continuous shaking, 1 ml of the cells were concentrated in a 1.5 ml Eppendorf microfuge tube. The medium was decanted off the top and the pellet resuspended in 0.1 ml of 10% formalin. Ten microliters of the cell suspension were placed onto a Neubauer hemacytometer counting chamber. From direct cell counts the cell number (number of organisms) was determined. In this experiment Tetrahymena growth was estimated by turbidity at $^{A}340$nm in a Beckman Spectronic 20 (Bell et al., 1987).

Preparation for Electron Microscopy

Stationary phase cells (3-day old) were selected for ultrastructural study. Two equal aliquots of cells were subjected to different fixation protocols. One sample was prefixed in 2.5% phosphate buffered glutaraldehyde (pH 7.0) for 2 hr at 4°C, then post-fixed in 4% OsO_4 in 0.2 M phosphate buffer (pH 7.0) for 1 hr. The second sample was prefixed as previously stated and post-fixed with a $KMnO_4$ solution. The samples went through 2-10 min buffer washes and then serial acetone dehydrations (30-100%). After overnight infiltrations in (1:1) acetone/Spurr's low viscosity embedding medium (firm) and 100% Spurr's overnight, the cells were then embedded in fresh Spurr's and cured for 8 hr at 70°C. Thin sections were obtained on an LKB ultramicrotome, counterstained in 1% uranyl acetate for 2.5 h followed by lead citrate (pH 12) for 30 min. The sections were then examined on a Joel-CX-100 electron microscope.

RESULTS AND DISCUSSION

Tetrahymena has been used as successful model in many research endeavors because it is a simplified but eukaryotic system (Hayes, 1972). Many of the nutritional requirements are the same as that of higher mammals and man. Tetrahymena pyriformis is a particularly valuable animal for many kinds of protozoan research. A primary reason for the success of Tetrahymena as an experimental subject is that it can be grown in large quantities on bacteria-free media or even on defined media.

There are at least fourteen species of Tetrahymena and the oval body shape and oral apparatus are characteristic to all of them. Variations may occur between the species as a result of life cycle phases. As members of the Subphylum Ciliophora, the abundant presence of cilia is the most obviously shared characteristic. Much work has been done on Tetrahymena ciliature by many investigators and in many laboratories worldwide. An appropriate description and classification of Tetrahymena cell surface has been provided by Dr. Joseph Frankel (personal communication). Dr. Frankel stated that in Tetrahymena the basal bodies and cilia are organized into two different systems (see Figure 1a). One the somatic system, consists of longitudinal rows of cilia, or ciliary meridians. These ciliary meridians extend from the anterior apex to the posterior end of the cell, except for two meridians which terminate anteriorly at the oral area. The meridians consist of much more than just cilia and basal bodies. Connected with the basal bodies are elaborate and very regular arrays of fibers. Some of these run longitudinal to the right of the ciliary meridian, collectively making up a structure known as the kinetodesmata (not shown on figure). Another name for the whole assembly (longitudinal row of cilia, basal bodies, and associated fibers) is kinety.

The second system of basal bodies and cilia is the oral system (oral apparatus). This is found, in Tetrahymena, near the anterior end of the cell. It consists of four sets of closely-connected basal bodies and cilia (hence the name Tetra-hymena, Latin for "four membrane"). The four compound membranes are disposed in two sets. One of these, on the animal's right margin of the oral apparatus is called the undulating membrane (Figure 1b), consisting of a single row of closely-spaced cilia with underlying basal bodies and fiber-systems. The second set consists of three membranelles, located within a shallow buccal cavity. Each membranelle consists of a triple file of closely-spaced cilia and basal bodies. The cilia of these membranelles beat in a coordinated fashion to set up water currents which sweep food down a shallow channel (the cytopharynx) to the mouth (or cytostome) at which place food vacuoles are formed.

Digestion takes place in the vacuole (Figure 1b); digestion products cross the vacuole membrane and reach the cytoplasm (Orias and Rasmussen, 1977). Indeed, the formation of the food vacuole appears essential in the standard media in common use for the laboratory culture of this ciliate (Rasmussen, 1976). It appears, however, that T. pyriformis can also take up dissolved amino acids via carrier-mediated transport systems located in membranes outside the food vacuole, presumably in the surface of cells and it thus becomes important to understand the relative contributions of the oral and surface uptake systems (Orias and Rasmussen, 1977).

Vacuole formation and presence is intimately associated with Tetrahymena growth. Tetrahymena can grow in a medium which is defined (Kidder and Dewey, 1951), which is undefined but axenic, or which contains bacteria as a food source. Media that are not free of bacteria are of little use to the biochemical protozoologist (Hill, 1972). Rasmussen and Kludt (1970) made an important observation in that particulate material is required for rapid cell multiplication. Growth in sterile-filtered proteose-peptone is improved by addition of insoluble materials such as talcum, quartz, clay, and calcium carbonate and by addition of water-soluble compounds that produce insoluble particles in aqueous solution Hill (1972). The previous authors surmised that some of the essential nutrients present in proteose-peptone must enter the cells via food vacuoles.

A typical cell population growth curve for T. pyriformis can be divided into the following phases: lag, acceleration, exponential (log), maximum stationary, decline, and logarithmic decline. The lag phase is characterized by an initial period of no increase in cell number or sometimes even a slight drop in cell number (Figure 2). This is followed by a phase of accelerated increase in cell number. The log phase is followed by a phase of decelerating increase in cell number, and then by a stationary phase in which cell number is maximal and rather constant for a time; with time the culture enters a decline phase which can be divided into a phase of accelerating cell death and one of logarithmic cell death (Elliott, 1973).

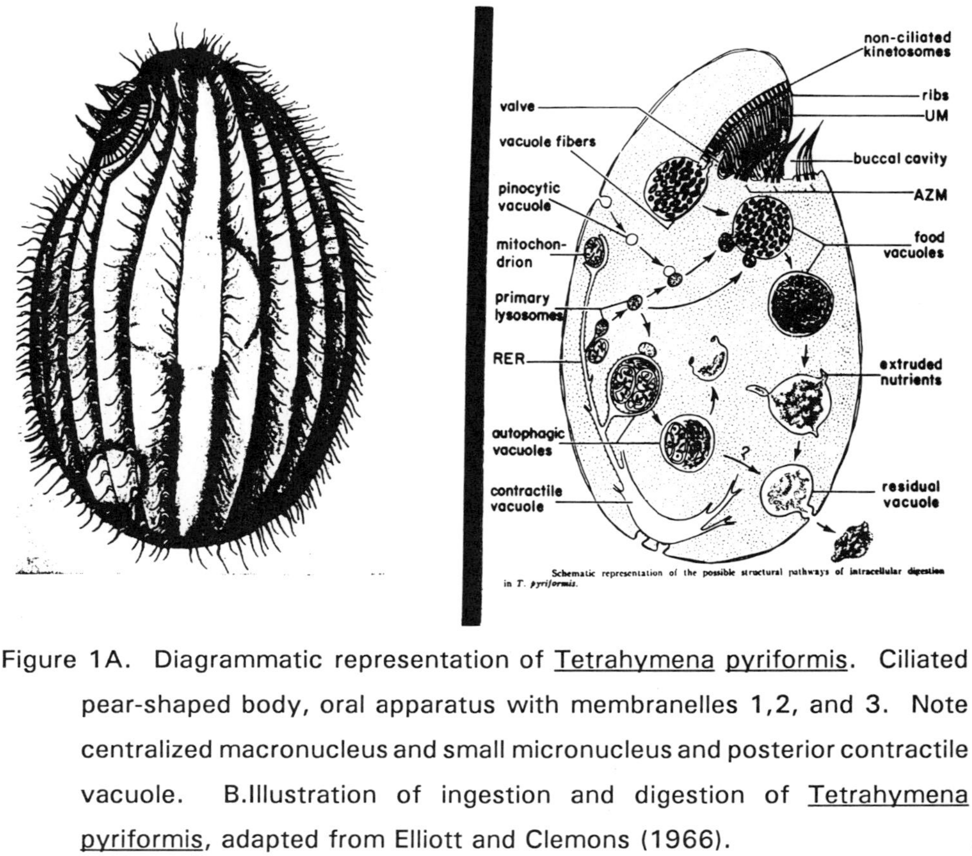

Figure 1A. Diagrammatic representation of <u>Tetrahymena</u> <u>pyriformis</u>. Ciliated pear-shaped body, oral apparatus with membranelles 1,2, and 3. Note centralized macronucleus and small micronucleus and posterior contractile vacuole. B.Illustration of ingestion and digestion of <u>Tetrahymena</u> <u>pyriformis</u>, adapted from Elliott and Clemons (1966).

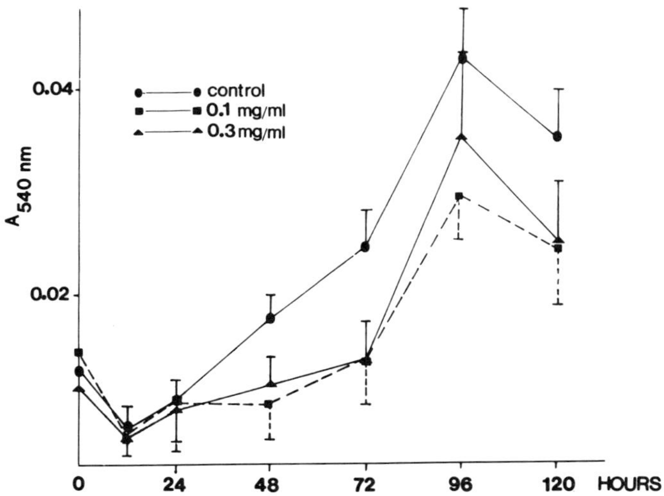

Figure 2. Effect of chlordiazepoxide hydrochloride on the growth of Tetrahymena pyriformis. ● , control; ■ , 0.1 mg 5 ml^{-2}; and ▲ 0.3 mg 5 ml^{-1}.

Tetrahymena has been and still is extensively utilized in pharmacological investigations. In recent years, these cells have yielded results of high validity. In 1959 Sanders and Nathan reported the employment of this protozoa in the pharmacological investigation of antihistamines (Forrest et al., 1964). The Tetrahymena cortex is a complex surface composed in many places of three membrane layers in close proximity to one another, as observed and demonstrated by Elliott and Levy, (1973) and the same can be seen in Figures 4-6. This arrangement may present a substantial barrier to entry of molecules from the outside (Prescott, 1972), and for a number of drugs the concentration required to achieve a given effect is sometimes much higher for Tetrahymena than mammalian cells (Durojaiye, 1979). Forrest et al., (1965) found that drug concentrations roughly comparable to human therapeutic doses are tolerated by suspensions of the protozoa in distilled water, and that tests could be completed within one hour.

An unusual property of chlordiazepoxide is its ability to stimulate the appetite of animals and man in subtoxic doses (Randall, 1961). Chlordiazepoxide and other benzodiazepines reliably increase food consumption in experimental animals primarily by increasing total eating duration (Cole, 1983). Chlordiazepoxide has shown unique taming effects first observed in animals and later confirmed in human subjects, including anticonvulsant activity and muscle relaxant effects at low dosages, appetite stimulating and sedative effects at high dosages (Randall, 1961). Tetrahymena cell growth was assessed in the presence of 0.1 mg 5 ml^{-1} and 0.3 mg 5 ml^{-1} of the benzodiazepine-chlordiazepoxide. Previous studies and data reported here (Figure 2) show that several points in the growth curve indicate that T. pyriformis exhibited stimulation of appetite for the medium (Bell et al., 1987). Results of motility studies revealed a proportional diminution in cell number with increasing drug concentration.

The cilium or eukaryotic flagellum is among the most important of the protistan organelles (Satir, 1984). Cilia as well as flagella also apparently have some sort of primitive sensory function. In the 1950s, as electron microscopy became an important tool of cell biological research, from examination of many diverse cilia and flagella, it became apparent that these organelles shared a common plan, the so called 9 + 2 microtubule-based axoneme surrounded by an extension of the cell membrane known as the ciliary membrane (Satir, 1984). In that same report Satir stated that this pattern was also shared by metazoan cilia and flagella. This would lead to the thought that many of our own sensory receptors have obviously evolved from cilia.

Previously the nature of the chemical composition of cilia and kinetosomes was much more elusive than their morphology (Elliott, 1972). With the development of ciliary isolation techniques (Child, 1959), (Child and Mazia, 1956) Watson et al., 1961) for Tetrahymena, analysis of cilia indicated that the bulk of the isolated material was protein (Child, 1959). The internal structure of cilia is composed of a common protein tubulin (Wallace et al., 1981). This is surely the protein the former investigators had retrieved from their isolations. Burgoyne (1986) isolated microtubules containing α and ß tubulin polypeptides of molecular weight 58,000 and 53,000. The cross section of a cilium from Tetrahymena can be seen in Figure 3. Allen's (1968) description was that of two central fibers surrounded by a central sheath, and spokes from this sheath connect with each of nine pairs of subfibers. One of each pair of subfibers has two arms protruding. High-resolution electron micrographs revealed that fibers of cilia of Tetrahymena pyriformis were composed of spherical subunits about 4 to 6 nm in diameter and arranged in straight longitudinal arrays to compose the walls of the fibers (Hill, 1973). Cilia move by contraction of fibrils which contain both ATP and proteins with ATPase activity, the movement being coordinated by a conductile component located in the ciliary membrane (Sovak et al., 1981). Sovak and his investigators then concluded that inhibition of ciliary movement must therefore, be a result of damage to motile or conductile components, or both. Tetrahymena axoneme has also been the focus of state of the art computer modelling. The computer model of Tetrahymena axonemes at macromolecular resolution is used to reconcile different interpretations of dynein arm structure and periodicity and to study aspects of axonemal dynamics, thus laying the foundations for the development of a complete functional model of the axoneme (Sugrue et al., 1991).

Tetrahymena ultrastructure has been thoroughly studied and described by Allen (1967). Figure 4 depicts different sectional views of Tetrahymena. Figures 4a and 4b are cross-sections through Tetrahymena and 4c and 4d are views through the oral apparatus region. Plates 4a and 4c were glutaraldehyde pre-fixed and osmium post-fixed. Note the osmophilic lipid droplet which are pronounced in this fixation method. Other typical cellular organelles can be seen

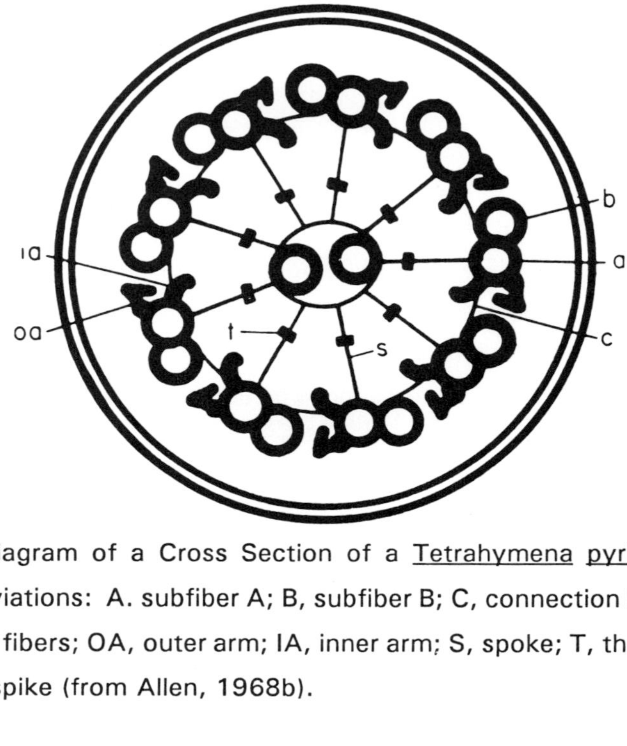

Figure 3. Diagram of a Cross Section of a Tetrahymena pyriformis cilium. Abbreviations: A. subfiber A; B, subfiber B; C, connection between outer double fibers; OA, outer arm; IA, inner arm; S, spoke; T, thickened region along spike (from Allen, 1968b).

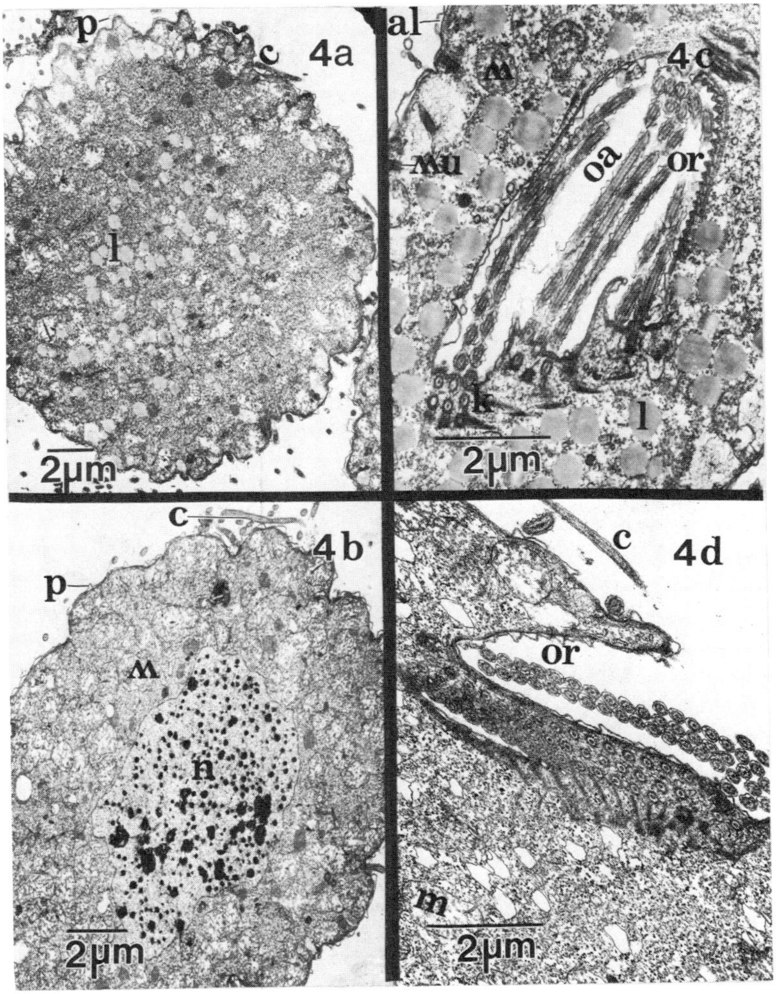

Figure 4: Cross-sections and oral regions of T. pyriformis cells. Plates 4a and 4b shows cilia (c), lipid droplets (1), mitochondria (m), nucleus (n) and the pellicular membrane (p). The oral apparatus (oa) displays the many basal bodies that comprise the membranelles and undulating membrane. Other structures are areolar sacs (al), oral ridge (or), kinetodesmal fibers (k), and mucosysts (mu).

and also the tripartite membrane of Tetrahymena. Micrographs 4b and 4d were prepared by the glutaraldehyde pre-fixed and $KMnO_4$ post-fixed protocol. Many of the organelles are not well preserved and excessive leaching is exhibited in plate 4d. These micrographs are of early stationary phase cells.

The micrographs in Figure 5 are of late stationary phase Tetrahymena cells. Many of the organelles earlier noted can be seen and various vacuoles are now apparent. Figure 5a had the traditional glutaraldehyde prefixed, osmium postfixed treatment. Mitochondria and other cytoplasmic particles are sequested into autophagic vacuoles. Residual vacuoles with digestive and degradative products can be seen. The bottom micrograph was produced as a result of glutaraldehyde/$KMnO_4$ treatment. Organelles are not readily visible and leaching is obvious.

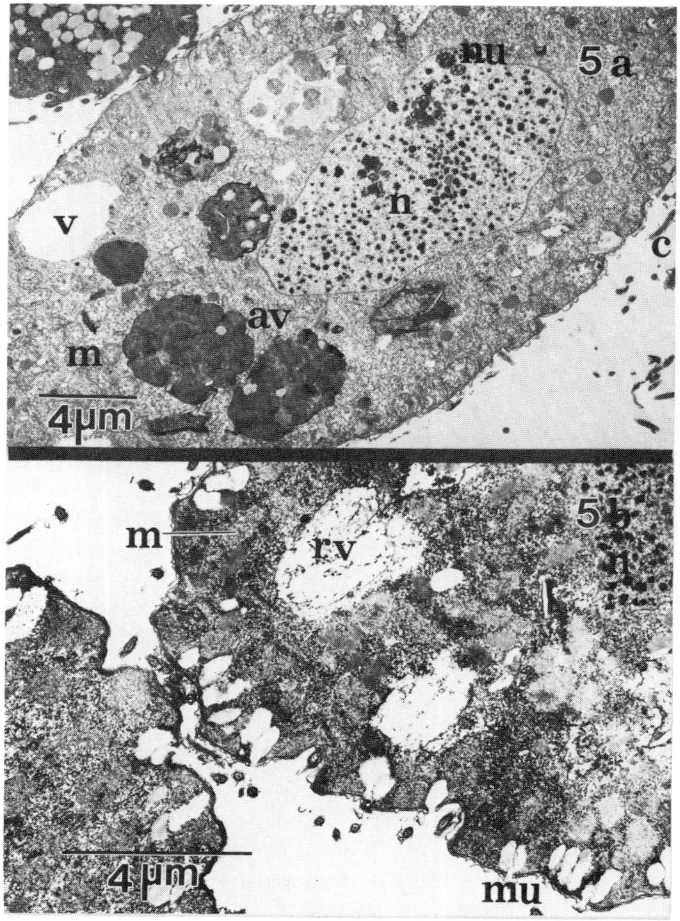

Figure 5. Late Stationary Phase Cells. Vacuole (v), autophagic vacuole (av), and residual vacuole (rv).

The highly magnified cortical area in Figure 6 shows the many types and arrangements of microtubules. The cortical region contains most of the major membrane proteins of Tetrahymena. The major microtubule bands found in ciliary meridians are the longitudinal, transverse, and post-ciliary bands (Ng and Williams, 1977). In their study electron-microscopic observations provided information about the positions and orientations of structures that are not visible in the light microscope, i.e. the kinetodesmal fibers, parasomal sacs and basal tubules.

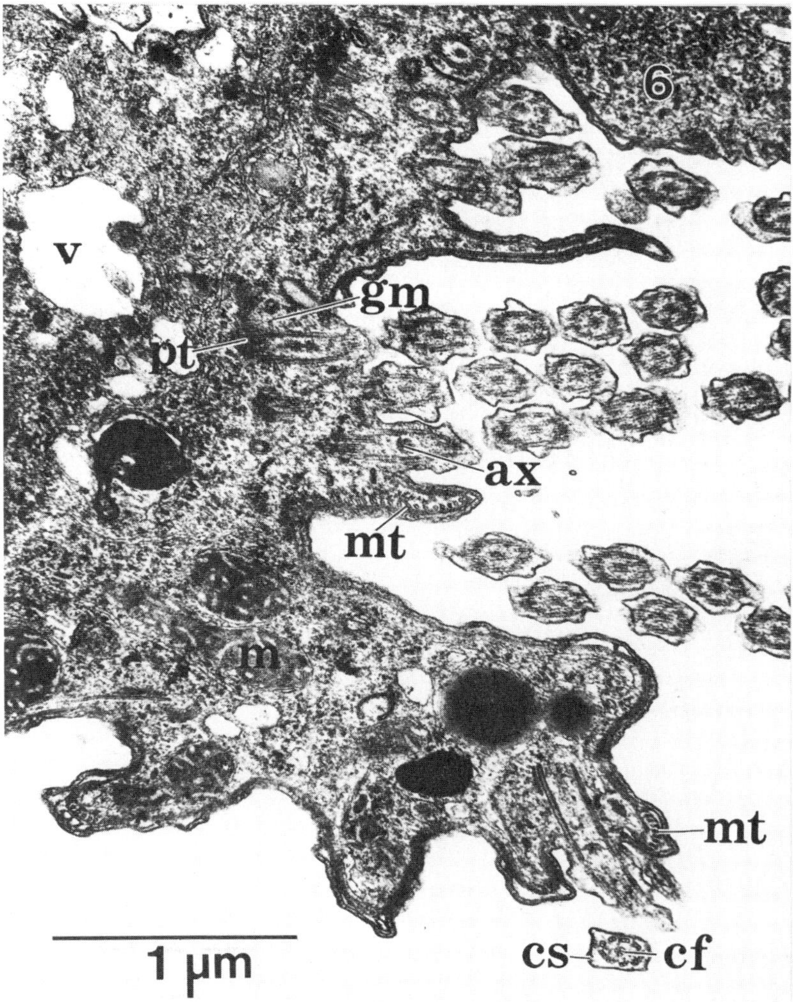

Figure 6. Cortical and Associated Structures. Microtubules (mt), posterior tubules, axosomes (ax), granular matrix (gm), ciliary sheath (cs), ciliary fiber (cf).

SUMMARY

The protozoan, Tetrahymena pyriformis, is a suitable organism to assess whether the anti-anxiety drug, chlordiazepoxide, can possibly bind to microtubules'tubulin rendering the organism immobile. Previous growth & motility investigations led to this hypothesis. Growth of T. pyriformis in liquid culture approximately simulated the lag, acceleration, exponential, decline phases of stationary & typical protozoan population growth curves. Although these phases were observed for both non-treated & drug-treated (0.1 mg 5 ml^{-1} & 0.3 mg 5 ml^{-1}) cultures, the drug readily produced growth curves reduced below that of non-treated cells. Hemacytometer motility cell counts revealed a proportional diminution in cell number with increasing drug concentration. Here is reported a comparative electron microscopic fixation analysis to establish the optimal preservation of non-treated T. pyriformis ultrastructure. Three-day old cultures grown in proteose-peptone medium were harvested & divided into two equal portions. One portion was pre-fixed 2 h in 2.5% glutaraldehyde & then post-fixed 1 h in 2% OsO_4, each buffered with 0.05M cacodylate, pH 7.4; the other was pre-fixed with glutaraldehyde & subsequently post-fixed 1 h in 2% aqueous $KMnO_4$. Cells were dehydrated through a graded acetone series & embedded in Spurr's. Thin sections of glutaraldehyde pre-fixed & osmium post-fixed T. pyriformis revealed abundant microtubules, mitochondria, ribosomes, lipid droplets, vesicles, & vacuoles, as well as a distinct pellicle & oral apparatus. In contrast, glutaraldehyde pre- and $KMnO_4$ post-fixations resulted in extensive leaching but well-preserved mitochondrial, vesicular, & pellicular membranes. These results suggest that glutaraldehyde pre- and osmium post-fixation is preferred for a subsequent comparative ultrastructural investigation of non-treated & drug-treated T. pyriformis as well as the drug's intracellular localization by high resolution radioautography & immunoelectron microscopy.

Acknowledgements

We thank Dr. Jerome Paulin for his patience, expert advice, and sharing of his works and facilities. We also thank Drs. David Porter, Barry Palevitz, and Charlie Keith for their collaborations. Utmost appreciation is extended to Ms. Doris Donnell for unselfishly giving of her time and clerical expertise. This work was supported by NSF-RIMI, Grant #RII-8912781 and the Amoco and Bush Foundations.

REFERENCES

Allen, R.D. (1967). Fine structure, reconstruction and possible function of components of the cortex of Tetrahymena pyriformis. J. Protozool. 14, 553-565.

Bell, B.J., Durojaiye, M.A. and Penn, J.H. (1987). Effects of chlordiazepoxide hydrochloride on Tetrahymena pyriformis. Microbios Lett. 34, 67-70.

Child, F.M. (1959). The characterization of the cilia of Tetrahymena pyriformis. Exptl. Cell. Res. 18, 258-267.

Child, F. and Mazia, D. (1956). A method for the isolation of the parts of ciliates. Expermentia. 12, 161-162.

Cole, S. (1983). Combined effects of chlordiazepoxide treatment and food deprivation on concurrent measures of feeding activity. Phar. Biochem., and Behav. 18, 369-372.

Dawes, C.J. (1990). Introduction to biological electron microscopy: Theory and techniques. Ladd Research Industries, Inc. Vermont.

DeRobertis, E.D.P., Nowinski, W.W. and Saez, F.A. (1970). Cell Biology. W.B. Saunders Company, Philadelphia.

Durojaiye, M. (1979). Ultrastructural studies and physiological effects of methyprylon on Tetrahymena pyriformis. Microbios Lett. 12, 117-120.

Elliot, A. and Levy, M. (1973). Biochemical and ultrastructural changes in Tetrahymena pyriformis during starvation. J. Protozool. 15, 208-222.

Elliott, A.M. (1973). Biology of Tetrahymena. Dowden, Hutchinson and Ross, Inc. Pennsylvania.

Elliott, A.M. and Clemmons, G.L. (1966). An ultrastructural study of ingestion and digestion in Tetrahymena pyriformis. J. Protozool. 13:311-323.

Forrest, I.S., Quesada, F., and Deitchman, G.L. (1964). Unicellular organisms as model systems for the mode of action of phenothiazine and related drugs. Pharm. Society. 7, 42-44.

Gill, T.S., Guram, M.S., and Geber, W.F. (1981). Comparative study of the teratogenic effects of chlordiazepoxide and diazepam in the fetal hamster. Life Sciences. 29, 2141-2147.

Hayes, A.W. (1977). Effect of the mycotoxin, penicillic acid, on Tetrahymena pyriformis. Toxicon. 15, 497-504.

Hill, D.L. (1972). The biochemistry and physiology of Tetrahymena. Academic Press. NY.

Kidder, G.W. and Dewey, V.C. (1951). The biochemistry of ciliates in pure culture. Biochem. and Physiol. of Protozoa. Vol. 1. Academic Press, NY.

Ng, S. and Williams, R.J. (1977). An ultrastructural investigation of 180°-rotated ciliary meridians in Tetrahymena pyriformis. J. Protozool. 24, 257-263.

Nilsson, J.R. (1977). Effects of dimethyl sulfoxide on Tetrahymena pyriformis G.L. Fine structural changes and their reversibility. J. Protozool. 24, 275-283.

Prescott, D. (1972). Methods in Cell Physiology. Academic Press. NY.

Randall, O. (1961). Pharmacology of chlordiazepoxide (librium). Dist. Nerv. Syst. 7 (Supple). 22, 7-15.

Rasmussen, L. (1976). Nutrient uptake in Tetrahymena. Carlsberg Res. Comm. 41, 143-167.

Rasmussen, L. and Kludt, T. (1970). Particulate material as a prerequisite for rapid cell multiplication in Tetrahymena cultures. Exp. Cell. Res. 59, 457-463.

Sabatini, D., Bensch, K. and Barnett, R.J. (1963). Cytochemistry and electron microscopy. J. Cell. Biol., 17, 19-58.

Satir, P. (1984). The generation of ciliary motion. J. Protozool. 31, 8-12.

Smith, J.D, King, J.C., and Holland, E. (1982). Phosphonoacetic Acid: Effect on and disposition by Tetrahymena. J. Protozol. 29, 103-105.

Sovak, M., Ranganathan, R., and Mutzel, W. (1981). Determination of lethal dose in a protozoa. Screen of contrast media toxicology. Invest. Radiol. 16, 438-442.

Spurr, A.W. (1969). A low viscosity epoxy embedding medium for electron microscopy. J. Ultrastruct. Res., 26, 31-43.

Sugrue, P., Avolio, J., Satir, P. and Holwill, M.E.J. (1991). Computer model of Tetrahymena axenomes at macromolecular resolution: Interpretation of electron micrographs. J. Cell. Sci. 98, 5-16.

Tingle, L.E., Parlat, W.A. and Cameron, I.L. (1973). Sublethal cytotoxic effect of mercuric chloride on the ciliate Tetrahymena pyriformis. J. Protozool. 20, 301-304.

Waldron, I. (1977). Increased prescribing of valium, librium, and other drugs - An example of the influence of economic and social factors on the practice of medicine. Int. J. Health Services. 7, 37-51.

Wallace, R., King, J. and Sanders, G. (1981). Biology - The Science of Life. Scott, Foresmanand Company, Glenview, IL.

Watson, M.R., Hopkins, J.M. and Randall, J.T. (1961). Isolated cilia from Tetrahymena pyriformis. Exp. Cell. Res. 23, 629-631.

HUMAN DERMAL MYIASIS CAUSED BY *CORDYLOBIA ANTHROPOPHAGA*: A CASE STUDY

TITUS A.(B.) OGUNNIYI, Department of Medical Microbiology and Parasitology, Faculty of Health Sciences, Obafemi Awolowo University, Ile-Ife, Nigeria
TENIOLA A. ADEWUNMI, Department of Medical Microbiology and Parasitology, Faculty of Health Sciences, Obafemi Awolowo University, Ile-Ife, Nigeria
BABATUNDE O. AMOLE, Laboratory of Parasitology, Morris Brown College, Atlanta, GA 30314, USA; Department of Medical Microbiology and Parasitology, Faculty of Health Sciences, Obafemi Awolowo University, Ile-Ife, Nigeria
BETTY R. JONES, Laboratory of Parasitology, Morris Brown College, Atlanta, GA 30314, USA

INTRODUCTION

Myiasis has been defined as the invasion of living vertebrate tissue by the larvae of certain species of Diptera (Vasallo, 1991). The infection has a cosmopolitan distribution. Only few cases of the disease have been reported in African subregions (Dipeolu and Olufarati, 1975; Ogunniyi, 1981; Edungbola, 1982; Lyko and Mutinga, 1988).

Reports of the epidemiology and documented cases studies in Nigeria are limited. Studies by Dipeolu and Olufarati (1975) and Edungbola (1982) described the first few cases of tumbu fly myiasis in Nigeria. One other study by Lyko and Mutinga (1988) reported cases of tumbu fly myiasis in a child in Kenya. As to our knowledge, no other studies have been reported of dermal myiasis. This paper describes a case study involving multiple furuncles in a seven month old infant.

Biodeterioration Research 4, Edited by G.C. Llewellyn *et al.*, Plenum Press, New York, 1994

CASE HISTORY

In July 1990, a seven month old female infant developed on parts of her body what her mother initially believed to be heat rashes. The third day of infection showed areas that appeared erythematous and coalesced into large focused nodules with suppurating appearances. On the fifth day, the infant appeared weak and inactive. On the seventh day, the infant appeared febrile with a persistent high temperature. Although a Pediatrician prescribed an antibiotic and an analgesic, the prescription was never administered. The high temperature persisted and on day ten the infant was treated with an antimalarial medication. At this time, movements were observed inside the nodules. After applying gentle pressure on one of the nodules a maggot (larva) was expelled. Subsequently, seven maggots were eventually expelled from the nodules. Anatomical distribution for these furuncular nodules were the abdomen, thigh, scalp, elbow, forearm and the back of the ear.

Four of the specimens are shown in Figure 1. Further microscopical observation revealed that the specimens were larvae of the tumbu fly Cordylobia anthropophaga as shown in Figure 2.

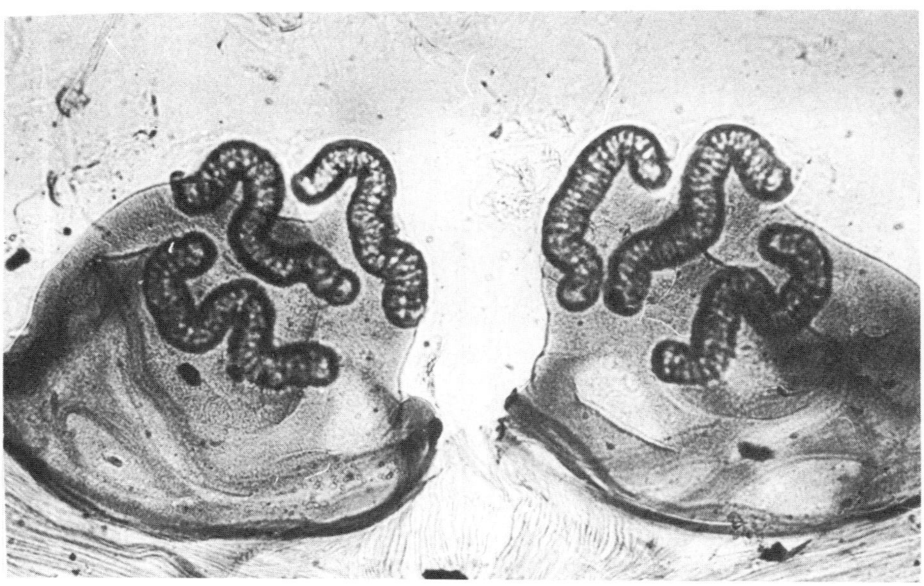

Figure 1. Four Maggots of Cordylobia anthropophaga Showing External Morphology (X 100).

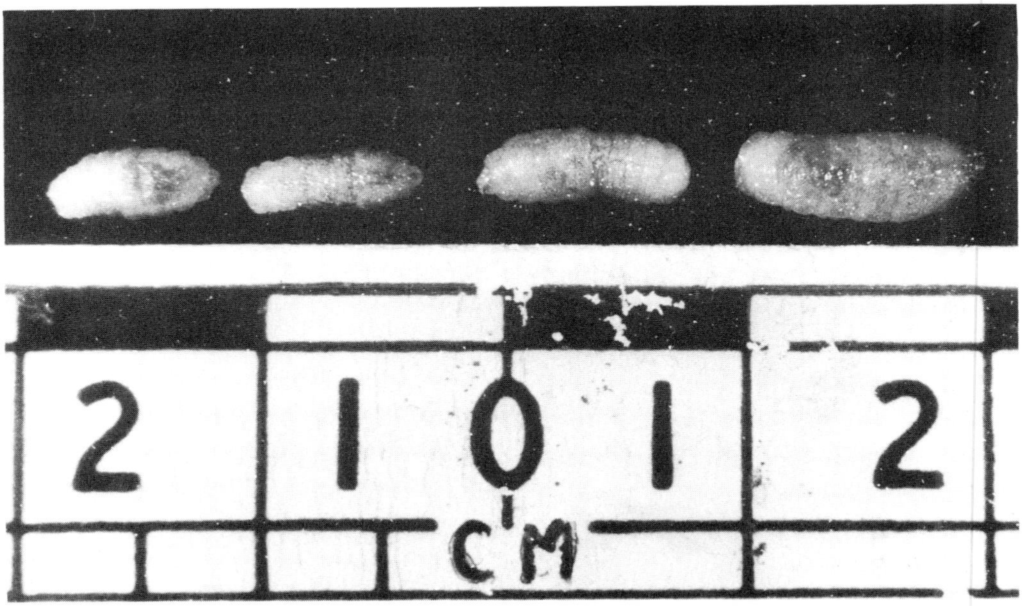

Figure 2. A pair of posterior spiracles of Cordylobia anthropophaga showing three slits on each side of the septum (X 100).

DISCUSSION

The causative agent, Cordylobia anthropophaga and the period of the year in which this infection took place are similar to other cases of myiasis which have been previously reported in Nigeria (Dipeolu and Olufarati, 1975; Ogunniyi, 1981; Edungbola, 1982). Although we cannot specifically determine the mechanism and mode of transmission in the present case, we would like to suggest that transmission may have taken place when a tumbu fly oviposited on the infant's linen and clothes. This is based on the occurrence of multiple lesions, since it is very unlikely that a fly could have extruded larvae on so many different anatomical locations during a single brisk visit. Thus, under conditions of warm body temperature, it is probable that the larvae hatched from the eggs lying in the infant's clothing and penetrated the skin to undergo further development.

Although, pyrexia in malaria holoendemic regions of the world usually suggest the possibility of a malarial infection, health workers in such regions should be aware of myiasis agents as also being possible causes of pyrexia.

Myiasis lesions have been found to self heal a few days post-larvae expulsion (Ogunniyi, 1981). Because of this rapid healing process, this has been suggested as a possible rationale for few reports in the literature in Nigeria (Dipeolu and Olufarati, 1975). It may be concluded based on available data about this disease and unreported cases that educational and training plans are needed to increase awareness of the disease and thereby enabling am implementation plan to be initiated.

We are of the opinion that this situation may be due to people's ignorance of the disease, especially its etiology and mode of transmission. Another factor that may be responsible is the very little importance associate with the disease, which rarely kills (except probably the internal form), hence considering it as not worth reporting. For example Ogunniyi, one of the authors of this paper, witnessed in 1981 a case of 6 larvae being brought by an indigenous Engineer to the Parasitology Laboratory of the University College Hospital in Ibadan, Nigeria. The Engineer made several previous visits to the outpatient clinic of the hospital with complaints of intermittent but irritating movements in his sinuses. According to the patient, his predicament did not attract any assistance from the hospital personnel. Instead, the larvae were procured through his nostrils by a traditional healer. Again in 1988, a fly larva was expelled from the forearm of a tutor Nurse and was presented to the same author. On both occasions the reporter did not stress the importance of the victims' sufferings and neither attempted identifying the larvae involved nor regarded the case worth reporting.

Cases of human myiasis may not be a very common occurrence, but it is doubtful that episodes receive adequate reporting.

SUMMARY

Epidemiologic studies of dermal myiasis are very limited. This may be due to the fact that very little importance is associated with the disease because it is rarely fatal. This paper described a case study involving dermal myiasis at multiple sites in a seven month infant. The disease was initially misdiagnosed because the symptoms, especially pyrexia were similar to that of malaria. This study highlights the need to increase awareness of the disease which may enable implementation plans to be initiated to alleviate the infection.

Acknowledgements

We wish to thank Mr. Charles Nelson, an undergraduate research associate working in the laboratory of Dr. Betty Jones in the area of parasitology for the word processing work associated with this manuscript and for making several revisions prior to initial review. We extend thanks to also to Dr. William Dashek, Associate Professor of Biology at Clark Atlanta University for further editing this manuscript. In addition, I extend my gratitude to Dr. Betty Jones for providing her time and support toward all aspects of this project.

REFERENCES

Dipeolu, O.O. and Olufarati, B. (1975). Furuncular myiasis caused by larva of Cordylobia anthropophaga in Ibadan, Nigeria. Trans. Roy. Soc. Trop. Med. Hyg. 69, 164-165.

Edungbola, L.D. (1982). Cutaneous myiasis due to tumbu-fly, Cordylobia anthropophaga in Ilorin, Kwara State, Nigeria. Acta. Trop. 39, 355-362.

Lyko, J. and Mutinga, M.J. (1988). Tumbu-fly myiasis in a child. Trans. Roy. Soc. Trop. Med. Hyg. 82, 345.

Ogunniyi, I.O. (1981). Furuncular swelling caused by larva of Cordylobia anthropophaga in Kaduna, Nigeria. Trans. Roy. Soc. Trop. Med. Hyg. 75, 752.

Vasallo, M.F. (1991). Human myiasis: A problem of today. Ann. R. Acad. Nac. Med. (Madr) 108, 153-179.

SECTION V

GENERAL BIOTOXINS, BIODEGRADATION AND BIODETERIORATION

GENERAL BIOTOXINS, BIODEGRADATION AND BIODETERIORATION

INTRODUCTION

In this section, we have included papers on many different topics. These topics range from the forensic sciences, bacterial standards and criteria, review articles on the influence of dissolved oxygen, and the evaluation of stainless steel susceptibility to microbiological corrosion. Additional articles on biocorrosion, biodeterioration of stone, enzyme degradation, and toxin-producing *Clostridium difficile* are also included in this section. And finally, a paper on the development of a Mycology Laboratory. Quite an interesting array of topics.

Textile fibers from fragments of exhumed fabric were examined using transmitted light and scanning electron microscopy. Wool fibers exhibited fungal tunneling, loss of cuticular scales and fiberillation, but the hemp fibers showed less evidence of biodeterioration. The forensic examination of hair indicated that environmental conditions may completely efface the cuticular scale patterns on hair, therefore, it may be impossible to determine the species of animal, from which the hair came. A paper on the development of an analytical procedure for delta-9-tetrahydrocannabinol as well as a paper on an historical sampling of marihuana and hashish, are also included in this section.

The establishment of a guide for sanitary water quality has made a significant impact on public health. However, more meaningful indicators are deemed necessary. In the realm of indicator bacteria, false positives, and pathogens, future methods of dealing with these problems are not able to be predicted at this time.

Review articles on the influence of dissolved oxygen on sulfite-reducing bacteria-related corrosion, and an evaluation of stainless steel susceptibility to microbiological-influenced corrosion have been included in this section. The first review highlighted the possible mechanisms of dissolved oxygen on the SRB-related corrosion in a closed industrial water system. The second review summarizes electro-chemical techniques, metallurgical and microbiological methods and surface analysis that are used in the study of microbiological-influenced corrosion of stainless steels. The final review, discusses the effects of biofilm surface coverage, water salinity, sunlight and the ennoblement of the Open-circuit Corrosion Potential (OCP) of passive metals and alloys immersed in natural waters. In addition to these reviews, included in this section is a paper discussing new trends and techniques in biocorrosion and biofouling monitoring.

Monuments, sculpture and stony objects-of-art exposed to the open-air have deteriorated throughout the centuries. In the gradual process of weathering, biological activity has played a key role due to its interactions with physical-chemical mechanisms. In addition, industrial and urban activities have been found to modify the composition of the atmosphere, thus effecting the colonization and growth pattern of certain microorganisms.

Two papers in this included in this section are on the key enzyme biogenic amines degradation in plasma. Also included, are new findings on the monoamine oxidase. Additionally, a paper on the detection and typing of toxin-producing *Clostridium difficile* is included.

A study on the disturbances in Tryptophan metabolism following chronic ingestion of a high-copper diet by male rats, is reported. The study showed a substantial reduced mean body weight gain in copper-treated animals. Testis and spleen weights were increased and the animals showed a slightly increased erythrocyte count and increased serum cholesterol. The urinary excretion of kynurenine and kynurenic acid were not altered, but Tryptophan 2,3 dioxygenase activity was slightly decreased. Also, some neuro-pathologic changes were noticed.

Finally, in this section, we have a paper concerning the development of a mycology laboratory. The report provides an overview of the major components needing consideration in the development of such a facility.

Charles E. O'Rear. Ph.D.
Department of Forensic Sciences
The George Washington University
Washington, DC 20052, USA

MICROSCOPIC EXAMINATION OF EXHUMED WOOL AND HEMP FIBERS FROM THE GRAVE OF THE VICTIMS OF ALFERD PACKER

JULIE B. KEMPTON, Department of Forensic Sciences, The George Washington University, Washington, DC 20052, USA
DOUGLAS DEGAETANO, Scientific Laboratories, Richmond, VA, 23210, USA
JAMES E. STARRS, National Law Center, The George Washington University, Washington, DC 20052, USA
WALTER F. ROWE, Department of Forensic Sciences, The George Washington University, Washington, DC 20052, USA

INTRODUCTION

In February 1874 a party of six gold miners set off from an Ute Indian camp near the present city of Delta, Colorado. They were bound east into the San Juan Mountains to the Los Piños Indian Agency (Figure 1). Only Alferd Packer completed the journey. According to his harrowing account, the other members of the party died of starvation or were killed when they attacked other members of the group. Packer admitted to killing one member of the party in self-defense; he also admitted surviving by eating flesh from the dead. In August 1874 an artist for Harper's Weekly came upon five mutilated bodies near the Lake Fork of the Gunnison River. Following an inquest held on the spot, a warrant was issued for Packer's arrest. He was not, however, to be brought to justice immediately: Before the warrant could executed Packer escaped from jail. He was to remain at large for nine years and this hiatus in justice was ultimately to save Packer's life. Brought to trial in 1883, Packer was convicted of premeditated murder and sentenced to hang. The Colorado Supreme Court overturned his conviction on the grounds that while Packer was a fugitive, the legislature had repealed the murder statute in effect in 1874 and enacted a new

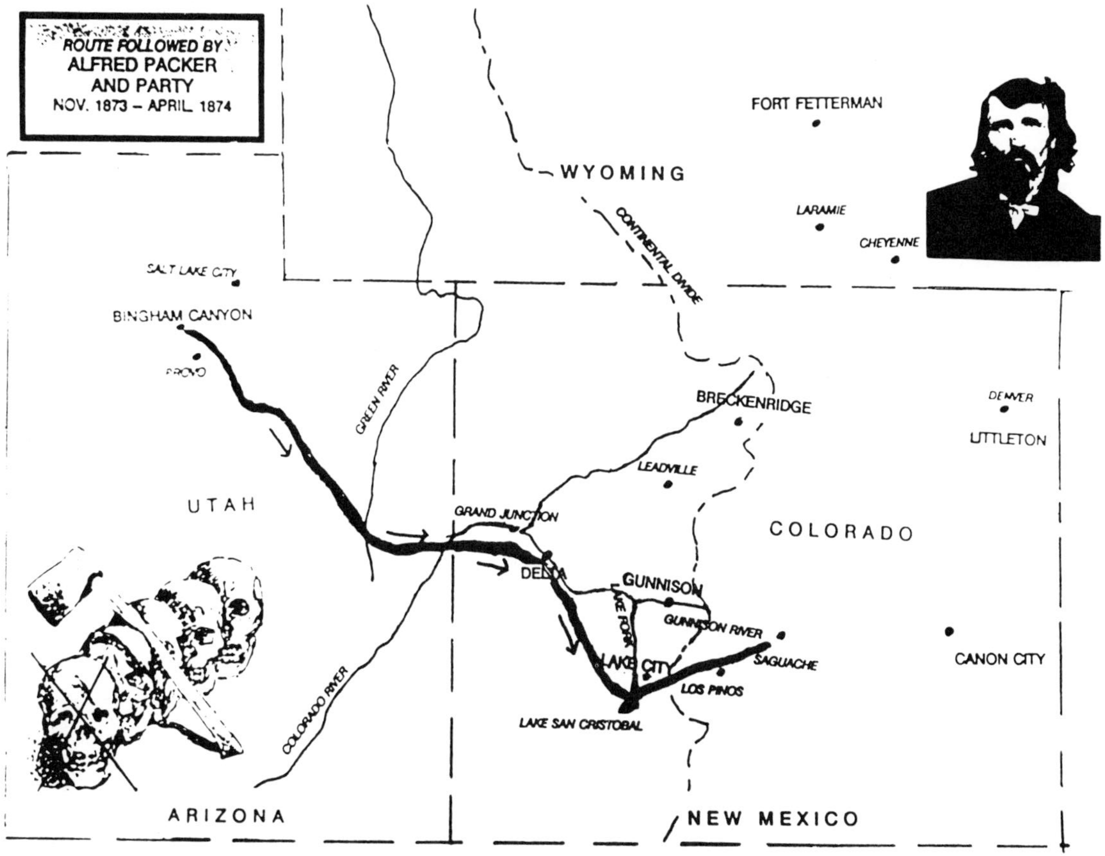

Figure 1. Route of Alferd Packer's Party, February 1874.

statute without including a grandfather clause. There was simply no statute under which Packer could be tried in 1883 for a crime committed in 1874. Packer was retried for voluntary man-slaughter and again convicted. Sentenced to forty years in prison, Packer was eventually paroled; he died in 1907. Even before his death Alferd Packer had already entered Western legend. According to one apocryphal story, the judge who sentenced Packer to hang accused him of eating five of only seven Democrats in Hinsdale County, Colorado. (Starrs, 1988) Students at the University of Colorado at Boulder have named their university cafeteria after Alferd Packer.

The graves of Alferd Packer's five victims were marked with five large stones topped with a boulder bearing a memorial tablet. Despite the clearly

Figure 2. The Bodies of Alferd Packer's Victims Following Exhumation.

marked grave site and the memorial inscription, some local historians doubted that the victims' bodies were actually buried there. Others maintained that even if the bodies were buried under the stones, the grave site was a secondary burial to which the bodies had been moved many years after the original burial. In 1989, one of the present authors (J.E.S.) lead an expedition to this site to exhume their remains for forensic examination. Excavation of the grave site revealed the skeletal remains of five male Caucasians (Figure 2), whose bones exhibited both blunt force and sharp force trauma. The bodies had been interred without coffins. The degree of articulation of the skeletal remains and the wealth of fragile artifacts recovered indicate that the grave site was the primary burial site, rather than a secondary burial. Bones of all five showed evidence of defleshing with a knife. Numerous artifacts were also recovered, including buttons and textile fragments. The latter artifacts were of particular interest because they afforded the opportunity to study the deterioration of textile fibers that had been buried 115 years.

MATERIALS AND METHODS

Numerous specimens of textile were recovered from the grave site. Samples of yarns, sewing thread and individual fibers were selected for examination with transmitted light microscopy. The samples were mounted on microscope slides using Permount (Fisher Scientific). The mounted fibers were then examined using a Ernst Leitz Dialux transmitted light microscope equipped with polarizing filters. Photomicrographs were taken with a Leica CL 35-mm camera attachment, using Kodak Tri-X Pan ISO 400 black and white film.

Additional fiber specimens were selected for examination with a scanning electron microscope. These fibers were carbon coated with a Hitachi HS5 Carbon Evaporator and examined with a CamScan Series II scanning electron microscope equipped with a Robinson backscattered electron (BSE) detector. X-ray spectra of the fibers were obtained using a EDAX 9800 Plus energy dispersive X-ray analyzer with a ECON IV light element detector.

RESULTS AND DISCUSSION

Many fragments of twill-weave cloth were found with brass eagle buttons still attached. Microscopic examination of yarns from this cloth revealed that it consisted of Z-twist woolen yarns. Many of the wool fibers had a greenish hue; a few retained the dark blue color that they had originally been dyed. The composition and weave of the cloth, the presence of military-style buttons, and the remaining blue-dyed wool fibers indicate that the fragments came from a U.S. Army great coat (Figure 3). The number of eagle buttons found is consistent with a single great coat. Both Alferd Packer and victim Wilson Shannon Bell served in the Union army; it is likely therefore that the fragments of twill-weave woolen cloth represent a U.S. Army greatcoat issued during the Civil War (1861-1865). One of the other victims, Israel Swan, served in the Mexican War (1846). However, the coat cannot have been issued during that war: One of the eagle buttons bore the backmark "Waterbury Button." According to Emilio (1911), the Waterbury Button Company came into existence in 1849 when the button department of Benedict and Burnham Manufacturing Company became a separate corporate entity.

In addition to the eagle buttons, the excavators found two-hole hard rubber poncho buttons, a variety of iron and brass trouser buttons, and a variety of small white glass buttons. The collection of buttons from the Packer grave site closely resembles the assemblage of buttons found on the Little Big Horn battlefield. (Scott et al., 1989). This is not surprising in that the Packer murders and the battle of the Little Big Horn occurred only two years apart.

Examination of wool yarns from the great coat fragments using transmitted light microscopy revealed that many of the wool fibers had developed transverse cracks (Figure 4). Some of the wool fibers also showed evidence of fungal tunneling (Figure 5). Fungal mycelia were observed with both transmitted light microscopy and scanning electron microscopy. A few heavily medullated hemp fibers (Figure 6) were found in one of the yarns from the great coat. The presence of hemp fibers is usually considered a sign of poor sheep breeding. Because medullated wool fibers do not dye as well as unmedullated fibers, their presence lowers the quality of the wool. (Von Bergen, 1947) The quality (or lack thereof) of clothing issued during the Civil War was a source of frequent complaint from Union Army personnel.(McPherson, 1988) The "shoddy" uniforms that dissolved in the first downpour have been elevated to legendary status. Nevertheless, this particular overcoat was evidently still in use over a decade after its probable date of issue.

Some fragments of the overcoat still had buttons attached. One small button still had thread attached and was still in its button hole. The button hole was edged with brown-colored thread. Originally, the threads used to sew on the button and the threads edging the button holes were dyed the same shade of blue as the woolen cloth. Even museum specimens of such garments show the same type of brown discoloration. Examination of the threads with a transmitted light microscope (Figure 7) and with a scanning electron microscope (Figure 8) revealed that they consist of hemp fibers. The transverse fissures shown in Figure 8 occur naturally in hemp fibers and are not the result of weathering. According to the 1864 U.S. Army Quartermaster Department specifications for overcoats, No. 30 linen thread was to be used throughout (Kloster, 1991). The hemp threads may represent a contractor's variance from the Army specifications or may be evidence of later repairs.

Figure 3. Civil War Era U.S. Army Great Coat for Mounted Men.

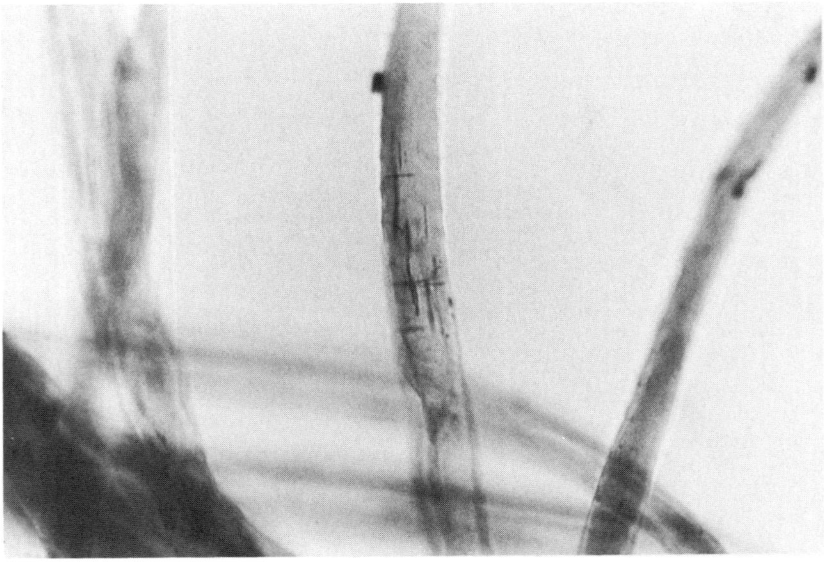

Figure 4. Wool Fibers from Great Coat, Showing Transverse Cracks.

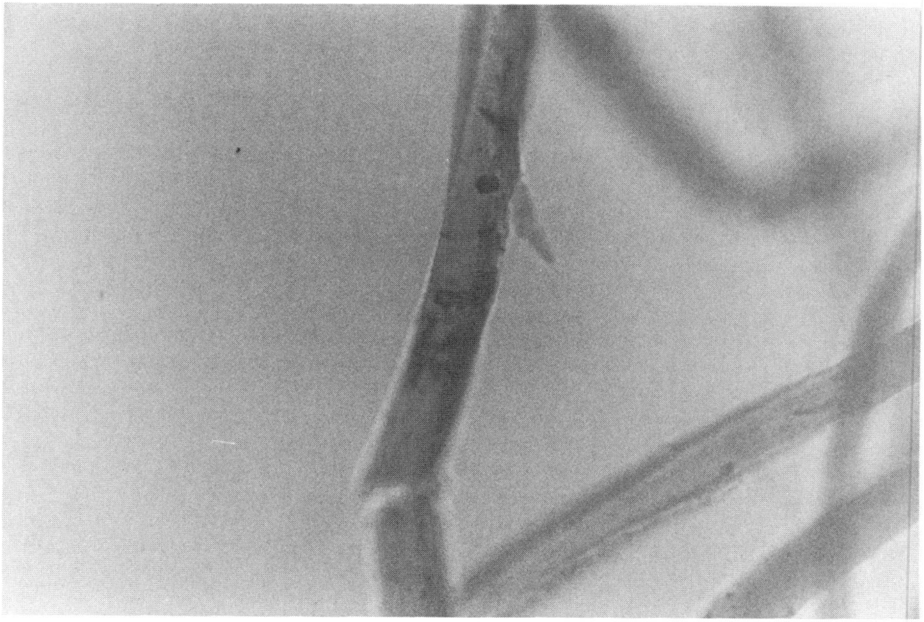

Figure 5. Fungal Tunnel in Wool Fiber from Great Coat.

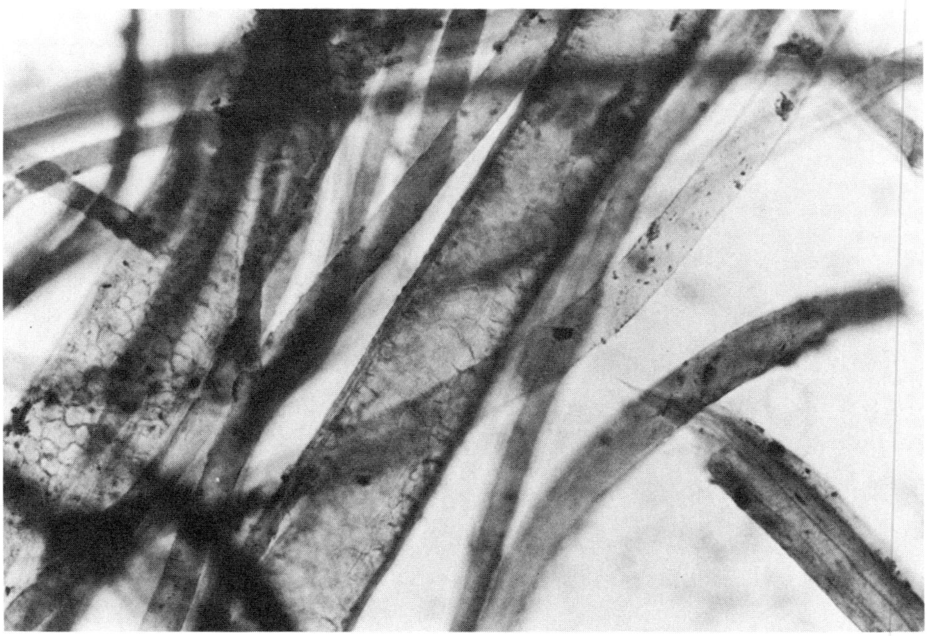

Figure 6. Kemp Fiber (Lattice Medulla) in Wool Yarn from Great Coat.

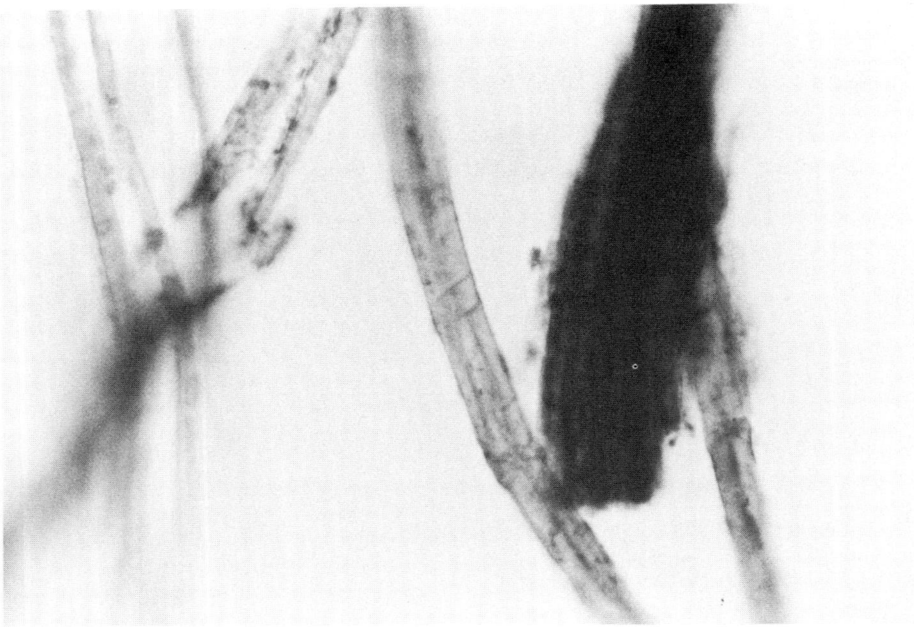

Figure 7. Hemp Fiber from Stitching on Button Hole on Fragment of Great Coat.

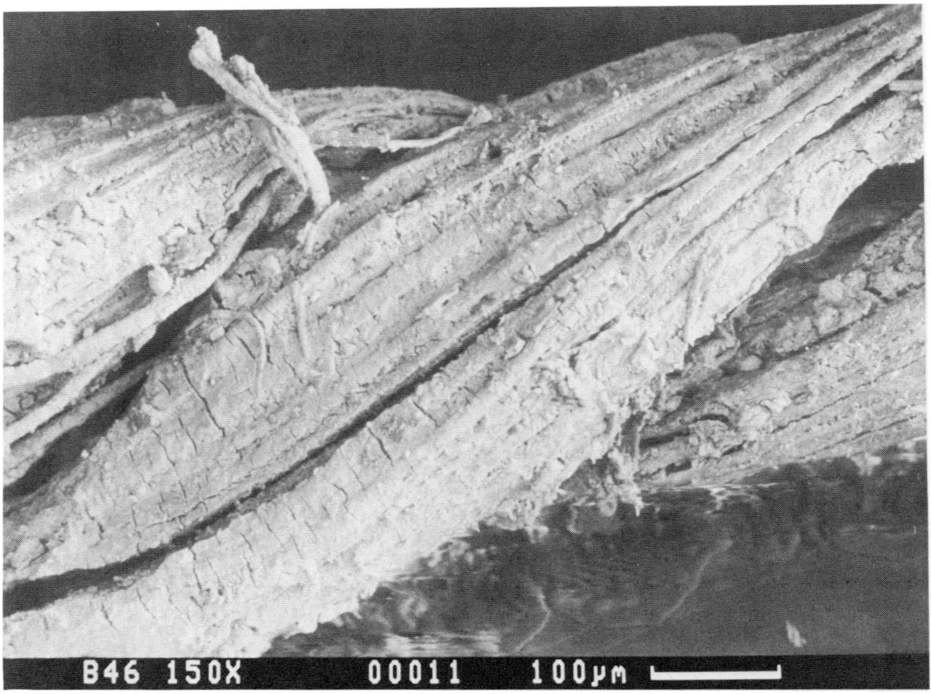

Figure 8. Hemp Sewing Thread from Great Coat. Backscattered Scanning Electron Micrograph.

Figure 9 shows a backscattered electron scanning electron micrograph of yarns from a fragment of brown plain weave cloth (perhaps from a pair of trousers). In this case the wool fibers have completely lost their cuticles, exposing the underlying fibril bundles. The lose of the cuticular layer may be the result of repeated freezing and thawing. Table 1 presents weather data for the period 1951-1980 for the two weather reporting stations closest to the Packer grave site. As may be seen, during a number of winter months the temperature cycles from above freezing during the day to below freezing at night. Ample moisture is also available, particularly during late winter or spring thaws. However, Ito et al. (1984) have found that freezing does not cause wool with intact scales to fibrillate. Once the wool has lost its scales, on the other hand, repeated freezing leads to separation of the wool fibrils. In the case of the clothing worn by Packer's victims the wool fibers could have lost their scale layers prior to burial through abrasion. The scales could also have been lost as result of microbial activity after burial.

Figure 9. Fibrillated Wool Fibers (Possibly from Trousers). Backscattered Scanning Electron Micrograph.

Very little research has been done on the effect of freezing on vegetable fibers like hemp. Batra (1985) demonstrated small decrements in the breaking strength and elongation at break of abaca fibers, which are bast fibers similar to hemp. Hemp fibers used in sewing thread or in textiles would be expected to be particularly resistant to biodeterioration. In the United States and Canada hemp fibers are prepared by allowing the mowed hemp plants (Cannabis sativa L.) to rot in the field to decompose gummy matter and free the fibers from the stalks. (Whitford, 1947a) Elsewhere, the hemp fibers are freed from the stalks by retting in water.

The textile fragments provided abundant evidence of microbial activity. Fungal mycelia (Figure 10) were frequently observed. Spherical fruiting bodies (Figure 11) were also numerous. Some of these were actually embedded in the weave of one of the woolen fabrics, indicating that they were not the result of contamination during the exhumation. Only one type of fungus could be tentatively identified: Figure 12 shows a possible conidiophore of Oncladium flavum (= Verticillium). (Ainsworth, 1961; Kendrick and Carmichael, 1973).

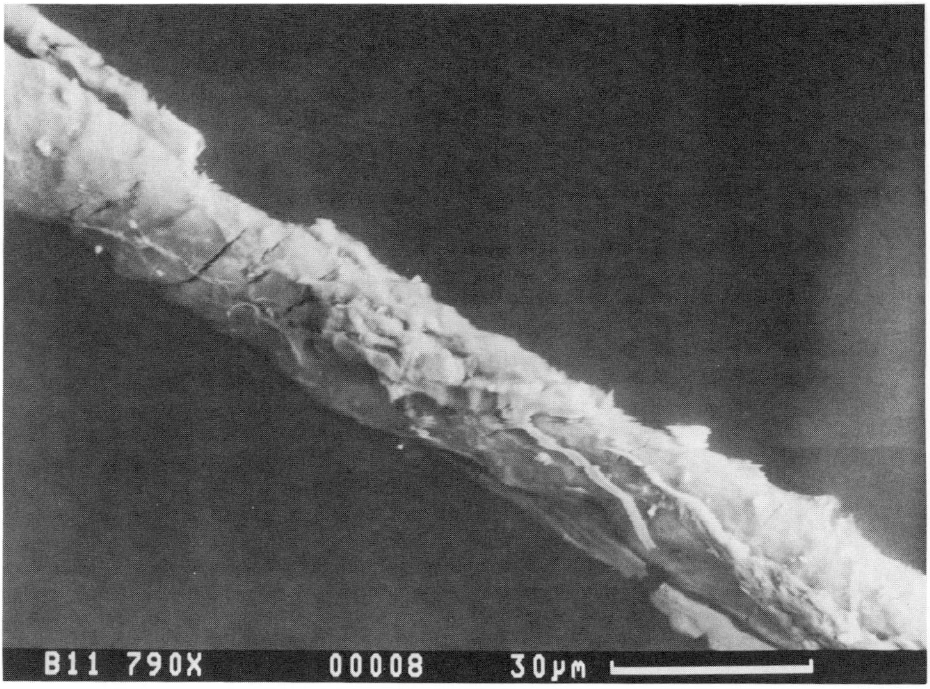

Figure 10. Fungal Mycelia on Wool Fibers. Backscattered Scanning Electron Micrograph.

Table 1. Temperature and precipitation at Gunnison and Montrose, Colorado (Ruffner, 1985).[a]

Gunnison, Colorado
Mean temperature

Maximum, °F

Jan	Feb	Mar	Apr	May	Jun	Jul	Aug	Sep	Oct	Nov	Dec
26	31	41	55	66	77	81	79	73	62	45	30

Minimum, °F

Jan	Feb	Mar	Apr	May	Jun	Jul	Aug	Sep	Oct	Nov	Dec
-7	-3	10	21	29	35	42	39	31	21	10	-3

Mean precipitation (inches)

Jan	Feb	Mar	Apr	May	Jun	Jul	Aug	Sep	Oct	Nov	Dec
1.0	0.8	0.8	0.7	0.7	0.5	1.5	1.5	1.0	0.9	0.7	0.9

Montrose, Colorado
Mean temperature

Maximum, °F

Jan	Feb	Mar	Apr	May	Jun	Jul	Aug	Sep	Oct	Nov	Dec
38	44	52	62	72	83	89	86	79	67	50	40

Minimum, °F

Jan	Feb	Mar	Apr	May	Jun	Jul	Aug	Sep	Oct	Nov	Dec
13	19	25	33	42	50	56	54	45	35	24	15

Mean precipitation (inches)

Jan	Feb	Mar	Apr	May	Jun	Jul	Aug	Sep	Oct	Nov	Dec
0.6	0.5	0.5	0.8	0.8	0.5	0.8	1.1	1.0	1.1	0.7	0.7

[a]Data collected between 1951-1980

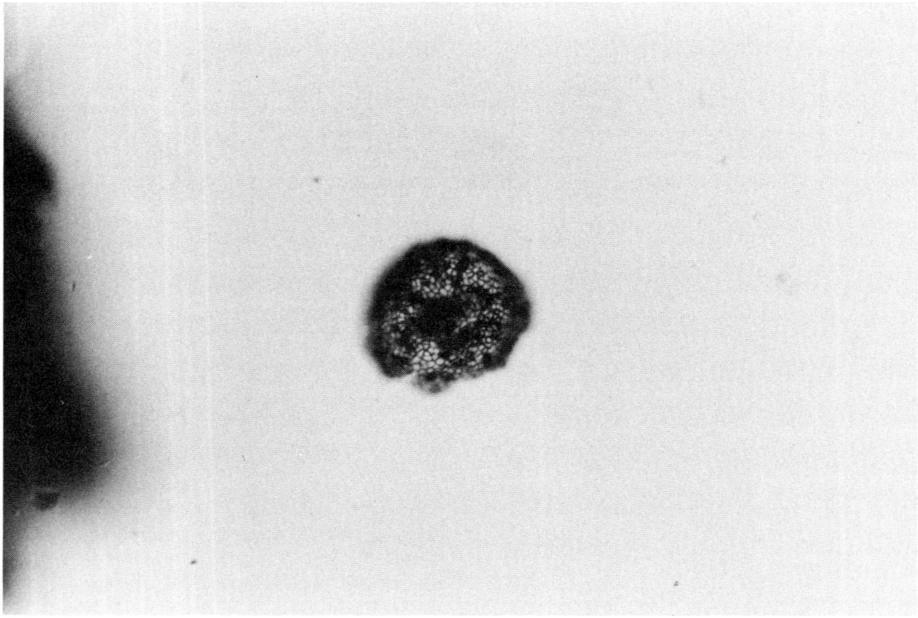

Figure 11. Fungal Fruiting Body.

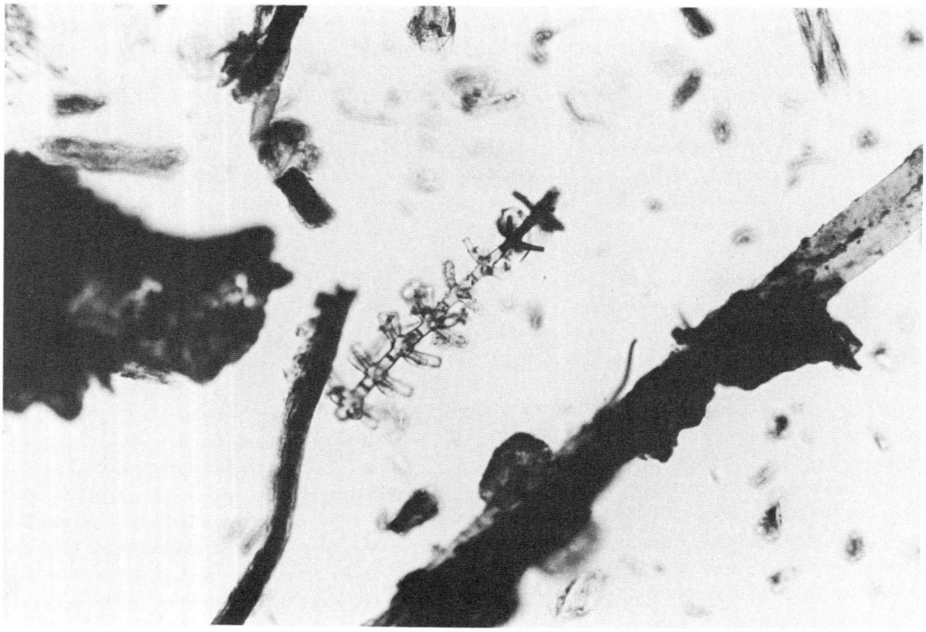

Figure 12. Possible Conidiophore of Oncladium flavum (= Verticillium).

SUMMARY

Textile fibers from fragments of fabric exhumed with the bodies of the victims of the notorious Western murderer and cannibal, Alferd Packer, were examined using transmitted light and scanning electron microscopy. Wool fibers exhibited fungal tunneling, loss of cuticular scales and fibrillation. The hemp fibers showed less evidence of biodeterioration. Fungal mycelia and fruiting bodies were found among the exhumed fibers.

REFERENCES

Ainsworth, G.C. (1961). Ainsworth and Bisby's Dictionary of the Fungi, Fifth Edition. p. 282, Commonwealth Mycological Institute, Kew, Surrey, p. 282.

Batra, S.K. (1985). Other long vegetable fibers. In: Handbook of Fiber Science and Technology (IV): Fiber Chemistry, pp. 728-807 (M. Lewin and E. M. Pearce, eds.), Marcel Dekker, NY.

Emilio, L.F. (1911). The Emilio Collection of Military Buttons: American, British, French and Spanish, with Some Other Countries and Non-Military in the Museum of the Essex Institute, Salem, Massachusetts. p. 29, The Essex Institute, Salem, Massachusetts.

Ito, H., Sakabe, H., Miyamoto, T., and Inagaki, H. (1984). Fibrillation of the cortex of merino wool fibers by freeze-thawing treatment. Textile Res. J. 54, 397-402.

Kendrick, W.B., and Carmichael, J.W. (1973). Hyphomycetes. In: The Fungi: An Advanced Treatise, Vol. IVA, pp. 323-509 (G.C. Ainsworth, F. K. Sparrow, and A.S. Sussman, ed.), Academic Press, New York.

Kloster, D. (1991). Military History Division, Smithsonian Institution, Washington, DC, personal communication.

Leeder, J.O. (1986). The cell membrane complex and its influence on the properties of the wool fiber. Wool Sci. Rev., 63, 3-35.

Lewis, J. (1981). Wool. In: Microbial Biodeterioration (Economic Microbiology vol. 6), pp. 81-129 (A. H. Rose, ed.) Academic Press, New York.

McPherson, J.M. (1988). The Battle Cry of Freedom: The Civil War Era, Oxford University Press, New York.

Nopitsch, M. (1953). Micro-organism attack on textiles and leather. Ciba Rev., 100, 3581-3591.

Ruffner, (1985). Climates of the States: National Oceanic and Atmospheric Administration Summaries, Tables and Maps with Overview of State Climatological Programs, 3rd Edition, Vol. 1. pp. 150-153, Gale Research Company, Detroit, MI.

Scott, D.D., Fox, R.A., Connor, M.A., and Harmon, H. (1989). Archeological Perspectives on the Battle of the Little Big Horn. p. 196, University of Oklahoma Press, Norman, Oklahoma.

Starrs, J.E. (1988). The distanced fascination of cannibalism: an Alferd Packer reprise. Scientific Sleuthing Newsletter, 12, 3-8.

Starrs, J.E. (1989). Words as true as bone: a report on the Packer victim's exhumation project," Scientific Sleuthing Review, 13, 1-9.

Von Bergen, W. (1947). Microscopical and physical properties of wool. In: Mathew's Textile Fibers: Their Physical, Microscopical and Chemical Properties, Fifth Edition, pp. 507-560 (H. R. Mauersberger, ed.), John Wiley and Sons, New York.

Whitford, A.C. (1947a). The bast fibers. In: Mathew's Textile Fibers: Their Physical, Microscopical and Chemical Properties, Fifth Edition, pp. 305-368 (H. R. Mauersberger, ed.), John Wiley and Sons, New York.

Whitford, A.C. (1947b). Structural or hard vegetable fibers. In: Mathew's Textile Fibers: Their Physical, Microscopical and Chemical Properties, Fifth Edition, pp. 369-397 (H. R. Mauersberger, ed.), John Wiley and Sons, New York.

SCANNING ELECTRON MICROSCOPE STUDIES OF THE BIODETERIORATION OF HUMAN HAIR BURIED IN SOIL AND IMMERSED IN WATER

TIMOTHY D. KUPFERSCHMID, RUTH VAN DYKE, and **WALTER F. ROWE***, Department of Forensic Sciences, The George Washington University, Washington, DC 20052. USA

INTRODUCTION

Forensic scientists are frequently called upon to examine minute bits of physical evidence that may be transferred between a criminal and the scene of his crime or the criminal and his victim. These minute bits of evidence are referred to as trace evidence. They are used to associate a suspect in a criminal case with a particular crime. Trace evidence includes glass particles, soil, paint chips and smears, textile fibers and hairs.

The forensic examination of hairs involves the following steps: Recognition of a fiber as an animal hair, identification of the hair as being from either a human being or other species of animal, the somatic origin of the hair (if human) and microscopic comparison of the evidentiary hair with known hair samples. The recognition of a fiber as an animal hair and the identification of the genus or species that the hair came from require the microscopic evaluation of an ensemble of morphological features. The most important of these are shown in Figure 1. A typical hair can been compared to a lead pencil: The paint on the outside of the pencil corresponds to the cuticle, a layer of flattened, overlapping cells between one and ten cells thick; the wooden shaft corresponds to the cortex; and the lead in the center corresponds to the medulla, a column of specialized cells running up the center of the hair shaft. The surface of a typical hair is covered with a layer of overlapping cuticular scales. (Hicks, 1977; Bisbing, 1983; Seta et al., 1988).

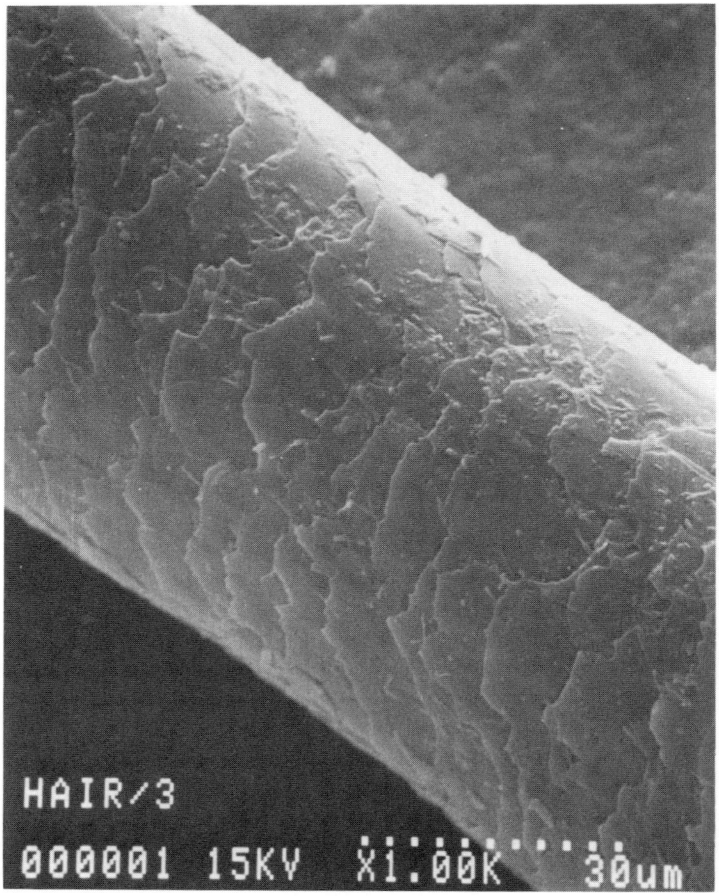

Figure 1. Diagram of Human Hair.

Three features of hairs are normally examined to determine the species of animal from which a hair came: Size, cuticular scale pattern and configuration of the medulla. Fur-bearing animals have three different sizes of hair: very large vibrissae (i.e. whiskers), large guard hairs and fine down hairs. Human hairs are intermediate in size between guard and down hairs. Cuticular scale patterns vary from very regular patterns resembling fish scales to irregular scale patterns found on human hairs. Medullas may appear as periodic, lattice-like or even absent. The width of the medulla relative to the width of the hair shaft (the so-called medullary index) is also useful in determining the species of animal that a hair came from. Human hairs have a medullary index of approximately 0.3, while most other animals have a medullary index greater than 0.5. (Hicks, 1977; Appleyard, 1978; Bisbing, 1983; Seta et al., 1988).

Forensic scientists encounter trace evidence that has been exposed to a wide range of environmental conditions for varying lengths of time. Hair is not easily destroyed by adverse environmental conditions. However, the probative value of hair evidence may decreased by biodeterioration. A number of studies have examined the effects of hair degradation caused by micro-organisms. (Davidson and Gregory, 1934; Page, 1950; Vanbreuseghem, 1952; Daniels, 1953; Barlow and Chattaway, 1955; Raubitschek, 1961; English, 1963; English, 1965; Keddie, 1967; English, 1969; Tosti et al., 1970; Yu et al., 1971; English, 1976; Deshmukh and Agrawal, 1982; Takatori et al., 1983; Serowick and Rowe, 1987; Hawks and Rowe, 1988; Kundrat and Rowe, 1989; and DeGaetano et al., 1992) While most of these studies have focussed on the effects of microbial attack on the ultrastructure of hair or the medical aspects of in vivo fungal infestation. Three stages have been identified in the fungal attack on hairs: Lifting of the cuticular scales (possibly caused by the growth of fungal mycelia along the surface of the hair), erosion of the cortex and tunneling of hair by specialized hyphae termed "penetrating organs." (Page, 1950; Vanbreuseghem, 1952; Daniels, 1953; Barlow and Chattaway, 1955; and English, 1963).

The studies of Serowik and Rowe (1987), Kundrat and Rowe (1989) and DeGaetano et al. (1992) specifically examined the implications of biodeterioration of human hair for forensic hair comparisons. Serowik and Rowe (1987) and Kundrat and Rowe (1989) used transmitted light microscopy to examine human hairs that had been buried in well-watered soil. DeGaetano et al. (1992)

reported on fungal tunneling of human head hairs that had been removed from the buried body of a murder victim. The emphasis of all of these studies was on the possible effects of biodeterioration on the microscopic comparison of human hairs.

Heretofore, there have been no studies on the effect of biodeterioration on the identification of fibers as hairs and on the determination of the genus or species of animal from which evidentiary hairs have come. The present study was conducted to determine the resistance of human head hair to the deleterious effects soil and aqueous environments. The scanning electron microscope (SEM) was used to examine the surface characteristics of the deteriorated hairs, especially the cuticular scale patterns. The SEM has proven to be a useful tool for examining the surface features of hairs. The SEM has been used to examine the surface characteristics of both normal (Riggott and Wyatt, 1983; Lindelöf at al., 1988; Hess et al., 1990), diseased (Lubach et al., 1982; Takatori et al., 1983; Forslind and Kaaman, 1984; Goldin et al., 1984; Prens et al., 1984; Forslind et al., 1985; Hebert et al., 1987; Shelley et al., 1987; Okuda et al., 1988) and treated (Choudry et al., 1983) human hair.

MATERIALS AND METHODS

For the burial study, forty-nine telogen hairs were collected from a Caucasian female. The hairs were obtained from the subject's comb after the she had washed and combed her hair. The only cosmetic treatment used by this subject on her hair was regular shampooing and occasional use of a creme rinse. Seven groups of seven hairs each were wound together into a ball and placed in nylon tulle mesh bags (approximately 2.5 cm X 2.5 cm), which were closed with metal staples. Six bags were buried in 10-cm diameter plastic flower pots using soil from an outdoor garden plot. Each pot was watered once a week with distilled water. The seventh bag was a control and was not buried. The buried hair samples were exhumed at intervals of thirteen, twenty, twenty-nine, forty-one, forty-nine and fifty-seven days. As each sample was retrieved from the soil, the nylon bag and its contents were rinsed under low pressure tap water to remove any adherent soil particles. The hairs were then air-dried. Finally, one hair from each exhumed sample was selected for SEM analysis.

Hair specimens for the immersion study were collected from a Caucasian male by combing the hair onto a clean piece of paper. This subject used no cosmetic treatment beyond shampooing on his hair. From the combings, eighty-one visually similar telogen hairs, each approximately 5 cm in length, were selected. These hair specimens were washed in a 1% solution of Triton X-100 (Fisher Scientific) and rinsed three times in distilled water. Test tubes were prepared by washing each once with soap and water and once with 3% hydrogen peroxide, and drying at 100°C. Three hairs were placed in each test tube. Twelve test tubes were filled with water collected from the Potomac River directly adjacent to the Lincoln Memorial in Washington, DC. An additional twelve test tubes were filled with distilled water. Three test tubes served as blanks. These contained hair samples, but no water.

All of the test tubes were placed in a Blue M dry type bacteriological incubator (Blue M Electric Company, Blue Island, Illinois) at 37°C, the optimum temperature for micro-organism activity. Every fourteen days, three tubes containing hair and Potomac River water, and three tubes containing hair and distilled water were removed. Incubation of samples continued for fifty-six days. The tubes were refrigerated until the hair samples could be examined microscopically.

At the end of the eight week period, the hairs were examined with a transmitted light microscope. From the original eighty-one hairs that were subjected to immersion, five were selected for examination with the scanning electron microscope. These five hairs represented typical hair morphology after four- and eight- week exposures in Potomac River and distilled water. A control hair specimen was selected from those in the control test tube.

Hairs selected for SEM examination were cut into 2-cm sections and attached by their ends to carbon discs with small drops of Duco cement. All discs and hairs were coated with gold using an Emscope sputter coater. The SEM analysis was conducted with a Hitachi S570 scanning electron microscope equipped with a secondary electron detector and Kevex EDX accessory.

RESULTS AND DISCUSSION

Figure 2 shows the scale pattern of one of the control hairs in the burial study. The scales are normal, with little or no lifting of the scale edges. The buried hairs showed progressive deterioration as the period of burial increased. The first two samples to be exhumed did not show significant loss of scales. The third sample showed some lifting of the cuticular scale edges (Figure 3), while the fourth sample showed some actual loss of scales. The hairs in the final sample to be exhumed were falling apart and fungal mycelia were evident on many of the hairs (Figure 4). The buried hairs clearly showed two of the, three stages of fungal attack on hair: Lifting of the cuticular scales and erosion of the cortex. (Page, 1950; Vanbreuseghem, 1952; Daniels, 1953; Barlow and Chattaway, 1955; and English, 1963) The scanning electron micrographs did not reveal tunneling of the hairs by the "penetrating organs" of the fungi. However, DeGaetano et al. (1992) reported that the openings of the tunnels on the hair surface are very small (and consequently easily overlooked) in the early stages of fungal attack.

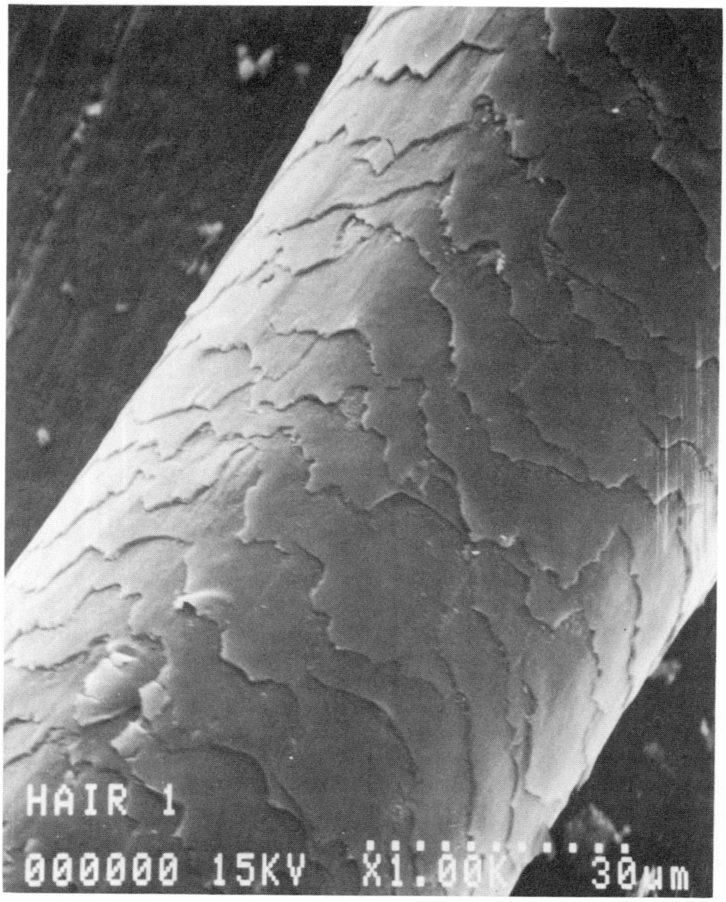

Figure 2. Human Head Hair Before Burial in Well-Watered Soil.

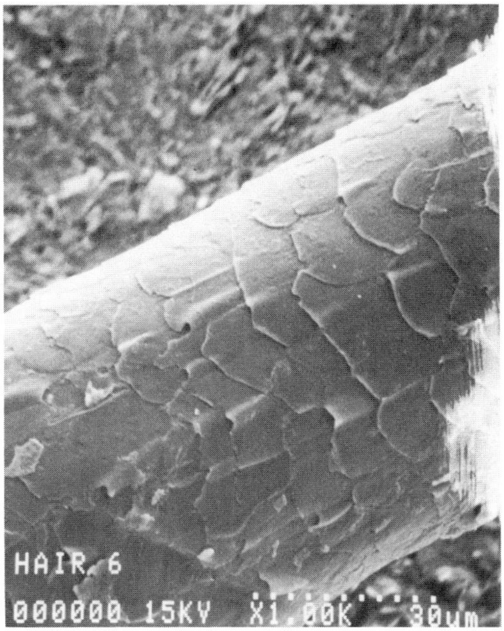

Figure 3. Human Head Hair After Burial for Forty-Nine Days in Well-Watered Soil.

Figure 4. Human Head Hair After Burial for Fifty-Seven Days in Well-Watered Soil.

Figure 5 shows one of the control hairs for the immersion study. As was the case with the burial study discussed above, the scale patterns are normal in all respects. The hairs immersed in water also displayed a progressive loss of cuticular scales (Figure 6). The edges of the scales became less and less pronounced as the period of immersion increased. The hairs immersed longest

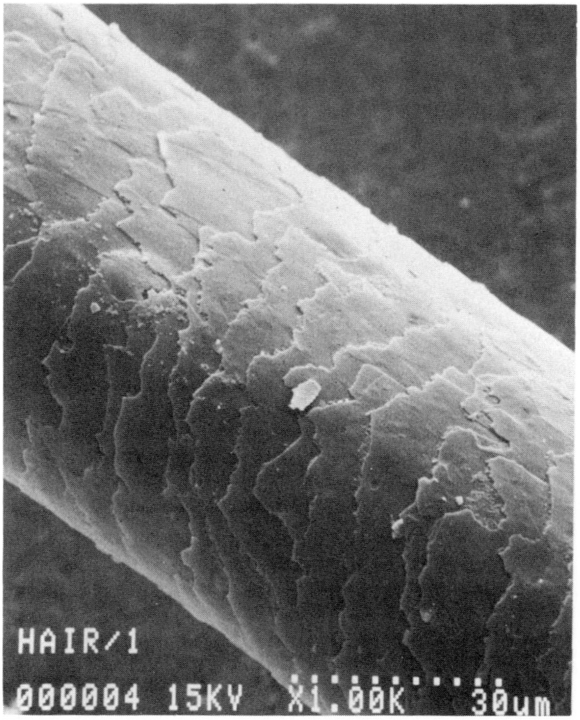

Figure 5. Human Head Hair Before Immersion in Water.

showed a complete loss of scales (Figure 7). These hairs were found to have developed lumpy surfaces, possibly as a result of the waterlogging of cuticular or cortical cells. This very unusual condition has not been previously described. No differences were noted between the hairs immersed in distilled water and those immersed in water from the Potomac River.

Figure 6. Human Head Hair After Immersion in Potomac River Water for Twenty-Eight Days.

Figure 7. Human Head Hair After Immersion in Potomac River Water for Fifty-Six Days.

The results of this study indicate that environmental conditions may completely efface the cuticular scale patterns on hairs. As a result, it may be impossible to determine the species of animal from which a hair came.

Acknowledgements

The authors would like to thank Dr. Frank Johnson, Chemical Pathology Division, the Armed Forces Institute of Pathology, for his aid in completing this project.

REFERENCES

Appleyard, H.M. (1978). Guide to the Identification of Animal Fibres, Wira, Leeds, UK.

Barlow, A.J.E., and Chattaway, F.W. (1955). The attack of chemically modified keratin by certain dermatophytes. J. Invest. Dermatol., 24, 65-74.

Bisbing, R.E. (1982). The forensic identification and association of human hair. In: Forensic Science Handbook, pp. 184-221 (R. Saferstein, ed.), Prentice-Hall, Englewood Cliffs, NJ.

Choudry, M.Y., Kingston, C.R., Kobilinsky, L., and DeForest, P.R. (1983). Individual characteristics of chemically modified human hairs revealed by scanning electron microscopy. J. Forensic Sci., 28, 293-306.

Daniels, G. (1953). The digestion of human hair keratin by Microsporum canis Bodin. J. Gen. Microbiol., 8, 289-294.

Davidson, A.M., and Gregory, P.H. (1934). In situ cultures of dermatophytes. Canadian J. of Res., 10, 373-393.

DeGaetano, D.H., Kempton, J.B., and Rowe, W.F. (1992). Fungal tunneling of hair from a buried body. J. Forensic Sci., 37, 1048-1054.

Deshmukh, S.K., and Agrawal, S.C. (1982). In vitro degradation of human hair by some keratinophilic fungi. Mykosen, 25, 454-458.

English, M.P. (1963). The saprophytic growth of keratinophilic fungi on keratin. Sabouraudia, 2, 115-130.

English, M.P. (1965). The saprophytic growth of non-keratinophilic fungi on keratinized substrata, and a comparison with keratinophilic fungi. Trans. Brit. Mycol. Soc., 48, 219-235.

English, M.P. (1969). Destruction of hair by Chrysosporium keratinophilum. Trans. Brit. Mycol. Soc., 52, 247-255.

English, M.P. (1976). destruction of hair by two species of Chrysosporium. Trans. Brit. Mycol. Soc., 66, 357-358.

Forslind, B., and Kaaman, T. (1984). Brittle hair in osteogenesis imperfecta: transmission and scanning electron microscopy and mycologic assay. Acta Derm. Venereol. (Stockholm), 64, 418-455.

Forslind, B., Thomsen, K., and Andersson, B. (1985). Spun glass hair: two cases investigated with SEM and TEM. Acta Derm. Venereol. (Stockholm), 65, 348-352.

Goldin, H.M., Bronson, D.M., and Fretzin, D.F. (1984). Woolly-hair nevus: a case report and study by scanning electron microscopy. Pediatric Dermatol., 2, 41-44.

Hawks, C.A., and Rowe, W.F. (1988). Deterioration of hair by airborne micro-organisms: implications for museum biological collections, In: Biodeterioration 7: pp. 461-465 (D.R. Houghton, R.N. Smith and H.O.W. Eggins, ed.) Proceedings of the Seventh International Biodeterioration Symposium, Elsevier Applied Science Publishers Ltd., Barking, Essex, UK.

Hebert, A.A., Charrow, J., Esterly, N.B., and Fretzin, D.F. (1987). Uncombable hair (pili trianguli et canaliculi): evidence for dominant inheritance with complete penetrance based on scanning electron microscopy. Am. J. Med. Genet., 28, 185-193.

Hess, W.M., Seegmiller, R.E., Gardner, J.S., Allen, J.V., and Barendregt, S. (1990). Human hair morphology: a scanning electron microscopy study on a male caucasoid and a computerized classification of regional differences. Scanning Microscopy, 4, 375-386.

Hicks, J.W. (1977). Microscopy of Hair: A Practical Guide and Manual, Federal Bureau of Investigation, Washington, DC.

Keddie, F.M. (1967). Further investigations of the electron microscopic examination of human hair naturally infected with Microsporum gypseum. Sabouraudia, 6, 1-4.

Kundrat, J.A., and Rowe, W.F. (1989). A study of hair deterioration in agricultural soil, In: Biodeterioration Research 2, pp. 91-98 (G.C. Llewellyn and C.E. O'Rear, ed.) Plenum Publishing Co., NY.

Lindelöf, B., Forslind, B., Hedblad, M., and Kaveus, U. (1988) Human hair form: morphology revealed by light and scanning electron microscopy and computer aided three-dimensional reconstruction. Arch. Dermatol., 124, 1359-1363.

Lubach, D., Glowienka, F., and Castellucci , M. (1982). Monilethrix: comparative scanning electron microscopic study of the hair in one family. Dermatologica, 164, 1-9.

Okuda, C., Ito, M., and Sato, Y. (1988). Fungus invasion into human hair in black dot ringworm: light and electron microscopy study. J. Invest. Dermatol., 90, 729-733.

Page, R.M. (1950). Observations on keratin digestion by Microsporum gypseum. Mycologia, 42, 591-602.

Prens, E.P., Peereboom-Wynia, J.D.R., De Bruyn, W.C., Van Joost, T., and Stolz, E. (1984). Clinical and scanning electron microscopic findings in a solitary case of trichorhinophalangeal syndrome type I. Acta Derm. Venereol. (Stockholm), 64, 249-253.

Raubitschek, F. (1961). Mechanical versus chemical keratolysis by dermatophytes. Sabouraudia, 1, 87-90.

Riggott, J.M., and Wyatt, E.H. (1983). Mensuration of scanning micrographs. A possible means of hair identification. J. Forensic Sci. Soc., 23, 155-160.

Serowik, J.M., and Rowe, W.F. (1987). Biodeterioration of hair in a soil environment, In: Biodeterioration Research 1, pp. 87-93 (G.C. Llewellyn and C.E. O'Rear, ed.) Plenum Publishing Co., NY.

Seta, S., Sato, H., and Miyake, B. (1988). Forensic hair investigations, In: Forensic Science Progress, Vol. 2, pp. 47-166 (A. Maehly and R.L. Williams, ed.) Springer-Verlag, Berlin.

Shelley, W.B., Shelley, E.D., and Burmeister, V. (1987). The infected hairs of tinea capitis due to Microsporum canis: demonstration of uniqueness of the hair cuticle by scanning electron microscopy. J. Am. Acad. Dermatol., 16, 354-361.

Takatori, K., Udagawa, S., Kurata, H., and Hasegawa, A. (1983). Microscopic observations of human hairs infected with Microsporum ferrugineum. Mycopathologica, 81, 129-133.

Tosti, A., Villardita, S., Fazzini, M.L., and Scalici, R. (1970). Contribution to the knowledge of dermatophytic invasion of hair: an investigation with the scanning electron microscope. J. Invest. Dermatol., 58, 123-134.

Vanbreuseghem, R. (1952). Keratin digestion by dermatophytes: a specific diagnostic method. Mycologia, 44, 176-182.

Yu, R.J., Harmon, S.R., Grappel, S.F. and Blank, F. (1971). Two cell-bound keratinases of *Trichophyton mentagrophytes*. *J. Invest. Dermatol.*, 56, 27-32.

A RAPID ANALYTICAL PROCEDURE FOR DELTA-9-TETRAHYDROCANNABINOL IN PLANT MATERIAL

CHARLES E. O'REAR and **GERALD C. LLEWELLYN**, Department of Forensic Science, The George Washington University, Washington, DC 20052, USA

INTRODUCTION

There are several methods of analysis available for detection of delta-9-tetrahydrocannabinol (delta-9-THC), a biologically active component of marihuana derived from Cannabis sativa L. These range from simple colorimetric tests (Beam 1911) to the Duquenois-Levine reaction which replaced the Beam test (Hahas, 1975). Analytical laboratories having available to them gas chromatography instrumentation utilize the procedures of Lerner (1986), Fetterman et al. (1971), Turner and Hadley (1973), or the DEA (1976) method when small quantities of material are evaluated.

The forensic laboratory is faced on a daily basis with a large number of samples, most of which represent confiscated plant material exceeding 10 g. The gas chromatography (GC) methods seem to meet most of the criteria of efficiency for such testing. Herein is described a modification to the GC method of Turner and Hadley (1973). This procedure is rapid, contains one quantitative extraction step, with extracts that contain low levels of plant waxes, eliminates most of the filtration and evaporation steps, and utilizes a straight chain hydrocarbon as an inexpensive internal standard.

MATERIALS AND METHODS

The source of the plant material (Cannabis sp.) selected for analysis was taken from confiscated, illicit materials known to contain delta-9 THC (delta-9,

Biodeterioration Research 4, Edited by G.C. Llewellyn *et al.*, Plenum Press, New York, 1994

tetrahydrocannabinol). These materials were originally illicit street samples submitted by law enforcement agencies to the Virginia State Laboratories.

Depending upon the state of the sample (loose and shredded, compacted or brick-like) the marihuana was chopped or crushed. The plant material was dried at room temperature to ambient humidity. To "manicure" the marihuana samples were passed through a No. 10, U.S. Standard sieve (2,000 micron openings) to remove stalks, stems and seeds. The extraction procedure follows: A random 5 g sample was placed directly into a 250 ml Erlenmeyer flask having a ground glass stopper. Next 50 ml of n-Hexane was added. The flask was stoppered tightly and placed on a magnetic stirrer for at least 30 minutes. After 30 minutes of natural sedimentation, 2 ml was removed and pipetted into a 25 ml culture tube. Then 4 ml of the internal standard solution was added and mixed thoroughly. This ratio of sample to standard (1/2; v/v works well for marihuana containing 0.5 to 50% THC.

The internal standard solution was prepared by weighing 250 mg of n-Triacontane (99% pure, Applied Science, State College, PA) to a 250 mg volumetric flask with n-Hexane. The standard dissolved and the solution q.s. (diluted to volume) with additional n-Hexane.

The THC standard solution (10 mg/ml-ethanol, United States Pharmacopeia, Rockville, MD) was removed quantitatively from the glass vial (broken at the neck). The 1 ml quantity and several washings (95% ethanol, USP) were added to a 10 ml volumetric flask (q.s. with ethanol). The solution was carefully evaporated to dryness by warming gently with a water bath while passing a fine stream of compressed air into the test tube. After dryness was achieved 9 ml of the internal standard solution was pipetted into the culture tube containing the delta-9-THC. The solution was mixed thoroughly.

The samples were tested utilizing a Bendix Model 3300 gas chromatograph equipped with hydrogen flame-ionization detectors and operated isotherically at 240°C. Glass columns, 2 mm i.d. x 1.83 m *6 ft.), were packed with 3% OV-17 (high purity polar phenyl methyl silicone, approx. 30,000 mol. wt.) on 60 to 80-mesh Gas-Chrom Q. Nitrogen was used as the carrier gas at a flow rate of 30/min. Attenuation was set at 1000. The hydrogen flame-ionization detector had a hydrogen gas flow rate of 40 ml/min and an air flow rate of 600 ml/min.

Injections of 4 μ1 (Hamilton microliter syringe, 801-RN, 10 μ11) of the resultant mixture (standard solution or sample solution both having internal standard) was made into the gas chromatograph. Mean values were derived from triplicate injections.

Calculations were based on the mean values for peak heights following the procedures outlined in Table 1.

Table 1. Calculation Steps for Determing Percentage of THC Based on Peak Heights.

$$\% \text{ THC} = \frac{{}^{C}\text{STD} \times R_2 \times 100}{R_1 \times W}$$

${}^{C}\text{STD}$ = Concentration of Standard (mg/ml)

$$R_1 = \frac{\text{Peak Height of Standard (THC)}}{\text{Peak Height of Internal Standard}}$$

$$R_2 = \frac{\text{Peak Height ofSample (THC)}}{\text{Peak Height of Internal Standard}}$$

V = Volume of Internal Standard Solution (ml)

W = Sample Weight (Calculate from original weight of marihuana and aliquots taken.)

RESULTS AND DISCUSSION

Although only a few sample results are shown in Table 2, this test procedure has been in exclusive use in our laboratory since January, 1976. To date more than 600 samples have been tested using this procedure. The data generated from these tests all are regarded as potential evidence for court cases. Some cases have been prosecuted based on the results of this procedure.

This method is a modification from that described by Lerner (1975), Fetterman et al. (1971) and Turner and Hadley (1973). It has consistently provided a satisfactory degree of precision and accuracy. The major changes in this method as compared to those listed above relate to the size of the sample to be tested and the quality of the extract. This procedure is actually a test for macro quantities using n-Hexane as the solvent with simplified and time-efficient methodology used.

Table 2. Test Results for THC Found in Illicit Samples Collected From Seven Different Geographic Areas of Virginia in 1976.

Sample No.	Location Confiscated	% THC [a,b]
1	A	9.3
2	B	1.1
3	C	2.0
4	D	4.2
5	D	3.7
6	E	1.6
7	E	1.5
8	F	4.6
9	G	0.6

[a]Mean from triplicate tests
[b]$X \pm SD$ - 3.18 ± 2.7*161.5

The forensic laboratory testing for marihuana normally has a substantial size of sample but there is a need to process a large number of samples. The modified method proposed herein follows more closely the needs of the forensic chemist. The sample is substantially larger in size which also may provide a more representative result. It also has a less complicated procedure allowing for a greater number of assays without significant loss of sensitivity. Many of the filtration and evaporation procedures have been eliminated. Hexane seems to be a better extraction solvent, apparently removing less plant waxes. It seems that extraction of 5 g of plant material with 50 ml of hexane, followed by refiltration and extraction with additional fresh hexane, showed little or no THC recovered. Petroleum ether used in the method proposed by the U.S. Drug Enforcement Administration (DEA) for a heated extraction, produced very unpleasant odors. Also the method reported herein follows their suggestion to use a straight-chain hydrocarbon as an internal standard. We used n-Triacontane because of its inexpensive nature and ease of purpose. The sensitivity range of the procedure rivaled but did not exceed those of Turner and Hadley (1973), Lerner (1986), Fetterman et al. (1971). Procedures have been tested recently, whereby a modification of this procedure can be used to assay for hashish and hashish oil.

SUMMARY

The development and testing of a rapid analytical procedure for delta-9-tetrahydrocannabinol, a biologically active component of marihuana was undertaken. It is a modified gas chromatography procedure having one quantitative extraction step. Results were found to be acceptable for plant materials have low levels of wax. The procedure is a test for macro quantities using n-Hexane as the solvent and n-Triacontane as the straight-chain internal standard. Additional hexane extractions of the 5 grams of plant material showed little or no additional recover of the active agent.

Acknowledgements

The authors wish to acknowledge the hospitality of the Virginia Bureau of Forensic Science during the period in which this work was done. We would also like to thank Eugene R. Harrison of the Virginia Division of Consolidated Laboratory Service for his assistance, comments and support which were so helpful in completing this study.

REFERENCES

Beam, W., Fourth Report, Wellcome Tropical Research Laboratory Chemistry Section Khartoum, No. 25., 1911.

Fetterman, P.S., Keith, E.S., Waller, C.W., Guerrero, O., Doorenbos, J.J., and Quimby, M.W., "Mississippi-Grown Cannabis sativa L.: Preliminary Observations on Chemical Definition of Phenotype and Variations in Tetrahydrocannabinol Content, Versus Age, Sex and Plant Part.: Journal of Pharmaceutical Science, 60, 1971. pp. 1246-1249.

Lerner, P., "The Precise Determination of Tetrahydrocannabinol In Marihuana and Hashish," Bulletin On Narcotics, Vol. 21, pp. 39-41, 1986.

Nahas, G.G., Marihuana - Deceptive Weed, Raven Press, NY, 1975, p. 93.

Turner, C.E., and Hadley, K., "Constituents of Cannabis sativa L. II; Absence of Cannabidiol in an African Varient," Journal of Pharmaceutical Science, Vol. 62, No. 2, 1973, pp. 251-255.

U.S. Drug Enforcement Administration. A Method of Analysis for Small Quantities of Marihuana, Washington, DC, 1976.

A HISTORICAL SAMPLING OF DELTA-9-TETRAHYDROCANNABINOL POTENCY OF MARIHUANA AND HASHISH CONFISCATED IN VIRGINIA

CHARLES E. O'REAR and **GERALD C. LLEWELLYN**, Department of Forensic Science, The George Washington University, Washington, DC 20052, USA

INTRODUCTION

Introduction with Cannabis, one of the oldest drugs known to man, is now widespread over the world, including Western countries where it had never been observed before. Cannabis sativa has been cultivated for centuries for the hemp in its stem, the oil in its seed, and the biologically active substance in its flowering tops. The principal biologically active ingredients include delta-9-tetrahydrocannabinol (delta-9-THC) (Nahas, 1975).

The plants cultivated in the USA for fibers during World War II have provided for much of the escaped plant growth in the Midwest. This wild variety seems not to have spread to the South. Also this variety usually has a low content of psychoactive materi:. (Doorenbos et al., 1971). According to Nahas, a psychoactive dose is approximately ten mg present in a one g cigarette. Haney and Bazzaz (1970) indicate that imported material is most commonly used in the USA.

Low grade marihuana including the wide variety found in this country may contain less than 0.5% delta-9-THC, while some Asian grades have more than 3%. It is estimated that when smoked with maximum efficiency, not more than 50% of the active agent in the cigarette is actually absorbed. Thus a one g cigarette containing 1% delta-9-THC would deliver five mg of the THC to the lungs (Goodman and Gilman, 1975). Rafaelsen et al. (1973) State that does equal to two cigarettes are needed before complex tasks such as driving are affected. Both Hollister (1971) and the Secretary of Health, Education and Welfare (HEW) report that delta-9-THC may be three times more potent when smoked as when taken by mouth (1972).

The Secretary's Report (1976) stated that typical marihuana used in the USA contains 1-2% delta-9-THC as compared with hashish which has as much as 10% delta-9-THC.

Marihuana use grew rapidly in the USA during the 1960's. For the year 1976 the greatest number of users continued to be among adolescents and young adults. The national survey conducted by Abelson and Fishburne (1976) found that 53% of this age group (18-25) has used marihuana at some time compared with 36% of adults in the age group (26-34) and 6% of adults over the age of 34. In the Sixth Annual Report to the U.S. Congress from the Secretary of HEW (1986), concern over the possible shift of Cannabis users towards the use of higher potency material was expressed. Among adolescents and young adults 29.2% reported ever using hashish (dried resinous exudate of the flowering tops, containing higher cannabinoid concentrations) and about 20% of those reporting use during the month preceding the survey.

The Drug Control Act of the Commonwealth of Virginia classifies marihuana and hashish as a Schedule I Substance (having no known and accepted medicinal uses). With the exception of experimental and analytical laboratory use, these substances are illicit.

In Virginia, marihuana, hashish, hashish oil and the associated paraphernalia of Cannabis abuse still constitute the major number of specimens submitted by law enforcement agencies to the forensic laboratory.

The majority of confiscated Cannabis specimens were comprised of manicured tobacco-like plant material normally associated with street distribution. Ranging in size from the "commercial" kilo (a 2.2 pound block which is compressed Cannabis leaves, stems and seeds) to the single "joint" (or cigarette). Marihuana specimens continue to be submitted to the laboratory from practically all law enforcement jurisdictions within Virginia. Occasionally, the laboratory received specimens of hashish oil for chemical identification. Hashish oil is a concentrated oily resin extracted from the marihuana plant. Existing as a liquid, this oil substance is spread on a cigarette and smoked by the abuser. One or two drops of hashish oil may be as potent as a marihuana cigarette.

It was the objective of this historical, state-wide study to evaluate the percent occurrence by weight of delta-9-THC as it related to geographic regions, monthly variations, and overall mean levels for the active agent in Cannabis and hashish. Were levels of delta-9-THC found in Virginia samples different from those for the country as a whole and is there a tendency for users to favor samples having higher concentrations of the psychoactive agent? Are "Virginia samples" psychoactive? Does the marihuana supply radiate from a common source within the State?

MATERIALS AND METHODS

Marihuana

Sources of Samples. The marihuana samples used in this study were illicit street samples weighing 10 or more grams each. A street sample represents the material being sold to the ultimate user. The latter person may use it in a variety of ways including smoking, ingestion, dilution, concentration or the addition of a second drug, i.e., phencyclidine. Samples that were seized by law enforcement agencies containing less than 10 grams were not included in this study. These smaller sample sizes were insufficient in weight for normal analytical tests, in addition to providing adequate evidence for prosecution. Therefore, this study does not include single dose samples, i.e., one or two cigarettes.

Geographical Regions. The two basic reasons for dividing the Commonwealth of Virginia into four geographic regions, Central, Northern, Tidewater, and Western are as follows: First, there were collection points for such illicit material since each contained a regional forensic state laboratory facility. Second, the population composition has different characteristics for each regional area.

The Central Virginia Region encompasses Richmond, the State Capitol; Hopewell and Petersburg, a chemical and industrial complex; and some rural communities. In contrast, Tidewater, Virginia contains the cities of Norfolk, Virginia Beach, Chesapeake, Hampton, Portsmouth, Williamsburg and Suffolk. These cities have large concentrations of military personnel, tourist, and recreation attractions as well as having one of the major seaports on the East Coast. Also Virginia's Eastern Shore is included as a rural area. Northern

Virginia is similar to Tidewater in that it contains a large population. However, it serves as a bedroom community for Washington, DC. Both Northern Virginia and Tidewater, Virginia have a constant flux in their populations. Western Virginia contains small cities and towns and large rural areas with its population being non-transient in nature.

Duration of Study. Marihuana is not exclusively used by college age students, but usage includes high school students as well as other young adults. With this in mind, to limit the variable in population shifts due to semester trends and the influx of summer tourists into the State, the spring semester was chosen for study. This included the months of January through June of 1976.

Sample Analyses. Each evidence sample was weighted upon arrival in the regional laboratory. When the sample exceeded 10 g the forensic chemist submitted a portion to the central laboratory for analysis. Analyses were completed by the same forensic chemist using the same techniques and instrumentation. Samples were prepared by air drying at room temperature to ambient humidity. The plant material was manicured by passing it through a No. 10, U>S> Standard Sieve (2000 micron openings). Manicuring removed any stems, stalks and sees. A random, five g sample was extracted with 50 ml of n-hexane and analyzed by a gas chromatographic method as modified from that described by Lerner (1969) and Turner and Hadley (1973). Analytical procedures for this internal standard method using n-Triacontane (Applied Science, State, PA) are described in work by O'Rear and Llewellyn presented in this same volume.

Hashish

The resin from this substance contains more THC than the plant material from which it was derived. Normally it is collected mechanically. This more potent form is considerably smaller in volume than the plant material but is used in much the same way. Hashish oil is even a more potent form, also being used in the same manner. it is a solvent extraction from either marihuana or hashish resin.

Two laboratory preparations of hashish oil were prepared during this segment of the study. Preparation A was derived from plant material seized in Northern Virginia. It contained 9.3% THC. The second source (Preparation B) was made from a composite sample from the same area but contained 1.3%

THC. Both preparations were made from 90 g of marihuana extracted using a Soxhlet apparatus for two hours. The volume of solvent n-hexane, was 200 ml. The extract was evaporated to dryness using a warm water bath and a small stream of air. The residue that remained was 14.9 g and 4.5 g respectively for Preparations A and B. This residue was analyzed for delta-9-THC using the same procedures reported above.

RESULTS AND DISCUSSIONS

Approximately 70% of the marihuana samples tested were confiscated in the Central and Tidewater Regions. The Central Region is thought by law enforcement agencies to be a North-South Traffic Corridor for the flow of illicit drugs. The Tidewater Region contains the international seaport having a large military population. As shown in Table 1, the Central Region had the highest mean level of delta-9-THC with the Northern Region having the lowest percentage, 1.66 and 1.26 respectively. The Central Region showed the greatest variability and had a standard deviation of 1.31. Not one of the four regions showed a percentage of delta-9-THC by weight that was significantly different from any of the other three regions. The greatest level of hashish as well as marihuana and their variations also related to the regions having the greatest number of samples. The Tidewater Region had $11.3 \pm 6.6\%$ hashish in the five samples tested. The Western and Northern Regions had no confiscated samples of hashish in the first six months of 1976.

Samples of marihuana for the month of January from the four regions ranged from 0.67 to 1.23% delta-9-THC and had a mean and standard deviation of 1.05 ± 0.63. February showed a mean value of 1.37; March, 1.51; May, 1.50, and June 1.66. The April value, 3.18, was the highest mean for the respective regions during the six month study. Also during the study, the Central Region, having the highest mean percentage of delta-9-THC in its samples, was about 24% higher than the lowest region, Northern Virginia. There was no significant difference between the means for any of the four regions. The overall mean on a state-wide basis was 1.49 ± 1.00 (Table 2).

When the samples of hashish were divided into oil and resin sample types (Table 3), the oil showed $14.5 \pm 9.4\%$ delta-9-THC whereas the resin contained $4.4 \pm 0.4\%$ delta-9-THC. As noted above, only two regions were represented and the total number of samples were very small.

Table 1. Marihuana Survey Showing Sample Distribution, Sample Number, Regional Percentages, and Mean Levels With Standard Deviations for THC and Hashish.

Regions of Virginia	# of Samples Tested	Percentage (%)	X ± SD of THC (% by Wt.)	S ± SD of Hashish (% by wt.)
Central	75	42	1.66 ± 1.31	7.3 ± 3.6
Tidewater	53	29	1.31 ± 0.61	11.3 ± 6.6
Western	36	20	1.51 ± 0.84	----
Northern	16	9	1.26 ± 0.44	----
Total (4)	180	100%	----	----

Table 2. The Mean and Standard Deviation for THC Levels (Percentage by Weight) Found in Illicit Street Samples Seized During the First Six Months of 1976 in the Four Geographic Regions of Virginia.

Geographic Region of Virginia	Months of the Year - 1976						Overall Regional X ± SD Six Months	
	January	February	March	April	May	June		
Central	1.23 ±0.52 (14)	1.29 ± 0.46 (17)	1.43 ± 0.56 (26)	3.78 ± 3.46 (5)	1.90 ± 1.77 (6)	1.84 ± 0.88 (7)	1.66 ± 1.31 (75)	
Northern	1.10 ± 0.92 (3)	1.42 ± 0.29 (6)	--	--	1.33 ± 0.49 (4)	1.32 ± 0.56 (3)	1.26 ± 0.44 (75)	
Tidewater	0.67 ± 0.59 (7)	1.41 ± 0.31 (15)	1.61 ± 0.73 (18)	--	1.11 ± 0.34 (7)	1.13 ± 0.56 (6)	1.31 ± 0.61 (53)	
Western	1.05 ± 0.71 (9)	1.40 ± 0.57 (8)	1.54 ± 0.74 (10)		2.43 ± 1.47 (4)	2.0 ± 0.71 (2)	2.60 ± 2.43 (3)	1.51 ± 0.84 (36)
Monthly and State-wide	1.05 ± 0.63 (33)	1.37 ± 0.41 (46)	1.51 ± 0.65 (54)		3.18 ± 2.70 (9)	1.50 ± 1.06 (19)	1.66 ± 1.13 (19)	1.49 ± 1.00 (180)[a]

[a]No. of samples state-wide for all months and areas studied.

For the laboratory preparations of hashish oils, a sample of marihuana that was 1.3% delta-9-THC and close to the state average potency, yielded a hashish oil of 29.9% delta-9-THC. For a second different preparation, made from a highly potent marihuana sample containing 9.3% delta-9-THC, the resultant oil was found to contain 48.5% delta-9-THC by weight. Even though the difference between delta-9-THC percentage in the starting material was approximately seven times greater for the higher of the two marihuana samples, the second hashish oil preparation yielded oil less than two time the potency for the lower sample.

Table 3. The Distribution of Hashish and Its THC Potency Found in Street Samples Seized in Virginia During the First Six Months of 1976.

Hashish	THC Level (% by Wt.)	Month of Seizure	Geographic Region
	9.8%	April	Central
	15.7%	June	Tidewater
Oil			
	17.7%	May	Tidewater
	14.6%	May	Tidewater
X ± SD	14.5 ± 3.4	-	-
	4.5%	February	Tidewater
Resin	3.9%	February	Tidewater
	4.7%	March	Central
X ± SD	4.4 ± 0.3	-	-

The report of the Secretary of HEW to Congress (1976) states that most users reported occasional use of stronger materials has increased. When this is compared to the state-wide mean for Virginia for the first six months of 1976, we find that 68% of the samples confiscated ranged from 0.5 to 2.5% delta-9-THC. It appears that a Virginia user wanting stronger materials would have had to turn to hashish which seems not to be readily available. Hashish samples represent 4% of the materials tested in this study. This report also claims that the typical marihuana in the USA contains 1-2% delta-9-THC as compared to

10% in hashish. These values compare similarly to those of our study period for Virginia when marihuana had 1.49 ± 1.00 and hashish had 10.3 ± 5.89% delta-9-THC.

According to Nahas (1975) there are two main types of Cannabis plants which are defined according to their concentrations of psychoactive substance (delta-9-THC). The fiber-type plant has a low delta-9-THC content (less than 0.2%) and the drug-type has a high delta-9-THC content of 3.,4 to 4.8%. A one g marihuana cigarette prepared from the drug-type plant contains about 34 to 48 mg of delta-9-THC, a sufficient amount to produce hallucinations and other psychotomimetic symptoms in a healthy person.

The number of samples in this study having less than 0.2% delta-9-THC was about 2%. It is thought that these materials, even if smoked would probably result in little if any response. The samples were distributed as follows: three samples in Tidewater, one in Central and one in the Western Regions. Four out of five occurred in the month of January, and the fifth was in February. January was the month having the lowest potency during our study.

Doorenbos et al. (1971) reports that Mexican Cannabis sativa L. grown in Mississippi provided about 3.7% of delta-9-THC whereas Turkish and other natural USA plants showed generally inactive levels of delta-9-THC. It appears that if this is true then the majority of the confiscated material in our study was imported rather than native material. If not imported, then it may have been produced from imported seeds.

If the same degree of enforcement occurs from region to region then the international seaport is not the main source but may supply 20-30%. The dilution effect on the active component is not observed as the distance from the seaport increases. Northern Virginia Region has a higher population than the other regions and there is a lower level of confiscated samples. This indicates that law enforcement agencies seem to vary greatly in their enforcement policies for the State of Virginia.

In general, it seems that nearly all street samples of marihuana confiscated in Virginia for this historical period, were psychoactive and contained about the national average for this active ingredient. The levels are similar from month to month and region to region, with similar high and low months found throughout the regions, possibly indicating a common supply source but not necessarily the international seaport.

SUMMARY

This historical study for the first six months of 1976 evaluated the occurrence of marihuana, its potency and distribution as based upon confiscated "street" samples taken in Virginia. On a state-wide basis, confiscated samples ranged from 0.5 to 2.5% active material. Virginia marihuana had 1.49% and hashish had 10.3% delta-9-tetrahydrocannabol. Samples collected during January showed the lowest potency. The levels were generally similar from month to month and region to region for this limited time-frame studied. Also, the level of active ingredient did not differ from that reported nationally. The potency of the active ingredient seemed to be routinely found at psychoactive levels.

Acknowledgements

The study was completed while C.E. O'Rear was Director of Forensic Science at the Virginia Division of Consolidated Laboratory Services, Richmond, Virginia. E.R. Harrison, Resident Forensic Chemist, contributed to the project.

REFERENCES

Abelson, H., and Fishburne, P.M., "Nonmedical Use of Psychoactive Substance."Response Analysis Corporation, Princeton, NJ, 1976, p. 30.

Doorenbos, N.J., Fetterman, P.S., Quimby, M.W., and Turner, C.C., "Cultivation, extraction and analysis of Cannabis sativa L.." Annals New York Academy of Science, Vol. 191, 1971, pp. 3-15.

Goodman, L.S., and Gilman, A., The Pharmacological Basis Therapeutics, Macmillan Publishing Co., Inc., NY, 1975, pp. 307-308.

Haney, A., and Bazzaz, F.A., "Some ecological implications of the distribution of hemp (Cannabis sativa L.) in the United States of America." In The Botany and Chemistry of Cannabis, C.R.B. Joyce and S.H. Curey (eds.), J. & A. Churchill, London, 1970, pp. 38-48.

Hollister, L.E., "Marihuana In Man: Three Years Later." Science, Vol. 172, 1971, pp. 21-29.

Lerner, P., "The Precise Determination of Tetrahydrocannabinol In Marihuana and Hashish." Bulletin On Narcotics, Vol. 21, 1969, pp. 39-42.

Nahas, G.G., Marihuana-Deceptive Weed, Raven Press, ny, 1975, pp. 66-69.

Rafaelsen, O.J., Bech, P., Christiansen, J., Christrup, H., Nyboe, J., and Rafaelsen, L., "Cannabis and Alcohol: Effects on simulated car driving." Science. Vol. 179, 1973, pp. 920-923.

Secretary of Health, Education and Welfare, Marihuana and Health: Second Annual Report to Congress, United States Printing Office, Washington, DC, 1972.

Secretary of Health Education and Welfare, Marihuana and Health: Sixth Annual Report to Congress, United States Printing Office, Washington, DC, 1976.

Turner, C.E., and Hadley, K., "Constituents of Cannabis sativa L. II: Absence of Cannabidiol in an African Variant." Journal of Pharmaceutical Science, Vol. 62, No. 2, 1973, pp. 251-255.

DIVERGENT INDICATOR BACTERIA STANDARDS AND CRITERIA FOR WATER QUALITY RELATED TO PUBLIC HEALTH

ERNST M. DAVIS*, Department of Environmental Science, University of Texas, School of Public Health, Houston, TX 77225, USA

RUFUS K. GUTHRIE, Department of Human Ecology, University of Texas, School of Public Health, Houston, TX 77225, USA

INTRODUCTION

Bacterial indicators of water quality have been developed over previous decades as tools to be used in decision-making processes for the protection of public health. These indicators serve the purpose of relating the degree of risk that a particular body of water holds regarding human contact, to the ingestion of the water or consumption of food species taken from the water. For the most part, they only suggest that pathogenic organisms may be present. In total, all of these can be considered "aquatic toxicants as biodeteriogens affecting human health." If present in high numbers the water may be judged to be unfit for its designated use. Nevertheless, no indicator group in use today can yield information with absolute certainty as to the sanitary quality of the water. To complicate matters further, there are several criteria and/or standards for assessing fresh water and marine water quality, depending on specific preestablished uses.

COLIFORM BACTERIA INDEXES

The first proposed federal criteria for water quality appeared in 1968 (U.S. Environmental Protection Agency, 1986) as a result of reevaluation of studies that were conducted in the late 1940's and early 1950's. The National Technical Advisory Committee (NTAC) of the Department of the Interior was

Biodeterioration Research 4, Edited by G.C. Llewellyn *et al.*, Plenum Press, New York, 1994

responsible for this development. The earlier results of epidemiological studies showed a highly significant correlation of illness in swimmers when the coliform geometric mean was 2,300/100 mL and no significant correlation when the count was 43/100 mL. This water quality index was refined into a fecal coliform index in the mid-1960's by using the ratio of fecal coliforms to coliforms from the same areas that were tested earlier. The NTAC recommended that the fecal coliforms were more "fecal specific" and less subject to variations than the total coliform group. Based on an 18% concentration of fecal coliforms in the total coliform mean of 2,300/100 mL, a level of 400/100 mL was established. Since a detectable risk was undesirable, one-half the density was proposed, or 200/100 mL.

On that basis, the contact recreational criterion was established at 200 fecal coliforms/100 mL by the most probable number (MPN) or membrane filter tests, with a minimum of five samples taken in not more than 30 days, and no more than 10% of the samples in the same period could exceed 400/100 mL. These were log mean values. The fecal coliform group includes species of the genera Escherichia, Enterobacter, Klebsiella, and Citrobacter (Standard Methods, 1989).

The State of Texas currently has both contact and non-contact fecal coliform criteria for surface waters (Texas Water Commission, 1991). For contact recreation waters the fecal coliforms shall not exceed a geometric mean of 200/100 mL with no fewer than 5 samples taken in a 30-day period, nor shall there be more than 400/100 mL in more than 10% of the samples. If ten or fewer samples are analyzed, no more than one sample shall exceed 400/100 mL. The non-contact recreation criteria read the same except the number for fecal coliforms is increased by one order of magnitude. Presumably, both the MPN and membrane filter techniques can be used for this evaluation although the Texas Water Commission (TWC) uses the membrane filter method for their evaluation of surface waters. The TWC defines fecal coliforms as, "that portion of the coliform group which is present in the intestinal tracts and feces of warm blooded animals." More discussion of this definition will follow.

Shellfish Criteria

The Texas Department of Health exercises jurisdiction over bay waters from which shellfish are harvested. Of primary importance are the oysters

(Thompson, 1991a,b). Prior to 1975, total coliforms were the indicator group of choice for determining whether oyster beds could be harvested. Since 1975, fecal coliforms have been used. The MPN method is used for this assessment. Based on decades of experience, closures or reopening designated areas is done based on rainfall duration and quantity data. This workable scheme also included decades of coliform data accompanying the rainfall data (Lowry, 1991). The agency can also close harvest areas due to unusual meteorological, biological or chemical pollutant spills. Closures by using rainfall data are based on the almost certainty that the fecal coliform count will exceed the criteria.

The criterion currently used for shellfish is a fecal coliform MPN density of 14/100 mL with no fewer than 5 samples taken per year in approved areas, or one sample per month in conditionally approved areas. This is to be a geometric mean of all samples. Fewer than 10% of all samples taken may exceed 43/100 mL. More than 15 samples must be taken at the same location in less than 3 years, and at least 5 samples must be taken in one year.

Classification of shellfish growing areas includes the areas that are approved, conditionally approved, or closed, due to persistently high coliform densities. Areas in estuaries are subjected to a four-part evaluation: 1) a pollution source identification based on types, quantities, points, and other parameters; 2) hydrographic factors such as volume, depth and current; 3) meteorological assessment of effects on pollutant sources; and 4) bacterial enumeration at selected stations to determine if the pollution extends beyond potential pollution areas identified in the first three categories above. A complete reevaluation of the entire bay system is done every 12 years with a classification reevaluation done every three years (Thompson, 1991b). This may appear to be somewhat like a speculative approach, especially since the currents, winds, and other factors change frequently in estuaries. Also, these changes can occur well within the time it takes to do the bacterial analyses. Nevertheless, the entire shellfish industry is subject to these restrictions. Judgment calls have had be based on prior experience, and the method has proven to be reliable.

Experience at determining pollution levels on both continuous and intermittent bases has most assuredly been a plus for public health protection. Recent counts of shellfish growing acreages show Texas, for example, to have 727,941 acres of approved (for harvesting) acres, 570,045 acres that are conditionally approved (criteria not met at all times), and 328,500 acres that are prohibited to harvesting (Broutman and Leonard, 1988). Seasonal variations of bacterial indicators and species also play an important part in oyster harvesting (Paille et al., 1987). Temperature of bay waters apparently plays a role in the predominant species present. _E. coli_ was the predominant species when the temperature was below 22°C whereas _Klebsiella spp._ constituted the higher numbers during warmer months. The _Klebsiella spp._ appeared to be of environmental rather than sewage origin. High levels of non-_E. coli_ fecal coliforms found in oysters during summer months from water which met the criteria "were not indicative of sewage pollution."

One of the problems inherent to the testing of shellfish growing waters is the MPN test itself. Whereas a plate count for fecal coliforms gives a number derived by counting the number of colonies on a plate, the MPN procedure may yield a higher number because it is a statistical evaluation (_Standard Methods_, 1989). Each MPN value also has a 95% confidence level. For example, for shellfish waters, an MPN of 14/100 mL can have an actual number between 6.0 and 35/100 mL. And an MPN of 170 has a range of 70-480/100 mL. Therefore, it is difficult to determine when the waters actually are in compliance.

Proposed and New Criteria and False Positives

The fecal coliform test has been criticized in recent years because at least one member of the coliform group, _Klebsiella_, can be from non-fecal sources (U.S. Environmental Protection Agency, 1986). The enterococci were shown to have a better correlation with swimming-associated illnesses: 0.75 versus -0.01 for fecal coliforms. The proposed criterion for enterococci for fresh recreational waters is 35/100 mL. Each is based on a geometric mean of at least five samples per 30-day period (_Standard Methods_, 1989; Kilgen, 1990). New drinking water coliform criteria (_Federal Register_, 1989) state that "public water systems need only determine the presence or absence of total coliforms; a determination of total coliform density is not required." This may complicate matters in the future because several species of coliforms that are of non-fecal origin can give false-positive results (Geldreich and Rice, 1987; Geldreich et al., 1989; U.S.E.P.A., 1986; Kilgen et al., 1988; Manville, 1991; Neilson, 1978).

A dilemma recently has plagued at least one regulatory agency. Recall the reference citation for the definition of fecal coliforms as having a warm-blooded animal source. Many species can be of environmental origin, especially some in the genera Klebsiella and Aeromonas (Geldreich and Rice, 1987). Klebsiella have been found on fruit and vegetables, wood pulp, cotton, tree tissues, in paper mills, textile plants and sugar processing. These probably were not of "warm-blooded animal feces" origin.

One unlikely source of high apparent fecal coliforms has been shown in effluent waters from catfish farms (Bennett and Gordon, 1982). Those were shown to have been Aeromonas, Klebsiella, Enterobacter and Citrobacter. It goes unsaid that catfish are not warm-blooded animals. Doland and Geldreich (1966) reported that channel catfish had a fecal coliform MPN range of 3,300 to 1,090,000 per gram of intestinal material and a fecal streptococci range of 3,300 to 240,000 per gram. Recent work by the senior author of this paper on water in a catfish farm in central Texas showed fecal coliform concentrations through the raceways from 5/100 mL to 5,900/100 mL and fecal streptococci from $<$10/100 mL to 3,500/100 mL.

Stormwater runoff is another source of high coliform concentrations (Field and Pitt, 1990). Relationships that have been developed, over years of research on this subject, between fecal coliform counts and pathogenic bacteria in other waters do not always hold for stormwater runoff. Pathogens such as Pseudomonas aeruginosa, not recorded by the fecal coliform test, can be found in runoff waters in high concentrations. And it is highly questionable whether all of the fecal coliforms and other species that may be pathogenic, i.e., Pseudomonas, are of fecal origin. Certainly, high humus soil can be a good source of many of these bacteria. Ubiquitous bacteria species that show positive on membrane filter fecal coliform plates somehow tend to defeat the purpose and intent of the sewage origin concept that led to the adoption of the fecal coliform test in the first place.

VIBRIO BACTERIA AND PUBLIC HEALTH

Only in recent years have the Vibrios been recognized as a real, not perceived, threat to the public health. Marine waters in particular, contain these pathogens. Recent outbreaks of shellfish-borne diseases have been related to Vibrio cholerae, V. vulnificus, and V. parahaemolyticus. These organisms have no correlation with fecal coliform bacteria counts (Broutman and Leonard, 1988). Vibrio mimicus has also caused fatal septicemia (Guthrie, 1991). Ingestion of uncooked shellfish may not be the only route of infection of Vibrio species. Swimmers with open wounds are also at risk.

The current wholesale market criterion for total aerobic plate counts in shellstock and processed raw oysters is ≤500,000 organisms per gram of meat. The MPN fecal coliform limit for oyster meat has been established at 230/100 grams (Kilgen et al., 1988). These authors also reported that when the vibrios were found in oysters and overlying waters, no Salmonella spp. were found. Salmonella spp. cases apparently are not reportable and estimates of the number of cases of this gastrointestinal disorder range to more than 2 million a year (Guthrie, 1991). Vibrios have been found in raw oysters even when those oysters have been taken from approved areas in Galveston Bay, Texas. Counts of these Vibrio spp. in overlying waters have ranged from 10 to 1,000/mL (Morgan and Guthrie, 1991). The problem with the existence of vibrios is not to be taken lightly. In Texas, during 1990, 25 culture-confirmed Vibrio infections were recorded with five deaths occurring among those 25 cases. Ten of the 25 cases were from seafood exposure (where four of the five deaths occurred). Nine cases out of the 25 were related to water exposure (swimming, windsurfing or fishing) and six cases were from other or unknown exposures. Vibrio species associated with the 25 cases were cholerae non-01, vulnificus, parahaemolyticus, fluvialis, and mimicus (Taylor, 1991). The neighboring state of Louisiana had 32 cases of vibrio recorded for 1990 (McFarland, 1991). Species associated with the 32 cases and number of cases per species were: V. cholerae 01 (3), V. fluvialis (1), V. hollisae (3), V. mimicus (5), V. cholerae non-01 (11), V. parahaemolyticus (6) and V. vulnificus (3). Apparently two additional cases occurred out of the state in persons who consumed Louisiana oysters. Of those two cases, one fatality occurred.

SUMMARY

Just exactly what all of these criteria mean is left up to the interpretation of the individual. The vast majority of the public has absolutely no idea that such things as these criteria exist. Apparently no regulatory agency in Texas is willing to post notices around bay areas known for persistent high fecal coliform counts or, for that matter, in areas where known treated sewage discharges exist. The establishment of the guide for sanitary water quality has, no doubt, made a significant positive impact on public health. Nevertheless, more meaningful indicators are needed for both freshwater and marine waters. Of course a fixed, firm and scientifically sound approach could resolve all of these problems. Simply stated, when it comes to seafood consumption, absolutely do not consume the product in the raw state. In the case of oysters the history of eating raw ones on the half-shell might take several decades to educate the public. And there are always some people who harvest them themselves and would never change their habits. Therefore, in the realm of indicator bacteria, false positives, pathogens that are not recorded by standard tests, and the like, the future is not predictable at this point in time.

Acknowledgments

A note of respect and appreciation for Dr. Rufus Guthrie, who passed away on November 5, 1992. Thanks are extended to Ms. Margaret Donnellan for manuscript preparation and patience with the senior author's writings, to Dr. John J. Mathewson, Center for Infectious Disease for manuscript review, and to Mrs. Gay Robertson for editorial assistance.

REFERENCES

Bennett, D., and Gordon, J. (1982). Sanitary significance of fecal coliforms in the effluents from catfish production raceways. Proc., 37th Annual Purdue Industrial Waste Conference, Lafayette, IN.

Broutman, M.A., and Leonard, D.L. (1988). The quality of shellfish growing waters in the gulf of Mexico. National Oceanic and Atmospheric Administration, U.S. Department of Commerce, Rockville, MD.

Davis, E.M. Unpublished data (1992). University of Texas, School of Public Health, Houston TX.

Doland, R.J., and Geldreich, E.E. (1966). Sanitary significance of fecal coliforms in the environment. U.S. Department of the Interior, Federal Water Pollution Control Administration, Cincinnati, OH.

Federal Register (June 29, 1989). Drinking water; national primary drinking water regulations (including fecal coliforms and E. coli) final rule. 54FR 27544. U.S. Environmental Protection Agency (1986).

Field, R., and Pitt, R.E. (1990). Urban storm-induced discharge impacts: U.S. Environmental Protection Agency Research Program Review. Wat. Sci. Tech. 22 (10/11), 1-7.

Geldrich, E.E., and Rice E.W. (1978). Occurrence, significance, and detection of Klebsiella in water systems. Res. Technol. J. AWWA, 74-80.

Geldrich, E.E., Goodrich, J.A., and Clark, R.M. (1989). Strategies for monitoring the bacteriological quality of water in distribution systems. Pres. AWWA Annual Conf., Los Angeles, June 18-22.

Guthrie, R.K. (1991). Personal communication. University of Texas, School of Public Health, Houston, TX.

Kilgen, M.B., Cole, M.T., and Hackney, C.R. (1988). Shellfish sanitation studies in Louisiana. J. Shellfish Res., 7(3), 527-530.

Kilgen, M.B. (1990). Personal communication. Proj. Dir., N.I.S. National Shellfish Indicator Study, Nicolls St. Univ., LA. Epidemiology Workshop Summary.

Lowry, M. (1991). Personal communication. Shellfish Sanitation Program, Texas Department of Health, Houston, TX.

McFarland, L. (1991). Personal communication. Epidemiology Section, Office of Public Health, Dept. of Health and Hospitals, New Orleans, LA.

Manville, D. (1991). Identification of bacterial populations which produce elevated MPN values in wastewater streams. Master of Science Thesis. University of Texas, School of Public Health, Houston, TX.

Morgan, G.R., and Guthrie, R.K. (1991). Unpublished data. University of Texas, School of Public Health, Houston, TX.

Neilson, A.H. (1978). The occurrence of aeromonads in activated sludge: isolation of Aeromonas sorbia and its possible confusion with E. coli. J. Appl. Bact. 44, 259-264.

Paille, D., Hackney, C., Reilly, L., Cole, M., and Kilgen, M. (1987). Seasonal variation in the fecal coliform population of Louisiana oysters and its relationship to microbial quality. J. Food Protection, 50(7), 545-549.

Standard Methods for the Examination of Water and Wastewater (1989). APHA, WPCF, AWWA. 17th Ed., Washington, D.C.

Taylor, J. (1991). Personal communication. Epidemiology In Texas, Annual Report. Texas Dept. of Health, Austin, TX.

Texas Water Commission (1991). Texas surface water quality standards. Texas Administrative Code, Chapter 307(307.2-307.10), Austin, TX.

Thompson, R.E. (1991a). History and status of the Texas Molluscan Shellfish Regulatory Program, In: Proc. Galveston Bay Characterization Workshop

(F. Shipley and R. Keisling, eds.), Galveston Bay National Estuary Program, GBNEP-6.

Thompson, R.E. (1991b). Personal communication. Texas Dept. of Health, Austin, TX.

U.S. Environmental Protection Agency (1986). Ambient water quality criteria for bacteria - 1986. EPA 440/5-84-002, Washington, DC.

REVIEW ARTICLE ON THE INFLUENCE OF DISSOLVED OXYGEN ON SULFATE-REDUCING BACTERIA RELATED CORROSION

WHONCHEE LEE* and **WILLIAM G. CHARACKLIS**, Center for Interfacial Microbial Process Engineering, Montana State University, Bozeman, MT 59717, USA

INTRODUCTION

Corrosion by sulfate-reducing bacteria (SRB) has been intensively studied during the last 40 years, but until now the importance of oxygen in SRB-related corrosion has rarely been emphasized (Hardy and Bown, 1984; Starkey, 1985; Hamilton, 1990; Hamilton, 1991). The impact of oxygen on SRB-related corrosion is attributed to a direct effect on the sulfur-related corrosion products rather than to any stimulation of SRB activity (Hamilton, 1990; Hamilton, 1991). Pitting corrosion is the characteristic mode of attack and deep pit is usually found underneath a porous corrosion products. However, the role that oxygen plays in the aerobic/anaerobic environments in relation to corrosion has not been clearly defined. The system is complex and dynamic. The role of SRB must be viewed in the context of biological consortia (biofilms) and/or mixed ecosystems. In addition to the biological factors, the chemical environments which influence corrosion are also complicated by the introduction of oxygen. The following statement is quoted from Starkey. " Factors that have been suggested or may be concerned with anaerobic corrosion relate particularly to the effect of ferrous sulfide, sulfur, ferrous hydrate and all other products of the corrosion process, differential aeration cells, and various combinations of all of those factors". In this review, we intended to focus on aspects of experimental systems that more accurately reflect those environmental conditions generally associated with corrosion in the field. Describing the role of dissolved oxygen on SRB-related corrosion, we will summarize the current published papers which describe the

Biodeterioration Research 4, Edited by G.C. Llewellyn
et al., Plenum Press, New York, 1994

corrosion of mild steel underneath aerobic biofilms containing SRB (Lee and Characklis, 1990; Lee et al. 1992). Finally, we will discuss various experimental approaches in an attemp to elucidate the true mechanism of SRB-related corrosion in aerobic environments.

THE SYSTEM

The experimental system is similar to the model proposed by Hamilton (1985) (Figure 1). The system consists of four compartments: 1) the gas, 2) the bulk water, 3) the biofilm, and 4) the substratum. Each compartment in the system may contain several phases. For example, the biofilm consists of at least four phases: interstitial water, the microorganisms, the extracellular polymer substance (EPS), and solid corrosion products. The substratum generally contains adherent corrosion products (iron oxides and iron sulfides) and bulk metal.

(1) Gas - The gas phase will supply gaseous components for microbial transformation such as oxygen. On the other hand, oxygen may accelerate corrosion of mild steel although this process is limited by the diffusion of oxygen to the metal surface.

(2) Bulk water - The bulk water contains almost all the nutrients needed for microbial growth in the biofilm. The bulk water also introduces corrosive species such as dissolved oxygen and chloride to maintain a corrosion current.

(3) Biofilm - The biofilm contains a number of microbial species in a matrix of extracellular polymeric materials and corrosion products. Thus, biofilms may substantially change the local chemistry of the adjacent metal due to microbial transformation and mass transfer processes (Characklis et al., 1990). This will result in concentration gradients and give rise to microenvironments and functional heterogeneities. These biofilm properties therefore, offer a ready explanation for the presence of SRB at the metal/biofilm interface, where the bulk water may be aerobic.

(4) Substratum - The substratum has at least one characteristic which may influence microbial corrosion: inclusions. It is known that inclusions are related to the initiation of localized corrosion under active SRB environments (Lee and Characklis, 1990). Mild steel is the major concern in this review due to its wide application in industrial water systems.

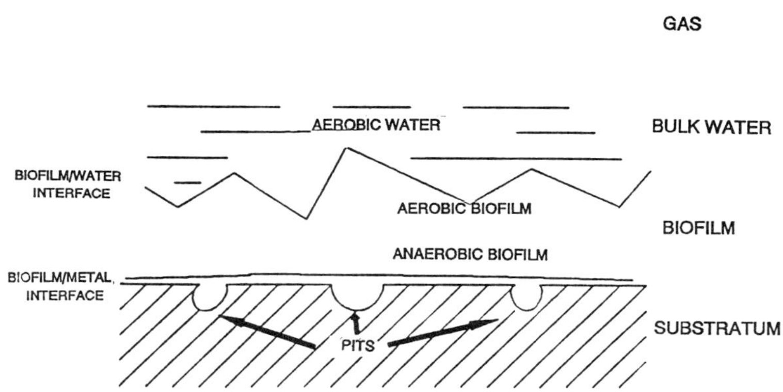

Figure 1. The Microbial Corrosion System Includes the Following Four Compartments: 1) Substratum, 2) Biofilm, 3) Bulk liquid, and 4) Gas.

THE ROLE OF DISSOLVED OXYGEN ON SRB-RELATED CORROSION

It has been demonstrated that SRB can proliferate within biofilm even when the bulk water is air-saturated (Lee et al.and Lee, 1992). The following events that occur on corroding steel during biofilm development. During the early stage of biofilm accumulation, aerobic process dominated corrosion and iron hydrates are the major corrosion products. The yellowish ferric hydrates are underlied by black sulfide-containing products after SRB developed. Elemental sulfur, pyrite, and mackinawite or greigite are detected within the fouling deposits in addition to iron hydrates. There is little corrosion under completely anaerobic conditions where there is an active development of SRB (Lee and Characklis, 1990). The role of dissolved oxygen on SRB-related corrosion is either indirectly through the corrosion products formation or directly where oxygen acts a cathodic depolarizer.

1. Ferrous or ferric hydroxides - The iron hydroxides are more easily converted into sulfide than the metal base. Ferrous sulfide, as mackinawite, cannot cause continued corrosion unless fresh sulfide is continuously produced and contact the metal surface. The mechanism of corrosion of mild steel by ferrous sulfide may be explained by 1) depolarization of cathodic areas by the adsorption of the polarizing hydrogen into the crystal lattice of the ferrous sulfide, or 2) the establishment of a galvanic cell, FeS/Fe, whereby the iron becomes anodic and hydrogen evolution occurs on the ferrous sulfide (King and Wakerley, 1973). The rate-controlling process is the hydrogen evolution reaction which is limited by the accumulation of atomic hydrogen within the iron sulfide lattice.

2. Elemental sulfur - The oxidation of H_2S by oxygen and/or iron oxides to produce elemental sulfur (Schmitt, 1991).

$$2H_2S + O_2 \rightarrow 2H_2O + 2S \qquad (1)$$

$$3H_2S + 2FeO(OH) \rightarrow 2FeS + S + 4H_2O \qquad (2)$$

Elemental sulfur can lead to considerable weight losses and shallow pitting of plain carbon steel. The corrosion rate caused by the elemental sulfur may be as high as several hundred mpy (Schmitt, 1991). The main features of the process were established by Kuster et al. (1984). Iron sulfide is formed as a corrosion product, and the rate-controlling process is the reduction of sulfur on the conductive iron sulfide

$$S + 2H^{+} + 2e^{-} \rightarrow H_2S \qquad (3)$$

Subsequently, Fe^{+2} from the anodic process reacts with the H_2S to form FeS

$$Fe^{+2} + H_2S \rightarrow FeS + 2H^{+} \qquad (4)$$

Thus the corrosion process is autocatalytic. Since the direct contact of elemental sulfur with iron sulfide is an essential prerequisite for the enhanced corrosion (MacDonald et al., 1978). The electrical conductivity of the fouling deposits is the key factor for this mechanism to be activated.

3. Pyrite - Oxygen may enhance the formation of pyrite within the fouling deposits. Pyrite has been shown to be more corrosive to steel than mackinawite in abiotic conditions by 1) a proton conduction mechanism; and/or by 2) a hydrogen absorption mechanism. It has been suggested that pyrite (FeS_2) formed at room temperature is a p-type (iron deficient) semiconductor while mackinawite ($Fe_{1-x}S$) is a n-type (sulfur deficient) semiconductor (Vaughan and Craig, 1978). Thus pyrite may make hydrogen ions more available to the metal surface through a proton conduction (Martin and Annand, 1981). On the other hand, Mara and Williams (1972) suggested that iron sulfides absorb hydrogen (evolved at cathodic sites on the mild steel surface) in proportion to the number of cationic defects in the iron sulfides. Pyrite absorbs hydrogen slowly as compared to mackinawite and thus causes more corrosion. The rate-controlling process is the hydrogen evolution reaction which is limited by the accumulation of atomic hydrogen within the iron sulfide lattice.

4. Oxygen - Oxygen acts as a cathodic depolarizer. In general, corrosion of mild steel in aerobic bulk water is completely governed by oxygen diffusion. However, local microbial activity (including SRB) may exert a decisive influence on localized corrosion, such as pitting and/or crevice corrosion. In this case, the cathodic area and hence the oxygen diffusion current is very large in relation to the active anode area. The localized corrosion rate is raised by an increase of the diffusion controlled rate of oxygen reduction.

In summary, direct contact of sulfide compounds with the metal surface is required to accelerate corrosion in a biofilm system. The rate-controlling process in all the suggested mechanisms is transport dependent.

SUMMARY AND FUTURE STUDIES

This review has highlighted the possible mechanisms of dissolved oxygen on the SRB-related corrosion in a system which is closed to the industrial water system. The detrimental effect of SRB will not show up during corrosion unless direct contact of sulfide compounds with the metal surface occurs. Obviously the physical and chemical properties of sulfide compounds play a key role in corrosion under aerobic/anaerobic conditions. Development or adaptation of advanced surface analytical techniques such as small spot x-ray photoelectron spectroscopy (XPS) and/or secondary ion mass spectroscopy (SIMS) is necessary to illustrate the spatial distribution of different sulfides within fouling deposits and their correlations with pits or crevice corrosion. Further studies such as the dynamic interaction of oxygen with iron sulfides within fouling deposits may complement the results from surface analysis and may identify the cathodic reactant(s). The role of SRB is also complex and must be viewed in the context of "patchiness" in terms of microbial activity. Their ability to grow by using hydrogen deserves more study in order to determine the true corrosion mechanism. In conclusion, the proposed studies shall determine the environmental conditions under which SRB can initiate and/or accelerate corrosion especially in the form of localized attack.

Acknowledgements

The support of the National Science Foundation, Engineering Research Center Program, and Center Industrial Associates is gratefully acknowledged.

REFERENCES

Characklis, W.G., Turakhia, M.H., Zelver, N., and Marshall, K.C. (1990). Process rates, In: Biofilms. pp. 193-341 (W.G. Characklis and K.C. Marshall, ed.) New York, John Wiley & Sons.

Hamilton, W.A. and Maxwell, S. (1986) Biological and corrosion activities of sulfate-reducing bacteria within natural biofilms. In: International Conference on Biologically Induced Corrosion, pp. 131-136, (S.C. Dexter, ed.) Houston, NACE.

Hamilton, W.A. (1991). Sulfate-reducing bacteria and their role in microbially influenced corrosion. In: Microbially Influenced Corrosion and Biodeterioration, pp. i-iv, (N.J. Dowling, M.W. Mittleman, and J.C. Danko, ed.), Knoxille, Tennessee.

Hamilton, W.A. (1991). Sulfate-reducing bacteria and their role in biocorrosion, In: Biofouling and Biocorrosion in Industrial Water Systems, pp. 187-195 (H.C. Fleming & G.G. Geesey, ed.), Berlin, Springer-Verlag.

Hardy, J.A. and Bown, J.L. (1984). The corrosion of mild steel by biogenic sulfide films exposed to air. Corrosion, 40, 650-654.

King, R.A. and Wakerley, D.S. (1973). Corrosion of mild steel by ferrous sulfide. Br. Corros. J., 8, 41-45.

King, R.A., Miller, J.D.A., and Smith, J.S. (1973). Corrosion of mild steel by iron sulfides. Br. Corros. J., 8, 137-141.

Kuster, K., Schlerkmann, H., Schmitt, G., and Steinmetz, D. (1984). Werkstoffe and Korrosion, 35, p.556.

Lee, W. and Characklis, W.G. (1990). Corrosion of mild steel under an anaerobic biofilm. Corrosion'90, Paper No. 126, Houston, NACE.

Lee, W., Lewandowski, Z., Okabe, S., and Characklis, W.G. (1992). Corrosion of mild steel underneath aerobic biofilms containing sulfate-reducing bacteria. Corrosion'92, Papaer No. 190, Houton, NACE.

Lee, W. (1992). Unpublished data. Center for Interfacial Microbial Process Engineering, Montana State University, Bozeman, MT.

MacDonald, D.D., Roberts, B., and Hyne, J.B. (1978). Corrosion Science, 18, p. 411.

Mara, D.D. and Williams, D.J.A. (1972). The mechanism of sulfide corrosion by sulfate-reducing bacteria. Biodeterioration of Materials, 2, 103-113.

Martin, R.L. and Annand, R.R. (1981). Accelerated corrosion of steel by suspended iron sulfides in brine. Corrosion, 36, 297-301.

Moosavi, A.N., Pirrie, R.S., Hamilton, W.A. (1990). Effect of sulfate-reducing bacteria activity on performance of sacrificial anodes. In: Microbially Influenced Corrosion and Biodeterioration, pp. 3-13, (N.j. Dowling, M.W. Mittleman, and J.C. Danko, ed.), Knoxville, TN.

Schmitt, G. (1991). Effect of elemental sulfur on corrosion in sour gas systems. Corrosion, 47, 285-308.

Starkey, R.L. (1986). Anaerobic corrosion - perspectives about causes. In: International Conference on Biologically Induced Corrosion, pp. 3-7, (S.C. Dexter, ed.) Houston, NACE.

Vaughan, D.J. and Craig, J.R. (1978). Electrical and magnetic properties of sulfides, In: Mineral Chemistry of Metal Sulfides, pp. 93-95 (D.J. Vaughan and J.R. Craig, ed.), Cambridge, Cambridge University Press.

EVALUATION OF STAINLESS STEEL SUSCEPTIBILITY TO MICROBIOLOGICALLY INFLUENCED CORROSION: AN UPDATED REVIEW

D.A. MORENO, J.R. IBARS, C. RANNINGER, Departamento de Ingeniería y Ciencia de Materiales, Escuela Técnica Superior de Ingenieros Industriales, Universidad Politécnica de Madrid, José Gutiérrez Abascal- 2, 28006 Madrid, Spain
M.F.L. DE MELE, H.A. VIDELA, Bioelectrochemistry Section, INIFTA, University of La Plata, Suc. 4, CC 16, La Plata 1900, Argentina

INTRODUCTION

Stainless steels (SS) are frequently used in a variety of industrial applications, where good corrosion resistance is needed. SS are usually in contact with waters of very dissimilar nature that generally, are favorable environments for microbial growth. Thus, they are susceptible to microbiologically influenced corrosion MIC (Pope et al., 1989).

Several disciplines, such as metallurgy, inorganic and organic chemistry, electrochemistry, biochemistry and microbiology, are involved in the study of MIC of SS. Among the various forms of corrosion induced by microorganisms, the most frequent are localized types of attack, such as pitting, crevice, stress corrosion cracking and corrosion fatigue (Videla, 1991). MIC mechanisms are very complex, and up to the present they are not fully understood. However, recent advances in corrosion studies allowed to reach practical solutions for several common industrial problems (Videla, 1989a).

Costly localized corrosion failures of SS components, caused by microorganisms, have occurred at dissimilar locations and environments, such

Biodeterioration Research 4, Edited by G.C. Llewellyn *et al.*, Plenum Press, New York, 1994

as waste water treatment plants, power generation and chemical processing plants (Pope, 1986; Borenstein, 1988; Zamanzadeh et al. 1989).

Different case histories of MIC, involving SS of different composition, in contact with different water environments have been reported (Stoecker and Pope, 1986; Borenstein and Lindsay, 1987; Scott and Davies, 1989). Although low chloride levels in water (20 ppm or less) should not be corrosive, it has been demonstrated (Stoecker and Pope, 1986) that, when microbial activity is present in such environments, corrosion could occur.

High alloy SS, which are thought to be relatively immune to aggressive environments, have been used to construct shell and tube heat exchangers for cooling sulfuric acid with brackish and seawater with excellent results. However, a case of UNS NO8904 (904 L) SS tube failures within two weeks of start up, accompanied by pitting attack of welds and crevice corrosion has been recently reported (Scott and Davies, 1989). Sulphate-reducing bacteria (SRB) were found, and appeared to alter the entire surface of the SS, etching the grain boundaries and creating small pock marks on the surface.

Another MIC case of widely used stainless steels, such as UNS S30400, S30403 and S31600 SS, corresponded to pipes from power plants showing pitting failures at or adjacent to welds (Borenstein and Lindsay, 1987; Borenstein, 1988). Generally, pits developed as large subsurface cavities, with minute pinholes at the surface. The action of iron-oxidizing bacteria typically results in brown or reddish-brown deposits of conical or hemispherical shape.

Nuclear power plants are particularly prone to MIC, as a result of their basic design. One of the major differences with other industries, is that SRB do not appear to be the most prevalent microbiological contaminant (Licina, 1989). When present, SRB produce black deposits of iron sulfide on steel and stainless steel (Sinha et al., 1991). A rounded gouge or pit, and black iron sulfide at the center, surrounded by a dark outer ring and a bluish inner ring with shiny metal beneath, are generally associated with SRB attack of SS (Licina, 1989). Iron-oxidizing bacteria typically result in brown or reddish-brown deposits. Through-wall perforations of SS caused by **Gallionella** or **Siderocapsa** generally result in rust colored streaks going up and down from the attacked area. Table 1 summarizes different studies on MIC of SS. In each case, the type of study, experimental technique used, kind of SS, kind of specimen , medium, microorganisms involved and corresponding references are cited.

A brief comment on the advantages and limitations of several electrochemical, metallurgical, microbiological and surface analysis techniques for assessing SS susceptibility to MIC is also presented.

ELECTROCHEMICAL TECHNIQUES TO EVALUATE SS SUSCEPTIBILITY TO MIC

The use of electrochemical techniques for assessing MIC has been recently reviewed (Dexter et al. 1991; Mansfeld and Little, 1991). According to the magnitude of the applied perturbation used, three general types of electrochemical techniques can be distinguished: a) without an external perturbation: corrosion potential vs. time evolution and reduction-oxidation potential; b) with minimal perturbations: polarization resistance methods or electrochemical impedance spectroscopy (EIS); c) with large perturbations: potentiostatic and potentiodynamic polarization, determination of pitting or breakdown potentials.

Electrochemical techniques without external perturbation to the system

The "corrosion potential" or "open-circuit potential", of a corroding metal can be measured by determining the voltage difference between the metal immersed in a corrosive medium and a proper reference electrode. The magnitude and sign of the corrosion potential are functions of the metal itself, and the composition, temperature and hydrodynamics of the electrolyte. It is important to indicate the sign of the potential and kind of reference electrode used. The corrosion potential may be measured using a potentiometric circuit, a high-impedance voltmeter or an electrometer, without drawing enough current to change the potential of either the reference electrode or the working electrode (Dexter et al., 1991). Several papers have been reported recently (Scotto et al., 1985, 1986; Johnsen and Bardal, 1986; Dexter and Gao, 1988; Videla et al. 1989; Little et al. 1990, 1991b) using variations in the corrosion potential with time to assess the effects of developing biofilms on the corrosion of SS immersed in seawater. The corrosion potential of a wide variety of SS, are often shifted in the positivedirection (through more noble potentials) when a biofilm is present on the metal surface (Dexter and Gao, 1988; Guezennec et al., 1988; Videla et al., 1989; Mollica et al., 1989), and in this way, SS become more susceptible to pitting attack. If some form of localized corrosion is in progress, the amount of scatter of the corrosion potential increase dramatically in the presence of the biofilm.

The "reduction-oxidation (redox) potential", is useful in predicting whether or not a given metal will suffer an increase in corrosion due to microorganisms (Dexter et al., 1991). This potential refers to the relative potential of an electrochemical reaction under equilibrium conditions where there is no net flow of electrical current. Thus, the redox potential of a certain aqueous environment, is actually the equilibrium potential of the oxygen reaction as measured on a platinum electrode immersed in the medium. For active-passive metals like SS, that frequently corrode by pitting, the combination of positive (noble) redox potential for the environment, with a noble corrosion potential and a negative (active) pitting potential (Ep) can indicate high aggressiveness of the medium for the alloy (Zhang et al. 1989; Ortiz et al., 1990).

Other electrochemical techniques that don't use external perturbation of the system like split-cell measurements (Little et al. 1986) or electrochemical noise (Dexter et al., 1991; Mansfeld and Little, 1991) have been successfully used to study MIC of different metals and alloys. Notwithstanding this, their use for assessing SS susceptibility to MIC still remains elusive.

Electrochemical techniques applying minimal perturbations to the system

The polarization resistance or linear polarization resistance technique is based on the nature of the linear relationship between changes in the applied potential and the resultant current density, when the applied potentials are within about +/- 10 mV of the corrosion potential (Dexter et al., 1991). The slope of the potential current curve is approximately linear and has units of resistance. The main advantage of this technique is the possibility to continuously monitor the instantaneous corrosion rate of a metal exposed to a corrosive environment (Mansfeld and Little, 1991). However, in practical situations, electrochemical systems are more complex than the simple model assumed by the polarization resistance method. For instance, metal surfaces immersed in aqueous environments are often totally or partially covered with biofilms. Under these circumstances, the interface behaves more like a combination of resistance and capacitance in parallel than a pure resistor, and there may be an uncompensated ohmic drop through the biofilm, leading to nonlinear polarization behavior. Despite these difficulties, several papers have been published using this method to assess different microbial effects on corrosion rates of SS (Nivens et al., 1986; Dowling et al., 1986, 1988).

Electrochemical impedance spectroscopy (EIS) uses impedance data as a function of the frequency of the applied signal, at a fixed potential and current point, of a polarization curve. In practice this working point corresponds to the corrosion potential. A large frequency range has to be applied to obtain a complete impedance spectrum. Many of the biofilms that adsorb on metallic surfaces immersed in natural aqueous environments are nonconducting and EIS techniques are potentially useful in their presence (Dexter et al., 1991). Several papers recently published, include the use of this technique for assessing SS susceptibility to MIC (Dowling et al., 1989, 1991; Feron, 1991; Mansfeld et al., 1990).

Electrochemical techniques applying large signals to the system

In these experimental techniques potential scans, which can range from several hundred millivolts to several volts, are used. Among these techniques it can be distinguished: i) Tafel polarization; ii) potentiostatic and potentiodynamic polarization; iii) determination of pitting (Ep) and breakdown (Eb) potentials.

i) The current-potential diagrams (Evans diagrams) are graphical methods for demonstrating the relationship between corrosion rate and the extent of polarization of anodic and cathodic reactions participating in the corrosion process (Videla, 1989a). Theoretically, both the cathodic and the anodic curves should be linear when the applied potential is plotted against the logarithm of the current density, and they should intersect at a point that represents the corrosion potential and the corrosion current density (Dexter et al., 1991). In practice, log current vs potential curves both deviate from linearity in the vicinity of the corrosion potential. Notwithstanding this, both anodic and cathodic curves often contain linear segments referred to as Tafel regions. Extrapolating these linear segments of either the anodic or cathodic curves to the corrosion potential yields the corrosion current density. Extrapolations from the linear segment of the cathodic curve are generally preferred in MIC studies, since it is less affected by corrosion products and biofilms. Tafel extrapolation has been mainly used in the literature to assess SRB effects on carbon steel. In the last years, some papers have been made to study SS susceptibility to MIC (Nivens et al., 1986; Dowling et al., 1986; Varjonen et al., 1991).

ii) potentiodynamic and potentiostatic polarization are most useful in characterizing the corrosion behavior for metal-electrolyte systems in which the metal is passivated by the formation of a passive film (e.g. SS). A polarization curve can provide an overview of several types of reactions that occur in a given metal/solution interface, such as charge transfer, diffusional processes, passivity, transpassivity and localized corrosion. For a given corroding metal, the corrosion potential and corrosion current will be determined by the point at which the cathodic curve intersects the anodic curve. The anodic and cathodic curves can be affected enough to disturb the passive behavior of a metal surface, by changes in solution chemistry, pH, temperature, hydrodynamics or passive film formation. Microorganisms, either in the form of biofilms or discrete colonies, can bring about such changes (Dexter et al., 1991). For instance, it has been reported that a transition from passive to active behavior can be caused by a biofilm forming on a SS immersed in seawater (Mollica et al., 1984, 1992; Dexter and Gao, 1988).

iii) Two useful parameters that can be determined either potentiodynamically (or potentiostatically) are the **Eb** and **Ep**. The former is defined as the potential at which the anodic polarization curve shows a marked increase in current, leading to pit initiation. The Ep is usually defined as the most noble potential at which the passive current density remains stable and pits do not nucleate on a metal surface. There have been several references in recent years on the use of large signal electrochemical techniques to assess SS susceptibility to MIC (de Cristofaro et al., 1986; Ringas and Robinson, 1988a, 1988b; Borenstein and White, 1989; Moreno et al., 1991; Salvago et al., 1991a; Webster et al. 1991). The anodic scan of different kinds of SS in sulfide and chloride mixtures have recently been analyzed to interpret the action of SRB on the breakdown of passivity (Moreno et al., 1992).

The disadvantage attributed to large signal polarization techniques is that they induce large deviations from steady state conditions, thus, changes in the pH at the interface are produced and can damage microorganisms.

MICROBIOLOGICAL, METALLURGICAL AND SURFACE ANALYSIS TECHNIQUES TO EVALUATE SS SUSCEPTIBILITY TO MIC

Microbiological methods are an important tool to identify and enumerate microbial species involved in MIC. The suspicion that corrosion is microbiologically induced, will begin when the predisposing environmental factors are noticed. If the surrounding environment is moist, and contains even minute amounts of any suitable food source, then MIC should be suspected. In most cases, MIC is accompanied by the formation of slimes or deposits, beneath which anaerobic conditions will prevail, and probably the SRB will be the causative agents (Gaylarde et al., in press).

In order to confirm MIC, identification of the organisms responsible is essential. This is not possible without using specialized microbiological techniques that seldomly can be carried out on site. Thus, good sampling technique and correct handling of samples once obtained, are of capital importance. A detailed description of microbiological techniques is out of the scope of this article, thus, a brief enumeration of the main microbiological techniques used in the literature on SS susceptibility to MIC will only be made. Direct microscopic examination is often the simplest and quickest of the tests available, this method is suitable for the detection of algae, fungi and, in clean samples, bacteria. Small bacterial cells can also be seen more readily by epifluorescence microscopy using stains such as acridine orange (Beech and Gaylarde, 1989).

Epifluorescence techniques allow the enumeration of total cells present in a sample, even those that would be unable to grow on the standard media in the laboratory. Counts obtained by this method are typically two orders of magnitude higher than those obtained by culture methods. This disadvantage may be overcome by the use of viable stains such as fluorescein diacetate. Several methods for the detection of microorganisms involved in MIC have been thoroughly reviewed in the literature (Gilbert and Herbert, 1987; Pope, 1990).

Conventional microbiological culture techniques provide a means of estimating the number of viable bacteria present in the biofilms and therefore give an indication of the extent of biofouling and the degree of MIC hazard. Recent years have seen the development of rapid methods for the specific identification and detection of environmentally important microorganisms. Some of these are available in kit form, but often they require a degree of expertise not normally found outside the specialist laboratory.

The dip-slide technique is a viable counting method which allows rapid enumeration of aerobic and anaerobic microorganisms without the necessity for making dilutions of the sample. This method cannot be used for very dilute cell suspensions, but is readily portable, and no special expertise is needed for either application or interpretation.

Activity measurements like radiorespirometric techniques (Maxwell and Hamilton, 1986), adenosine triphosphate (ATP) measurements (Prasad, 1988); electron transport system activity (Scotto et al. 1986), have been recently used in MIC studies. For the specific detection of SRB, total numbers of bacteria may be determined within a day by the extremely sensitive enzyme-linked immunoassay (ELISA) method (Gaylarde and Cook, 1987).

A modified version of the ELISA test, involving the use of filters, has been developed by Pope (1990) and a commercial kit based on the ELISA principle is now available for SRB detection (Tatnall, 1991). Although not yet available for known corrosion-causing microorganisms, the use of gene probes could be a powerful technique for the detection of specific organisms causing MIC.
Attempts to simulate service conditions (diffusional characteristics, water quality, microbial contaminants, etc.) have been made in laboratory studies when assessing SS susceptibility to MIC. Different aerobic and anaerobic microorganisms, some of them isolated from industrial environments, were inoculated into various types of culture media in laboratory experiments. Thus, continuous and batch conditions, and pure and mixed cultures, have been used (Juanas et al., 1988; Ringas and Robinson, 1988b; Weimer and Ng, 1988; LLoret et al., 1988; Scott and Davies, 1989; Ortiz et al. 1990).

Microstructural characteristics and chemical composition data are rarely reported in research works on SS susceptibility to MIC. However, metallurgical factors are of paramount importance when the corrosion performance of a SS is under study. Welding heat-affected zones are frequently the center of microbial attack (Borenstein, 1988, 1991; Kerns and Borenstein, 1991). In the weld deposits either the austenite or ferrite phase can be preferentially attacked, depending on whether the environmental conditions are either oxidizing or reducing, respectively (Stein, 1991). According to this author, the susceptibility to MIC presented by AISI 300 series of SS, is not related to sensitization, but to the microstructure produced during the manufacturing process. Conversely, recent results (Mele et al. 1991; Moreno, 1990) clearly show that sensitization heat treatments lower pitting corrosion resistance of the 304 SS alloy. These treatments are not only detrimental in relation to intergranular corrosion but to pitting corrosion (the most frequent form of attack in MIC) as well. An increasing

Table 1. Analysis of Publications on Stainless Steels MIC Research

*UNS Numbers are Listed in Metals and Alloys in the Unified Numbering System, published by the Society of Automotive Engineers (SAE) and co-sponsored by the American Society for Testing and Materials (ASTM).

PLACE	EXPERIMENTAL METHODS	MATERIAL	PRODUCT	MEDIUM	MICROORGANISMS	REFERENCE
Field	Test installation for monitoring systems -Bacterial analysis	UNS* S30400	Small vessels Corrosion coupons Welded pipe	Waters and aqueous waste systems in chemical process industry (USA)	Sulfate reducers, sulfur oxidizers, slime forming bacteria (Pseudomonas), filamentous iron bacteria (Gallionella, Sphaerotilus)	Tatnall (1981)
Lab Field	Model heat exchanger Recirculating systems -Bacterial counts Trough systems -Algal count	UNS S30403	Heat exchanger tubes Tubing	Hudson river water Additions of: NaCl Cloride Biocides Cooling water systems from the Savannah river Nuclear Power Plant (USA)	Bacterial populations Algae	Pope et al. (1982)
Lab	Immersion in a recirculation system -Electrochemical (OCP) -Weight loss -Bacterial activity	UNS: S30400, S31600 21Cr-3Mo 18Cr-2Mo	Piping coupons: 20 mm diameter x 50 mm length	Natural and artificial seawater (ASTM)	Bacterial populations	Scotto et al. (1985)
Lab	Immersion test of coupons with preformed biofouling -Weight loss -Pitting corrosion (ASTM G-48-76)	UNS S30400	Test coupons	Demineralizated water Standard corrosion testing water Cooling Tower water from Usina Santa Elisa System (Brazil)	Iron bacteria and fungi	Silva et al. (1986)
Field	Immersion test -Weight loss -SEM, EDAX	UNS S30400	Welded and unwelded plates: 78 x 38 x 2 mm	Natural seawater from Lake Chilka (India)	Hyphomicrobium vulgare	Das and Mishra (1986)
Lab	Measurement of oxygen concentration with a platinum electrode	UNS S31600	Machined test cells from bars: 25.4 mm diameter x 25.4 mm length	Artificial and natural seawater	Marine aerobic bacteria	Dexter et al. (1986)
Lab	Immersion test in batch cultures -SEM, EDAX -FT-IR -Bacterial analysis Electrochemical test -Tafel constants -Polarization resistance	UNS S30400	Discs: 1.59 cm Polished to 600 grit	Marine broth (2216 Difco)	Vibrio natriegens Vibrio anguillarum	Nivens et al. (1986)
Lab	Immersion in both batch and continuous cultures -SEM, GC/MS Electrochemical test -Tafel plots and polarization resistance on coupons with biofilm	UNS S30400	Test coupons	Marine broth (2216 Difco)	Vibrio natriegens, Desulfobacter postgatei 2a c9	Dowling et al. (1986)

(continued)

Table 1 (continued)

PLACE	EXPERIMENTAL METHODS	MATERIAL	PRODUCT	MEDIUM	MICROORGANISMS	REFERENCE
Lab Field	Immersion test in a PVC-box (lab) and in a circulation loop (field) -OCP -Cathodic polarization	UNS: S31254, N08028 SAF 2205	Specimens: 0.1 x 0.1 m in circulation loop 110 mm diameter in PVC-box	Artificial seawater (ASTM D 1141-52) NaCl 3% Natural seawater from Trondheimsfjord or Byfjorden (Bergen)	------	Johnsen and Bardal (1986)
Lab Field	Electrochemical test -Potentiodynamic polarization -OCP Immersion test -OCP -SEM, EDAX	UNS S30400	Discs: 15 mm diameter Working electrode area: 0.38 cm² Polished to 1 μm	Filtered artificial and natural seawater or NaCl 0.5 M Natural seawater in a Thermal Power Plant, Mar del Plata Harbor (Argentine)	------ Rods and coccoid species	Videla et al. (1987)
Lab	Immersion in a recirculation test loop -Visual and optical examination	UNS S30403	Piping: 25.4 mm diameter x 100 mm length Sensitizing heat treatments and "as-received" state	ASTM D-993	Sulfate reducing bacteria	Juanas et al. (1988)
Lab	Electrochemical test -Polarization -Pitting potential -Potentiostatic Immersion test in semicontinuous culture -Weight loss -Pit density -SEM, EDAX	UNS: S30403, S31603, S31600, S43000, S40900 Sandvik-1802 SX 3CR12	Coupons exposed area: 1 cm² Passivated in 40% HNO_3 1/2 h Specimens: 50 x 20 x 3 mm Polished to 120 grit and passivated in 40% HNO_3	SRB media, with and without Fe	<u>Desulfovibrio desulfuricans desulfuricans</u>, <u>Desulfovibrio vulgaris vulgaris</u>, and mixed culture	Ringas and Robinson (1988a, 1988b)
Lab Field	Electrochemical test -Cathodic and anodic polarization (on Polarit 757-316-material) Immersion test -OCP -TEM	UNS: S31254 S31703 S31603 S30403 S44635 S31803 N08904	Sheet: 100 x 30 x (1-3) mm Polished and passivated with HNO_3	Fresh brackish water Brackish water in Baltic Sea (Gulf of Finland)	Biofilm from field Different types of bacteria	Varjonen et al. (1988)
Lab Field	Immersion in a test loop -Bacterial analysis -OCP	20Cr-18Ni-6Mo	Tubes	Natural seawater from/in Genova (Italy)	Aerobic and anaerobic bacteria	Guezennec et al. (1988)
Lab	Electrochemical test -Pitting potential	UNS S41000	Bars: 9 mm diameter Polished to 800 grit With different heat treatments (annealed, quenched, quenched & tempered)	Postgate C with chlorides, sulphides and metabolites	<u>Desulfovibrio desulfuricans</u> (DSM 642)	Ibars et al. (1988)

PLACE	EXPERIMENTAL METHODS	MATERIAL	PRODUCT	MEDIUM	MICROORGANISMS	REFERENCE
Lab	Immersion test in semicontinuous culture -Visual and optical examination	UNS: S30300, S30400, S30403, S31603, S42000	Pipe, plate and bar Superficial grinding to 180 grit With different heat treatments (sensitization, quenched & tempered, etc.)	Beckwith	Sulfate reducing bacteria	Lloret et al. (1988)
Lab	Electrochemical test -Potentiodynamic polarization Immersion test -SEM, EDAX, XPS, AES	UNS S30400	Bars: 5 mm diameter Polished to 1μm	Postgate C with and without chlorides, and artificial seawater	SRB from Urumea river (Spain)	Erauzkin (1988)
Lab	Immersion test in continuous cultures -Zero resistance ammetry	UNS: S30400, S31600	"Sandwich" style coupons: 38.1 x 12.7 x 1.6 mm	BTZ - 4	*Desulfovibrio desulfuricans* (G100 A)	Weimer and Ng (1988)
Field	Immersion test -Visual inspection	UNS: S30400, S31600, S43000, S44600	Test coupons	Natural seawater in Galveston, Texas (USA)	Fouling	Griffin et al. (1988)
Lab Field	Immersion test in batch cultures -SEM Immersion test -SEM -OCP	UNS S30400	Discs: 15 mm diameter Rods: 5 mm diameter Polished to 600 grit	Sterilized seawater from a Thermal Power Plant Natural seawater in a Thermal Power Plant (Argentine)	*Pseudomonas sp.* (marine) Bacterial cells, diatoms and protozoans	Videla (1989b)
Field	Immersion test in a seawater circuit -Cell counts	UB6	Test coupons Polished to 1 μm	Natural seawater in Brest Bay (France)	Rod-shaped and coccoidal bacteria	Fera et al. (1988)
Lab	Immersion test -SEM-EDAX Electrochemical test -Cathodic polarization	UNS N08366	Test coupons from metal sheet	Artificial seawater with variyng pH through acetic acid additions	*Acetobacter aceti* (ATCC 15.983)	Little et al. (1988)
Lab Field	Electrochemical test -OCP -Cathodic polarization Immersion test -SEM -OCP	UNS S31600	Rod: 0.635 cm diameter x 5 cm length Polished to 320 grit and passivated in 20% nitric acid, room temperature for 20 min	Filtered natural seawater from South of the Mouth of the Delaware Bay (USA) Natural seawater in Delaware Bay (USA)	Biofilm from field Marine bacteria, ciliates and protozoans	Dexter and Gao (1988)

(continued)

Table 1 (continued)

PLACE	EXPERIMENTAL METHODS	MATERIAL	PRODUCT	MEDIUM	MICROORGANISMS	REFERENCE
Lab	Immersion test -OCP -Potentiodynamic polarization -Polarization resistance -EIS -Surface analysis -SEM -Bacterial counts	TYPE 316L + As,Cu,Se	Test coupons Polished to 800 grit	Natural and artificial seawater (ASTM)	Aerobic and anaerobic	Hernández et al. (1988)
Lab Field	Immersion test -Weight loss Immersion test -SEM, EDAX	UNS: S30300, S30403, S30800, S31603, S32100, S42000, S43100	Split tubes and sheets Tube: 100 x 24 mm As-received and oven cooling (700°C-1h)	Beckwith Slime from a Hydraulic Power Station Natural water in two Hydraulic Power Stations (Spain)	SRB ------- Sulfate and sulfite reducing bacteria, Gallionella, Sphaerotilus, Leptothrix, Siderocapsa, Fragilaria	OCIDE (1988)
Lab	Recirculating water system -OCP -Potentiostatic polarization -EIS -Biochemical analysis	Eight different SS	Rectangular specimens 75 x 25 x 2 mm 250 x 25 x 1.5 mm	Natural seawater	Bacteria and diatoms	Mollica et al. (1989)
Field	Immersion in a flow channel -OCP -Cathodic polarization -Bacterial activity and counts	UNS: S31254, S31600, S32750	Specimens exposed area: 60 cm² Etched in 4% HF+ 20% HNO_3	Natural seawater from the Trondhein fjord (Norway) with and without chlorination	Bacteria	Gundersen et al. (1989)
Lab Field	Immersion test -Pitting potential -OCP Immersion test -Visual inspection -Radiography	UNS: S30403, S31603	Pipe coupons Polished to 9 μm and passivated in 50% HNO_3 1/2 h Welded with (three percentages of ferrite content): ER 308 (low), ER 308 (med) and ER 312 (high)	Hutners medium modified Natural water (USA)	Bacillus, Pseudomonas, Acinetobacter Mixed	Borenstein and White (1989)
Lab	Immersion test -Visual inspection -SEM	UNS N08904	Coupons from heat exchanger tubing	Autoclaved seawater with nutrient additions	Desulfotomaculum sp., Desulfovibrio desulfuricans	Scott and Davies (1989)
Lab	-EIS -Small amplitude cyclic voltammetry (SACV) -Electron surface chemical analysis (ESCA)	UNS S31603 welded with E-308	Welding coupons Polished to 9 μm and passivated in 50% HNO_3	Culture medium	Unselected marine bacteria	Dowling et al. (1989)

PLACE	EXPERIMENTAL METHODS	MATERIAL	PRODUCT	MEDIUM	MICROORGANISMS	REFERENCE
Lab Field	Immersion test -OCP Electrochemical test -Pitting potential Immersion test -OCP -SEM	UNS: S30403, S43000 AISI 316T	Discs: 15 mm diameter Polished to 1 μm	Artificial seawater filtered Natural seawater in Mar del Plata Harbor (Argentine)	------- Bacteria, protozoa, algae, etc.	Videla et al. (1989)
Lab	Immersion test -Pitting potential -OCP -SEM	UNS: S30403, S31603, S43035, N08367 E-308 deposited weld metal	Test coupons: 1.5 cm² x 0.5-1 cm Passivated in 50% HNO_3 Polished to 600 grit	Synthetic culture medium	A6F Bacteria	Zhang et al. (1989)
Lab Field	Immersion test in batch cultures -SEM Immersion test -OCP -SEM -Bacterial counts	UNS S30400	Discs: 5.8 or 15 mm diameter Polished to 600 grit	Postgate C Air conditioning cooling water recirculation Thermal Power Plant seawater cooling system (Brazil)	<u>Pseudomonas fluorescens</u> Aerobic bacteria, iron-oxidizing bacteria, anaerobic bacteria	Silva and Videla (1989)
Lab Field	Electrochemical test -EIS Immersion test -SEM, EDAX -OCP -Bacterial counts	UNS: S30400, S31600, N08366	Electrode area: 1 cm² Rectangular 75 x 10 x 1 mm 100 x 6 x 1 mm	Seawater Natural seawater in Pacific Ocean Port Hueneme, California (USA)	Aerobic heterotrophic bacteria, sulfate reducing bacteria	Mansfeld et al. (1990)
Lab	Immersion test -OCP -Microbiological analysis	UNS S30400 welded with E-308	Rod and pipe	Artificial seawater and natural seawater from Pacific Ocean (USA)	<u>Lepthothrix</u> and acid producing bacteria (APB)	Licina et al. (1990)
Lab Field	Electrochemical test -Pitting potential Exposed in cooling water systems -SEM, EDAX -Bacterial counts	UNS S31600	Coupons area: 0.5 cm² Polished to 300-600 grit	Water of the recirculating system Water in two cooling towers (Brazil)	------- Fungi, aerobic bacteria, <u>Pseudomonas sp.</u>, anaerobic bacteria, iron related	Videla et al. (1990)
Lab Field	Immersion test -OCP -Bacterial counts Immersion test -OCP -SEM	UNS: S30400, S31600, N08366	Electrode area: 1 cm² Polished to 600 grit Rectangular 100 x 6 x 1 mm	Natural and sterile sea water from Gulf of Mexico, 3.5% NaCl and artificial seawater Natural seawater in Pacific Ocean, Port Hueneme, California (USA)	Aerobic heterotrophic bacteria and anaerobic SRB from natural seawater Diatoms	Little et al. (1990)

(continued)

Table 1 (continued)

PLACE	EXPERIMENTAL METHODS	MATERIAL	PRODUCT	MEDIUM	MICROORGANISMS	REFERENCE
Lab	Electrochemical in a recirculating seawater system -OCP -SEM, EDAX -Bacterial counts	UNS: N10276	Sheet: 20 x 20 x 2 mm Polished to 600 grit Exposed area: 1 cm^2	Filtered natural seawater	Biofilms from lab (Vibrio Harveyi) and field (bacterial from estuarine seawater, Delaware Bay, USA)	Dexter and Lin (1990)
Lab	Electrochemical test -OCP -EIS ASTM G48-76	UNS: S31254, S31600, N08367	Sheet or tube area: 3.5 cm^2 Polished to 600 grit	Widdel and Pfennig	Bacterial consortium (Bacillus sp., Pseudomonas sp., SRB, etc.)	Dowling et al. (1991)
Lab	Immersion test in semicontinuous cultures -SEM, EDAX Electrochemical test -Pitting potential	UNS: S30300, S30400, S30403, S31600, S31603, S42000	Split tubes and sheets Polished to 120 and 800 grit, rectangular polished to 800 grit Working electrodes were passivated in 25% nitric acid, 50°C for 20 min With different heat treatments (sensitization, quenched & tempered, etc.)	Beckwith Postgate C + NaCl (3% w/v) NaCl (3.56% w/v) Filtered culture from Desulfovibrio desulfuricans (DSM 1926)	Desulfotomaculum sp. Desulfovibrio desulfuricans (DSM 1926)	Moreno (1990)
Lab	Immersion test in batch cultures -SEM -Visual inspection Electrochemical test -OCP -EIS -Pitting potential	UNS: S31603, S43036	Biocoupons: 18 cm^2 Plate: 1.0 mm Electrode area: 1 cm^2 (test with flash) or 18 cm^2 (test with fermenter)	Standard culture medium with or without sodium sulphide or ascorbic acid	Desulfovibrio vulgaris (NCIMB 8303) Desulfovibrio gigas (NCIMB 9332)	Feron (1991)
Lab	Immersion test in an apparatus simulating the moderator heat treatment -Cyclic polarization -Zero resistance ammetry -OCP	UNS N08800	-------	Water from lake Ontario (Canada) with sulfate (70 ppm) and chloride (30 to 230 ppm)	SRB and thermophilic SRB obtained from a moderator heat exchanger	Gendron et al. (1991)
Lab	Tubular flow system Electrochemical test -OCP Cell counts	UNS S31603	Plate	Bacterial culture media	Vibrio natriegens, Desulfovibrio vulgaris	Benbouzid-Rollet et al. (1991)
Lab	Electrochemical test -OCP -EIS -Anodic polarization	UNS: S31600, welded with ER308	Weld metal with different surface conditions (as welded, ground and polished)	------	Bacterial consortium	Buchanan et al. (1991)
Lab	Immersion test Electrochemical test -Tafel polarization	UNS: S30403, S31603, S31703	15 mm and 30 x 15 cm diameter plates Polished to 600 grit	White water (paper machine environment)	Aerobic Bacteria	Marmo et al. (1991)

PLACE	EXPERIMENTAL METHODS	MATERIAL	PRODUCT	MEDIUM	MICROORGANISMS	REFERENCE
Lab Field	Immersion in a test loop -Cathodic polarization	UNS S31254	Bars: 5 mm diameter x 15 mm length Polished to 1000 grit	Lab: artificial seawater (ASTM D1141) Field: Natural seawater from Arenzano, Genoa (Italy)	Aerobic bacteria	Salvago et al. (1991a)
Lab Field	Immersion test -SEM-EDAX Electrochemical test -EIS -OCP	UNS: S30400, S31600, N08366	Test coupons	Synthetic seawater Natural Pacific Ocean Water	Biofilms	Little et al. (1991a)
Lab Field	Immersion -Crevice corrosion tests Electrochemical -OCP -Cathodic polarization	UNS: S30400, S31254	------	Artificial seawater (lab) and natural seawater from Arenzano, Genoa (Italy)	------	Salvago et al. (1991b)
Lab	Electrochemical test -Pitting potential -Transient current Immersion in batch cultures -SEM, EDAX	UNS S41000	Bar: 9 mm diameter Electrodes area: 0.30 cm^2 Polished to 600 grit With different heat treatments (annealed, quenched and quenched & tempered)	Modified Postgate C, with Cl- and/or S^{2-} in several concentrations	_Desulfovibrio vulgaris_	Moreno et al. (1991)
Lab	Electrochemical test -Potentiodynamic polarization -EPR Immersion in batch cultures -SEM	UNS S30400	Rod: 6 mm diameter Electrodes exposed area: 0.30 cm^2 Polished to 600 grit Solution annealing and sensitization heat treatments	Modified Postgate C	_Desulfovibrio vulgaris_	Mele et al. (1991)
Lab Field	Electrochemical test -Pitting potential Immersion test -OCP -Bacterial counts Immersion test -SEM, EDAX	UNS: S30400, S31600, N08366	Electrode area: 1 cm^2 Polished to 600 grit	Natural and sterile seawater from Gulf of Mexico, 3.5% NaCl and artificial seawater Natural seawater in Gulf of Mexico water (USA)	Aerobic heterotrophic bateria and anaerobic SRB from natural seawater Diatoms	Little et al. (1991b)
Lab	Immersion test in batch cultures -SEM	UNS S31603, S44660, S31254, N06625, N08028, N08904, N08925, N08366, N08367, N08800, N08825	20 mm cut from heat exchanger tubes	A propietary medium for growth of SRB and other bacteria	_Desulfovibrio desulfuricans_, _Desulfovibrio salexigens_, _Desulfotomaculum orientis_, _Desulfotomaculum nigrificans_, _Desulfotomaculum oceanicum_, _Desulfotomaculum sp._, _Hyphomicrobium indicum_, _Hyphomicrobicum sp._ and _Thiobacillus ferroxidans_	Scott et al. (1991)

(continued)

Table 1 (continued)

PLACE	EXPERIMENTAL METHODS	MATERIAL	PRODUCT	MEDIUM	MICROORGANISMS	REFERENCE
Field	Immersion test -SEM	UNS S31600	Coupons area: 0.50 cm²	Seawater injection system (Argentine)	Rod-shaped and filamentous bacteria	Videla et al. (1991)
Lab	Electrochemical test -Zero resistance ammetry (ZRA) -OCP -Potentiostatic test	15Cr-10Ni	Hot-rolled to 4 mm thickness Electrode rectangular area: 1 cm² Polished to 600 grit	Postgate-type	Desulfovibrio vulgaris	Webster et al. (1991)
Lab	Immersion in a simulated service water test loop -Visual, SEM and metallographic examinations -Bacterial analysis	UNS: S30400, S31600, S31603	Pipe material: 10 in. Welded, wrought and furnace sensitized condition	Lake water	SRB, APB, anaerobes and aerobes	Stein (1991)
Lab	Microautoradiographic techniques	UNS S30400	Test coupons	Bacterial culture media	Pseudomonas, sp., Hafnia, sp., Desulfovibrio gigas	Franklin et al. (1991)
Field	Immersion test -OCP -Bacteriological analysis	UNS: S30403, S31603, S31703, S31725, S31803, S32900	Specimens: 200 x 100 x 2-3 mm As received and welded	Different paper machire environments (Finland)	Aerobic bacteria: Bacillus sp., Xanthomonas sp., Pseudomonas sp., Clavibacter sp.	Varjonen et al. (1991)
Lab	Electrochemical test -Tafel polarization	UNS S30403	Discs: 15 mm diameter Polished to 600 grit		Biofilm from LWC-machine II	
Field	Immersiont test -Polarization potential -Cathodic protection -Visual and optical examinations -Bacterial counts	UNS S30403 welded with E-308	Pipes and condenser water boxes Polished to 120 grit	Untreated seawater from the Gulf of Mexico (USA)	SRB and APB	Nekoksa and Gutherman (1991)
Field	Immersion in a test loop -Potentiodynamic scans -OCP	UNS: S30400, S31600, S44660, N08367	Electrodes as cylinders Polished to 600 grit	Low pressure service water system of the Catawba Nuclear Station (USA)	-------	Licina et al. (1991)
Field	Immersion in a test loop -Bacteriological analysis -Weight loss -Microscopic examination -EDAX	UNS: S30400, S30403, S31600, S44600, N08367	Service water piping and tubing material Polished to 120 grit Welded with types 308 (UNS S30400), 308L (UNS S30403) and C-276 (UNS N08367)	Low pressure service water system of the Catawba Nuclear Station (USA)	Sulfate reducing, iron oxidizing and slime bacteria	Ward and Anderson (1991)
Field	Immersion test -Visual, radiographic and metallographic evaluations -Bacteriological analysis	UNS: S30403, S31603	Pipe Welded with ER 308L (UNS S30403) and E 312 (UNS S31603)	Service water systems in a Nuclear Power Plant (USA)	Pseudomonas, Runella, Acinetobacter, Alcaligenes	Kearns and Borenstein (1991)

awareness of the importance of metallurgical studies is necessary at present, to reach a better understanding of SS susceptibility to MIC.

Surface analysis techniques, complemented with different kinds of electron microscopy are also very useful appliances to investigate MIC of SS. Scanning electron microscopy (SEM), complemented with Energy Dispersion X - Ray Analysis (EDAX), X-ray Photoelectron Spectroscopy (XPS) and Auger Electron Spectroscopy (AES) have been recently used to provide valuable data on the modifications of base metal composition caused by SRB corrosion (Erauzkin, 1988). The impressive advances of electron microscopy, since the beginning of the 90's, offer to MIC studies very powerful tools such as the scanning tunneling microscope (STM); the environmental scanning electron microscope (ESEM), able to operate at atmospheric pressure and using humid samples (Wagner, 1992); and the Atomic Force Microscope (AFM) the highest resolution microscope developed up to the present and already used for MIC studies (Geesey and Bremer, 1992).

SUMMARY

Electrochemical techniques, metallurgical and microbiological methods and surface analysis that have been used in the study of microbiologically influenced corrosion of stainless steels are succinctly reviewed. A sole technique can not provide the complete information needed to understand all the biological and electrochemical processes influencing the metal attack. Advances in the field have been reported by several researchers, during the last years. Sixty-one of these reports on MIC of stainless steels, are reviewed in this article, and their experimental conditions are briefly analyzed. The uses and limitations of several experimental techniques are also discussed.

Acknowledgements

D.A.M, J.R.I. and C.R. are grateful to Comisión Interministerial de Ciencia y Tecnología of Spain, for financial support (Grant MAT90-0124).

REFERENCES

Beech, I.B., and Gaylarde, C.C. (1989). Adhesion of *Desulfovibrio desulfuricans* and *Pseudomonas fluorescens* to mild steel surfaces. J. Appl. Bacteriol., 67, 201-207.

Benbouzid-Rollet, N.D., Guezennec, J., Conte, M., and Prieur, D. (1991). Experimental biofilms containing sulphate-reducing and fermentative bacteria on stainless steel in a laboratory tubular flow-system. In: Microbially Influenced Corrosion and Biodeterioration, 3/ pp. 47-54 (N.J. Dowling, M.W. Mittleman and J.C. Danko, eds.), The University of Tennessee, Knoxville.

Borenstein, S. (1988). Microbially influenced corrosion failures of austenitic stainless steels welds. Mat. Perf., 27 (8), 62-66.

Borenstein, S.W. (1991). Microbiologically influenced corrosion of austenitic stainless steel weldments. Mat. Perf., 30 (1), 52-54.

Borenstein, S.W., and Lindsay, P.B. (1987). Planning and conducting microbiologically influenced corrosion failure analyses. Corrosion/87, paper no. 381, Houston, NACE.

Borenstein, S.W., and White, D.C. (1989). Influence of welding variables on microbiologically influenced corrosion of austenitic stainless steel weldments. Corrosion/89, paper no. 183, Houston, NACE.

Buchanan, R.A., Xiao Zhang, Ping Li, Stansbury, E.E., Dowling, N.J.E., Hall, T., and Lindberg, A. (1991). Effects of surface condition on susceptibility to microbially influenced corrosion: stainless steel weldments and carbon steel. In: Microbially Influenced Corrosion and Biodeterioration, 3/ pp. 99-102 (N.J. Dowling, M.W. Mittleman and J.C. Danko, eds.), The University of Tennessee, Knoxville.

Cristofaro, N.B. de, Acosta, C.A., Salvarezza, R.C., and Videla, H.A. (1986). The effect of sulfides on the electrochemical behavior of AISI 304 stainless steel in neutral buffered solutions. Corrosion, 42 (4), 240-242.

Das, C.R., and Mishra, K.G. (1986). Biological corrosion of welded steel due to marine algae. In: Biologically Induced Corrosion, pp. 114-117 (S.C. Dexter, ed.), NACE-8, International Corrosion Conferences Series, Houston, NACE.

Dexter, S.C., and Gao, G.Y. (1988). Effect of seawater biofilms on corrosion potential and oxygen reduction of stainless steel. Corrosion, 44 (10), 717-723. Discussion on this paper by F. Mansfeld and B. Little in Corrosion, 45 (10), 786-789.

Dexter, S.C., and Lin, S. (1990). Effect of marine bacteria on calcareous deposition. Corrosion/90, paper no. 157, Houston, NACE.

Dexter, S.C., Lucas, K.E., and Gao, G.Y. (1986). The role of marine bacteria in crevice corrosion initiation. In: Biologically Induced Corrosion, pp. 144-153 (S.C. Dexter, ed.), NACE-8, International Corrosion Conferences Series, Houston, NACE.

Dexter, S.C., Duquette, D.J., Siebert, O.W., and Videla, H.A. (1991). Use and limitations of electrochemical techniques for investigating microbiological corrosion. Corrosion, 47 (4), 308-318.

Dowling, N.J.E., Guezennec, J., and White, D.C. (1986). Facilitation of corrosion of stainless steel exposed to aerobic seawater by microbial biofilms containing both facultative and absolute anaerobes. In: Microbial Problems in the Offshore Oil Industry, pp. 27-38 (E.C. Hill, J.L. Shennan and R.J. Wilkinson, eds.), Institute of Petroleum, London.

Dowling, N.J.E., Guezennec, J., and White, D.C. (1988). Methods for insight into mechanisms of microbially influenced metal corrosion. In: Biodeterioration 7, pp. 404-410 (D.R.Houghton, R.N.Smith and H.O.W.Eggins, eds.), Elsevier Applied Science, London.

Dowling, N.J.E., Franklin, M., White, D.C., Lee, C.H., and Lundin, C. (1989). The effect of microbiologically influenced corrosion on stainless steel weldments in artificial seawater. Corrosion/89, paper no. 187, Houston, NACE.

Dowling, N.J.E., White, D.C., Buchanan, R.A., Danko, J.C., Vass, A., Brooks, S., and Ward, G.L. (1990). Microbiologically influenced corrosion of 6% molybdenum stainless steels and AISI 316: comparison with ferric chloride testing. Corrosion/90, paper no. 532, Houston, NACE.

Erauzkin, E. (1988). Biocorrosion of AISI stainless steel by sulphate reducing bacteria. In: Microbial Corrosion 1, pp. 412-430 (C.A.C. Sequeira and A.K. Tiller, eds.), Elsevier Applied Science, London.

Fera, P., Siebel, M.A., Characklis, W.G., and Prieur, D. (1988). Seasonal variations in bacterial colonization of stainless steel, aluminum and polycarbonate surfaces in a seawater flow system. Biofouling, 1 (3), 251-261.

Feron, D. (1991). Chemical and electrochemical aspects of the corrosion of stainless steels in the presence of sulphate reducing bacteria. In: Microbially Influenced Corrosion and Biodeterioration, 5/ pp.73-83 (N.J. Dowling, M.W. Mittleman and J.C. Danko, eds.), The University of Tennessee, Knoxville.

Franklin, M.J., Guckert, J.B., White, D.C., and Isaacs, H.S. (1991). Spatial and temporal relationships between localized microbial metabolic activity and electrochemical activity of steel. Corrosion/91, paper no. 115, Houston, NACE.

Gaylarde, C.C., and Cook, P.E. (1987). ELISA techniques for the detection of sulphate-reducing bacteria. In: Immunoassay Techniques for the Detection of Bacteria, pp. 231-244 (J.M. Grange, A. Fox and N.L. Morgan, eds.), Blackwell Scientific Publications, Oxford.

Gaylarde, C.C., Edyvean, R.G.J., and Videla, H.A. (1992). Biologically Induced Corrosion, Research Studies Press, London, in press.

Geesey G.G., and Bremer P.J., (1992). Properties of a bacterium that causes pitting corrosion. In: Proceedings of the International Marine Biotechnology Conference '91, Society for Industrial Microbiology, Annandale, Va, in press.

Gendron, T.S., Cleland, R.D., and Lavoie, P.A. (1991). The role of chloride and SRB in underdeposit corrosion of alloy 800 in fresh water. In: Microbially Influenced Corrosion and Biodeterioration, 2/ pp. 1-7 (N.J. Dowling, M.W. Mittleman and J.C. Danko, eds.), The University of Tennessee, Knoxville.

Gilbert, P.D., and Herbert, B.N. (1987). Monitoring microbial fouling in flowing systems using coupons. In: Industrial Microbiological Testing, pp. 79-98 (J.W.Hopton and E.C.Hill, eds.), Blackwell Scientific Publications, Oxford.

Griffin, R.B., Cornwell, L.R., Seitz, W., and Estes, E. (1988). Localized corrosion under biofouling. Corrosion/88, paper no. 400, Houston, NACE.

Guezennec, J., Scotto, V., and Alabiso, G. (1988). Biocorrosion of stainless steels in natural seawater comparison tests in field and in lab. In: Microbial Corrosion 1, pp. 183-187 (C.A.C. Sequeira and A.K. Tiller, eds.), Elsevier Applied Science, London.

Gundersen, R., Johansen, B., Gartland, P.O., Vintermyr, I., Tunold, R., and Hagen, G. (1989). The effect of sodium hypochlorite on the electrochemical properties of stainless steels and on the bacterial activity in seawater. Corrosion/89, paper no. 108, Houston, NACE.

Hernandez, G., Lemaitre, C., Guezennec, J., Audouard, J.P., and Beranger, G. (1988). Etude de l'influence d'elements poisons sur la resistance a la corrosion bacterienne d'aciers inoxydables en milieu marin. In: Proc. 7th Intl. Congr. on Marine Corrosion and Fouling, Valencia, Spain, in press.

Ibars, J.R., Moreno, D.A., Lloret, M., and Ranninger, C. (1988). Microstructural effects on the electrochemical behaviour of AISI 410 in microbiological media. In: Microbial Corrosion 1, pp. 287-289 (C.A.C. Sequeira and A.K. Tiller, eds.), Elsevier Applied Science, London.

Johnsen, S., and Bardal, E. (1986). The effect of a microbiological slime layer on stainless steel in natural sea water. Corrosion/86, paper no. 227, Houston, NACE.

Juanas, F., de la Cruz, M., and Ranninger, C. (1988). Corrosión por bacterias de acero inoxidable AISI 304L. Rev. Iberoamer. Corros. Prot., 19 (2), 84-85.

Kearms, J.R., and Borenstein, S.W. (1991). Microbially influenced corrosion testing of welded stainless alloys for nuclear power plant service water systems. Corrosion/91, paper no. 279, Houston, NACE.

Licina, G.J. (1989). An overview of microbiologically influenced corrosion in nuclear power plants. Mat. Perf., 28 (10), 55-60.

Licina, G.J., Pytel, M.L., and Hirota, N.S. (1990). An innovative method for mitigation of MIC in stainless steel weldments. Corrosion/90, paper no. 114, Houston, NACE.

Licina, G.J., Anderson, S., Maner, M.K., and Ward, G.L. (1991). In-plant electrochemical studies of service water system materials. Corrosion/91, paper no. 277, Houston, NACE.

Little, B.J., Wagner, P.A., Gerchakov, S.M., Walch, M. and Mitchell, R. (1986). The involvement of a thermophilic bacterium in corrosion processes. Corrosion, 42 (9), 533-536.

Little, B.J., Wagner, P., and Duquette, D. (1988). Microbiologically induced increase in corrosion current density of stainless steel under cathodic protection. Corrosion, 44 (5), 270-274.

Little, B., Ray, R., Wagner, P., Lewandowski, Z., Lee, W.C., Characklis, W.G., and Mansfeld, F. (1990). Electrochemical behavior of stainless steels in natural seawater. Corrosion/90, paper no. 150, Houston, NACE.

Little, B., Wagner, P., and Mansfeld, F. (1991a). An electrochemical evaluation of biofilms and calcareous deposits formed in natural seawater. In: Microbially Influenced Corrosion and Biodeterioration, 5/ pp. 9-15 (N.J.

Dowling, M.W. Mittleman and J.C. Danko, eds.), The University of Tennessee, Knoxville.

Little, B., Ray, R., Wagner, P., Lewandowski, Z., Chee Lee, W., Characklis, W.G., and Mansfeld, F. (1991b).Impact of biofouling on the electrochemical behaviour of 304 stainless steel in natural seawater. Biofouling, 3 (1), 45-59.

Lloret, M., Moreno, D.A., APARICIO, M.L., and Ranninger, C. (1988). Stainless steel corrosion analysis in semicontinous culture of sulphate-reducing bacteria. In: Microbial Corrosion 1, pp. 335-360 (C.A.C. Sequeira and A.K. Tiller, eds.), Elsevier Applied Science, London.

Mansfeld, F., Tsai, R., Shih, H., Little, B., Ray, R., and Wagner, P. (1990). Results of exposure of stainless steels and titanium to natural seawater. Corrosion/90, paper no. 109, Houston, NACE.

Mansfeld, F., and Little, B. (1991). A technical review of electrochemical techniques applied to microbiologically influenced corrosion. Cor.Sci., 32 (3), 247-272.

Marmo, S.A., Nurmiaho-Lassila, E.L., Varjonen, O., and Salkinoja-Salonen, M.S. (1991). Biofouling and microbially induced corrosion on paper machines. In: Microbially Influenced Corrosion and Biodeterioration, 4/ pp. 33-38 (N.J. Dowling, M.W. Mittleman and J.C. Danko, eds.), The University of Tennessee, Knoxville.

Maxwell, S., and Hamilton, W.A. (1986). Modified radiorespirometric assay for determining the sulfate reduction activity of biofilms on metal surfaces. J. Microbiol. Met., 5 (2), 83-91.

Mele, M.F.L. de, Moreno, D.A., Ibars, J.R., and Videla, H.A. (1991). Effect of inorganic and biogenic sulfide on localized corrosion of heat-treated type 304 stainless steel. Corrosion, 47 (1), 24-30.

Mollica, A., Trevis, A., Traverso, E., Ventura, G., Scotto, V., Alabiso, G., Marcenaro, G., Montini, V., de Carolis, G. and Dellepiane, R. (1984). Interaction between biofouling and oxygen reduction rate on stainless steel in seawater. In: Proc. 6th Intl. Congr. on Marine Corrosion andFouling, p 269-281. Athens, Greece.

Mollica, A., Trevis, A., Traverso, E., Ventura, G., Carolis, G. de, and Dellepiane, R. (1989). Cathodic performance of stainless steels in natural seawater as a function of microorganism settlement and temperature. Corrosion, 45 (1), 48-56.

Mollica, A., (1992). Biofilm and corrosion on active-passive alloys in seawater. In: International Biodeterioration, in press.

Moreno, D.A. (1990). Análisis de la susceptibilidad a la corrosión microbiana de los aceros inoxidables en diferentes estados microestructurales. Ph.D. Dissertation. Universidad de Alcalá de Henares, Madrid, Spain.

Moreno, D.A., Mele, M.F.L. de, Ibars, J.R., and Videla, H.A. (1991). Influence of microstructure on the electrochemical behavior of type 410 stainless steel in cloride media with inorganic and biogenic sulfide. Corrosion, 47 (1), 2-9.

Moreno, D.A., Ibars, J.R., Ranninger, C., and Videla, H.A. (1992). Technical note: Use of potentiodynamic polarization to assess pitting of stainless steels by sulfate-reducing bacteria. Corrosion, in press.

Nekoksa, G., and Gutherman, B. (1991). Test results from electrochemical exposure racks at the Crystal river nuclear power plant. Corrosion/91, paper no. 275, Houston, NACE.

Nivens, D.E., Nichols, P.D., Henson, J.M., Geesey, G.G., and White, D.C. (1986). Reversible acceleration of the corrosion of AISI 304 stainless steel exposed to seawater induced by growth and secretions of the marine bacterium *Vibrio natriegens*. Corrosion, 42 (4), 204-210. OCIDE. (1988). Corrosión bacteriana de aceros inoxidables. Final Report, Proj. no. 120.11. Madrid, Spain.

Ortiz, C., Guiamet, P.S., and Videla, H.A. (1990). Relationship between biofilms and corrosion of steel by microbial contaminants of cutting oil emulsions. International Biodeterioration, 26 (5), 315-326.

Pope, D.H. (1986). A study of microbiologically influenced corrosion in nuclear power plants and a practical guide for countermeasures. Electric Power Research Institute. Final Report EPRI NP-4582s. Palo Alto, CA.

Pope, D.H. (1990). Development of methods to detect sulfate-reducing bacteria - Agents of microbiologically influenced corrosion; Publication no. 37, Materials Technology Institute of the Chemical Process Industries, Inc., St. Louis, MO.

Pope, D.H., Soracco, R.J., and Wilde, E.W. (1982). Studies on biologically induced corrosion in heat exchanger systems at the Savannah River Plant, Aiken, SC. Mat. Perf., 21 (7), 43-50.

Pope, D.H., Duquette, D., Wayner, P.C. and Johannes, A.H. (1989). The case for microbiologically influenced corrosion of stainless steels. In: Microbiologically Influenced Corrosion: A State-of-the-Art Review, second edition, Publication 13, Materials Technology Institute of the Chemical Process Industries, Inc. Columbus. Ohio.

Prasad, R. (1988). Pros and cons of ATP measurement in oil field waters, Corrosion/88, paper no. 87, Houston, NACE.

Ringas, C., and Robinson, F.P.A. (1988a). Corrosion of stainless steel by sulfate-reducing bacteria. Electrochemical Techniques. Corrosion, 44 (6), 386-396.

Ringas, C., and Robinson, F.P.A. (1988b). Corrosion of stainless steel by sulfate-reducing bacteria - Total immersion test results. Corrosion, 44 (9), 671-678.

Salvago, G., Taccani, G., and Fumagalli, G. (1991a). Electrochemical approach to biofilms monitoring. In: Microbially Influenced Corrosion and Biodeterioration, 5/ pp. 1-7 (N.J. Dowling, M.W. Mittleman and J.C. Danko, eds.), The University of Tennessee, Knoxville.

Salvago, G., Taccani, G., Fumagalli, G., and Galelli, L. (1991b). Acidification effects on the electrochemical behaviour of stainless steel in seawater. In: Microbially Influenced Corrosion and Biodeterioration, 8/ pp.5-11 (N.J. Dowling, M.W. Mittleman and J.C. Danko, eds.), The University of Tennessee, Knoxville.

Scott, P.J.B., and Davies, M. (1989). Microbiologically influenced corrosion of high molybdenum austenitic stainless steel. Corrosion/89, paper no. 186, Houston, NACE.

Scott, P.J.B., Goldie, J., and Davies, M. (1991). Ranking alloys for susceptibility to MIC - A preliminary report on high-Mo alloys. Mat. Perf., 30 (1), 5557.

Scotto, V., di Cintio, R., and Marcenaro, G. (1985). The influence of marine aerobic microbial film on stainless steel corrosion behaviour. Cor. Sci., 25 (3), 185-194.

Scotto, V., Alabiso, G., and Marcenaro, G. (1986). The behaviour of stainless steels in natural seawater. In: Bioelectrochemistry and Bioenergetics, 16, 347-355.

Silva, A.J.N., Tanis, J.N., Silva, J.O., and Silva, R.A. (1986). Alcohol industry biofilms and their effect on corrosion of 304 stainless steel. In: Biologically Induced Corrosion, pp. 76-82 (S.C. Dexter, ed.), NACE-8, International Corrosion Conference Series, Houston, NACE.

Silva, R.A., and Videla, H.A. (1989). A corrosao microbiológica e sua relacao com a presenca de biofilmes en metais e ligas de uso industrial. Proc. 3rd Iberoamerican Congress of Corrosion and Protection - Brazilian Corrosion Congress/89, pp. 1580-1592. Río de Janeiro, Brazil.

Sinha, U.P., Wolfram, J.H., and Rogers, R.D. (1991). Microbially influenced corrosion of stainless steels in nuclear power plants. In: Microbially Influenced Corrosion and Biodeterioration, 4/ pp.51-60 (N.J. Dowling, M.W. Mittleman and J.C. Danko, eds.), The University of Tennessee, Knoxville.

Stein, A.A. (1991). Metallurgical factors affecting the resistance of 300 series stainless steels to microbiologically influenced corrosion. Corrosion/91, paper no. 107, Houston, NACE.

Stoecker, J.G., and Pope, D.H. (1986). Study of biological corrosion in high temperature demineralized water. Mat. Perf., 25 (6), 51-56.

Tatnall, R.E. (1981). Fundamentals of Bacteria Induced Corrosion. Materials Performance, 20 (9), 32-38.

Tatnall, R.E. (1991). New perspectives on testing for sulfate reducing bacteria. In: Microbially Influenced Corrosion and Biodeterioration, 5/ pp.17-32 (N.J. Dowling, M.W. Mittleman and J.C. Danko, eds.), The University of Tennessee, Knoxville, TN.

Varjonen, O., Hakkarainen, T., Nurmiaho-Lassila, E., and Salkinoja-Salonen, M. (1988). The brackish water biofilm on stainless steels: an electrochemical and morphological study. In: Microbial Cor. 1, pp. 164-178 (C.A.C. Sequeira and A.K. Tiller, eds.), Elsevier Applied Science, London.

Varjonen, O.A., Hakkarainen, T.J., and Marmo, S.A. (1991). Observations on biofilms and corrosion resistance of stainless steels in paper machine environments. Corrosion/91, paper no. 190, Houston, NACE.

Videla, H.A. (1989a). Metal dissolution/redox in biofilms. In: Structure and Function of Biofilms, pp. 301-320 (W.G.Characklis and P.A.Wilderer, eds.), John Wiley & Sons, Ltd, Chichester, UK.

Videla, H.A. (1989b). Biological corrosion and biofilm effects on metal biodeterioration. In: Biodeterioration Research 2, pp.39-50 (C.E. O'Rear and G.C. Llewellyn, eds.), Plenum Press, New York.

Videla, H.A. (1991). Microbially induced corrosion: an updated overview. In: Biodeterioration and Biodegradation 8, pp. 63-88 (H.W.Rossmoore, ed.), Elsevier Applied Science, London.

Videla, H.A., Mele, M.F.L. de, and Brankevich, G. (1987). Microfouling of several metal surfaces in polluted sea water and its relation with corrosion. Corrosion/87, paper no. 365, Houston, NACE.

Videla, H.A., Mele, M.F.L. de, and Brankevich, G. (1989). Biofouling and corrosion of stainless steel and 70/30 copper - nickel samples after several weeks of immersion in seawater. Corrosion/89, paper no. 291, Houston, NACE.

Videla, H.A., Mele, M.F.L., Silva, R.A., Bianchi, F., and Gonzales Canales, C. (1990). A practical approach to the study of the interaction between biofouling and passive layers on mild steel and stainless steel in cooling water. Corrosion/90, paper no. 124, Houston, NACE.

Videla, H.A., Guiamet, P.S., Pardini, O.R., Echarte, E., Trujillo, D., and Freitas, M.M.S. (1991). Monitoring biofilms and MIC in an oilfield water injection system. Corrosion/91, paper no. 103, Houston, NACE.

Ward, G.L., and Anderson, S. (1991). MIC and corrosion of service water system materials. Corrosion/91, paper no. 278, Houston, NACE.

Wagner, P. (1992). Investigations of MIC using environmental scanning electron microscopy. Corrosion/92, paper no. 185, Houston, TX.

Webster, B.J., Newman, R.C., and Kelly, R.G. (1991). SRB-Induced localized corrosion of stainless steels. Corrosion/91, paper no. 106, Houston, NACE.

Weimer, P.J., and Ng, T.K. (1988). Use of two-stage continuous culture to assess biocorrosion by sulfate-reducing bacteria. Corrosion/88, paper no. 90, Houston, NACE.

Zamanzadeh, M., O'Connor, K., and Bavarian, B. (1989). Case histories of corrosion problems in waste-water facilities. Mat. Perf., 28 (9), 43-48.

Zhang, X., Buchanan, R.A., Stansbury, E.E., and Dowling, N.J.E. (1989). Electrochemical responses of structural materials to microbially influenced corrosion. Corrosion/89, paper no. 512, Houston, NACE.

BIOFOULING EFFECTS ON CORROSION OF STAINLESS ALLOYS IN SEAWATER

S. C. DEXTER*, **H-J. ZHANG** and **P. CHANDRASEKARAN**, College of Marine Studies, University of Delaware, Lewes, DE 19958, USA

INTRODUCTION

The purpose of this paper is to present a brief review of how microbial fouling films affect the corrosion behavior of passive metals and alloys in fresh, brackish and sea waters. It has often been reported that microbial films change the open circuit corrosion potential (OCP) of passive metals immersed in natural waters (Mollica and Travis, 1976; Johnsen and Bardal, 1985, 1986; Scotto, et al., 1985; Dexter and Gao, 1988; Gallagher, et al., 1988; Scotto, 1989). The change has usually been in the noble (electropositive) direction, and it has been called, "ennoblement." The significance of this effect lies in it's influence on localized corrosion initiation and propagation. In chloride bearing waters, the initiation of pitting and crevice corrosion is statistical, with the probability of initiation increasing directly with chloride ion activity and OCP. Thus, at a given chloride level, the probability of localized corrosion initiation is increased by anything (such as a biofilm and its metabolic products) that causes the OCP to shift in the noble direction.

The phenomenon of ennoblement was first documented for passive alloys in full strength seawater by Mollica and Trevis (1976). They and the other investigators cited above found that the OCP of an alloy resistant to chloride initiation of localized corrosion, such as NO8367 (6XN), usually became shifted to a steady noble value between +350 and +500 mV vs. the saturated calomel electrode (SCE) as a microbial biofilm formed at the metal/water interface. In contrast, the OCP of the same metal in the same water ranged from +100 to

Biodeterioration Research 4, Edited by G.C. Llewellyn
et al., Plenum Press, New York, 1994

+150 mV SCE if the microbial biofilm was inactivated or prevented from forming (See the review by Dexter, 1992). On alloys with less resistance to localized corrosion (eg., S30400), the OCP was first shifted noble, but as soon as pitting or crevice corrosion initiated, the current supplied by the active corrosion polarized the potential back to a more active value (Dexter and Zhang, 1990, 1991). Thus, it has been found that potential ennoblement can only be sustained for long periods of time (weeks to months) in the absence of corrosion initiation. It has been established by many authors (see Scotto, 1989; Dexter, 1992) that ennoblement of the OCP in the presence of a microbial film is a direct result of the metabolic activity of microorganisms in the film. Dexter (1992) has reviewed the types of data and arguments used to establish that the microorganisms cause the effect and that an actively metabolising film, as opposed to a dead film is required.

Other citations in the literature report no ennoblement effect. Mansfeld et al. (1990, 1992) showed that ennoblement did not occur under conditions in which the film consisted of a scatter of individual cells. Little et al., (1990) published data showing no ennoblement when a very thick film formed rapidly. Thus, the question arises as to what factors promote or suppress the effect.

In an effort to shed some light on that question, we will proceed in the following sections to review the effects of environmental parameters such as: surface coverage by the biofilm, salinity of the water and the presence or absence of natural sunlight on the ennoblement phenomenon. We will also review the mechanisms that have been proposed to account for ennoblement of the OCP. Finally, we will consider the effect of ennoblement on the initiation and propagation of corrosion itself.

EFFECT OF ENVIRONMENTAL PARAMETERS ON ENNOBLEMENT

Surface Coverage

Biofilms formed from natural aqueous environments are noted for providing spotty, rather than continuous coverage of solid surfaces. This is particularly true for films that have not yet reached a steady state thickness in dynamic equilibrium with the hydrodynamic flow conditions in which they are growing. Figure 1 shows that the extent of surface coverage does have an effect on the degree to which a biofilm is able to change the OCP. The various levels of nominal coverage were achieved by first growing a mature film from a natural assemblage of microorganisms from coastal ocean water on a one cm^2 coupon of alloy N10276 (C-276) and then coupling it to various sized bare panels of the same alloy. The results show that 30 to 40 percent coverage was required before the OCP of the couple rose above the +150 mV SCE level. Thus, complete coverage is not necessary for substantial ennoblement to occur, but the scatter of individual cells reported by Mansfeld et al. (1990, 1992) would not be expected to have much effect.

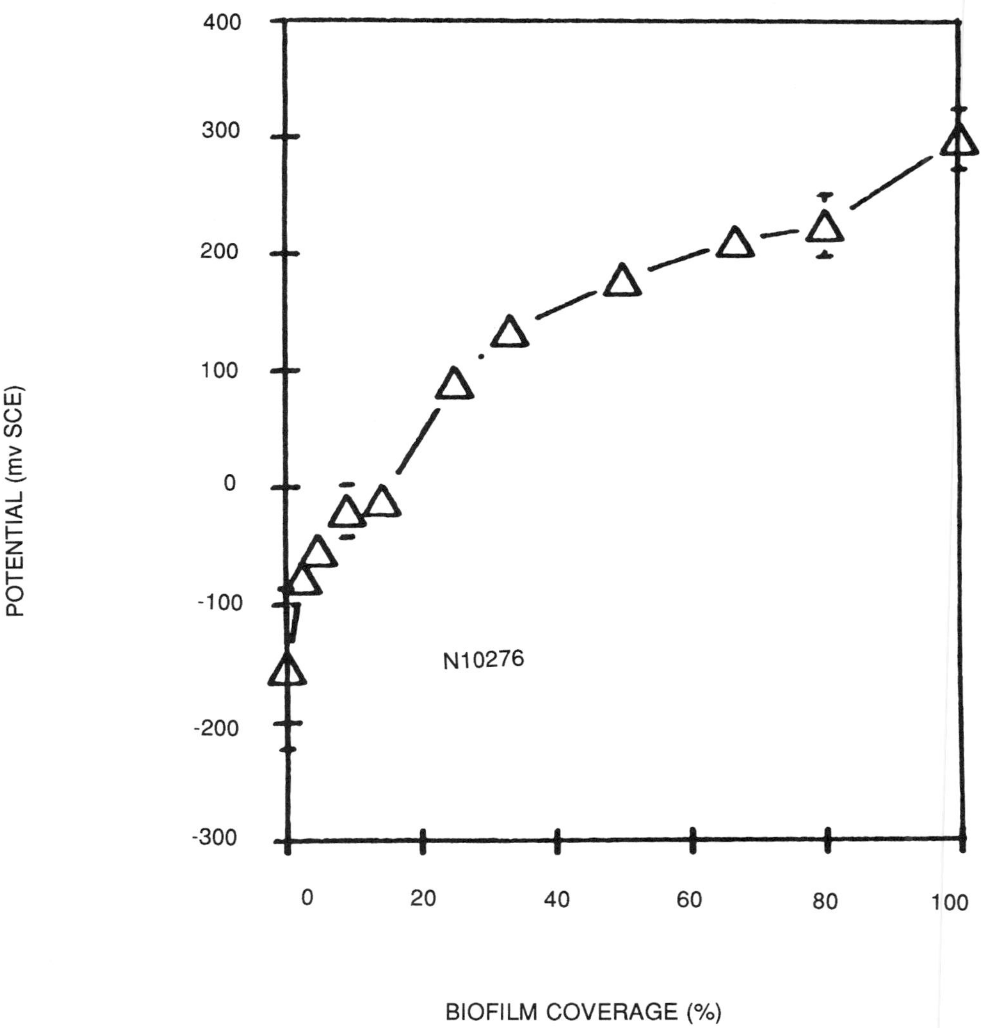

Figure 1. OCP Ennoblement as a Function of Persent Surface Coverage by a Predominantly Bacterial Biofilm on Passive Alloy N10276.

Salinity

Many coastal structures that use passive alloys are located on the tidal portions of rivers and bays in which the salt content of the water varies from nearly fresh to over 30 parts per thousand (ppt). Nominally full strength open ocean water is 35 ppt salinity. Microbial biofilms are known to form in natural waters of all salinities, thus, it is of interest to know how the change in OCP is affected by salinity.

Dexter and Zhang (1991) have studied the change in OCP produced by biofilms grown on a variety of passive alloys in waters ranging from 0.02 to 25 ppt salinity. They have shown that biofilms grown at all salinities are able to ennoble the OCP. Both the most noble potentials (about +475 mV SCE) and the largest amount of ennoblement (over 500 mV) were found in fresh water, and both decreased with increasing salinity. For instance, the most noble potential reached on alloy N08367 decreased from +460 mV in fresh water to +400 mV at 25 ppt salinity, while the average amount of ennoblement decreased from 475 to 250 mV over the same salinity range.

Sunlight

Stainless alloys are used in shipping and coastal structures under conditions ranging from complete darkness inside a heat exchanger to full sunlight and the daily light-dark cycle. Dexter and Zhang (1991) tested the effect of light on OCP under natural assemblage biofilms on the highly corrosion resistant alloy N10276. They found as shown in Figures 2 and 3 that ennoblement was the most persistent and reproducible under darkened, or continuously low light level conditions, where photosynthetic microorganisms, even if present in the film, would not be producing oxygen (see the dashed curve in Figure 2). When photosynthetic algae were present and the experiment was exposed to the day-night cycle (solid curve, Figure 2), there were large fluctuations in the OCP. On many, but not all samples, the fluctuations followed the light-dark cycle. Figure 3 is a magnified view of region "A" in Figure 2, and it shows that the most noble potentials (solid data points) were measured at night, while the potentials of the same sample in the day time (open data points) were usually about 150 mV less noble. They rationalized this behavior in terms of the photosynthetic activity of algae in the biofilm as explained in the next section.

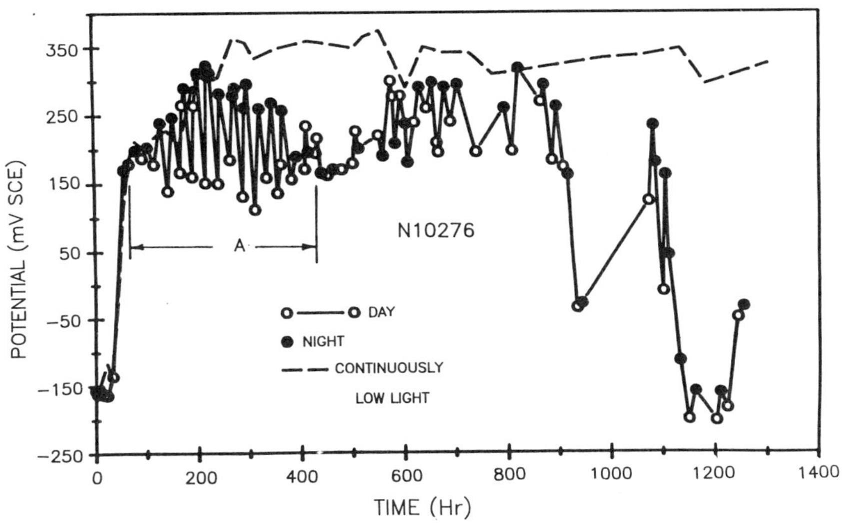

Figure 2. Effect of Sunlight on OPC of Alloy N10276 in Coastal Seawater. Dashed Line Represents Data Under Low Light Conditions. Open Circles - Data Takem in Daylight; Solid Circled - Data Taken at Night.

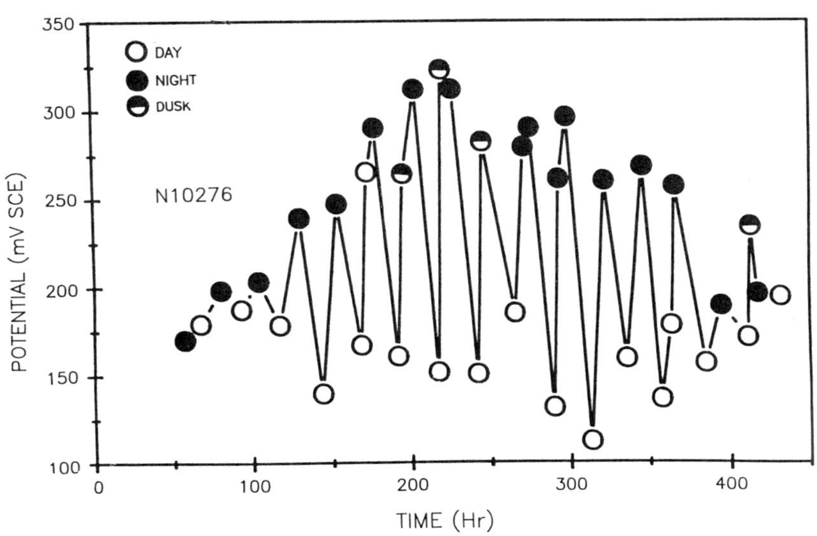

Figure 3. Details of OPC Variations Within Time Zone "A" of Figure 2.

PROPOSED MECHANISMS

Thermodynamic and Kinetic Effects

Ennoblement of the OCP as described above could come about from an influence of the biofilm on either the thermodynamics or kinetics of the prevailing cathode reaction. In a recent discussion of this phenomenon (Mansfeld, Little and Dexter, 1989; see also Dexter, 1992) it was shown how an ennoblement of the OCP could arise from either a thermodynamic effect on the reversible potential of the cathode reaction or a kinetic effect on the current density or Tafel slope of that reaction.

A biofilm on a metal surface can affect electrochemical parameters at the metal surface through both its physical presence and its metabolic activity. Physically, the film acts as a diffusion barrier, tending to concentrate chemical species produced at the metal-film interface and to retard diffusion of species from the bulk water toward the metal surface. Aller (1983) measured a 60 to 90 % reduction in the diffusion coefficients of inorganic solutes within organic materials produced by 8 species of marine invertebrates as compared to their values in pure water. LaMotta (1976) found that the effective diffusivities of glucose within a porous biological slime were 19 to 50 % of their molecular diffusivities in water. It is difficult to apply these numbers, which were measured in bulk materials, to the case of a thin biological film due to variations in film properties along the metal surface as mentioned above. The reduced diffusivities in the film, as compared to those in bulk solution, however, will contribute to a decrease in the average rate of transport of chemical species perpendicular to the metal surface.

Effect of pH

One effect of reduced diffusivities perpendicular to the metal surface is to concentrate acidic by-products from the metabolism of microorganisms in the biofilm. It is well known that the reversible oxygen potential shifts about 60 mV in the noble direction for each unit decrease in pH (Pourbaix, 1974). Reduced diffusivity in the biofilm, however, also makes it harder to maintain a positive oxygen tension at the metal surface, where oxygen is being consumed by both the electrochemical reaction and the respiratory activity of organisms. Dexter and Lin (presented at 7th Intl. Congr.on Marine Corrosion and Fouling, Valencia, Spain, 1988, but unpublished) used the Nernst Equation to calculate that the ennobling effect of a decrease in pH at the metal surface more than balanced the decrease in oxygen concentration within the film as long as conditions at the metal surface were not completely anaerobic. The results of their calculations are shown in Table 1.

Table 1. Variations in EO_2 with pH and PO_2 at 25°C.

pH	PO_2 (atm)	EO_2 (mV SCE)	Change in EO_2 (mV SCE)
8	0.20	+505	----
5	0.02	+668	+163
3	0.02	+786	+281

The value of +505 mV SCE at pH=8 represents that a bare metal surface, while the values at pH=5 and 3 represent possible conditions at a metal surface under a biofilm, where the oxygen partial pressure is expected to be low. This thermodynamic data predicts that the reversible oxygen potential will be shifted in the noble direction by the change in pH and oxygen tension expected under a biofilm, and that the maximum amount of change will be about 280 mV for pH=3. Since the oxygen electrode is not strictly reversible, even on platinum, one would expect the actual noble shift to be smaller than the calculated value.

Open circuit potential measurements on platinum in aerated seawater are shown as the lower curve in Figure 4 along with the thermodynamic data (upper curve). The pH of the solution was varied by additions of 0.1 N HCl and NaOH. The data show that the OCP of the oxygen electrode on platinum actually shifted about 170 mV noble for a pH change from 8 to 3. Similar measurements in deaerated solutions showed that the rate of potential change with decrease in pH was the same as in Figure 4 within experimental error. Equivalent data on Stainless alloy S31600 in aerated seawater showed a noble shift of 250 ± 25mV (220 mV in aerated 3% NaCl plus $NaHCO_3$ solution). This agrees closely with data for alloy S31600 in synthetic seawater reported by Little, et al. (1991). These data show that the often observed ennoblements of 250 mV or less on common stainless steels in full strength natural seawater can be explained by the pH mechanism alone if it is possible for the biofilm to establish a pH of about 3 at the metal surface. The larger amounts of ennoblement often observed on the high molybdenum stainless steels and superalloys in seawater and on all alloys in lower salinity waters, however, require some mechanism in addition to that of pH.

The effect of pH on ennoblement also provides a basis for understanding the effect of sunlight on OCP ennoblement. During the dark hours of the day, all the organisms in the biofilm undergo respiration, taking in oxygen and producing carbon dioxide. The production of carbonic acid would drive the pH of the water trapped in the biofilm in the acid direction (Fogg, 1975; Dexter and Culberson, 1980) adding to acidification from the direct production of other acidic by-products (Pope et al., 1988) by bacteria in the film. The acidic pH would have an ennobling effect on the OCP as described. In the daylight hours, however, photosynthetic organisms (bacteria, algae and diatoms) in the film use CO_2 in the production of oxygen. This would drive pH in the basic direction, perhaps reaching values above 10 (Edyvean and Moss, 1986), which are sufficient to interfere with the metabolism of acid producing bacteria in the film (Gaudy and Gaudy, 1980). Thus, ennoblement of the OCP would tend to decrease in the day time. The daily light-dark cycle, then would lead to a corresponding fluctuation in the amount of ennoblement as reported by Dexter and Zhang (1991).

Recent work by Dowling et al. (1992) showed that the large fluctuations in potential reported above could not be due to an influence of photosynthesis on oxygen concentration alone. They found that changes in oxygen concentration with variations in illumination in a photosynthetic biofilm brought about changes in OCP of 15 mV or less. Their measured changes in OCP agreed closely with those predicted by the Nernst Equation for oxygen concentration variations at constant pH.

Bacterially Produced Enzymes

Scotto (1989) has proposed that oxygen reduction at the cathode surface might be enhanced by bacterially produced enzymes. One possibility for this mechanism involves the suite of respiratory enzymes that bacteria in the biofilm produce to control toxic oxygen derivatives. Hydrogen peroxide, superoxide and hydroxyl radical are all toxic by-products of oxygen reduction (or respiration). The concentrations of these by-products must, therefore, be controlled by the organisms for survival. The enzyme, superoxide dismutase, catalyses the destruction of superoxide free radical (O_2^-) according to:

$$2O_2^- + 2H^+ \text{---->} O_2 + H_2O_2$$

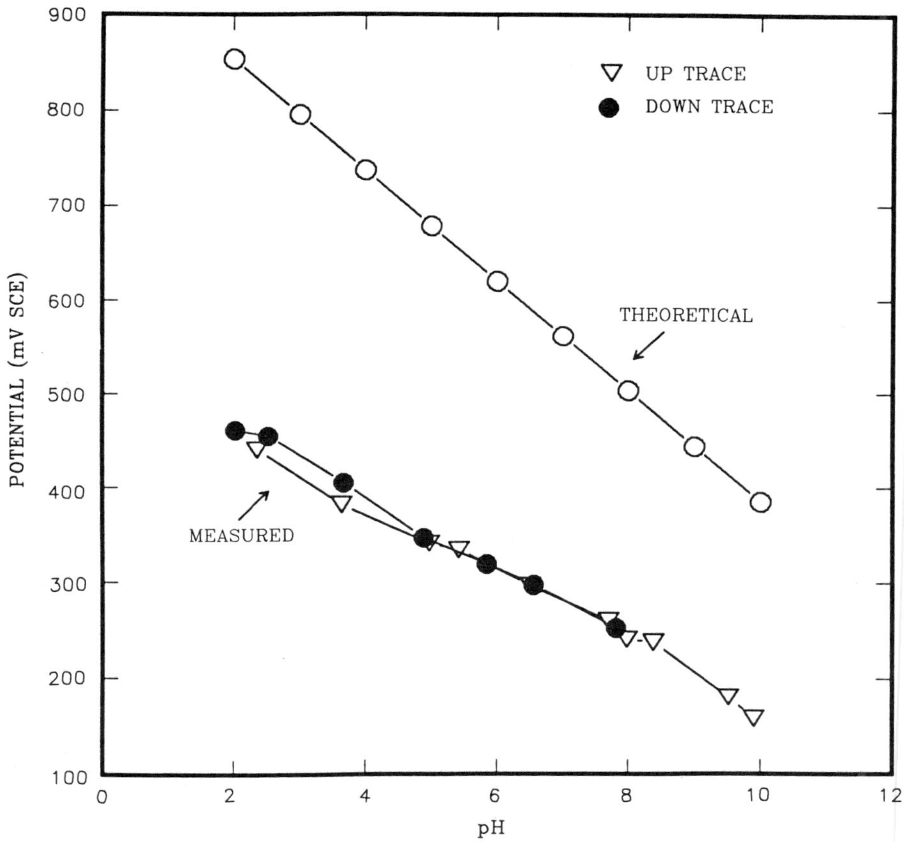

Figure 4. Effect of pH Variations in Seawater on the OPC of the Oxygen Electrode on Platinum. pH was Controlled by Adding 0.1N HCl and NaOH.

Superoxide dismutase, however, is inactivated by the build up of H_2O_2. Therefore, it is usually associated with two other enzymes, catalase and peroxidase, which break down hydrogen peroxide to water and oxygen according to:

$$2H_2O_2 \text{ ----> } 2H_2O + O_2$$

A possible mechanism for ennoblement may involve the ability of biofilm organisms to maintain a balanced concentration of oxygen and peroxide within the film. Peroxide has a more noble redox potential (1.76 V NHE) than oxygen (1.23 V NHE), which may contribute to the ennoblement phenomenon. These effects are currently under investigation.

Organo-Metallic Catalysis

Johnsen and Bardal (1985) and Dexter and Gao (1988) have proposed that the mechanism of potential ennoblement might involve the catalytic action of bacterially produced organo-metallic complexes on the oxygen reaction at the metal surface. Presumably, this catalysis would effect the exchange current density of the oxygen electrode (a kinetic effect), rather than the reversible potential (thermodynamic). The scientific basis for this mechanism comes from extensive use of such compounds to catalyze reduction of oxygen to water in fuel cells (see the review by Van den Brink et al., 1980).

These fuel cell processes emulate the natural reduction of oxygen in biochemical systems in which oxygen is both transported and reduced by organic complexes of transition metal ions. The transition metal ion is chelated in a square planar complex by a macrocyclic organic molecule having four nitrogen atoms as ligands (Van den Brink et al., 1980; Hancock and Martell, 1989). The fuel cell research has utilized organic compounds of the phthalocyanin, porphyrin and tetraazaannulene types, but only the porphyrins are known to be produced by marine bacteria. In that work, substantial catalysis of oxygen reduction has been achieved by organo-complexes with the heavy metals: Fe(II) and Fe(III), Co, Mn and Ni. Most of the stainless alloys and superalloys shown to be ennobled by biofilms contain one or more of these heavy metals as alloying elements, thus, providing a ready source of them at the metal-film interface.

One objection to this mechanism (see Mansfeld, Little and Dexter, 1989) has been that ennoblement is also observed on titanium and platinum, which do not contain the required heavy metals. Microorganisms in the biofilm, however require most of these heavy metals as micronutrients (Brock and Madigan,

1988). Hence, in a mature biofilm the needed heavy metals, as well as the porphyrins, may be provided by the organisms themselves as they die and release their cell contents to the biofilm matrix. Another objection to the oxygen catalysis mechanism is that the oxygen tension at the metal surface under a mature biofilm can fall to zero, as shown by microelectrode measurements (Lewandowski et al., 1988). Even this potentially serious problem might be overcome by the biochemical system as long as oxygen reaches the outer layers of the biofilm (ie as long as the ambient environment is aerobic). This would require biochemical methods of electron transport to connect the metal surface with the aerobic layer of the biofilm.

EFFECT ON CORROSION

The question now is, what effect does the observed ennoblement of the OCP by a distributed biofilm have on corrosion of passive metals and alloys in marine environments? In the following sections, the effect on localized corrosion in the form of pitting and crevice corrosion will be discussed, followed by a consideration of the effects expected on galvanic corrosion and cathodic protection.

Pitting

Total penetration perpendicular to the surface by pitting is a function of both initiation and propagation. Initiation of pits on passive alloy surfaces in chloride containing media depends on the chloride ion concentration and the electrode potential. The more electropositive (or noble) the potential of the metal the easier it will be for chloride ions to penetrate the passive film at local weak spots and create small anodic areas. Thus, anything that causes the OCP to change in the positive direction (ennoblement) increases the probability of pit initiation at a given chloride ion concentration.

Evidence has been presented above for the ability of biofilms to ennoble the OCP. This means that the formation of biofilms is expected to increase the likelihood of pit initiation. If one could measure a meaningful pitting potential at the metal-biofilm interface, then it would be possible to predict whether or not a biofilm would cause pit initiation just by measuring the OCP. The difficulty in doing this is that very little detailed information is yet known about the chemistry at the metal-film interface, or its variability in the presence of a spotty film. Dexter and Zhang (1991) tried comparing the maximum measured ennobled potential to the critical pitting potential for stainless alloys S30400 and S31600 as a function of water salinity.

From this data Dexter and Zhang (1991) concluded that the maximum salinities at which these alloys should be resistant to pit initiation in the absence of crevices should be reduced by the ennobling action of a biofilm. Their general conclusion is correct, but their results cannot be used for predicting localized corrosion behavior in practical applications for at least two reasons. First, they considered only chloride ion activity when determining the critical pitting potentials, ignoring other chemical changes (eg., pH or production of inhibitors) within the biofilm. There is currently some confusion about the effect of pH on the critical pitting potential. Little et al., (1991) reported that the critical pitting potential for Type 304 stainless steel measured potentiodynamically in 3.5% NaCl solution fell from about 0.3 V SCE at pH 8 to about zero V SCE at pH 2. Little et al., (1991), however, did not evaluate their findings against other data taken in the 1960s showing that pH had little influence on the critical pitting potential of 18Cr-8Ni stainless steel in 0.1N NaCl in the acid pH range (Leckie and Uhlig 1966; Fokin et al., 1965; Kolotyrkin et al., 1963). Thus, it is probably fair to say that we do not yet fully understand how the chemical changes under a microbial biofilm influence the critical pitting potential.

The second reason why Dexter and Zhang's results are difficult to apply is that stainless steels in chloride bearing marine environments are almost always found to undergo crevice corrosion rather than pitting, and crevice corrosion can initiate over a wide range of potentials. The effect of biofilms on crevice corrosion will be considered in the next section. Propagation of corrosion in pits and crevices after initiation will also be dealt with in the next section.

Some of the newer stainless alloys and superalloys such as N08367 and S44735 have critical pitting potentials more positive than +800 mV SCE. These alloys can withstand ennoblement to +400 - 500 mV SCE produced by a biofilm without pit initiation, so it can be stated with certainty they will suffer no direct consequence of biofilm produced ennoblement, at least in the absence of crevices.

Crevice Corrosion

Under the actual conditions in which stainless alloys are used in industrial service, crevices are always present. Any surface film that is not continuous, including a biofilm, can create crevice like conditions, under which localized corrosion initiation can take place over a wide range of potentials. Johnsen and Bardal (1985) said that OCP ennoblement should increase the tendency for crevice corrosion initiation and propagation as well as that for pitting. Dexter et al. (1986) showed that bacteria in the crevice solution could contribute to the depletion of oxygen, potentially decreasing crevice initiation times. Using the remote crevice assembly technique, Valen (1986) found that the crevice corrosion current density for alloy S31603 in natural seawater was two orders of magnitude higher than in synthetic seawater even though the cathode-to-anode area ratio was larger in the synthetic seawater tests. Gallagher et al. (1988) used the multiple crevice assembly technique to study crevice corrosion of stainless alloys in natural, synthetic and transported seawaters. They found that crevice corrosion was dramatically less in artificial than in natural seawater, and they attributed the difference to microorganisms in the natural water.

Recently, Zhang and Dexter (1992) have used traditional crevice assembly and remote crevice assembly tests to show that both initiation and propagation of crevice corrosion are affected by biofilms on alloys with low resistance (S30400 and S31603) and intermediate resistance (NO8904, S31703 and S31803) to crevice corrosion in coastal seawater. Crevice initiation times in their experiments were 2 to 5 times shorter in natural seawater than in a control water made by filtering the same water to remove most of the film forming microorganisms. On many alloys they also measured increases in the: percent of crevice area that was attacked, maximum and average depths of attack, corrosion current and weight loss for the biofilmed samples versus controls.

Zhang and Dexter (1992) concluded that the acceleration of crevice corrosion by the action of biofilms in their coastal waters was more severe and more consistent on the intermediate resistance alloys, N08904 and S31703, than it was on the low resistance alloys, S30400 and S31603, and on alloy S31803. The biofilm had no effect in their tests on alloy N08367, but that does not mean this alloy is immune to microbial acceleration of crevice corrosion.

The evidence is that ennoblement itself, and hence any accelerating effects that has on crevice corrosion and pitting, is often reduced when the corroding metal has access to sunlight (Dexter and Zhang, 1990). Thus, the most severe condition is expected to occur under the darkened conditions characteristic of many industrial systems such as heat exchangers. It is under those conditions that the most benefit to corrosion should be realized in biofilm formation on the metal surface, provided of course, that the fouling control method does not itself accelerate corrosion.

Galvanic Corrosion

There is no direct evidence yet on the effect of biofilms on galvanic corrosion. Since the rate of galvanic corrosion, however, is often controlled by the reaction on the cathodic member of the couple, any increase in kinetics of the cathodic oxygen reaction on passive metals has the potential to increase the severity of these metals as cathodes in galvanic couples. Work summarized by Scotto (1989) showed that the upper (or noble) portion of the cathodic polarization curve was shifted toward higher current densities, while the lower portion, dominated by the limiting diffusion current for oxygen diffusion, was affected to a lesser extent. The cathodic polarization data suggest that the rate of galvanic corrosion on the anodic member of the couple should be increased by the action of a biofilm if the mixed potential of the couple is within that noble portion of the cathodic polarization curve.

For a series of galvanic couples with various anodes but the same passive alloy as the cathode, one would expect the accelerating effect of a biofilm on the cathode to decrease as the couple potential decreased. In the absence of actual data, then, the prediction is that galvanic corrosion of a copper alloy would be increased by the action of a biofilm on a stainless alloy to which it was connected because the couple potential would be quite noble. The acceleration of corrosion on steel connected to the same passive alloy would be less because the couple potential would be lower. An important question is whether the consumption of sacrificial anodes (usually aluminum, zinc or magnesium) is increased by the action of biofilms on the cathode being protected. The prediction based on the cathodic polarization data cited above is that the effect on these active materials would be small because the couple potential is low. This prediction is currently being tested by actual galvanic corrosion experiments.

Cathodic Protection

The economics of cathodic protection depend on the amount of current required to polarize the structure to the protected potential. The required current in turn depends on the exposed area of bare metal surface and the electrochemical properties of that surface. We have already documented that a biofilm can enhance the cathode reaction on passive metal surfaces. The main consequence to cathodic protection of such an effect on the cathode reaction is an increase in the applied current needed to protect the metal. Johnsen and Bardal (1985, 1986) showed the influence of biofilms on the applied current density needed to polarize various passive metals to potentials of -400 and -1100 mV vs. the saturated calomel electrode. The current for holding the metals at -400 mV, corresponding to the current for protecting stainless steels from pitting in seawater, started below 10 mA/m^2, but increased rapidly to over 300 mA/m^2 as a biofilm formed between 4 and 8 days of exposure. For the alloys held at -1100 mV, corresponding to a slight overprotection of steel, there was no such increase in the applied current. Dexter and Lin (1992) published similar data, and they offered an explanation (Dexter and Lin, 1991; 1992) based on interactions between a biofilm and the change in pH at the metal surface caused by the applied current. They argued that the applied current necessary to hold the metal at -1100 mV was large enough to shift the pH at the metal surface above 9, a value high enough to interfere with the metabolism of bacteria in the biofilm and negate any biological enhancement of the cathodic reaction. In contrast, the smaller applied current for holding the metal at -400 mV would not give an appreciable increase in surface pH. Thus, at the lower current density, the organisms would still be able to enhance the cathodic reaction and increase the required current density.

Even though high applied current density levels reduce or eliminate the ennobling effect of biofilms, the presence of biofouling is widely regarded (see Edyvean, 1984; Edyvean, et al., 1985; Edyvean and Moss, 1986) as increasing the applied current density for complete protection of steel on offshore structures. The reasons for this given in the literature are partly related to the usual spotty coverage of the metal surface by the fouling, leading to oxygen, pH and chemical concentration cells, and partly due to the influence of sulfate reducing bacteria (SRB). The distribution of SRB in shielded niches, such as under deposits and the edges of disbonded coatings, makes it harder to supply cathodic protection current to the corroding area, thus, necessitating a higher applied current density.

Partially counteracting these above negative effects of biofilms on cathodic protection is the effect of biofilms on the distribution of calcareous deposits. Dexter and Lin (1991, 1992) showed that a biofilm formed on the metal surface before the application of cathodic protection caused the calcareous deposit resulting from the applied current to be distributed more evenly on those areas of the metal surface covered by the biofilm. This effect is expected to make cathodic protection more efficient.

SUMMARY

The effects of biofilm surface coverage, water salinity and sunlight on ennoblement of the OCP of passive metals and alloys immersed in natural waters have been reviewed. Ennoblement is largest and most persistent in low salinity waters under darkened conditions.

The mechanism of ennoblement is not completely understood. Most of the effect observed in full strength seawater can be explained on the basis of a decrease in pH at the metal surface under the biofilm. However, there are questions about how low a pH can be maintained under the film and what effect the low pH has on the potential at which localized corrosion initiates. Other mechanisms for ennoblement could involve catalysis by organo-heavy metal complexes or bacterially produced enzymes. The respiratory enzyme system of marine microorganisms may also stabilize a high enough peroxide concentration at the metal-film interface to make peroxide reduction a significant contribution to the cathode reaction in addition to oxygen reduction.

Ennoblement of the OCP has several effects on corrosion of stainless alloys. It has been well demonstrated that ennoblement of the OCP decreases the initiation time for both pitting and crevice corrosion of susceptible alloys. Ennoblement also increases the propagation rate of crevice corrosion. It is suspected of increasing the severity of stainless steels as the cathodes of galvanic couples, although no data has yet been published, and there is some question as to whether the consumption of anodes more active than zinc would be affected. Finally, ennoblement can be either beneficial or detrimental to the process of cathodic protection.

Acknowledgements

The work on which this paper is based was supported by NOAA Office of Sea Grant, U.S. Deptartment of Commerce, under Grant No. NA86AA-D-SG040.

REFERENCES

Aller, R.C. (1983). The importance of the diffusive permeability of animal burrow linings in determining marine sediment chemistry. J. Marine Res. 41, 299-322.

Brock, T.D. and Madigan, M.T. (1988). Biology of Microorganisms, 5th Ed., Prentice Hall.

Dexter, S.C. (1992). Role of microfouling organisms in marine corrosion. Biofouling, in press.

Dexter, S.C. and Culberson, C. (1980). Global variability of natural seawater. Materials Performance, 19, No. 9, 16-28.

Dexter, S.C., Lucas, K.E. and Gao, G.Y. (1986). The role of marine bacteria in crevice corrosion initiation. In: Biologically Induced Corrosion, pp. 144-153 (S.C. Dexter, ed.,), National Assoc. of Corrosion Engineers, Houston, TX.

Dexter, S.C. and Gao, G.Y. (1988). Effect of seawater biofilms on corrosion potential and oxygen reduction of stainless steel. Corrosion, 44, 717-723.

Dexter, S.C. and Lin, S-H. (1991). Effect of marine bacteria on calcareous deposition. Materials Performance, 30 No. 4, 16-21.

Dexter, S.C. and Lin, S-H. (1992). Effect of marine bacteria on cathodic protection. Intl. Biodeterioration and Biodegredation, 29, 231-249.

Dexter, S.C. and Zhang, H-J. (1990). Effect of biofilms on corrosion potential of stainless alloys in estuarine waters. Proc. 11th Intl. Corrosion Cong. Florence, Italy, Vol. 4, 333-340.

Dexter, S.C. and Zhang, H-J. (1991). Effect of biofilms, sunlight and salinity on corrosion potential and corrosion initiation of stainless alloys. EPRI NP-7275, Final Report on Project 2939-4, Electric Power Research Inst., Palo Alto, CA.

Dowling, N.J.E, Guezennec, J., Bullen, J., Little, B. and White, D.C. (1992). Biofouling, 5, 315-322.

Edyvean, R.G.J. (1984). Interactions between microfouling and the calcareous deposit formed on cathodically protected steel in seawater. Proc. 6th Intl. Congr. on Marine Corrosion and Fouling, Marine Biology, Athens, Greece, 469-483.

Edyvean, R.G.J., Terry, L.A. and Picken, G.B. (1985). Marine fouling and its effects on offshore structures in the North Sea - A Review. Intl. Biodeterioration, 21, 277-284.

Edyvean, R.G.J. and Moss, B.L. (1986). Microalgal communities on protected steel substrata in seawater. Estuarine, Coastal and Shelf Sci., 22, 509-527.

Fogg, G.E. (1975). Primary Productivity, In: Chemical Oceanography, 2nd Ed., Vol 2, pp. 385-453 (J.P. Riley and G. Skirrow, eds.), Academic Press, New York.

Fokin, M.N., Kurtepov, M.M. and Bochkareva, V.I. (1965). Sbornik po Korozii, Moscow.

Gallagher, P., Malpus, R.E. and Shone, E.B. (1988). Br. Corrosion J., 23, No. 4, 229.

Gaudy, A. and Gaudy, E. (1980). Microbiology for Environmental Scientists and Engineers. p. 183, McGraw-Hill Book Co., New York.

Hancock, R.D. and Martell, A.E. (1989). Chem. Rev., 89, 1875.

Johnsen, R. and Bardal, E. (1985). Cathodic properties of different stainless steels in natural seawater. Corrosion, 41, 296-304.

Johnsen, R. and Bardal, E. (1986). The effect of microbiological slime layer on stainless steel in natural seawater. Paper No. 227, Presented at Corrosion/86, NACE:Houston, TX.

Kolotyrkin, Ya.M., Golovina, G.V. and Florianowich, G.M. (1963). Dokl. Akad. Nauk SSSR, 148, 1106.

Lamotta, E.J., (1976). Internal diffusion and reaction in biological films. Env. Sci. and Tech., 10, 765-769.

Leckie, H.P. and Uhlig, H.H. (1966). Environmental factors affecting the critical potential for pitting in 18-8 stainless steel. J. Electrochem. Soc., 113, 1262-1267.

Lewandowski, Z., Lee, W.C. and Characklis, W.G. (1988). Dissolved oxygen and pH microelectrode measurements at water immersed metal surfaces. Paper No. 93, Presented at Corrosion/88, NACE:Houston, TX.

Little, B., Ray, R., Wagner, P., Lewandowski, Z., Lee, W., Characklis, W., and Mansfeld, F. (1990). Paper No. 150, Presented at CORROSION/90, NACE:Houston, TX.

Little, B., Ray, R., Wagner, P., Lewandowski, Z., Lee, W., Characklis, W., and Mansfeld, F. (1991). Biofouling, 3, 45-59.

Mansfeld, F. and Little, B. (reply by Dexter, S.C.). (1989). Discussion on effect of seawater biofilms on corrosion potential and oxygen reduction on stainless steel. Corrosion, 45, 786-789.

Mansfeld, F., Tsai, R., Shih, H., Little, B., Ray, R. and Wagner, P. (1990). Paper No. 109, Presented at CORROSION/90, NACE:Houston, TX.

Mansfeld, F., Tsai, R., Shih, H., Little, B., Ray, R. Wagner, P. (1992). Corrosion Science, 33, No. 3, 445-456.

Mollica, A. and Trevis, A. (1976). Correlation entre la formation de la pellicule primaire et la modification de la cathodique sur des aciers inoxydables experimentes en eau de mer aux vitesses de 0.3 A 5.2 m/s. Proc. 4th Intl. Cong. Marine Corrosion and Fouling, Juan-Les-Pins, Antibes, France, 351-365.

Pope, D.H., Zintel, T.P., Kuruvilla, A.K. and Siebert, O.W. (1988). Organic acid corrosion of carbon steel. Paper No. 79, Presented at CORROSION/88, NACE:Houston, TX.

Pourbaix, M. (1974). Atlas of electrochemical equilibria in aqueous solutions. National Assoc. of Corrosion Engineers, Houston, TX, 97.

Scotto, V. (1989). Electrochemical studies of biocorrosion of stainless steel in seawater. Proc. EPRI Workshop, Microbial Corrosion: 1988, Electric Power Research Institute, Palo Alto, CA, B-1 to B-36.

Scotto, V., DiCintio, R. and Marcenaro, G. (1985). The influence of marine aerobic microbial film on stainless steel corrosion behaviour. Corrosion Science, 25, 185-194.

Valen, S. (1986). Thesis, Department of Materials and Processes, Norwegian Institute of Technology, Trondheim, Norway.

Van den Brink, F., Barendrecht, E. and Visscher, W. (1980). Recueil. J. Royal Netherlands Chemical Soc., 99, No. 9, 253.

Zhang, H-J. and Dexter, S.C. (1992). Effect of marine biofilms on crevice corrosion of stainless alloys. Paper No. 400, Presented at CORROSION/92, NACE:Houston, TX.

NEW TRENDS IN BIOCORROSION/BIOFOULING MONITORING TECHNIQUES

HECTOR A. VIDELA, Consultant, Marketing Department, Aquatec Quimica S.A., Rua Sampaio Viana 425, Sao Paulo, SP. 04004, Brazil

INTRODUCTION

An industrial plant contains several environments where corrosion and fouling processes are potentially troublesome, such as cooling water systems, storage tanks, water and wastewater treatment facilities and piping (Characklis, 1986). Corrosion is one of the undesirable results of fouling, in addition to energy losses owing to increased heat transfer and frictional resistances. A recirculating cooling water system provides a good example of an industrial environment where corrosion and fouling hazard is particularly critical and needs to be kept under strict control.

Until the 80's, inorganic corrosion and inorganic fouling (scaling) have been the two major objectives of cooling water treatments. A typical treatment program generally consisted in the use of a synergistic mixture of anodic and cathodic corrosion inhibitors such as chromates and zinc, seldomly combined with phosphates and polymers, to alleviate the aggressive effects of concentrated water. Scale was controlled by the addition of sulfuric acid to transform insoluble bicarbonates into the more soluble sulfates. The possible deleterious effects of the low pH (6.0 - 7.0), used in these water treatment programs, was properly counterbalanced by the aforementioned corrosion inhibitors, while microbial contaminants were maintained at low levels, through the use of chlorine, occasionally combined with pentachlorophenol or quaternary ammonium compounds (Lamont, 1988).

Biodeterioration Research 4, Edited by G.C. Llewellyn *et al.*, Plenum Press, New York, 1994

Since the beginning of the 80's, a drastic change was produced in water treatment strategies as a consequence of more strict environmental concerns and increases regulatory pressures. The generalized use of chromates has been eliminated and, high pH values (7.0 - 9.0) have been adopted. Organic phosphonates are now generally used, in association with polycarboxilates to achieve scale control. The undesirable result of these new strategies is an increase of suspended solids, greater numbers and diversity of microorganisms, and elevate amounts of organic and inorganic deposits.

Thus the major concern now is to improve methods for maintaining heat transfer efficiency by controlling biological deposits, while still preserving corrosion inhibition and avoiding scale deposition.

Fortunately, a better understanding of biocorrosion and biofouling processes as a consequence of an increasing intellectual and technical cross-fertilization of ideas between researchers from different disciplines, has provided the industry with new strategies for monitoring biocorrosion and biofouling to counterbalance their undesirable effects.

This paper will review different microbiological and electrochemical methods, as well as several surface analysis and microscopic techniques frequently used to monitor biofouling/ biocorrosion in the field or in the lab.

A new non-conventional reusable appliance used as sampling device, will be described in this article, jointly with a realistic strategy for monitoring biofouling and biocorrosion hazard in industrial water systems.

NATURE OF FOULING DEPOSITS

A metal surface exposed to a natural or industrial water environment undergoes a sequence of biological and inorganic changes that lead to fouling and corrosion. The biological changes are due to the irreversible adhesion of different types or microorganisms (bacteria, microalgae, fungi and protozoa) on the metal surface, through the production of extracellular polymeric substances (EPS). This permanent bonding results in the development of a biofilm consisting mainly of water, bacterial cells and their EPS. Thus a biofilm is the result of an accumulation process at the interface which is not necessarily uniform in time or space (Characklis and Marshall, 1990).

Another deposition process of inorganic nature, usually designated as scaling, occurs when a crystalline growth of insoluble salts or oxides is produced on a heat exchanger surface. The rate of this process is a complicated function of many variables such as temperature, concentration of scale-forming species, pH, water quality and hydrodynamic conditions.

To complete the sequence of inorganic changes, a process of metal dissolution (corrosion) and corrosion products formation (passivation) occurs simultaneously with biofilm formation. Although corrosion and biofouling on metal surfaces take place in aqueous media within the same time scale, both processes occur in opposite directions (Videla, 1989). Corrosion and subsequent accumulation of corrosion products are directed from the metal towards the solution, whereas biofouling and scaling are produced in the opposite direction. Thus, a complex biologically and electrochemical conditioned interface is produced as a result of the interactions between inorganic and biological processes. The consequent corrosion behavior of the metal substratum will vary according to the extent of this reciprocal interaction (Videla, 1991). An adequate understanding of the effects of biofilms on the corrosion process is needed to guarantee effective water treatment in cooling water systems. The interrelationship between inorganic corrosion and biofouling of heat transfer equipment must be controlled if long-term reliability is to be achieved.

MONITORING BIOCORROSION/ BIOFOULING: WHAT TO MEASURE AND HOW TO DO IT

According to the different elements participating in the complex fouling and corrosion processes at a metal/ solution interface, different kinds of monitoring techniques are needed to properly cover all the variables affecting the behavior of the metal substratum.

Microorganisms influence the corrosion reaction by several mechanisms operating simultaneously or consecutively, but in very seldom cases (e.g. acid corrosion by sulphur oxidizing bacteria), a single cause can account for the microbial effects on corrosion.

Electrochemical conditions at the metal surface are changed by microbial activity. These changes may have many effects, ranging from the induction of localized corrosion, to a change in the rate of general corrosion or to corrosion inhibition (Videla, 1991). In each case the corrosion process is electrochemical and electrochemical techniques can be useful, in the investigation of biofouling and biocorrosion interactions.

Microscopic observations are significantly valuable to assess the degree and type of metal attack, to verify the morphological characteristics of microbial films, and to evaluate the amount and variety od sessile microorganisms at the interface.

Surface analysis will supply the information needed to determine the chemical composition of biological and inorganic deposits formed during biofouling and biocorrosion processes. Surface analysis techniques can be essential to correlate the chemical composition of corrosion products and biofilms with their causative microorganisms.

Microbiological Methods

Traditionally, the quantitative evaluation of microorganisms in industrial water systems has largely bee restricted to the planktonic population. However, monitoring of the sessile population has several advantages (Gilbert and Herbert, 1987): i) biofilms represent more properly the microbial population of a system in terms of numbers or biomass; ii) microbial species which predominate in the bulk phase may be different from those present in the biofilm; iii) susceptibility of sessile microorganisms to antimicrobial agents markedly differs from that of equivalent microorganisms dispersed in the bulk water phase, because of the protective action of their EPS.

Direct evaluation of sessile population will be described within microscopic methods. Indirect microbial methods for sessile bacteria quantification include: a) viable counts; b) biomass assessments and c) activity measurements.

a) Viable count techniques provide an estimation of the number of viable bacteria present in a biofilm without requiring specialized equipment. However, some of the disadvantages of these techniques are: i) they require a prior removal of the biofilm from the coupon surface; ii) they do not intricate properly the number of bacteria in terms of activity (Gilbert and Herbert, 1987).

b) Biomass assessment can be accomplished indirectly by means of ATP measurements (Prasad, 1988), fatty acid profiles (White et al., 1986) and different serological techniques such as enzyme linked immunosorbent assay (ELISA), (Cook and Gaylarde, 1987).

c) Activity measurements have been recently considered as the most important criterion for sessile bacteria quantification (Hamilton, 1985). Among these methods, radiorespirometric techniques provide a sensitive and rapid means of determining microbial activity in the biofilms. This method has been mainly used to determine the activity of specific groups of bacteria, like SRB in the oil production industry.

Electrochemical Methods

The uses and limitations of electrochemical techniques for investigating biocorrosion have been thoroughly reviewed by Dexter st al. (1991). Thus, only a brief overview of the electrochemical methods most frequently used for monitoring biocorrosion/ biofouling will be presented here.

The corrosion potential or open-circuit potential, because of its simplicity has been widely used in biocorrosion studies for many years. Corrosion potential variation with time can be measured by determining the voltage difference between a metal immersed in a liquid medium and an appropriate reference electrode (generally, the standard calomel electrode). The magnitude and sign of the corrosion potential are functions of the metal itself and of the composition, temperature and hydrodynamics of the electrolyte. To measure the corrosion potential a reference electrode, an high impedance voltmeter and in most cases a suitable recording device are needed. Several papers have been recently published (Scotto et al., 1985; Dexter and Gao, 1988) using variations of the corrosion potential with time to assess the effects of biofilms on the corrosion behavior of stainless steel in seawater.

Another simple electrochemical parameter, useful in predicting whether a certain metal will be corroded or not in a given biologically conditioned environment, is the reduction-oxidation potential or redox potential. At the redox potential, the rates of the oxidation and reduction reactions at the platinum surface are assumed to be equal, and the redox potential itself is a function of the oxygen concentration, temperature and pH of the environment. The use of the redox potential jointly with the corrosion potential, can be of special importance to determine whether the oxidation conditions of the environment can reach the values needed to initiate localized corrosion.

Direct current polarization has been used in biocorrosion since the beginning of the 1960's for assessing SRB effects on the anaerobic corrosion of iron. Among these methods, the current-potential diagrams (Evans diagrams) are useful graphical methods to demonstrate the relationship between corrosion rate and the extent of polarization of anodic and cathodic reactions participating in the corrosion process. Several limitations should be kept in mind when applying these methods to biofilmed metal surfaces: i) the large polarizations needed may change the electrochemical conditions at the surface and may be deleterious to microorganisms in the biofilm; ii) these measurements alone do not give information on the relative contribution to corrosion of the biofilm.

Potentiodynamic-sweep techniques are more useful in characterizing and predicting the corrosion behavior when the metal shows a passive behavior due to the presence of a protective (passive) film. Two electrochemical parameters obtained through these methods have been used to assess biofilm/passive layers interactions in the corrosion behavior of a certain metal alloy: i) the pitting potential (Ep), i.e. the potential below which the metal remains passive and above which pitting nucleates on the surface (Szklarska-Smialowska, 1986); ii) the repassivation potential (Er), i.e. the potential associated with a drop in current caused by the repassivation of pits. The potential range between the Er and the Ep in a polarization curve, and its variation between sterile and biologically conditioned media, can be useful for determining biofouling effects on the corrosion behavior of the base metal (Videla, 1991).

Polarization resistance measurements are based on the nature of the linear relationship between changes in the applied potential and the resultant current density, when the applied potentials are within about $\pm$ 10 mV of the corrosion potential. The slope of the potential-current curve is approximately linear and has units of resistance. A number of commercial instruments are now readily available (e.g. the Corrater™), for monitoring instantaneous corrosion rates in industrial plants through linear polarization measurements.

Electrical resistance probes (e.g. the Corrosometer™) are widely used as industrial monitoring techniques jointly with the classical gravimetric methods using corrosion coupons. A this metallic wire or strip sensing element is immersed in the corrosive medium and its electrical resistance is monitored with time. A constant temperature, corrosion of the probe, which causes a decrease in cross-sectional area, can be monitored by periodically measuring the electrical resistance. The average corrosion rate is determined from the increase in resistance between any two readings. The electrical resistance method is excellent to indicate a change in the general corrosion rate, but the results are harder to interpret in the presence of localized forms of corrosion-like pitting.

During the last years, several electrochemical techniques used in inorganic corrosion have been applied to biocorrosion studies. Some of these techniques are: i) electrochemical impedance spectroscopy (EIS); ii) electrochemical noise and iii) split-cell measurements. A discussion on the advantages and disadvantages of those techniques for monitoring biocorrosion/ biofouling is out of the scope of this article and can be found in the reference of Dexter et al. (1991).

Microscopical Methods

Examination of biofilms and quantification of sessile population can be undertaken by direct observation through different microscopical methods.

In epiflourescence (incident light) microscopy, bacteria are stained with a fluorescent dye and later observed under an incident light source and filter combined to produce fluorescence of the stained bacteria. Epiflourescence techniques allow the enumeration of total cells present in a sample, even those that would be unable to grow on the culture media used in the laboratory. Thus, bacterial counts obtained with this method are higher than those obtained by cultural methods. The inconvenience can be overcome by using viable stains (e.g. fluorescein diacetate). An important improvement of this technique is achieved through the use of immunofluorescence methods that provide a means of identifying specific microbial species within the biofilm (Pope, 1986).

Scanning electron microscopy (SEM) has been one of the most frequently used techniques to characterize sessile bacteria in biofouling/ biocorrosion studies, during the last decade. A disadvantage of this technique is the sample preparation prior to biofilm examination, which includes fixation, desalination (if required), dehydration and critical-point or freeze drying. The presence of "artifacts" or distorted biological structures is sometimes due to this complex preparation procedure. Recently, it became possible to expand SEM to high gas pressures (including atmospheric pressure) through the development of the environmental SEM technology. This improvement in the instrumentation allows wet specimens like biofilms, to be observed in their natural state avoiding structure-destroying and laborious preparations of the samples (Wagner, 1992).

High-resolution electron microscopy, like scanning tunneling microscope (STM) and atomic force microscopy (AFM) are promissory tools for the study of biocorrosion and biofouling, although in a preliminary stage their use is restrained to the research field.

Surface Analysis Techniques

Surface analysis techniques, complemented with some kind of electronic microscopy, may be very useful in monitoring the chemical nature of biological and inorganic deposits. Energy dispersion X-ray analysis (EDXA), X-ray photoelectron spectroscopy (XPS) and Auger electron spectroscopy (AES) are increasingly used to provide valuable data on the modification of base metal composition caused by biocorrosion and biofouling interactions. Several papers using these techniques have been recently published (Erauzkin, 1988; Lee et al., 1992).

A NEW SAMPLING DEVICE FOR MONITORING BIOCORROSION/ BIOFOULING IN INDUSTRIAL WATER SYSTEMS

Sampling devices for monitoring biofouling can be simultaneously used to assess corrosive attack after the removal of biological and inorganic deposits, offering a more ample and realistic scope of the problem causality (Videla, 1991).

Sampling devices fall into two main types: i) directly implanted in the system; ii) side-stream implanted. In both cases, metal coupons of known surface area, to enable an accurate count of sessile bacteria/cm^2 after biofilm detachment, are mounted in holding assemblies which are then inserted in the pipework of the laboratory or industrial system. Flow characteristics have a major influence on biofouling development (Characklis, 1990), consequently, flush mounting is indispensable to ensure that metal coupons undergo a similar flow regime to the internal pipe wall. Generally, the entire holding assembly or individual coupons can be withdrawn at intervals to proceed with further examination.

Directly implanted samplers, used to monitor sessile bacteria in oil waterflood systems, are designed to be compatible within high-pressure fittings to avoid partial shut-down and depressurization.

Side-stream samplers, mounted in parallel to the main system, receive a proportion of its flow, and can be isolated prior to sampling. One of the advantages of these devices is the number of coupons of identical or different composition that they hold, allowing more versatile applications than directly implanted samplers. Additionally, side-stream samplers can be isolated from the main system and subsequently connected to laboratory flow loops for biocide testing.

The pioneer side-stream sampler has been the Robbins device (Ruseska et al., 1982) originally constructed from Admiralty brass. Several disadvantages such as, time-consuming and complex construction, different sampling material than the actual,pipe surface, very small sampling area, have led to this sampling losing favor.

A new multi-purpose side-stream sampling device, than can be fitted in any conventional corrosion rack, has been recently developed (Videla et al.,1990) and used to monitor biofouling and biocorrosion effects in recirculating cooling water systems and in different oil field injection lines. The sampler, identified as RNEAprobe™ (**RE**usable **N**on-conventional **A**ppliance), consists of a Teflon™ holding rod, alternately drilled on opposite faces to hold 8 round metal coupons (of 0.5 cm^2 area each) distributed four on each face (Figure 1). The sampler can be directly inserted in non-pressurized areas of the flow system or generally, side-stream implanted in a conventional corrosion rack for corrosion coupons (ASTM D2688-83).

A NEW MONITORING PROGRAM FOR BIOCORROSION/ BIOFOULING IN INDUSTRIAL WATERS

A complete monitoring program for biofilms and corrosion in industrial water systems has been recently presented in three different publications (Amaral et al., 1990; Videla et al., 1990; Videla et al., 1991) each one describing a particular practical case.

The monitoring program includes parallel field and laboratory investigations using the RNEAprobe™ as the main sampling device.

Field measurements consist of corrosion coupons for loss in weight procedures, the use of polarization resistance or electrical resistance probes, and redox potential measurements. According to each type of system, valuable data like water quality, dissolved oxygen, sulfide evaluations, etc. were also provided.

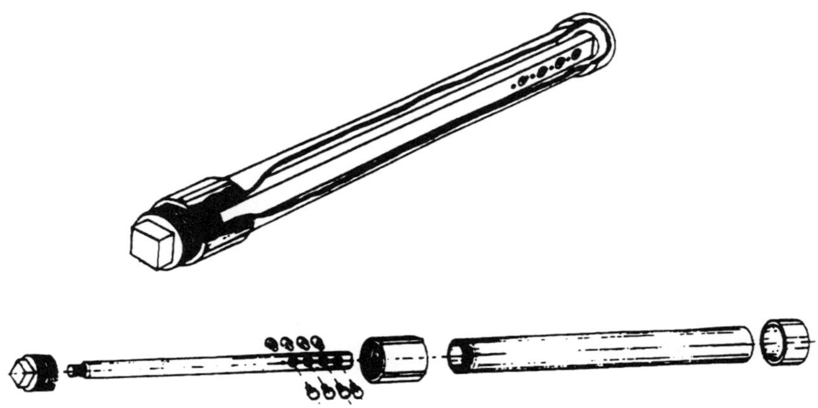

Figure 1. RNEAprobe™ Device and container.

Laboratory measurements include: i) electrochemical techniques such as potentiodynamic polarization (Ep and Eb evaluation) and corrosion potential vs. time registers; ii) microbiological methods form-assessing planktonic and sessile population, identification of the main groups of bacteria (SRB, slime-formers, iron-oxidizing bacteria, etc.); iii) microscopical techniques like epifluorescence counts, SEM observation of the biofilm, SEM and optical observations of the morphology and intensity of the metal attack; iv) surface analysis of biological and inorganic deposits and corrosion products by means of EDXA or electron microprobe.

To accomplish with all these measurements, after each exposure period (generally from 1 day to 1 week), the coupon holder was withdrawn and coupons were distributed as follows: One (1) stainless steel coupon (generally of the 316 AISI type for biofilm observation by SEM; Two (2) coupons (generally made with system structural material (SSM) for sessile bacterial counts (by duplicate);
Two (2) SSM coupons for biofilms/ corrosion products observations by SEM; One (1) SSM coupon for SEM observation of the corrosive attack; One (1) SSM for EDXA of biological and inorganic deposits and One (1) SSM coupon for corrosion potential vs. time measurements. In each case two (2) RNEAprobes™ may be installed per station to allow duplication.

For sessile bacterial counts, the entire device is taken to the laboratory within a period of 12 hours, in a closed receptacle filled with system water (Figure 1). Biofilms are generally detached by scraping or by using an ultrasonic device. For SEM observations metal coupons are firstly washed with sterile system water, fixed, washed with buffer, dehydrated through an acetone series to 100% and finally critical point dried. All these procedures were avoided for SEM observation of the corrosive attack, although in all cases samples were metal coated prior to examination by sputtering a carbon or palladium/ gold coating layer.

Results obtained with this monitoring program have been successful for modifying or optimizing biocide strategies, for improving corrosion inhibition treatments, and eventually for establishing a proper diagnosis of the biodeterioration causes.

SUMMARY

New trends in monitoring biodeterioration effects (biocorrosion and biofouling) in industrial water systems require a multi-disciplinary approach comprising different microbiological, electrochemical and microscopic measurements as well as several surface analysis techniques. A brief critical review of these techniques is complemented with the description of a new multi-purpose sampling device, the RNEAprobe™. A complete monitoring program based in laboratory and field measurements is also described. This approach has been successfully applied in different practical cases referred to in this article.

REFERENCES

Amaral, M. C., Silva, R. A., Canales, C. G., and Videla, H/ A. (1991). Field and laboratory evaluation of a new multipurpose sampling device for monitoring microbially influenced corrosion and biofilms in recirculating cooling water systems, In: Biodeterioration and Biodegradation 8, pp. 592-594 (H. W. Rossmoore, ed.), Elsevier Applied Science, London.

Cook, P. E., and Gaylarde, C. C. (1987). Rapid techniques for the detection and quantification of sulfate-reducing bacteria, In: Microbial Problems in the Offshore Oil Industry, p. 245 (E. C. Hill, J. L. Shennan and R. J. Watkinson, eds.), John Wiley & Sons, Chichester.

Characklis, W. G. (1986). Influence of microbial films on industrial processes,d Proceedings of Argentine-USA Workshop on Biodeterioration (CONICET-NSF), pp 181-217 (H. A. Videla, ed.), published by Aquatec Quimcia S.A., San Paulo, BRAZIL.

Characklis, W. G. (1990). Biofilms processes, In: Biofilms, pp. 195-231 (W. G. Characklis, and K. C. Marshall, eds.) John Wiley & Sons, New York.

Characklis, W. G., and Marshall, K. C. (1990). Biofilms: A basis for an interdisciplinary approach, In: Biofilms, pp. 3-15 (W. G. Characklis, and K. C. Marshall, eds.), John Wiley & Sons, New York.

Dexter, S. C., and Gao, G. Y. (1988). Effect of seawater biofilms on corrosion potential and oxygen reduction of stainless steel. Corrosion, 44, 717-23.

Dexter, S. C., Duquette, D. J., Siebert, O. W., and Videla, H. A. (1991). Use and limitations of electrochemical techniques for investigating microbiological corrosion. Corrosion, 47, 308-318.

Erauzkin, E. (1988). Biocorrosion of AISI stainless steel by sulfate-reducing bacteria, In: Microbial Cor. I, pp. 412-430 (C. A. C. Sequeria, and A. K. Tiller, eds.), Elsevier Applied Science, London.

Gilbert, P. D., and Herbert, B. N. (1987). Monitoring microbial fouling in flowing systems using coupons, In: Industrial Microbiological Testing, pp. 79-98 (J. W. Hopton, and E. C. Hill, eds.), Blackwell Scientific Publications, Oxford.

Hamilton, W. A. (1985). Sulfate-reducing bacteria and anaerobic corrosion. Ann. Rev. Microbiol., 39, 195-217.

Lamont, J. E. (1988). Role of boicides in controlling microbial corrosion, In: Microbial Cor., pp. 224-234 (C. A. C. Sequeira, and A. K. Tiller, eds.), Elsevier Applied Science, London.

Lee, W., Lewandowski, Z., Ikabe, S., and Characklis, W. G. (1992). Corrosion of mild steel underneath aerobic biofilms containing sulfate-reducing bacteria. Corrosion 92, paper no. 190, Houston, NACE.

Pope, D. H. (1986). MIC in U.S. Industries. Detection and prevention, In: Proceedings of Argentine-USA Workshop on Biodeterioration (CONICET-NSF), pp. 105-118 (H. A. Videla, ed.), published by Aquatec Quimica S.A., San Paulo, Brazil.

Prasad, R. (1988). Pros and cons of ATP measurement in oil field waters, Corrosion 88, paper no. 87, Houston, NACE.

Ruseska, I., Robbins, J., Costerton, J. W., and Lashen, E. S. (1982). Biocide testing against corrosion-causing oil-field bacteria helps control plugging. Oil, Gas J., 80, 253-264.

Scotto, V., DiCintio, R, and Marcenaro, G. (1985). The influence of marine microbial film on stainless steel corrosion behavior. Cor. Sci., 25, 184-194.

Szklarska-Smialowska, Z. (1986). Measurement of susceptibility to pitting, In: Pitting Corrosion of Metals, pp. 39-67, Houston, NACE.

Videla, H. A. (1989). Metal dissoulton/redox in biofilms, In: Structure and Function of Biofilms, pp. 301-320, (W. G. Characklis, and P. A. Wilderer, eds.), John Wiley & Sons, Chichester.

Videla, H. A., deMele, M. F. L., Silva, R. A., Bianchi, F., and Gonzales Canales, C. (1990). A practical approach to the study of the interaction between biofouling and passive layers on mild steel and stainless steel in cooling water. Corrosion 90, paper no. 124, Houston, NACE.

Videla, H. A., Guiamet, P. S., Pardini, O. R., Echarte, E., Trujillo, D., and Freitas, M. M. S. (1991). Monitoring biofilms and MIC in an oilfield water injection system. Corrosion 91, paper no. 103, Houston, NACE.

Videla, H. A. (1991). Microbially induced corrosion: an updated overview, In: Biodeterioration and Biodegradation *, pp. 63-88 (H. W. Rossmore, ed.), Elsevier Applied Science, London.

Wagner, P. (1992). Investigations of MIC using environmental scanning electron microscopy. Corrosion 92, paper no. 185, Houston, NACE.

White, D. C., Nivens, D. E., Nichols, P. D., Mikell, A. T., Kerger, B. D., Henson, J. M., Geesey, G. G., and Clarkmem, C. K. (1986). Role of anaerobic bacteria and their extracellular polymers in the facilitation of corrosion: use of Fourier transforming infrared spectroscopy and "signature" phospholipid fatty acid analysis, In: Biologically Induced Corrosion, pp. 233-243, (S. C. Dexter, ed.), Houston, NACE.

BIODETERIORATION OF STONE IN HISTORIC BUILDINGS AND MONUMENTS

C. SAIZ-JIMENEZ, Instituto de Recursos Naturales y Agrobiologia, Apartado 1052, 41080 Sevilla, Spain

INTRODUCTION

Microbial geochemical cycles have been going on since life on earth began. This activity results in chemical and mineralogical modification of the different types of rocks and minerals within the biosphere and involves essentially the solubilization processes of major mineral elements from silicates, carbonates, phosphates, oxides, etc. While the weathering of minerals in natural environments results in the destruction of rocks, thus contributing to soil formation processes, it has a deleterious effect when these rocks are part of historic buildings and monuments.

The monuments, sculptures and stony objects of art exposed to open air have deteriorated, through the centuries, due to natural causes. Sun, frost, wind, rain, etc. contribute to the gradual process of weathering, in which biological activity plays a key role because its interactions with physico-chemical mechanisms are considered central to the understanding of long-term deterioration.

The weathering of building stones is not a recent phenomenon but it was well-known in antiquity. This process can be considered a result of natural environmental factors. However, industrial and urban activities modify the composition of the atmosphere, resulting in a more agressive environment. Obviously, these anthropogenic compounds may influence to some extent the colonization and growth pattern of microorganisms in polluted urban

Biodeterioration Research 4, Edited by G.C. Llewellyn *et al.*, Plenum Press, New York, 1994

environments when compared with the growth of the same or similar organisms in rural environments.

Although atmospheric pollution is generally recognized as a significant factor in the deterioration of cultural properties, stonework biodeterioration deserves more attention that has received. In fact, a great variety of organisms colonizing stones in urban environments have been reported (Griffin et al., 1991). They can be found both on and beneath surfaces.

In this paper recent findings in biodeterioration of stoneworks in historic buildings and monuments are reviewed.

BACTERIA AND FUNGI

Most bacteria and fungi are heterotrophic organisms that consume, not produce, organic matter. However, a small group of photosynthetic and chemoautotrophic bacteria fix carbon dioxide and are therefore classified as autotrophs. These organisms, of which sulphur and nitrifying bacteria are representatives, are rare in nature, and their share in the total production of organic matter is small. On the contrary, most of the microorganisms living in rocks, and stoneworks are heterotrophic bacteria and fungi. Both types appear to be involved in the weathering and formation of minerals and rocks.

Microbiological analyses of weathered stone have shown the presence of microorganisms which utilize sulphur and thiosulphates as sources of energy and which give rise to sulphate as a product of metabolism. The facultative chemolithotrophs species Thiobacillus novellus and T. intermedius were isolated by Gugliandolo and Maugeri (1988) from an Italian church.

Bacteria not only can be involved in sulphur oxidation but also reduction. In this connection, it is interesting reporting that the calcium ions obtained by release from gypsum when Desulfovibrio desulfuricans reduces sulphate, react with carbon dioxide resulting in the formation of calcite (Atlas et al., 1988). Marble crusts treated with bacterial broth showed the cleaning and some neomineralization carried out by the bacterium consuming a large portion of the available gypsum in the crust.

Meincke et al. (1989) studied the population of nitrifying bacteria from German sandstone monuments. The population analysis indicated that the most abundant ammonia-oxidizing organism in the sandstones belongs to the genus Nitrosovibrio, although other ammonia-oxidizing bacteria of the genera

Nitrosomonas and Nitrosospira were detectable. Nitrobacter was the only detectable genus of the nitrite-oxidizers. Both groups of nitrifying bacteria live endolithically in the sandstone of different monuments. The pH tolerance of Nitrosovibrio and the resistance to high concentrations of salts might contribute substantially to the selective enrichment of Nitrosovibrio in the sandstone of the monuments. Bock et al. (1990) identified a new species of Nitrobacter, N. vulgaris isolated from stone samples from the Cologne cathedral, which was also found in soils, concrete and freshwaters.

Colonization and biodeterioration on stoneworks are usually linked to environmental conditions. The most significant parameters affecting microbial growth are represented by physical factors, mainly moisture, temperature, and light, as well as by the chemical nature of the substratum. Thus, Tayler and May (1991) reported that the quantitative and qualitative nature of the heterotrophic bacterial population on sandstone varies seasonally and can be related to moisture and temperature changes associated with a temperate climate. The potential of these populations to cause damage to sandstone also varies with season. In this respect, Lewis et al. (1988) suggested that bacteria on stone can be extremely versatile and could maintain their activity during nutrient perturbations, operating at low nutrient levels and utilizing what the environment has to offer. As a consequence, bacterial populations may be able to maintain their involvement in the process of stone deterioration during periods of nutrient flux.

A primary colonization of chemoautotrophic bacteria favours a settling in of heterotrophic bacteria, fungi, and specific actinomycetes populations. The activity of heterotrophic bacteria on stone is governed by the availability of suitable organic matter. It is likely that most stoneworks contain sufficient organic material from soil and dust or dead autotrophic and phototrophic organisms to maintain growth and activity of large population of bacteria. Most heterotrophic bacteria are able to use various organic compounds as a source of carbon and energy for growth.

Warscheid et al. (1988) determined chemoorganotrophic bacteria from surfaces of sandstones monuments. It was shown that most of the isolated bacteria used a wide range of different carbohydrates, amino acids, fatty acids, and hydrocarbons. About 40 % of the strains analyzed were shown to be potential acid producers, whereas the capability of manganese and iron oxidation was only sporadically found in the bacteria living on the rock. Kerosene, as a representative mixture of different hydrocarbons detectable in polluted atmosphere, was well metabolized by 70 % of the analyzed bacteria.

Giacobini et al. (1988) observed that actinomycetes grow on frescoes producing a whitish powdery mycelium or colourles patinas. They isolated 200 strains from which only 60 were identified. The actinomycetes thrive preferentially in caves and mural paintings, although it also have been recorded in limestone monuments (Saiz-Jimenez and Samson, 1981; Saiz-Jimenez, 1984).

It has been shown that in addition to bacteria, fungi can break down minerals mechanically or chemically. The fungi isolated from weathered stones are usually ubiquitous saprophytes whose conidia are readily distributed through the atmosphere. Most of the isolated species are commonly included in floristic listings. Aspergillus, Penicillium and Cladosporium are among the most common airborne fungi and thus have a worldwide distribution. Also, the fungi have a remarkable saprophytic and polysaccharide-degrading potential.

Petersen et al. (1988) found that Penicillium, Phoma and Exophiala species colonized most examined German sandstone monuments. Furthermore, they stressed the biodeteriorating activities of the genera Cladosporium, Penicillium and Trichoderma, which were the most abundant on the monuments investigated. The principal genera isolated from weathered sandstone from the cathedral of Salamanca, Spain, were Penicillium, Phoma, Cladosporium, Trichoderma and Fusarium (De la Torre et al. 1991) and an association between fungi and algae was found.

The presence of an organism on decayed material does not necessarily imply that it has caused the damaged observed. The activity of microorganisms in promoting stone deterioration is largely dependent upon the production of corrosive metabolites which can solubilize minerals in a manner similar to chemical agencies. Probably, low-frequency isolation cannot be directly correlated with metabolic activity, as the fungi isolated in culture media may be dormant (spores) and are not necessarily the ones which are functioning in the ecosystem.

The ability of fungi to decompose mineral matter could arise, probably only in a microhabitat, when there is an accumulation of organic matter which can be metabolized. In addition, many bacteria and fungi derive much of their carbon from organic materials of which algal products may represent an appreciable proportion (e.g. mucilages and algal cell wall polysaccharides) and can give rise to organic acids. This can be interpreted as a response to the rising dampness from the ground creating a richer nutrient environment through dissolved salts and organic matter, where green algae flourish.

An efficient organic acid-producer fungus was A. niger which is, by far, the most effective oxalic acid synthesizer from mono- and polysaccharides (De la Torre et al., 1991). It is well-known that organic acids are produced in fungal cultures. These acids might play a role in mineral weathering, due to their ability to form complexes with cations, promoting the removal of the latter.

The extensive corrosion ability of fungi on limestone was studied by Koestler et al. (1985). After only five weeks in culture the fungal hyphae penetrated at least 1 mm into the crystals. It is apparent that the fungus corroded channels along and in front of its path. Trichotecium spp. are known to secrete oxalic acid, citric acid, and gluconic acid among other products. These secretion could chemically attack the dolomite or calcite and account for the observed etching. The channels created by the fungus provided new avenues of attack deep inside the crystal, and a vastly increased surface area susceptible to attack by water and/or dissolved acids in the water, such as nitric and sulphuric acids from air pollutants.

CYANOBACTERIA AND ALGAE

Although cyanobacteria (formerly blue-green algae) and algae are organisms commonly associated with aquatic ecosystems, they are also widespread over terrestrial substrata. Their occurrence at the surface of soils, stones, rocks and trees, particularly in moist areas, is well-known. In building stones they grow in humid places, on cornices, in holes and crevices or beneath crusts, where water is retained and evaporation is slow due to protection against winds or direct sunshine. Furthermore, growth may also be due, for example, to leaking or badly sited roof guttering, inadequate drainage of flat areas, frequent wetting by wind-blown rain, or from adjacent watercourses. Their growth is rarely uniform, frequently forming streaks that follow the areas of dampness. Their presence is more apparent on the north faces of buildings than on the south, because the latter dry out more readily.

Only water and a minimal supply of mineral salts are required for colonization. It is the duration of the period of wetness that is crucial, rather than the frequency of wetting, in predisposing a surface to colonization (Grant, 1982). Inoculation is more rapid where there is adjacent or overhanging vegetation from which cyanobacteria and algae can be brought by wind and rain, and development is further accelerated if bird droppings, agricultural fertilizers and pollution-derived nitrogen oxides introduce additional nitrogen and phosphorous. Rough or porous surfaces facilitate attachment of airborne propagules and the accumulation of nutrients. Although cyanobacteria and algae are phototrophic organisms, deriving their energy for growth from light, they can also be found in very poorly illuminated places inside buildings (Caiola et al., 1987).

Extensive taxonomical and ecological studies on the cyanobacteria and algae present in some European monuments were reported by Anagnostidis et al. (1983), and Ortega-Calvo et al. (1992). Saiz-Jimenez et al. (1990) found an endolithic microbial community developed as a green layer 1 mm below the surface of a limestone from the basilica of Tongeren, Belgium. This community was dominated by unicellular cyanobacteria. An endolithic filamentous cyanobacterium (Phormidium spp.) was also found growing under black sulphated crusts developed on limestones in the cathedral of Seville, Spain (Saiz-Jimenez, 1990a; Saiz-Jimenez et al., 1991a).

Microclimatic differences, probably including pH, porosity (allowing a higher degree of water retention) and richer substratum, can influence the algal distribution. Algae may grow in abundance on some stone blocks and yet be totally absent on another blocks of different nature but of the same age directly beside it. In fact, a different species distribution was observed by Ortega-Calvo et al. (1992) between granite and adjacent mortar in the cathedral of Toledo (Spain).

A unfavourable microclimate in the stone may induce the formation of resistant algal forms, as was seen by Pietrini et al. (1985) on marble remains in Rome. This is a common mechanism in survival strategy. The reddish-orange coloured patina present on the marble was produced mainly by the presence of cysts of Haematoccus pluvialis. Their formation is accompanied by a change in colour from green to red-orange due to the production of large quantities of cytoplasmatic carotenoids, responsible for the colour alteration.

Endolithic algae have also been described on historic buildings. Endolithic growth of Stichococcus bacillaris has been described in granite from the cathedral of Toledo (Spain) by Ortega-Calvo et al. (1992), and from the Custom House of Dublin, Ireland (unpublished observations).

The occurrence of phototrophic microorganisms on stone surfaces does not automatically imply a destructive action. The aggressive action of cyanobacteria and algae in relation to the substratum where they develop has been considered negligible by some authors, although it is still a matter of controversy. Apart from the unaesthetic appearance, evident in most of the reports on cyanobacteria and algae on historic buildings, there are references in the literature that point to direct decay mechanisms. Danin (1983) reported the weathering induced by coccoid cyanobacteria in limestone walls and monolithic tombs, up to 2600 years old, in Jerusalem. By decreasing the coherence of rock crystals around their colonies, the cyanobacteria accelerate detachment of rock particles, leading to the weathering process.

Cyanobacteria and algae may also have a role in the mechanical deterioration of the stone, exerting considerable force through repeated shrinking and relaxation when they are going through cycles of drying and moistening. This is mainly attributed to the mucilage formed by the sheaths of the cyanobacteria and algae, which adheres them to the sustrate and suffers deep changes in volume due to its water retention properties. The sheath allows the establishment of a self-contained aquatic environment within which the cells can attain adequate irradiation and gaseous exchange. In addition, the presence of this water reservoir with a high water potential ensures that the cells remain essentially aquatic, thus permiting the cyanobacteria and algae to overcome drought periods. Through sheath contraction and expansion, they can loosen rock grains, contributing to the gradual destruction of the rocks (Ortega-Calvo et al., 1991 a, b). In addition, the formation of cyanobacterial and algal patinas result in a longer moisture retention at the surface of the stone, increasing the mechanical damage produced by the freezing and thawing of water present in the pores of the stone.

LICHENS AND BRYOPHYTES

For some years, studies have been published on the different aspects of deterioration of stone materials. However, the role of lichens and bryophytes in the deterioration of building materials of historic buildings and monuments has been dealt with in a few cases (i.e. Casas Sicart and Saiz-Jimenez, 1982; Nimis and Monte, 1988; Seaward and Giacobini, 1988; Garcia-Rowe and Saiz-Jimenez, 1991). This process is an extension of the weathering of rocks in nature, in which the same or related species are involved.

Lichen growth is of universal occurrence on rocks, both in nature and when used as building stones, but they also extent to most other building materials including cement, concrete, mortar, and glass. Due to a certain number of common characteristics, both the lichens and mosses usually exploit the same ecological niche. Subsequent colonization by vascular plants is enabled by the activity of these organisms, basically rock degradation and soil formation.

Lichens, as pioneer organisms in rock colonization, occupy environments normally hostile to the majority of other forms of life. These lower plants are resistant to dessication and extreme temperatures, have long life, low growth rate, and are efficient in accumulating nutrients from their environment. The response of lichens to a potentially colonizable substratum depends on their ecological requirements and their tolerance to that substratum, since each species has its own needs and physiological limitations. Important limiting factors for this colonization include texture, moisture-retaining capacity of the substratum, its stability, pH, and the quantity of nutrients (Hale, 1983).

In urban environments the presence of a lichen mosaic, sometimes aesthetically pleasing, is a characteristic feature of southern European cathedrals with respect to those from north Europe, which usually display monotonous green patches of algae. Colonization of different stone materials by lichens is helped by the contribution of the organic nitrogen from large-bird excrement, such as storks, pigeons, and kestrels. This gives rise to a characteristic ornithocoprophilous flora (Saiz-Jimenez, 1981). Some species of lichens, such as _Ramalina_ spp., are characteristic of rainy climates and excrement-rich habitats, and are thus found in the sandstones of the cathedral of Salamanca, Spain. On the contrary, _Lecanora muralis_ is distributed in all the substrata and climates studied (Garcia-Rowe and Saiz-Jimenez, 1991).

Crustose lichens colonize the stones, taking particular advantage of their ability to resist drought and temperatures above 50°C in sunny exposed habitats. The colonization of stones by lichens originates a biogenic crust, covering the stone surface. Most lichen species are innocuous and may also afford protection to the stone against pollutant gases. However, Seaward (1988) suggested that changes in environmental pollution levels may conduce to increasing detrimental invasion by certain lichen species due to lack of competition from other species. This was the case of L. muralis causing blistering in Roman terracotta pots. This is a highly successful lichen in urban environments, recognized by its high tolerance to pollution so that it readily adapts to man-made substrata and spreads towards the urban center.

The effect of lichen attack has been studied in a variety of building materials, including different types of stones, bricks, and mortars (Jones and Wilson, 1985; Seaward, 1988; Seaward et al., 1990). These substrata generally support different communities. This was particularly investigated in the Archaeological Site of Italica, a natural laboratory for stone colonization and biodeterioration studies (Saiz-Jimenez, 1990 b; Saiz-Jimenez et al., 1991 b; Saiz-Jimenez and Garcia-Rowe, 1992). Since the first quarter of this century, the exposure of most of the mosaics after excavation has resulted in colonization of the tesserae and mortars by pioneering species of lichens, and subsequent invasion of less pioneering species of lichens, and mosses. However, the classic plant succession concept, starting with pioneering crustose lichens, which are replaced by foliose lichens and/or mosses cannot be generally applied to mosaics.

The mortars are subjected to mechanical disintegration caused by endolithic and epilithic crustose lichens, and mosses. Sometimes foliose lichens appear later. Among the lichens colonizing the mortars, two well-defined strategies stand out; one, of those which colonize the mortars, from which they invade the stones, and the other, of those which colonize only the mortars (Garcia-Rowe and Saiz-Jimenez, 1988).

Lichens adhere to, penetrate and digest the minerals of which the stones are made. It is well-known that lichen attack is both mechanical and chemical. Although it is rather difficult to obtain direct evidence of the mechanical deterioration of stones by microscopic examination, it is well established that lichens are able to withstand prolonged periods of desiccation, reabsorbing water and swelling quickly once it becomes available again. This is particularly true in xerophytic species such as those identified in most Southern Europe areas. The process can result in the deterioration of stone, particularly by succesive expansion and contraction of the lichen on wetting and drying. Furthermore, it has been reported (Seaward, 1988) that many lichen species growing on ancient monuments create microclimatic effects at the thallus/substratum interface, which undoubtedly lead to mechanical damage to stonework. Lecania erysibe, also caused damage by pulling away and scaling important limestone fragments in the Sevilla's cathedral belfry (Saiz-Jimenez, 1981).

Lichens produce a wide range of secondary metabolites which are often referred to as lichen acids. Most of these are aromatic polyketides with orsellinic acid and its homologues as principal structural unit, condensed as di-, tri-, and tetracyclic compounds. Syers and Iskandar (1973) demonstrated that orcinol and ß-orcinol depsides and depsidones are sufficiently soluble in water to function as metal-complexing agents in rock weathering. In fact, lecanoric acid formed a reddish-yellow complex when it was allowed to react with biotite, and significant amounts of Ca, Mg, Fe and Al were released from silicates by solutions of lichen acids.

In addition, oxalic acid is recognized as an active lichen metabolite in the weathering of rocks in nature. Calcium oxalates (whewellite and weddellite) were encountered as patinas on marble and limestones of statues, monuments, and historic buildings as a consequence of the presence of lichens, which provide the acid required to form the oxalates (Del Monte et al., 1987).

SEM observations made upon the stone surface, after lichen removal, showed areas heavily damaged by characteristic holes, varying of diameter. The bigger holes were shown to be formed by association of many small holes. This phenomenon is also known as pitting. Higher magnifications revealed the appearance of a distinctive etching pattern, which has been previously reported near the border of lichen colonies, and assumed to be produced from chemical action of the prothallus, around the margin of crustose lichens (Saiz-Jimenez and Garcia-Rowe, 1992). The effects of the lichen activity through such metabolites were clearly seen from microscopic examination of limestone by Saiz-Jimenez and Garcia-Rowe (1992), and by Ascaso et al. (1990) who studied by X-ray diffraction the formation of new materials, beneath the thalli, not found in the parent rocks.

Pitting is common in many marble monuments, such as the Trajan column in Rome, the terminal column of the Via Appia in Brindisi (Italy), the walls of the cathedral of Cadiz (Spain), or the Teseion in Athens (Greece). This type of decay has been attributted to the action of endolithic lichens by Krumbein et al. (1989), and Del Monte (1991). The latter author considers that the hole morphology and the dense canalization due to fungus hyphae leave no doubt in this regard.

However, both the origin of oxalate patinas and pitting were recently discussed (Alessandrini and Realini, 1990). Lazzarini (1990) considers that pitting formation is related with the water action on surfaces with structural or textural defects, resulting in microkarstic erosion visible at the micro-, meso- or macrolevel. This hypothesis, according to Lazzarini, is substantiated by the fact that the pitting is not observed in areas not washed out.

Casas Sicart and Saiz-Jimenez (1982) identified 25 species of bryophytes growing on the roofs and walls of the cathedral of Seville. These species were restricted to mortars, and stone fissures on shady, protected, and wet places. Although characteristically few species can thrive in polluted urban areas, those identified in the cathedral of Seville were not poleophiles.

In Southern Europe, in contrast to the smooth, hard surfaces of stones, which avoids biological colonization of mosses, the mortars of most buildings and monuments, with a higher water-retention capacity and lower cohesion of material, allow a more efficient colonization, firstly of algae, and then mosses. This is especially true in wet and shading environments from adjacent buildings or trees canopies, while in sunny ones the mortars are partially colonized by xerophytic, calcicolous lichen communities. In both cases, the penetration and eventual disgregation of mortars is clear, but strongly accelerated by the annual action of mosses which provide a good input of organic matter for heterotrophic microorganisms.

The effect of mosses on the mortars is important. Optic and electron microscopy reveal the profuse network of rhizoids in the areas of mortar. Rhizoids may penetrate the mortar down to the base allowing easy access of water to deep levels. However, in the case of mosaics, rhizoids are not capable of penetrating crystalline limestone tesserae, but extensively invade the mortar in wet seasons, breaking the cohesion between tessera and mortar, causing cracks and disgregation. The mosses contribute to organic enrichment of the substratum and to the formation of humus, opening the way to invasion by vascular plants (Garcia-Rowe and Saiz-Jimenez, 1991).

VASCULAR PLANTS

Biodeterioration of building materials in European cathedrals and monuments is not very significant if compared with deterioration caused by atmospheric phenomena (frosts, rain, wind, etc.) and pollution (weathering, deposits of dirt, formation of sulphated crusts, etc.). However, in southern countries colonization by lichens and bryophytes is one of the first stages in a succession which gives rise to the later invasion by vascular plants, whose roots cause important deterioration in the structure of the buildings, above all in the roofings, tiles, gutterings, drains, etc., giving rise in the long term to their ruin. All this is increased by the difficulty in access to the roofs, and the traditional lack of cleaning and maintenance which is common in these great monuments.

Weed settling in urban monuments can be attributed to biological and anthropogenic causes. In fact, pioneer colonizators prepare the substratum for vascular plant invasion, following the typical plant succession in nature. The process is accelerated by stone weathering due to air pollution, lack of maintenance and accumulation of debris in roofings, faults in drainage or roof guttering, and in general by the frequent wetting of surfaces.

Garcia-Rowe et al. (1991) catalogued 91 species of vascular plants in three Spanish cathedrals. Fifty seven per cent of the species develop in Seville's cathedral, 41 % in Salamanca and 32% in Toledo. Asteraceae (Compositae) is the best represented family (from a total of 29 catalogued) with 21 species. Poaceae (Gramineae) presents 16 species with a high biomass, overcoming that of Compositae. Brassicaceae (Cruciferae), with a high biomass as well, is represented by 9 species.

Biological spectra show a high percentage of therophytes (79 %) whereas another life-forms are comprehensively less represented (11 % hemicryptophytes, 7 % phanerophytes, 2 % chamaephytes and 1 % geophytes). Caneva et al. (1990) considered that on the basis of the life cycle and degree of lignification, the chamaephytes and phanerophytes clearly represent the most danger species, whereas hemicryptophytes and therophytes are less damaging to the substratum.

The action of plants in stone deterioration may be mechanical, through the thrust exerted by the growth of roots, and chemical, through the acidity of rootlets and through exudates. Caneva and Altieri (1988) stated that the chemical role of plant roots is produced by the acidity of the rootlets and by excretion of inorganic and organic compounds, with aggresive or chelating capabilities, the secretion of organic acids being remarkable. Lewin and Charola (1981) reported that plant growth produce substances that deposit on and penetrate into the stone. Ivy, in addition to whewellite deposition, exude organic substances, among which were strongly chelating agents, that infuse limestone producing stone coloration.

Caneva and Altieri (1988) collected ruderal plants from Roman monuments and measured the root acidity, which usually ranged from pH 5.3 to 6.4 in mashed rootlets. Leyval and Berthelin (1991) suggested that the release of organic acids by pine roots may explain the weathering of phyllosilicates into vermiculite, which is basically a cation-exchange process.

Extensive growth of vascular plants (up to 57 different plants were recorded in the area) result in decay of mortars, loosening and removal of tesserae, and ultimately in the complete destruction of the mosaics of Italica, as was recorded by Saiz-Jimenez et al. (1991 b) for mosaics excavated around 1920s. The vascular flora colonizing the mosaics and brick walls is typically nitrophilic, comparable to that found in the surroundings (very altered due to the impact of human activity), and in the uncultivated surrounding land (notably nitrophilic due to grazing), with a large number of small mammals (mice, moles, and rabbits), both in the already excavated places and the unexcavated ones.

The destructive action of vascular plants can be observed in any historic building or monument lacking of maintenance, elimination and/or removal of plants. This invasion represents the last step in the biological stone decay process.

Acknowledgements

This work has been supported by the C.E.C. through contract STEP-CT90-0107, C.I.C.Y.T. (PAT91-1056-C03), and the Junta de Andalucia.

REFERENCES

Alessandrini, G. and Realini, M. (1990). Note conclusive sul convegno Le Pellicole ad Ossalati: Origine e Significato nella Conservazione. Arkos, 9/10, 19-32.

Anagnostidis, K., Economou-Amilli, A. and Roussomoustakaki, M. (1983). Epilithic and chasmolithic microflora (Cyanophyta, Bacillariophyta) from marbles of the Parthenon (Acropolis-Athens, Greece). Nova Hedwigia, 38, 227-287.

Ascaso, C., Sancho, L.G. and Rodriguez-Pascual, C. (1990). The weathering action of saxicolous lichens in maritime Antarctica. Polar Biol., 11, 33-39.

Atlas, R.M., Chowdhury, A.N. and Gauri, K.L. (1988). Microbial calcification of gypsum-rock and sulfated marble. Stud. Conserv., 33, 149-153.

Bock, E., Koops, H.-P., Moller, U.C. and Rudert, M. (1990). A new facultatively nitrite oxidizing bacterium, Nitrobacter vulgaris sp. nov. Arch. Microbiol., 153, 105-110.

Caiola, M.G., Forni, C. and Albertano, P. (1987). Characterization of the algal flora growing on ancient Roman frescoes. Phycologia, 26, 387-396.

Caneva, G. and Altieri, A. (1988). Biochemical mechanisms of stone weathering induced by plant growth. In: VIth International Congress on Deterioration and Conservation of Stone, pp. 32-44 (J. Ciabach, ed.), Nicholas Copernicus University, Torun.

Caneva, G., Dinelli, A., De Marco, G. and Vinci, M. (1990). Halophylous vegetation in the deterioration of stone monuments in coastal environments. In: La Conservazione dei Monumenti nel Bacino del Mediterraneo, pp. 231-234 (F. Zezza, ed.), Grafo, Brescia.

Casas Sicart, C. and Saiz-Jimenez, C. (1982). Los briofitos de la catedral de Sevilla. Collec. Bot., 13, 163-175.

Danin, A. (1983). Weathering of limestone in Jerusalem by cyanobacteria. Z. Geomorphol., 27, 413-421.

De la Torre, M.A., Gomez-Alarcon, G., Melgarejo, P. and Saiz-Jimenez, C. (1991). Fungi in weathered sandstone from Salamanca cathedral, Spain. Sci. Total Environ., 107, 159-168.

Del Monte, M. (1991). Trajan's column: lichens don't live here any more. Endeavour, 15, 86-93.

Del Monte, M., Sabbioni. C. and Zappia, G. (1987). The origin of calcium oxalates on historical buildings, monuments and natural outcrops. Sci. Total Environ., pp. 67, 17-39.

Garcia-Rowe, J. and Saiz-Jimenez, C. (1988). Colonization of mosaics by lichens: the case study of Italica. Stud. Geobot., 8, 65-71.

Garcia-Rowe, J. and Saiz-Jimenez, C. (1991). Lichens and bryophytes as agents of deterioration of building materials in Spanish cathedrals. Int. Biodeter., 28, 151-163.

Garcia-Rowe, J., Aparicio Martinez, A. and Saiz-Jimenez, C. (1991). Weeds settling in Spanish cathedrals (Salamanca, Seville and Toledo). In: Science, Technology and European Cultural Heritage, pp. 497-500 (N.S. Baer, C. Sabbioni and A.I. Sors, eds.), Butterworth-Heinemann, Oxford.

Giacobini, C., De Cicco, M.A., Tiglie, I. and Accardo, G. (1988). Actinomycetes and biodeterioration in the field of fine art. In: Biodeterioration 7, pp. 418-423 (D.R. Houghton, R.N. Smith and H.O.W. Eggins, eds.), Elsevier, London.

Grant, C. (1982). Fouling of terrestrial substrates by algae and implications for control - A review. Int. Biodet. Bull., 18, 57-65.

Griffin, P.S., Indictor, N. and Koestler, R.J. (1991). The biodeterioration of stone: a review of deterioration mechanisms, conservation case histories and treatment. Int. Biodeter., 28, 187-207.

Gugliandolo, C. and Maugeri, T.L. (1988). Biodeterioration of stone by sulphur bacteria. In: La Conservazione dei Monumenti nel Bacino del Mediterraneo, pp. 221-224 (F. Zezza, ed.), Grafo, Brescia.

Hale, M.E. (1983). The Biology of Lichens, Edward Arnold, London.

Jones, D. and Wilson, M.J. (1985). Chemical activity of lichens on mineral surfaces -A review. Int. Biodeter., 21, 99-104.

Koestler, R.J., Charola, A.E., Wypyski, M. and Lee, J.J. (1985). Microbiologically induced deterioration of dolomitic and calcitic stone as viewed by scanning electron microscopy. In: Ve Congrés International sur l'Alteration et la Conservation de la Pierre, pp. 617-626 (G. Felix ed.), Presses Polytechniques Romandes, Lausanne.

Krumbein, W.E., Petersen, K. and Schellnhuber, H.-J. (1989). On the geomicrobiology of yellow, orange, red, brown and black films and crusts developing on several different types of stone and object of art. In: Le Pellicole ad Ossalato: Origine e Significato nella Conservazione delle Opere d'Arte, pp. 337-347, Centro CNR "Gino Bozza", Milano.

Lazzarini, L. (1990). Tavola Rotonda, Trascrizione. Convegno Internazionale Le Pellicole ad Ossalati: Origine e Significato nella Conservazione. Arkos, 9/10, 20-36.

Lewis, F.J., May, E. and Bravery, A.F. (1988). Metabolic activities of bacteria isolated from building stone and their relationship to stone decay. In: Biodeterioration 7, pp. 107-112 (D.R. Houghton, R.N. Smith and H.O.W. Eggins, eds.), Elsevier, London.

Lewin, S.Z. and Charola, A.E. (1981). Plant life on stone surfaces and its relation to stone conservation. Scanning Electron Microsc., I, 563-568.

Leyval, C. and Berthelin, J. (1991). Weathering of a mica by roots and rhizospheric microorganisms of pine. Soil Sci. Soc. Am. J., 55, 1009-1016.

Meincke, M., Krieg, E. and Bock, E. (1989). Nitrosovibrio spp., the dominant ammonia-oxidizing bacteria in building sandstone. Appl. Environ. Microbiol., 55, 2108-2110.

Nimis, P.L. and Monte M. (1988). The lichen vegetation of the cathedral of Orvieto (Central Italy). Stud. Geobot., 8, 77-88.

Ortega-Calvo, J.J., Hernandez-Marine, M. and Saiz-Jimenez, C. (1991 a). Mechanical deterioration of building stones by cyanobacteria and algae. In: Biodeterioration and Biodegradation 8, pp. 392-394 (H.W. Rossmoore, ed.), Elsevier, London.

Ortega-Calvo, J.J., Hernandez-Marine, M. and Saiz-Jimenez, C. (1991 b). Biodeterioration of buildings materials by cyanobacteria and algae. Int. Biodeter., 28, 165-185.

Ortega-Calvo, J.J., Sanchez-Castillo, P.M., Hernandez-Marine, M., and Saiz-Jimenez, C. (1992). Isolation and characterization of epilithic chlorophyta and cyanobacteria from two Spanish cathedrals (Salamanca and Toledo). Nova Hedwigia, in press.

Petersen, K., Kuroczkin, J., Strzelczyk, A.B. and Krumbein, W.E. (1988). Distribution and effects of fungi on and in sandstones. In: Biodeterioration 7, pp. 455-460 (D.R. Houghton, R.N. Smith and H.O.W. Eggins, eds.), Elsevier, London.

Pietrini, A.M., Ricci, S., Bartolini, M. and Giuliani, M.R. (1985). A reddish colour alteration caused by algae on stoneworks. Preliminary studies. In: Ve Congrés International sur l'Alteration et la Conservation de la Pierre, pp. 653-662 (G. Felix, ed.), Presses Polytechniques Romandes, Lausanne.

Saiz-Jimenez, C (1981). Weathering of building materials of the Giralda (Seville, Spain) by lichens. In: 6th Triennial Meeting ICOM, Commitee for Conservation, Ottawa, paper 81/10/4.

Saiz-Jimenez, C. (1984). Weathering and colonization of limestones in an urban environment. In: Soil Biology and Conservation of the Biosphere, Vol. 2, pp. 757-767 (J. Szegi, ed.), Akademiai Kiado, Budapest.

Saiz-Jimenez, C. (1990 a). Analytical approaches for the study of organic compounds in weathered building stones. In: Advanced Workshop on Analytical Methodologies for the Investigation of Damaged Stones, 22 p. (F. Veniale and U. Zezza, eds.), Pavia.

Saiz-Jimenez, C. (1990 b). The mosaics of Italica. European Cultural Heritage, Newsletter on Research, 4, 34-38.

Saiz-Jimenez, C. and Samson, R.A. (1981). Microorganisms and environmental pollution as deteriorating agents of the frescoes of Santa Maria de la Rabida, Huelva, Spain. In: 6th Triennial Meeting ICOM, Commitee for Conservation, Ottawa, paper 81/15/5.

Saiz-Jimenez, C. and Garcia-Rowe, J. (1992). Biodeterioration of marbles and limestones in Roman paviments. In: 2nd International Symposium for the Conservation of Monuments in the Mediterranean Basin, in press.

Saiz-Jimenez, C., Garcia-Rowe, J., Garcia del Cura, M.A., Ortega-Calvo, J.J., Roekens, E. and Van Grieken, R. (1990). Endolithic cyanobacteria in Maastricht limestone. Sci. Total Environ., 94, 209-220.

Saiz-Jimenez, C., Hermosin, B., Ortega-Calvo, J.J. and Gomez-Alarcon (1991 a). Applications of analytical pyrolysis to the study of cultural properties. J. Anal. Appl. Pyrol., 20, 239-251.

Saiz-Jimenez, C., Garcia-Rowe, J. and Rodriguez-Hidalgo, J.M. (1991 b). Biodeterioration of polychrome Roman mosaics. Int. Biodeter., 28, 65-79.

Seaward, M.R.D. (1988). Lichen damage to ancient monuments: a case study. Lichenologist, 20, 291-295.

Seaward, M.R.D. and Giacobini, C. (1988). Lichen-induced biodeterioration of Italian monuments, frescoes and other archaeological materials. Stud. Geobot., 8, 3-11.

Seaward, M.R.D., Capponi, G. and Giacobini, C. (1990). Biodeterioramento da licheni in Puglia. In: La Conservazione dei Monumenti nel Bacino del Mediterraneo, pp. 243-245 (F. Zezza, ed.), Grafo, Brescia.

Syers, J.K. and Iskandar, I.K. (1973). Pedogenetic significance of lichens. In: The Lichens, pp. 225-248 (V. Ahmadjian and M.E. Hale, eds.), Academic Press, New York.

Tayler, S. and May, E. (1991). The seasonality of heterotrophic bacteria on sandstones of ancient monuments. Int. Biodeter., 28, 49-64.

Warscheid, T. Oelting, M. and Krumbein, W.E. (1991). Physico-chemical aspects of biodeterioration processes on rocks with special regard to organic pollutants. Int. Biodeter., 28, 37-48.

SOME NEW FINDINGS ON MONOAMINE OXIDASE "The Key Enzyme of Biogenic Amines Degradation"

HENRY M. ZEIDAN, Chemistry Department, Clark Atlanta University, Atlanta, GA 30314, USA, University of Texas, Chemistry Department, El-Paso, TX 79968, USA, and Life College, Marietta, Georgia 30060, USA

INTRODUCTION

Monoamine oxidase (MAO) is postulated to play an important role in controlling the levels of certain neurohormones and other pharmacologically important amines (Schacht and Leven, 1977; Von Korff, 1979). This involvement has stimulated considerable interest in the physiochemical and enzymatic properties of the enzyme.

The enzyme has a molecular weight of 104,000 per protomer and is made up of two subunits, each of which has a molecular weight of 52,000 (Chaung et al., 1976) and there are no disulfide linkages between the subunits. There appears to be one mole of FAD per two subunits, which is covalently attached to the protein (Orleand, 1971). From the effect of pH on the kinetic properties of the enzyme (Igaue at al., 1967), it was concluded that a cysteine residue in the enzyme is catalytically important in the cleavage of the C-N bond in the substrate. MAO has been characterized by amine substrate specificities and response to inhibitors as having two independent catalytic sites (Tipton et al., 1981; Jani, 1977) or as two distinct enzyme forms (A, B) embedded in the same mitochondrial membrane (White and Tanik, 1979; White and Stine, 1982; Bach et al., 1988). Other authors postulated that MAO-A and MAO-B may be the same enzyme in different lipid environments (Haung, 1980; White and Stine, 1982; Fowler and Ross, 1984). The MAO-B which preferentially deaminates

Biodeterioration Research 4, Edited by G.C. Llewellyn *et al.*, Plenum Press, New York, 1994

benzylamine is sensitive to deprenyl inhibition. Tyramine and tryptamine are substrates for both MAO-A and MAO-B enzymes (Tipton et al., 1981; Oreland and Eksted, 1976). The potential significance of the enzyme in neuropharmacology has promoted us to explore the nature of the active site by Fluorescence probes and spin Label techniques (Zeidan et al., 1980; Zeidan, 1984; Zeidan, 1988; Zeidan, 1990; Zeidan and Buckanan, 1990). As a continuation of our efforts to verify the nature of the active site of MAO A versus B, we have embarked upon a specific study to probe the active site of MAO A and B by using a suitably chosen labeled substrate, trytamine. The ESR spectra of the label in the presence of an enzyme provides a convenient method to characterize structural features and differences of the active site in vitro and to provide a clue as to the nature of the MAO's active sites. Although there have been many speculations in the literature that the two forms of MAO represent different active sites on the same protein or different lipid complexes of one enzyme, none of these notions was persuasive (Singer, 1985, 1986). In 1988, Bach et al., reported the complete amino acid sequences of both proteins. The A type enzyme, which deaminates serotonin, is sensitive to clorgyline inhibition. Tyramine and tryptamine are substrates for both the A and B enzymes (Tipton et al., 1981; Orleand and Eksted, 1976). Many of the complex problems surrounding the mechanism of action, regulation and substrate specificity of mitochondrial MAO have resisted definitive resolution when explained by traditional enzy- mological approaches. In our laboratory, we have been exploring the utility of spin labels in studying these problems. Of particular interest has been the controversial question of the function and structural basis of the enzyme's multiplicity of substrate specificities (Chaung et al., 1974; White and Wu, 1975). As a first step toward a three-dimensional model, Zeidan et al., (1979) embarked upon a specific study of the environment of the essential sulfhydryl residues using conventional ESR spin labeling techniques. The details of this aspect of MAO research have been elaborated elsewhere (Zeidan et al., 1979). As a continuation of the efforts to understand the structure-function relationship of the enzyme, we used fluorescence, spin probes and NMR techniques to verify the nature of the active site of MAO-B (Zeidan, 1988, 1990; Zeidan and Buchanan, 1990). In the present investigation, spin label tryptamine was used as a structural probe to provide a clue as to the nature of the active site of MAO-A and to detect the structural differences at the three dimensional level between MAO-A and MAO-B.

MATERIALS

Highly purified MAO A and B were obtained following the procedures of Weyler and Salach (1985) and Salach (1981), respectively. The highly purified either MAO A or B were electrophoretically homogeneous in SDS- polyacrylamide gel electrophoresis. Spin-labeled tryptamine (N-(1-oxyl-2, 2, 5, 5-tetramethylpyrroline-3-yl)- N(3- B- ethyl) indolyl-carbamide (TrySL) and Spin labeled tryptophan/Methyl-N-/1- oxyl-2, 2, 5, 5- tetramethyl- pyrroline-3-aminocarbonyl were purchased from Medimpex (Budapest, Hungary) (Figure 1). Spin labeled pargyline was a generous gift from Drs. Buckman and the Eiduson at University of California, Los Angeles. The other chemicals used were purchased from standard chemical companies.

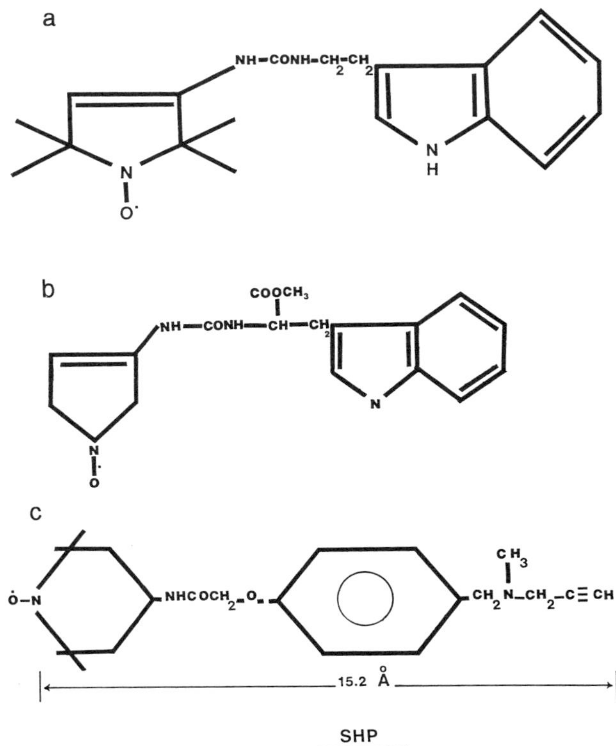

Figure 1. a. Structure of Spin Labeled Tryptamine (N-(1-oxyl-2, 2, 5, 5-tetramethylpyrrolin-3-yl)-N-[-3-B-ethyl) indolyl]-carbamide TrySL).
b. Structure of Spin Labeled Tryptophan (TrpSL).
c. Structure of Spin Labeled Pargyline (SHP).

METHODS

The enzyme was assayed by the spectrophotometric procedure of Tabor et al. (1951) using benzylamine as a substrate. Electron paramagnetic resonance (EPR) measurements were recorded at room temperature by a Varian EPR spectrometer. The field setting was 3415 G; the microwave frequency was 9.5 Hz; and modulation amplitudes of 4.0 and 2.0 G were used. MAO-TrySL interaction was monitored by the addition of TrySL dissolved in 0.01 M phosphate buffer (pH 7.5) directly to MAO in the same buffer. Titrations of MAO with TrySL were performed by successive addition of 0.005 ml increments of 5 x 10^{-3} M TrySL. The MAO concentration was 3.8 x 10^{-5} M. The concentration of spin-labeled tryptamine was determined by computer assisted double integration of spectra and comparison with a standard known concentration. In the case of a low concentration of MAO-bound label, quantitative measurements were performed from the decrease of the amplitude of the free label after addition of substrate analogue. The concentration of paramagnetic molecules was used in all calculations of dissociation constants. MAO-TrySL and MAO-SHP interactions were monitored as described above for MAO-TrySL interaction. (Figure 2) All solutions contained NaCN to retard extraneous nitroxide reduction.

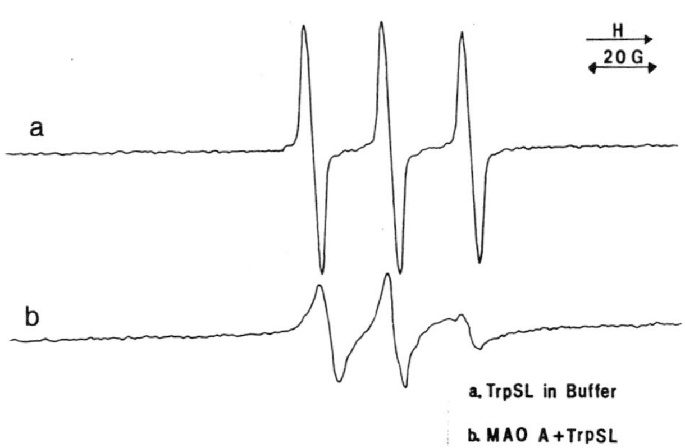

Fig. 2. a. ESR Spectrum of Spin Labeled Tryptamine (TrySL) Free in Ssolution (1.0 x 10^{-5}M).

b. ESR Spectrum of the Ssubstrate Sspin Labeled Ttryptamine (TrySL) in the Presence of MAO-A. The enzyme (2.0mg in 1.0 ml of 0.01M potassium phosphate buffer (pH 7.5) was reacted with 1 x 10^{-5}M (final concentration) TrySL under anaerobic conditions at pH 7.5 and room temperature.

RESULTS AND DISCUSSION

MAO-A/MAO-B dilemma: One or Two enzymes?

MAO-A-TrySL Interaction: Classification of MAO into two forms, A and B, first suggested by Johnston (1968), is based on the selective inhibition of the activity of these forms by clorgyline and deprenyl. During the period of 1968-1988 there has been a debate as to whether this is due to the presence of either one enzyme with two active sites or two separate enzymes. Bach et al. (1988) reported the amino acid sequences of both proteins. They concluded that MAO-A and MAO-B are probably separate enzymes. In the present investigation, we have tried to explore the active site of MAO-A and MAO-B at the three dimensional level by spin labeling techniques. The specificity of binding between TrySL and either MAO A or B was carefully investigated. On mixing TrySL with highly purified MAO-A, the observed changes were indicative of binding. This was manifested as selective broadening of the high and low magnetic field lines in the three-line ESR spectrum (Figure 1). Such broadening can be attributed to decreased motional freedom of the bound state for the interaction of a spin label with a protein. The spectrum clearly suggests a relatively fast and nearly isotropic motion of the spin label, suggesting one type of motional freedom. The bound tryptamine analogue appears to be in a "partial immobile state" suggesting that the catalytic sites are close to the surface of the enzyme (Zeidan et al., 1988; Zeidan et al., 1980; Zeidan et al., 1986; Zeidan, 1990).

Calculation of the correlation time, which reflects the environment of the active site of the enzyme was straight forward, since there is only one type of motional freedom (Zeidan et al., 1986; Zeidan, 1988; Zeidan, 1990) where H is the peak to peak width of the central field (in G), and H_0/H_{-1} are peak to peak heights of the central and high field lines, respectively. The calculated rotational correlational time was found to be 1.0 x 10^{-9} seconds. The corresponding $\tau/_c$ for TrySL in buffer would be of the order of 10^{-10} S.

$$\frac{1}{\tau_c} = \frac{3.6 \times 10^{-9}}{\left(\frac{h_0}{h_{-1}}-1\right)\Delta H_0}$$

Selective binding of spin-labeled tryptamine to MAO was examined by using other proteins such as bovine serum albumin, ribonuclease and alkaline phosphatase at concentrations comparable to that of MAO-A. The signal indicated no binding of spin labeled tryptamine to any of these proteins. That, the MAO bound spin labeled tryptamine reversibly was clearly indicated by the lack of a residual observable ESR signal after a mixture of MAO-spin label complex was dialyzed overnight against 0.01 M potassium phosphate, pH 7.5. The enzyme was titrated with TrySl to determine the dissociation constant for the radical enzyme complex and the number of moles of TrySL bound per mole of enzyme. A Schatchard plot of the data is shown in Figure 3 Extrapolation of

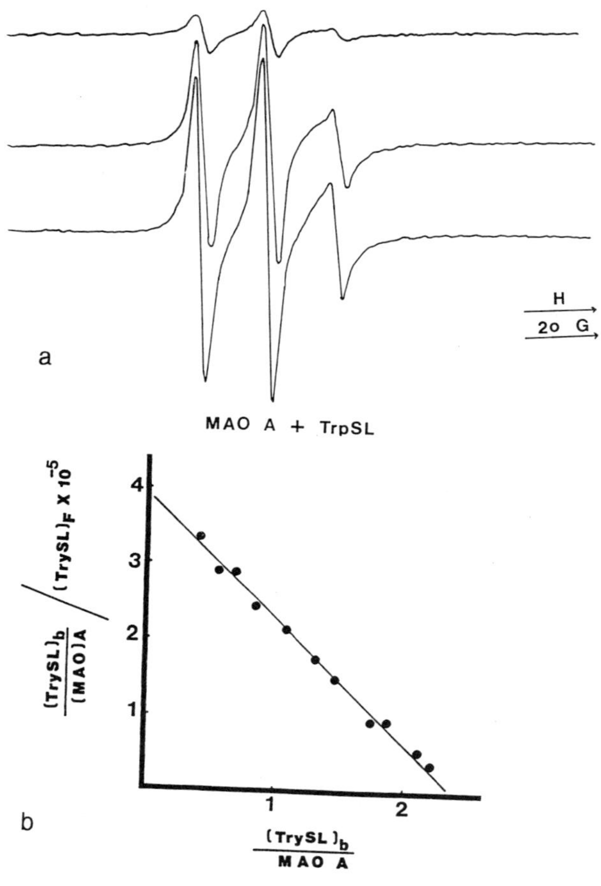

Fig. 3. a. Tritation of MAO-A with TrySL, i) 1 x 10^{-6}M TrySL, ii) 5 x 10^{-6}M TrySL, iii) 1 x 10^{-5}M TrySL.

b. Scatchard Plot of the Paramagnetic Resonance Data for TrySL Binding to MAO. The line intersecting the abscissa at (TrySL)b/Et = 2.2 was determined by a least squares fit of the data.

the line to (TrySL)b/Et = 2.4 fits the data with the dissociation constant, Kd, for the radical enzyme complex 5.2 10^{-5} M.

MAO-B-TrySL Interaction: The electron paramagnetic resonance spectra obtained by spin labeling MAO-B from bovine liver with TrySL is presented in Figure 4a. The observed changes were indeed indicative of binding. Again, this manifested as selective broadening of the high and low magnetic field lines. Again, this broadening is attributed to decreased motional freedom of the bound state for the interaction of TrySL with MAO-B. However, the nitroxide radical in this state appeared to be located in a more restricted environment. Increasing immobilization of the nitroxide resulted in a more asymmetrical ESR spectrum. The rotational freedom exhibited by MAO-B-TrySL interaction (Figure 4) is in accordance with the prediction that the microenvironment of the MAO-B active site is not identical to MAO-A. The spectrum in Figure 4 suggests that the nitroxide is located in a more hindered environment than in Figure 2b, however, still relatively fast and nearly isotropic, suggesting that the active site of MAO-B is more hydrophobic than the active site of MAO-A. The catalytic sites of MAO-B are still most likely located near the surface of the enzyme. However, the geometry of both MAO-A and B are different as judged by the existence of a hindered environment for the MAO-B active site.

Selective binding of TrySL to MAO-B was examined by using other proteins such as bovine serum albumin, ribonuclease and alkaline phosphatase at concentrations comparable to that in the highly purified MAO-B preparations. The ESR spectrum indicated no binding of TrySL to any of these proteins (Figure 4b, other spectra not shown). There was no detectable interference with the assay from Triton X-100 which was present at a very low concentration in an enzyme preparation during the determination in either MAO-A or MAO-B. That MAO-A bound TrySL reversibly was clearly indicated by the lack of a residual observable ESR signal after a mixture of MAO-B complex was dialyzed overnight against 0.01 M potassium phosphate, pH 7.5. The dissociation constant for the radical enzyme complex and the number of moles of TrySL per mole of enzyme was determined as described with MAO-A. A Scatchard plot of the data is shown in Figure 5. Extrapolation of the line to $(TrySL)_b$ = 2.2 fits the data with a dissociation constant, Kd, for the radical enzyme complex 3.9 x 10^{-9} M. There were no indications of secondary, weaker and/or nonspecific binding.

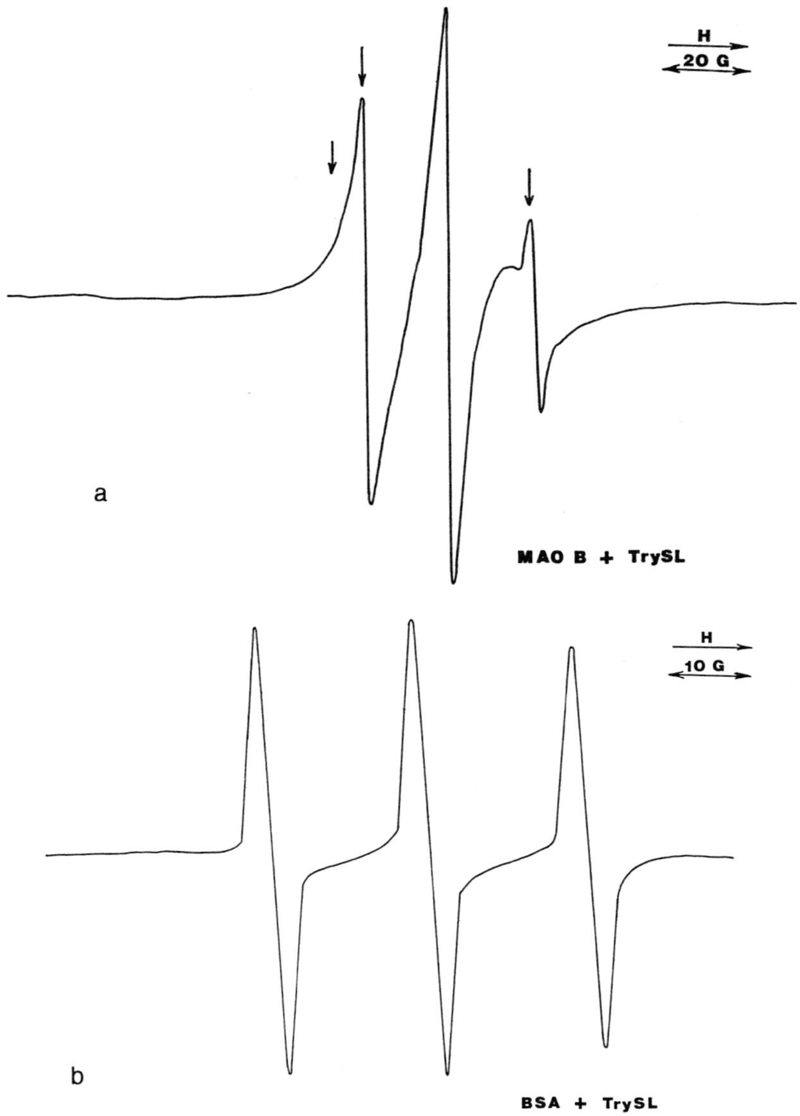

Fig. 4. a. ESR of TrySL in the Presence of MAO-B. The enzyme (2.0mg in 1.0ml of 0.01M potassium phosphate buffer (pH 7.5) was reacted with 1.0×10^{-5}M (final concentration) TrySL under anaerobic conditions at pH 7.5 and room temperature.

b. ESR Spectrum of TrySL in the Presence of Albumin. The same experimental methods as a.

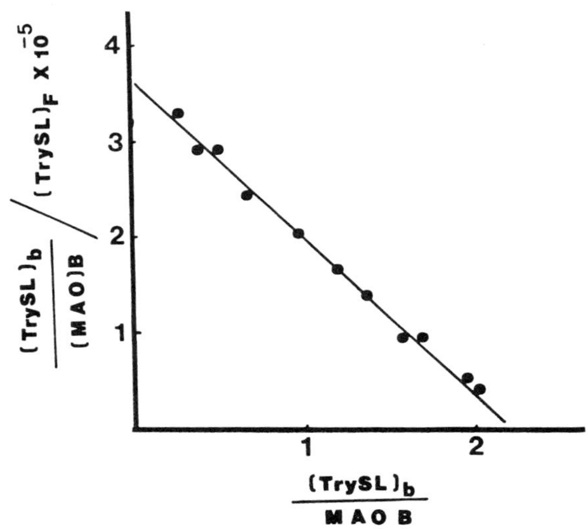

Figure 5. Schatchard Plot of the Paramagnetic Resonance Data for TrySL binding to MAO-B. The line intersecting the abscissa at (trySL)b/Et = 2.2 was determined by a least squares fit of the data.

When increasing concentrations of tryptamine were added to a solution of TrySL and MAO, the bound TrySL was progressively displaced from the enzyme. This release of tryptamine bound label demonstrated unambiguously that it binds mainly at the active site. Tryptamine appeared to be competing for the same site on the enzyme. This con- clusion is further supported by the finding that degradation of tryptamine was competitively inhibited by TrySL. Interaction of MAO with Spin Labeled Tryptophan: A number of MAO inhibitors have been shown to exhibit true antidepressant activity and some have found their way into clinical use Youdim and Finberg (1982), Tipton, et al., (1981), and Finberg, and Youdim (1985). Animal behavioral studies have shown that treatment with the amine precursors L-tryptophan, L-5- hydroxytrptophan, or L-dihydrophenyalanine following MAO inhibition can result in overt behavioral syndromes that can be attributed to the accumulation and release of amine neurotransmitter Tipton, et al., (1981). It was of interest to us to examine MAO-spin labeled tryptophan (TrpSL) interaction to permit a detailed study at the molecular level, to understand the detailed binding geometrics of selective potent drugs which may exhibit site specific inhibitory activity, and to elucidate the mode of action of psychotropic and hypotensive drugs. The MAO-TrpSL interaction is presented in Figure 6. As the spectra suggest that the mode of binding of TrpSL to either MAO-A or MAO-B is identical to TrySL interaction with the A and B forms of the enzyme, respectively. The EPR spectra adds suggestive evidence for differences in the hydrophobicity of the active site of MAO. MAO-B appeared to be more hydrophobic than MAO-A. The binding sites of MAO-B appeared to be in a more restricted environment. In line with these suggestions, it is possible to envision an MAO-A selective inhibitor which is able to bind to MAO-A, but is sterically hindered from binding to the active center of MAO-B. On the other hand, an inhibitor of MAO-B, which must have a high affinity for the hydrophobic site, is not sterically hindered from binding to MAO-A and thus cannot attain the same degree of sensitivity as the MAO-A selective inhibitors. Other investigators contend that compounds which the aromatic nucleus is separated by 1 or 2 carbons from the oxidizable N, react preferentially with MAO-B. Typical examples are benzylamine, phenylethylamine, deprenyl and pargyline. When the distance is equivalent to three or more carbons, the substrate or inhibitor bind preferentially to MAO-A. This was interpreted to suggest that the active sites of the two forms differ by the distance between the binding sites for the N and of the aromatic ring Finberg and Youdim (1985). However, some substrates (e.g., noradrenaline), do not seem to follow this pattern. The results of this investigation, along with previous investigations suggest that the active site of MAO-A is close to the surface of the enzyme and that MAO-B is more hydrophobic.

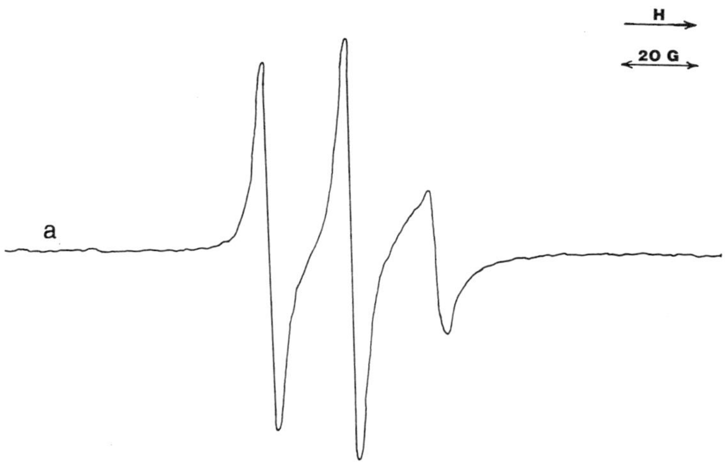

MAO A + TrpSL

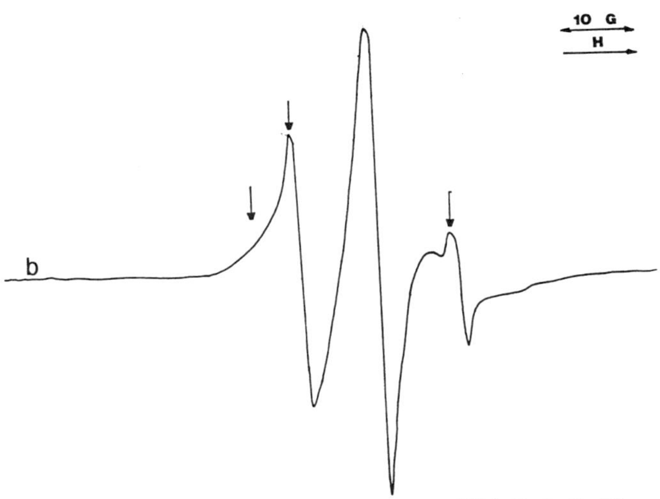

MAO B + TrySL

Fig. 6. a. ESR Spectrum of Spin Labeled Tryptophan (TrpSL) in the Presence of MAO-A. The same experimental methods as Figure 4a. b. ESR Spectrum of TrpSL in the Presence of MAO-B. The same experimental methods as Figure 2b.

The Effect of Proteolytic Hydrolysis by Trypsin on MAO:

We have attempted to study the effect of trypsin on Spin labeled either MAO-A or MAO-B with TrySL and the results of this investigation are presented in Figure 7. As can be seen, proteolysis for up to 30 min at room temperature did not reverse the MAO-A induced nitroxide signal loss, however, the EPR signal

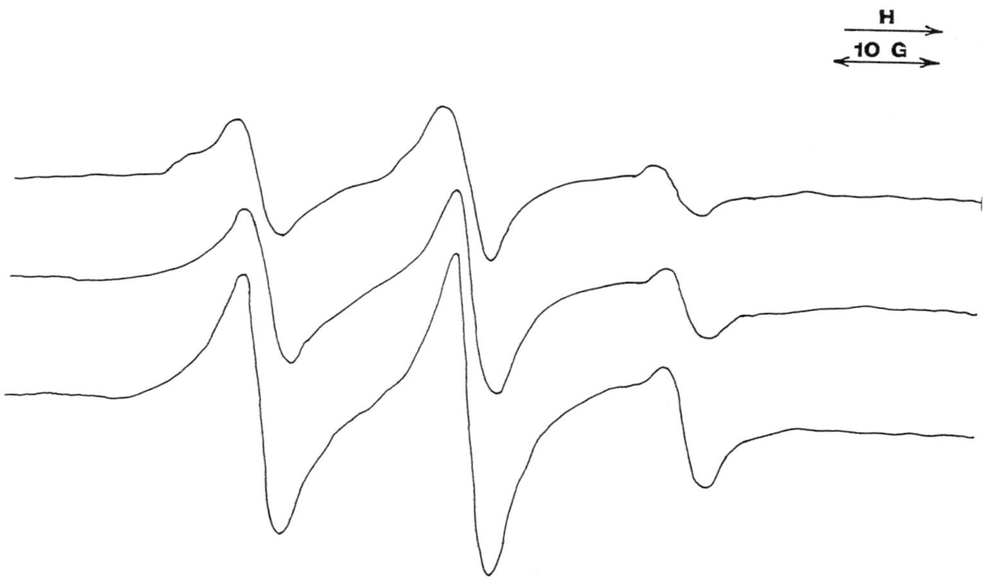

Figure 7. ESR Spectrum of Spin Labeled MAO-B with TrySL After Addition of Trypsin. MAO-B-TrySL derivative was prepared as in Figure 4b then 1×10^{-6}M trypsin was added to the labeled enzyme. The same experimental conditions as in Figure 4.

was restored upon addition of trypsin to MAO-B suggesting a major conformational change in the active site of spin labeled MAO-B enzyme. Again, these spectral data highly suggest structural differences in the microenvironment of the active site of both MAO-A and B.

MAO-Pargyline Interactions

Pargyline behaves as a normal suicide inhibitor with a variety of soluable MAO preparations Buckman and Eiduson (1980). It was of interest to us to examine the effect of pargyline on spin labeled MAO with TrySL. As seen in Figure 8, addition of pargyline increased the motional freedom of the nitroxide radical in spin labeled MAO-B, however, a very slight effect was observed with spin labeled MAO-A. Again, these results are consistent with the structural differences between MAO-A and MAO-B. Binding of spin labeled pargyline (SHP) with MAO-B was examined and the results are shown in Figure 9. On mixing SHP with MAO-B there were some changes in the ESR spectra indicative of binding. The observed broadening can be attributed to decreased motional freedom of the bound state for SHP interaction with MAO-B. The addition of pargline reversed the EPR signal loss and the addition of SHP to spin labeled MAO-B with TrySL had the same effect as pargyline. An increase in the motional freedom of the nitroxide radical was observed and the total concentration of the partially immobile and free label increased (Figure 10).

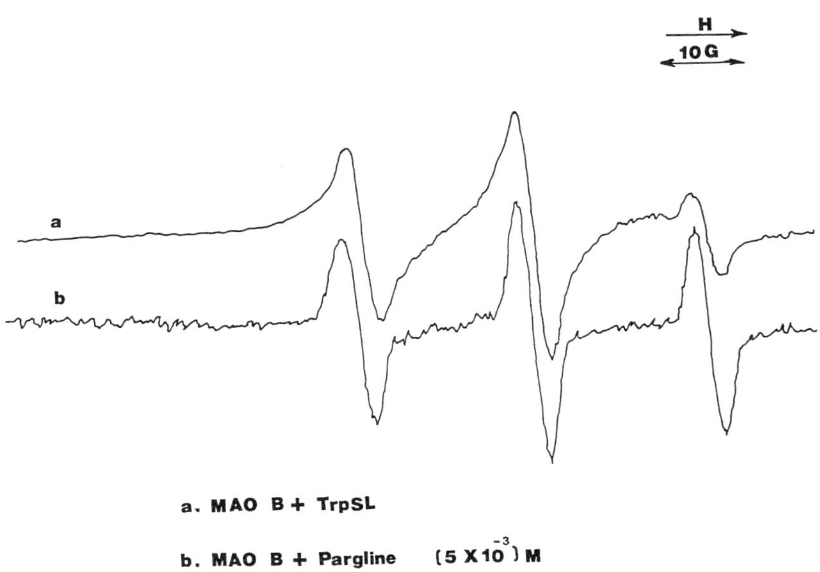

Fig. 8. a. ESR Spectrum of Spin Llabeled MAO-B with TrySL (same as Figure 4a).

b. Same as Figure 4a, then 1.5 x 10^{-5}M Pargyline Was Added.

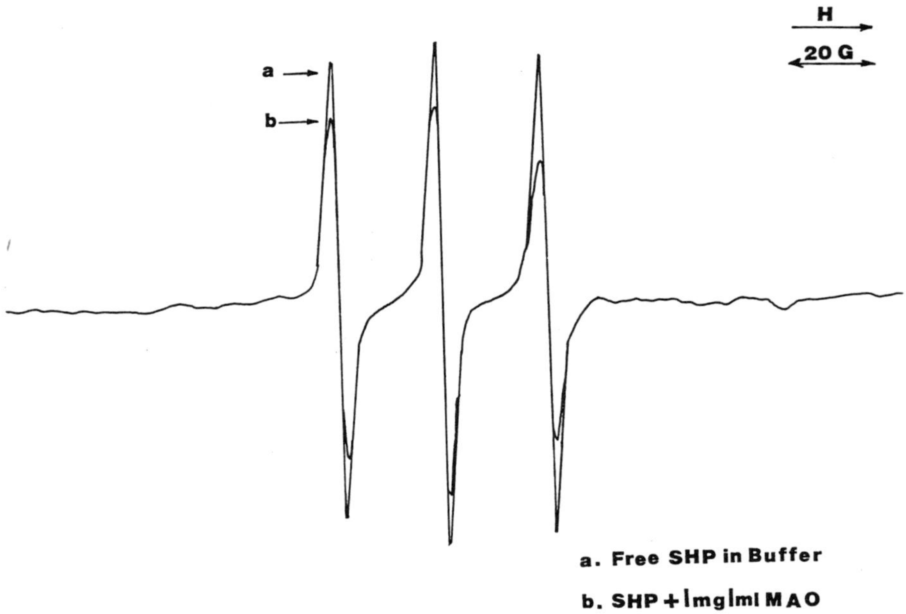

Figure 9. a. ESR Spectrum of Spin Labeled Pargyline in Buffer (1×10^{-5}M). b. ESR Spectrum of SHP-MAO-B. A total concentration of SHP at 1×10^{-5}M was added to 2mg/ml MAO-B.

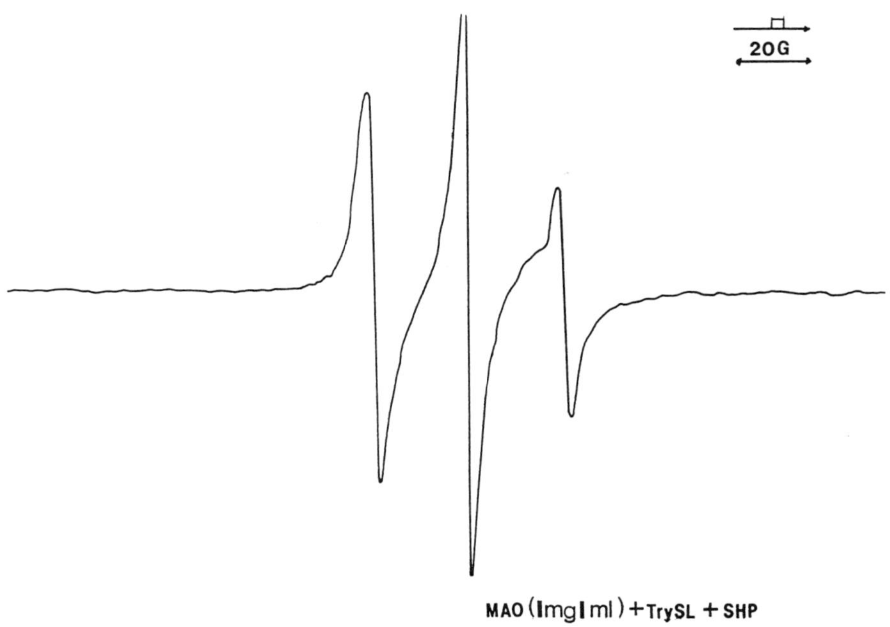

Figure 10. ESR Spectrum of MAO-B-TrySL-SHP. MAO-B-TrySL derivative was first prepared as in Figure 4a, then 1×10^{-5}M SHP was added to the derivative.

All of the above data can be used to make certain tentative conclusions about the MAO-active site structure. These are summarized below:

1) The spin-label substrate, tryptamine, can be used as a structural probe to explore the structural differences between the active site of MAO-A and MAO-B.

2) Binding of the probe is highly specific for MAO, with a dissociation constant of 42 μM and 2.4 mole per enzyme dimer for MAO-A and 3.6 x 10^{-5} and 2.2 mole bound per enzyme dimer for MAO-B.

3) From an analysis of the motional freedom of the nitroxide radical, it appears that the radical is located in a more hindered environment in MAO-B suggesting structural differences in the active site of MAO-A and MAO-B. MAO-B is most likely more hydrophobic than MAO-A.

4) Spin-labeled MAO-B is more sensitive to trypsin hydrolysis than MAO-A suggesting structural differences in the active site.

5) Spin labeled MAO-B is more sensitive to pargyline than MAO-A, again, suggesting structural differences in the active site.

Acknowledgements

We wish to thank Ms. Tondaleyo Conley for typing this manuscript. This work was supported by NIH, RCMI Grant #1G12-RR03062 and NIH Grant #R08247 to HZ. The author gratefully acknowledges Dr. William Dashek, Biology Department, Clark Atlanta University, for his critical reading of this manuscript. The technical support of Emory University is also highly appreciated. The author also acknowledges Mr. Khaled Abboud for his outstanding graphic work. The author also is grateful to the assistance of his graduate students and all who contributed to the amine oxidases research in our laboratory.

REFERENCES

Bach, A.W.J., Lan, N.C., Johnson, D.L., Abell, C.W., Bembenk, M.E., Kwan, S., Seeburg, P.H. and Shih, J.C. (1988), Proc. Natl. Acad. Sci., 85, 4934-4938.

Buckman, T. and Eiduson, S. (1980) J. Neurochem., 34, 1594-1602.

Chaung, H.Y.K., Pateck, D.R. and Hellerman (1974) J. Biol. Chem., 249, 2384-2391.

Finberg, J.P.M. and Youdim, M.B.H., Monamine Oxidases, in Handbook of Neurochemistry (1985), Lajath. A. Eds., Plenum Press, NY.

Haung, H.K. (1981) Functions and Regulations of Monoamine Enzymes (Usdin, E., Weiner, N. and Youdim, M.P.H., eds), pp. 489-501, MacMillan, New York.

Haung, H.K. (1980) Mol. Pharmacol., 17, 192-198.

Igaue, I, Gomes, B., and Yasunobu, K. (1967), Biochem. Biophys. Res. Commun., 29, 562-570.

Jain, M. (1977) Monoamine Oxidase, 1925-1933, Life Science Co.

Johnston, J.P. (1968) Biochem. Pharmacol., 16, 285.

Kamijo, K. Koelle, G.B. and Wagner, H.H. (1956) J. Pharmacol. Exp. Ther., 117, 213.

Minamuria, N. and Yasunobu, K.T. (1978) Arch. Biochem. Biophys., 189, 481-489.

Murphy, D.L., Garrlick, N.A., and Cohen, R.M. (1982) Monoamine Oxidase. Biochemical and Physiological Aspects Revelant to Human Psychopharmacology in Antidepressants, Burrows, N., and Davies, D. Eds., Elsevier, Amesterdam. 209.

Orleand, L. and Eksted, T.B. (1976) Biochem. Pharmacol., 25, 119-124.

Schact, V. and Levien, M. (1977), Depressive Disorders Synposium, Rowe, Publisher F.K. Schattaner Verlag, Stutgart, NY, 21-40.

Severina, I.S. (1973) Eur. J. Biochem., 38, 239.

Shih, J.C. and Eiduson, S. (1973) J. Neurochem., 21, 41-49.

Singer, T.P. Von Korff, R.W. and Murphy, D.L. (1979) (eds.) Monoamine Oxidase, Structure, Function and Altered Functions, 1-57, Academic Press, NY.

Singer, T.P., Mitochondrial monoamine oxidase (1986) (eds.) in Advances in Biochemical Pharmacology, 1, Vessey, D.A. and Zakim, D., eds., John Wiely and Sons. NY.

Singer, T.P. (1985) (eds.) Demise and ressurection of misconceptions about monoamine oxidase in Neuropharmacology, Kelemen, K., Magyar, K., and Vizi, E.S., eds., Akademiai Kiado, Budapest, 29.

Tabor, G.E., Tabor, H. and Rosenthal, S.M. (1951) J. Biol.Chem., 193, 265-271

Tipton, K.F., Housley, M.D. and Mantle, T.J. (1981) in Monoamine Oxidase and its Inhibition. Ciba Foundation Symp. 39, 5-31, Elsevier, Amesterdam.

Tipton, K.F. (1968), Eur. J. Biochem., 4, 103.

Usidin, E. (1976) (eds.) Neuropysychopharmacology of Monoamines and Their Regulatory Enzymes, 1-6, Raven Press, NY.

Von Korff, R.W. (1979), Monoamine Oxidase; Structure, Function and Altered Functions, (Singer, T.P., Von Korff, R.W. and Murphy, D.L., eds.), Academic Press, NY, 1-6.

White, H.L. and Tansik, T.L. (1979) Monoamine Oxidase Structure, Function and Altered Functions (Singer, T.P.; Von Korff, R.W.; and Murphy, D.L., eds.), pp. 129-144, Academic Press, NY.

White, H.L. and Stine, D.k. (1982), J. Neurochem., 38, 1429-1436.

White, H.T. and Wu, J.C. (1975) Neurochemistry, 25, 21-26.

Yagi, K. and Naoi, M. (1982) Biochem. Int., 4, 457.

Youdim, M.B.H. and Finberg, J.P.M. (1982) Monoamine Oxidase Inhibitor Antidepressants in Psychopharmacology I, Graham-Smith, D.G., Hippius, H., and Winoker, G., eds., Elsevier, Amesterdam, 38.

Zeidan, H., Watanabe, K.T., Piette, L.H. and Yasunobe, K.T. (1979) Frontiers in Protein Chemistry, 133-148, Elsevier/North-Holland, Amesterdam.

Zeidan, H. (1988), Biochem. Biophys. Acta., 955, 111-118.

Zeidan, H., Watanabe, K., Piette, L.H. and Yasunobu, K.T. (1980) J. Biol. Chem., 255, 7621-7626.

Zeidan, H. (1988) Biochem. Biophys. Acta, 957, 266-271.

Zeidan, H., Pearson, K., Brown, S., and Han, P. (1986) Biochem. Biophys. Acta, 870, 141-147.

Zeidan, H. (1990), Biodeterioration Research, 3, 555-564, G.C. Llewellyn and C.E. O'Rear, (eds.), Plenum Press, N.Y.

Zeidan, H. and Buchanan, S. (1990), Biodeterioration, 577-585, G.C. Llewellyn and C.E. O'Rear, (eds.), Plenum Press, NY

THE ACTIVE SITE OF PLASMA AMINE OXIDASE "THE KEY ENZYME OF BIOGENIC AMINES DEGRADATION IN PLASMA" AS SEEN BY FLUORESCENCE INVESTIGATIONS

HENRY M. ZEIDAN*, and **S. OYOUGA**, Clark Atlanta University, Chemistry Department. Atlanta, GA 30314, USA, University of Texas, Chemistry Department, El-Paso, TX 79968, USA, and Life College, Marietta, GA 30060, USA.

INTRODUCTION

The role of Plasma Amine Oxidase (diamino: 0_2 oxidoreductase; EC 1.4.3.6) in the catabolism of various amines found in association with DNA and its function in one of the catabolic pathways of histamine and other pharmacologically important amines have led to considerable interest in this enzyme. The enzyme catalyzes the oxidation of certain amines and diamines according to the following reaction:

$$RCH_2NH_2 + 0_2 + H_20 \text{ ----- } RCH0 + NH_3 + H_20_2$$

Physicochemical studies of bovine plasma amine oxidase indicated that the enzyme is made up of two identical subunits and has a molecular weight of 170,000 KD (Yasunobu et al., 1976). The enzyme contains one essential histidine residue per monomer (Hiramatsu, et al., 1975). There are two gram-atom Cu/ mole protein which are essential for enzyme activity (Yasunobu et al., 1976). It has been reported that the copper does not change its valence state during catalysis either under aerobic or anaerobic addition of substrate. The only variance to the no valence change was the work reported by Mondovi et al., 1976 who reported that copper changed its valence state during substrate oxidation. It was concluded acid analysis was performed on the native enzyme and the reduced modified enzyme. The concentration of the protein was determined by a biuret method and absorbance was read at 280nm (R = 13). N-(1-pyrene) maleimide was purchased from Molecular Probes. The other common reagents used were of reagent grade and were purchased from standard companies.

METHODS

Fluorescence Labeling of the Essential Liberated Sulfhydryl Groups

The native PAO was assayed prior and after labeling of the enzyme with NPM. The activity of the enzyme did not change. This result confirms our previous investigation and others and adds a suggestive evidence that the native oxidized form of the enzyme, the sulfhydryl groups is essential for activity.

When 12.5 umoles of benzylamine at pH 7.6 was added to 1ml of PAO and the reduced enzyme was reacted with NPM to form the corresponding succinimido PAO adduct (Figure 1), the enzyme lost 95% of its activity. Modification of the enzyme with NPM was carried out by applying 5 x 10^{-4} moles NPM in 1.0ml of dimethyl sulfoxide to Whatman filter paper in a test tube which was then dried by a stream of nitrogen. Then, 1.2 x 10^{-5} moles of reduced PAO in 1,0ml of 0.05M potassium phosphate buffer, pH 7.5 was added to the test tube. The reaction was allowed to proceed with gentle stirring at room temperature and under anaerobic conditions.

Amino acid analyses were performed on 24 hour acid hydrolyzates (110°C), after the samples were passed through a 1.5 x 10cm column of Sephadex G-25, on a Beckman Model 121 MB automatic amino acid analyzer as described by Tarr (1985). Both the native and the reduced modified enzyme were analyzed for amino acid composition.

Spectral Measurements

Fluorescence excitation and emission spectra were recorded with a Perkin-Elmer MPF-4 fluorescence spectrophotometer using a 1.0cm path cuvette. Excitation was at 330nm for NPM fluorescence. Emission spectra were recorded at room temperature unless otherwise stated. The experiments are described in the legends.

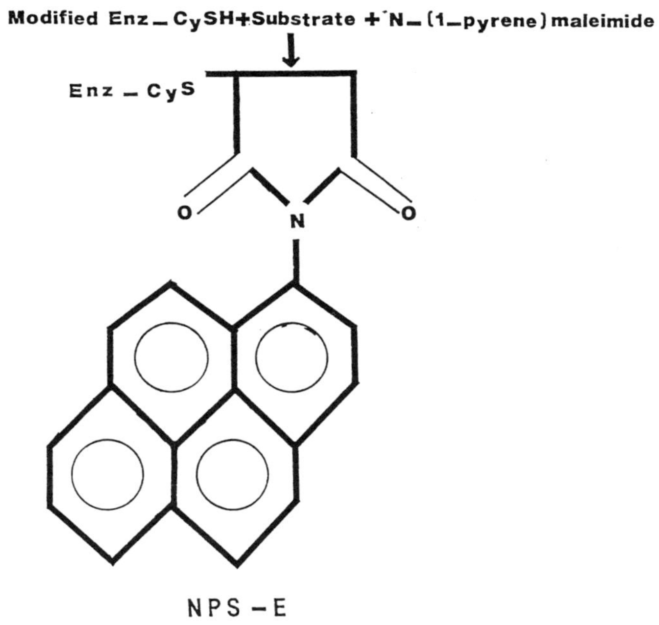

Fluorescent Labeling of the essntial sulfhvdrvl group on PAO

Figure 1. Reaction Scheme of NPM With a Cysteine Residue.

RESULTS AND DISCUSSION

NPM-PAO Interaction

Under the conditions described in the experimental section and by Zeidan et al., 1980, PAO was first reduced by addition of 12.5 umoles of benzylamine at pH 7.6 to 1.0ml of the enzyme. The reduced enzyme was then reacted with fluorescent label NPM. This then attached the label on the essential liberated sulfhydryl group per monomer. It was then exhaustively dialyzed against 0.05M potassium phosphate buffer, pH 7.6. After dialysis to remove excess reagent, the fluorescence spectrum of NPM- PAO adduct was recorded (Figure 2). The activity of the enzyme was checked prior to recording the spectrum, it was found that the enzyme lost 95% of its activity after its reduction with benzylamine and reaction with NPM fluorescence probe. NPM alone did not fluoresce or form a fluorescent product in solution alone. The fluorescence spectrum of NPM-PAO exhibits two major vibronic bands at about 375 and 395nm and a shoulder at 415-420nm. It closely resembles the spectra of several NPM- labeled proteins reported by Weltman et al. (1973); Zeidan and Buchanan (1990), but is shifted slightly to the blue with respect to them. When native PAO was treated with NPM, the fluorescence intensity was diminished significantly suggesting that addition of the substrate liberated an essential sulfhydryl group per monomer which reacted with NPM. The native enzyme did not lose any activity when it reacted directly with NPM. The activity studies along with the fluorescence spectra adds a suggestive evidence to the liberation of an essential sulfhydryl group upon addition of the substrate.

The pyrenemaliemide adduct with reduced PAO exhibit two distinctive fluorescent components at 375 and 395nm. Pyrene maleimide is quite sensitive to one or more specific factors which define the microenvironment about the protein bound chromophore. We are presently analyzing these factors. The two spectroscopic components may be a reflection of the two ways in which the pyrene may rest on the protein surface, that is with either the bulk of the pyrene ring buried and the succinoamide moiety exposed to the water that copper forms a tetraagonal complex with the enzyme. From the effect of the pH on the kinetic properties of the enzyme, it was concluded that the enzyme contains four sulfhydryl groups per mole which are neither essential for activity nor chelated to the copper present in the enzyme (Wang and Yasunobu, 1968). The two subunits of the enzyme are linked by S-S bridges (Achee, et al, 1968). It has been suggested by many investigators that there is another unknown cofactor X involved in catalysis which was identified tentatively by some investigators as pyridoxal phosphate (Davidson, 1958). Several investigators have disclosed that addition of substrate liberates two sulfhydryl groups per enzyme dimer (Suva and Abeles, 1978; Zeidan et al., 1980). It was postulated by these investigators that the liberated sulfhydryl groups are covalently linked to the unknown cofactor in the oxidized enzyme. It was reported by Inamasu and Yasunobu, 1974 that Plasma amine oxidase is not a pyridoxal phosphate dependent enzyme; furthermore, a few investigators suggested that CU- amine oxidase did not contain any other prosthetic group other than copper (Allison et al., 1973; Bentiz and Allison, 1974). Highly purified enzyme contains 4.6% hexoses bound to the protein (Watanabe et al., 1972). As a first step toward a three-dimensional model of PAO, Zeidan et al., 1980, embarked upon a specific study of the environment of the liberated sulfhydryl groups and the active site of the enzyme by conventional ESR- Spin labeling techniques. The details of this aspect of PAO research have been elaborated elsewhere (Zeidan et al., 1980).

As a continuation of the efforts to characterize the local environment of the liberated essential sulfhydryl groups, one per monomer, this laboratory has begun a specific study to probe the nature of the active site by using an extrinsic pyrene probe. In the work described here, we have used an extrinsic covalently attached label, N-(1-pyrene) maleimide, a specific sulfhydryl reagent that becomes fluorescent only after reacting with a sulfhydryl group (Wu et al., 1976). This fluorophore offers the advantages of a high quantum yield for easily dectable signal. Clear excitation and emission spectra, and a sufficiently long fluorescent life time. This probe has been used in this study to monitor any conformational changes which took place in the local environment of the essential liberated sulfhydryl groups during the pH change, urea denaturation and increasing ionic strength.

MATERIALS AND METHODS

Materials

Highly purified enzyme was isolated following the procedure of Zeidan et al., 1980. The enzyme activity was assayed according to Yasunobu et al., 1976. One enzyme unit catalyzes the formation of 1 umole of benzyaldehyde per minute. Specific activity of the highly purified enzyme was 3500μ/mg. The purity of the enzyme was checked by SDS polyacrylamide gel electrophoresis. Only a single band appeared at 85KD. Amino is buried or exposed would be due to the specific environment of each sulfhydryl group, which would differ for each protein. We have previously demonstrated by using a spin label approach that the liberated sulfhydryl group, one per monomer are located in a deep pocket (Zeidan et al., 1980). The use of pyrenemaleimide fluorescent label seems in part support our previous EPR findings and further characterizes the local environment of the liberated essential sulfhydryl groups.

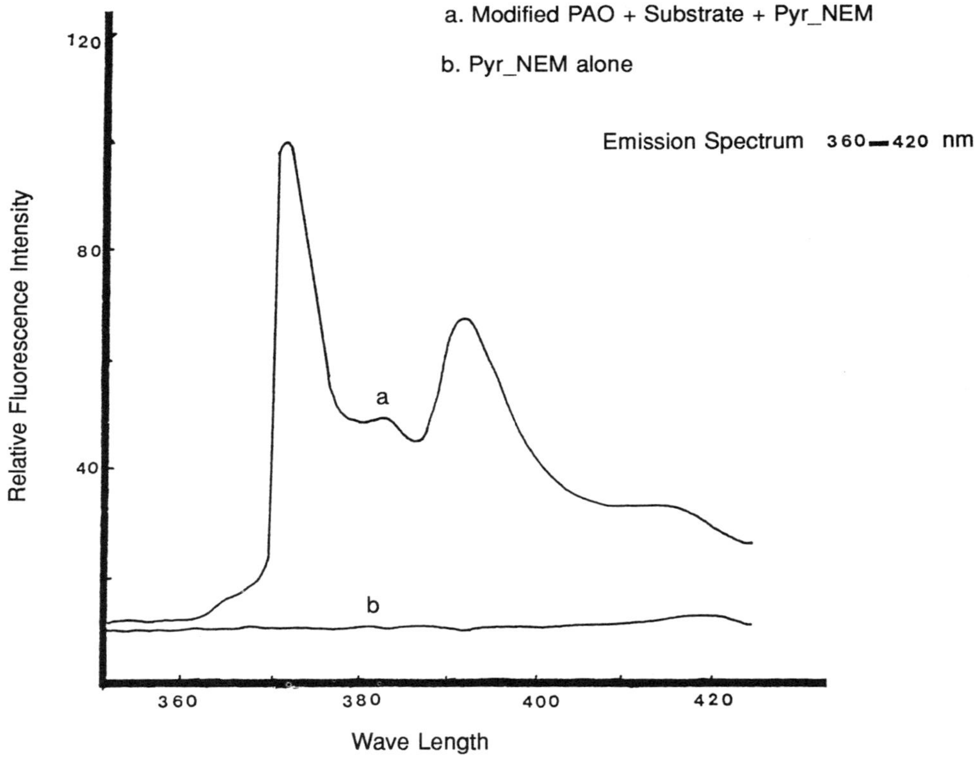

Figure 2. Fluorescence Emission Spectrum of NPM in the Absence and Presence of PAO. Fluorescence was recorded in 1.0ml of 0.01M KPB, pH 7.5. Both the excitation and emission slits were set at 7mm. The sample sensitivity was 30. The protein concentration was 0.087mg/ml. The NPM labeled enzyme was dialyzed against 0.01M KPB, pH 7.5, for 24 hours before recording the spectra.

Influence of Ionic Strength On The Pyrenemaleimide-PAO

We have examined the effect of increasing the concentration of potassium chloride on pyrenemaleimide- PAO fluorescence spectrum and PAO activity. The purpose of this experiment was to detect smaller conformational changes, closer to those which may take place under physiological conditions. Increasing ionic strength appeared to have a small effect on either the fluorescence spectrum of the adduct (Figure 3) or the activity of the enzyme.

Influence of pH on the Pyrenemaleimide- PAO Spectrum

The effect of varying the pH solutions of pyrenemaleimide- PAO derivative are shown in Figure 4. As seen in the spectrum, there is a gradual change in the spectrum from pH 7 to 11. Activity studies have demonstrated that the enzyme maintains 80% of its activity within the range of 7-9 and loses about 95% of its activity at pH 11.5. Thus, loss of activity with pH change parallels loss of the pyrenemaleimide- PAO derivative signal. As has been previously reported that addition of a substrate liberated two sulfhydryl groups per enzyme dimer (Suva and Abeles, 1978; Zeidan et al., 1980), the fluorescence spectra reported here with the activity studies clearly support the suggestive evidence of the significance of the sulfhydryl groups and characterizes the environment of the essential liberated sulfhydryl groups where the residues are located in the enzyme.

Influence of Urea on the Pyrenemaleimide-PAO Spectrum

As previously stated NPM by itself is not fluorescent, but becomes so after reacting covalently with the sulfhydryl group of a large or small molecule (Weltman et al., 1973). Emission spectra of both kinds of NPM are shown in Figures 5 and 6. The large molecule being the PAO derivative and the small, 2-mercaptoethanol (Figure 5). Differences appeared when 6M urea was present in the medium. The emission of NPM-labeled PAO derivative was markedly quenched (Figure 6). The major band at 375-6nm was reduced to less than half its former intensity; the 395-7nm band and the shoulder at about 385nm were also quenched, but less. In contrast, the emission of NPM adduct of mercaptoethanol was not quenched by 6M urea. The quenching by urea of PAO-NPM fluorescence spectrum cannot, therefore, be due only to solvent-fluorphore interaction, but must be partly due to urea-induced conformational changes in the NPM-PAO protein. The conformational change induced by urea was irreversible at 6M urea. Thus, unfolding and exposure to urea causes massive

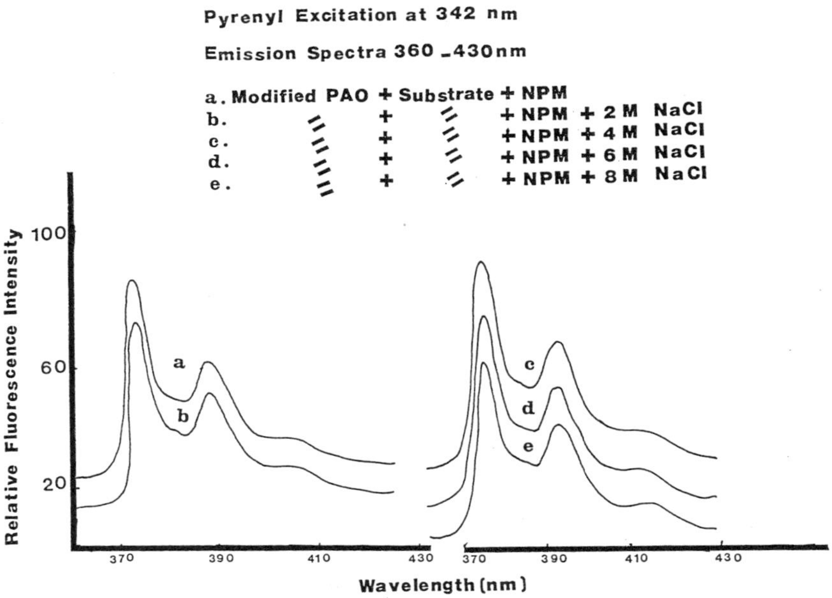

Figure 3. Effect of Varying Ionic Strength on PAO-NPM Fluorescence. The enzyme was treated with sodium chloride and modified with NPM as described in the experimental section. The NPM labeled enzyme was dialyzed against 0.01M KPB, pH 7.5, for 24 hours before recording the spectra. Both excitation and emission spectra were set at 7mm.

structural changes in the PAO. This fluorescence spectra was supported by activity studies (not shown). Thus, loss of activity by denaturation paralleled loss of pyrenemaleimide fluorescent signal.

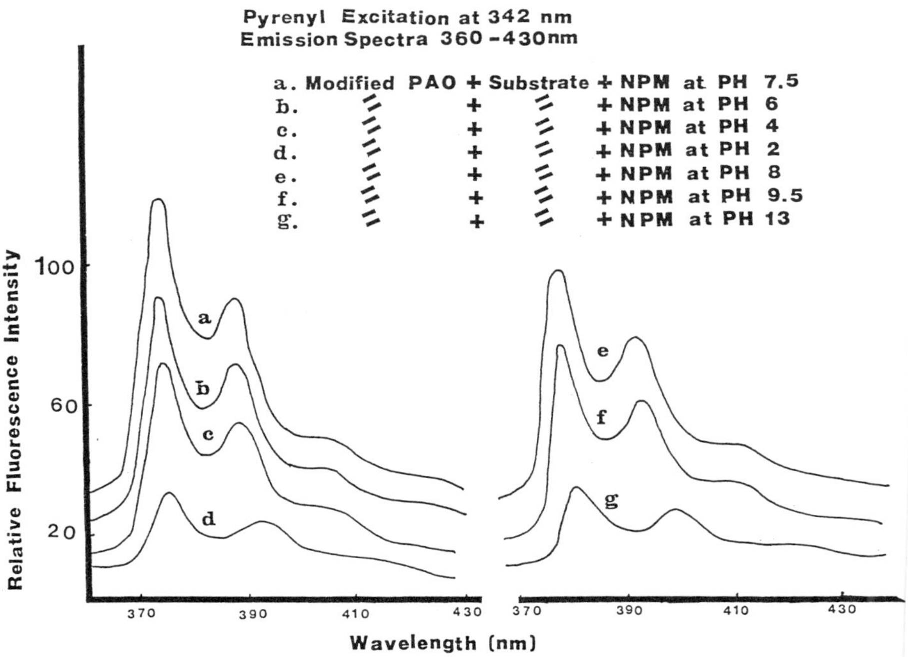

Figure 4. Effect of pH on PAO-NPM Fluorescence. The enzyme was modified and labeled at the indicated pH as described in the experimental section. The NPM labeled enzyme was dialyzed against 0.01M KPB, pH 7.5 for 24 hours before recording the spectra. Both excitation and emission spectra were recorded at 7mm.

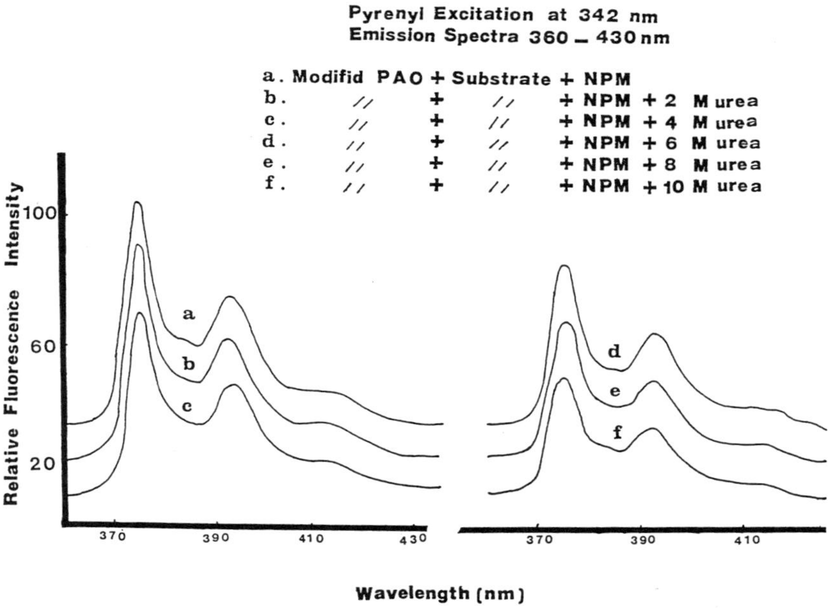

Figure 5. Effect of Urea Concentration on PAO-NPM Fluorescence. The enzyme was modified as described in experimental methods section. The NPM labeled enzyme was dialyzed against 0.01M KPB, pH 7.5, for 24 hours before recording the spectra.

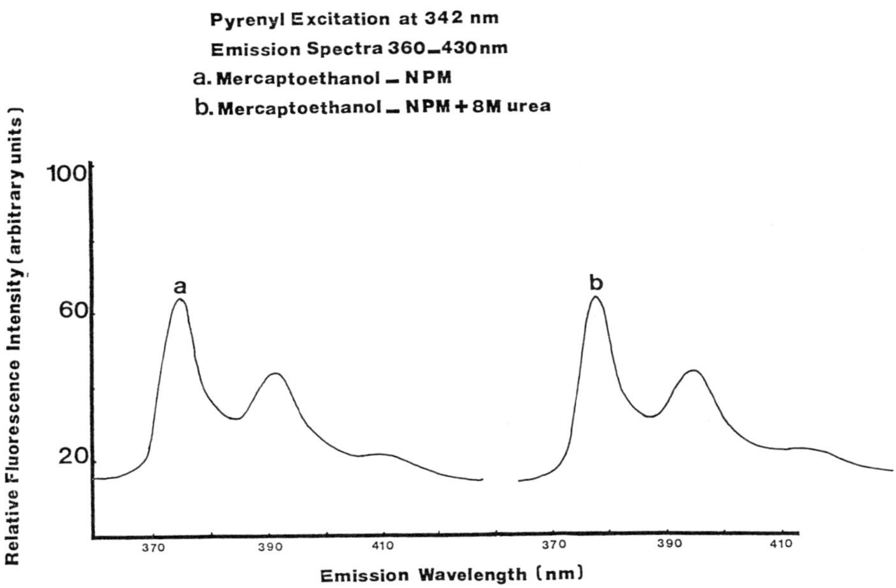

Figure 6. Effect of 8M Urea on the Fluorescence Emission Spectrum of Mercaptoethanpl. Excitation and emission slits were set at 7mm. The sample sensitivity was set at 30.

SUMMARY

A pyrenemaleimide attached to the essential liberated sulfhydryl groups by the substrate of plasma amine oxidase has been used to probe the structure of the active site of the enzyme. The label is nonfluorescent in aqueous solution but forms a strongly fluorescent adduct with the liberated essential sulfhydryl groups of the enzyme. The extrinsic probe was found to be sensitive to a limited and fully reversible transition that took place when the enzyme was incubated under conditions that changed its activity in vitro. This highly suggests that N-(1-pyrene) maleimide label may be used to probe both the environment of liberated essential sulfhydryl groups of the enzyme and the conformational changes that occur under conditions consistent with activity.

Therefore, we conclude that the extrinsic sensitive NPM probe could be used as an effective detector to monitor conformational changes at the active site. Furthermore, it can be used to probe the microenvironment of the proteins that carry the bound NPM molecules.

Acknowledgments

We wish to thank Ms. Tondaleyo Conley for typing this manuscript. This work was supported by NIH, Grant IG#12-RR03062, NIH Grant R #08247 to Henry Zeidan. The author highly acknowledges Dr. William V. Dashek for reviewing this manuscript, Mr. Khaled Abboud for his outstanding graphic work, the assistance of his graduate students and all who contributed to the Amine Oxidases Research in our laboratory. The technical assistance of Emory University is also highly appreciated.

REFERENCES

Achee, F.M., Chervenka, C.H., Smith, R.A. and Yasunobu, K. (1968), Biochemistry, 128, 7, 4329-4335.

Davidson, A.N. Physiol. Rev. (1958), 38, 729-747.

Hiramatsu, A., Tsurushin, S., and Yasunobu, K.T. (1975), Eur. J. Biol. Chem., 57, 587-593.

Inamasu, M., Yasunobu, K.T., and Konig, W.A. (1974), J. Biol. Chem., 249, 5265-5268.

Mondovi, B., Costa, M.T., Argo, A.F., and Rotilo, G., (1976), Arch. Biochem. Biophys., 119, 373-381.

Suva, R.H., and Abewles, R.H. (1978), Biochemistry, 17, 3538-3546.

Watanabe, K., Smith, R.A., Inamasu, M., and Yasunobu, K.T. (1972), Adv. Biochem. Psychopharmacol., 5, 107-117.

Weltman, J.K. (1973)., J. Biol. Chem., 248, 3173.

Wu, Cheng-Wen, Yarbough, L. and Wu, F. (1976), Biochem., 15, 2863-2867.

Yasunobu, K.T., Minamiura, N., and Ishizaki, H. (1976), Mol Cell. Biochem., 13 3-29.

Zeidan, H., Watanabe, K., Piette, L.H., and Yasunobu, K. (1980), J. Biol. Chem., 255, 7621-7626.

Zeidan, H., and Buchanan, S. (1990), Biodeterioration Research, 3, 565-576, G Llewellyn and C. O'Rear (eds.), Plenum Press, N.Y.

POLYMERASE CHAIN REACTION ANALYSIS, DETECTION AND TYPING OF TOXIN PRODUCING CLOSTRIDIUM DIFFICILE

DAVID E. McMILLIN*, Department of Biology, Clark Atlanta University, Atlanta, GA 30314, USA

LYCURGUS L. MULDROW, Department of Biology, Clark Atlanta University, Atlanta, GA 30314, USA

INTRODUCTION

Clostridium difficile, the primary cause of pseudomembranous colitis (PMC) in humans, produces two toxins; toxin A, an hemorrhagic enterotoxin and toxin B, a potent cytotoxin. Both toxin proteins appear to act synergistically to induce colitis and PMC. Size estimates of the toxin proteins have varied widely; however recent DNA sequence information estimated the size of toxin A to be 307,972 daltons (Dove et al. 1990) and toxin B to be 269,696 daltons (Barroso et al. 1990). These large bio-toxins are only found in pathogenic strains of C. difficile, but not found in non-pathogenic strains. In addition to the studies of the toxin proteins, a number of labs have been investigating the molecular biology of these toxin genes.

Several research groups have reported isolating part of the toxin A gene. Muldrow et al. (1987) first reported on the cloning of a 0.3kb DNA fragment of toxin A in the expression vector λ gt11. Price et al. (1987) soon thereafter reported the cloning of a 4.7kb region of the toxin A gene. In addition, Wren et al. (1987) and Eichel-Streiber et al. (1989) cloned overlapping fragments of the toxin A gene. Recently, the entire toxin A gene has been cloned and expressed (Phelps et al. 1991), and toxin A (Dove et al. 1990, Sauerborn and von Eichel-Streiber 1990) and toxin B (Barroso et al.1990) have been sequenced. The toxin

Biodeterioration Research 4, Edited by G.C. Llewellyn *et al.*, Plenum Press, New York, 1994

DNA sequence information in conjunction with amino acid sequence information, have allowed the molecular characterization and detection of the toxin genes using the polymerase chain reaction (PCR).

PCR is a rapid *in-vitro* method for amplification of target DNA. The target DNA that is amplified can be of any origin, and in picogram concentrations. DNA is incubated in a thermocycler in the presence of short oligonucleotide primers, a reaction buffer containing Mg^{++}, nucleotides, and heat stable Taq DNA polymerase. Each cycle of the polymerase chain reaction consists of three steps, denaturation of the target DNA, annealing of the PCR primers to the target DNA, and extension of DNA between the PCR primers. This process of denaturation, annealing, and extension is repeated 15-45 times resulting in the logarithmic amplification of target DNA.

This paper will summarize the use of PCR to characterize and detect toxic and non-toxic strains of C. difficile and to generate DNA profiles for future epidemiological studies. The approach used to study the C. difficile toxin genes could be used for any bio-toxin producing bacteria.

ORGANIZATION OF THE TOXIN GENES

The DNA sequence information for toxin A (Dove et al. 1990, Sauerborn and von Eichel-Streiber 1990) and toxin B (Barroso et al. 1990) have shown that the toxin genes are linked with an open reading frame of unknown function (Figure 1). Toxin B is 7,098bp, toxin A is 8,133bp, while the open reading frame is 500bp in length. PCR has been used in our laboratory to amplify the entire toxin B gene, and the open reading frame, as well as portions of the toxin A and B genes (Figure 1).

AMPLIFICATION OF THE TOXIN B GENE

Prior to the publication of the toxin B DNA sequence, the entire toxin B gene was amplified by McMillin et al. (1991) using N-terminal amino acid sequence information (Meador and Tweten 1988) and the published DNA sequence which flanked the 3′ end of the toxin B gene (Figure 2). PCR primer BP-3, used for amplification of the toxin B gene, contained mixed nucleotide sequences which were derived from the toxin B N-terminal sequence and the codon usage frequency of C. difficile. The codon usage frequency was determined using the published DNA sequences of toxin A (Dove et al. 1990) and the CAT-D gene (Wren et al. 1989). The other primer (BP-4) was derived from the 3′ sequence of toxin B (Figure 1, Table 1). A 7.1kb DNA fragment was amplified, and after obtaining sequence information it was confirmed to be the toxin B determinant by Southern blot analysis (McMillin et al. 1991).

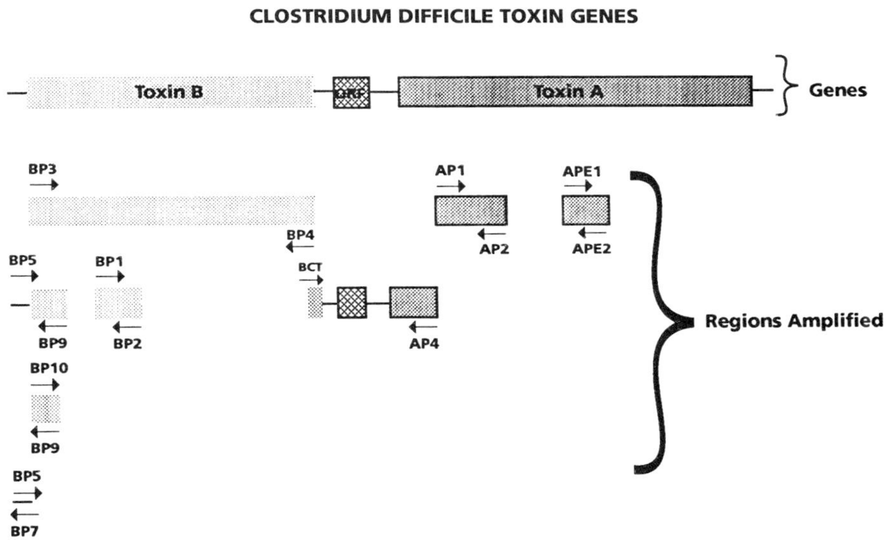

Figure 1. C. difficile Toxin Genes and PCR Amplified Fragments.

Amplification of the entire gene also resulted in the appearance of some lower molecular weight DNA fragments (Figure 2). These lower weight fragments seemed to be reproducible in each of the toxic strains and may represent partially amplified toxin B gene fragments or sequences unique to toxin producing strains. No amplification was observed in non-toxic strains when the BP-3 and BP-4 primers were used.

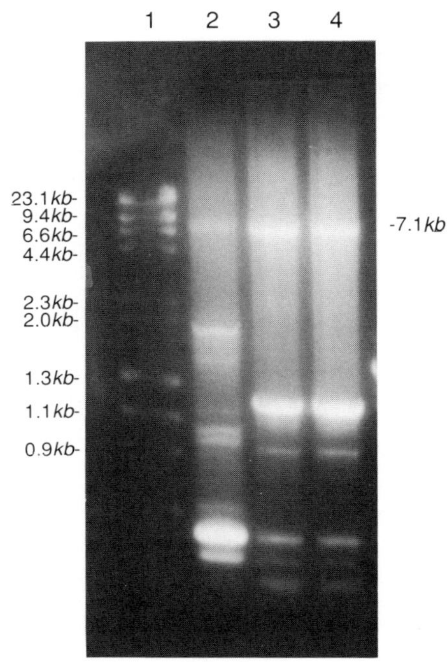

Figure 2. PCR Amplification of the Entire C. difficile Toxin B Gene. Lane 1, Hind III digested λ DNA and Hae III digested φX174 DNA Markers. Lanes 2-4 represent the toxic strains 74, 9689, and 56.

Table 1. Primer Sequences Used For Amplification of C. difficile DNA.

Primer	Sequence 5'-3'	Region of Gene	Reference
AP1	GGAAATTTAGCTGCA GCATCTGAC	toxin A 784-807	McMillin et al. 1991, 1992
AP2	TCTAGCAAATTCGCT TGTGTTGAA	toxin A 1977-2001	McMillin et al. 1991, 1992
AP4	GTCAGATGCTGCAG CTAAATTTCC	toxin A 807-784	Unpublished
AEP1	CTAGACGAACATGAC CCATTA	toxin A 3151-3171	Unpublished
AEP2	TTCCTGCTCCATCAA ATGAATAAGA	toxin A 3976-3949	Unpublished
BP1	GGTGATATGGAGGCA TCACCACTAG	toxin B 1030-1054	McMillin et al. 1991, 1992
BP2	TCCAGGATAAGTCTC CTCTACGTTG	toxin B 2115-2139	McMillin et al. 1991, 1992
BP3	GAAAAAATGGC(T) or(A)AATGT	toxin B 26-43	McMillin et al. 1991
BP4	AGGTGAACTACTGTG CATTC	toxin B 7179-7160	McMillin et al. 1991
BP5	TGAGCTTAAAGAGAT A	toxin B 305-289	Unpublished
BP7	TCTTGAGTACGAAAT C	toxin B 53-47	Unpublished
BP9	TCTAACATCATTTCC AC	toxin B 707-723	Unpublished
BCT	GATCCTGATACAGCT CAATTAGTG	toxin B 7006-7030	Unpublished
BP10	GATTTCGTACTCAAG A	toxin B 47-53	Unpublished
ARP-1	TAGCTCAGTGG	Random	McMillin and Muldrow, 1992
ARP-2	TCACGATGCA	Random	McMillin and Muldrow, 1992
ARP-3	GCGGCTATTAT	Random	McMillin and Muldrow, 1992
ARP-4	AGTGACTGGG	Random	McMillin and Muldrow, 1992
ARP-5	TAGCCAACGC	Random	McMillin and Muldrow, 1992
ARP-6	AGTAAAGGAG	Random	McMillin and Muldrow, 1992

Twenty one strains of C. difficile were examined using these primers; fifteen toxic strains and six non-toxic strains. All toxic strains examined amplified the entire toxin B gene, while most non-toxic strains did not (McMillin et al. 1991). One strain was found which did amplify the toxin B gene but was non-toxic. This strain is being analyzed further. The above reported data suggest that toxic strains possess the toxin B gene while most non-toxic strains do not. It also demonstrated the ease by which one can amplify a large gene for cloning or detection using mixed PCR primers if there is partial amino acid sequence information available.

CHARACTERIZATION OF THE TOXIN GENES

Numerous publications demonstrated the synergistic activity of toxin A and B in the induction of PMC, and that these proteins were only found in C. difficile pathogenic strains. To determine if the toxin genes are always present simultaneously and if the lack of toxicity is always correlated with the lack of the toxin genes, strains were characterized using PCR and dot blot analysis. Primers specific for toxin A (AP1 and AP2) were used to amplify a 1.2kb fragment, while a 1.0kb toxin B fragment was amplified using primers BP-1 and BP-2 (Figures 1,3). Amplified fragments from selected strains were labelled and used for dot blot analysis of other C. difficile strains. Both techniques were used to characterize 37 toxic strains and 10 non-toxic strains. All toxic strains gave positive hybridization or PCR results for the presence of both toxin genes, but nine of the non-toxic strains did not demonstrate the presence of either toxin gene (McMillin et al. 1991). However, one of the non-toxic strains gave positive results which demonstrated the presence of both toxin genes. This was the same strain that amplified the entire toxin B gene. These results indicated that in the toxic strains tested both toxin genes are present simultaneously, while the toxin genes are absent from most of the non-toxic strains.

Similar results have been observed by other investigators for toxin A when various regions of this gene were analyzed. Kato et al. (1991) found that 35 toxic strains amplified a toxin A fragment, but none of the 26 non-toxic strains showed amplification. Wren et al. (1990b) amplified a 63bp toxin A fragment from a repeated region in 62 toxic strains but not in 23 non-toxic strains. Dot blot analysis using a 4.5kb DNA fragment of toxin A gene revealed that 58 toxic strains hybridized with this probe and 17 non-toxic strains did not (Wren et al.

1990a). When the results of these studies are combined with our findings, all toxic strains possessed both toxin A and B, while 75 out of 76 non-pathogenic strains did not contain either toxin A or toxin B. The non-toxic strain found in our studies which amplified both toxin A and B could represent a mutant. This strain is being investigated further. Another mutant strain has been described by Borriello et al (1992), which was cytotoxic in animals but lacked the toxin A gene. The molecular analysis of these mutant strains could provide important information concerning the regulation of the bio-toxins. Other studies, in our laboratory are currently being conducted to PCR characterize the 5′ flanking region of the toxin B gene (Figure 1). This region may be responsible for varying toxicity levels observed among toxic strains and could provide important information concerning the regulation of toxin production in C. difficile.

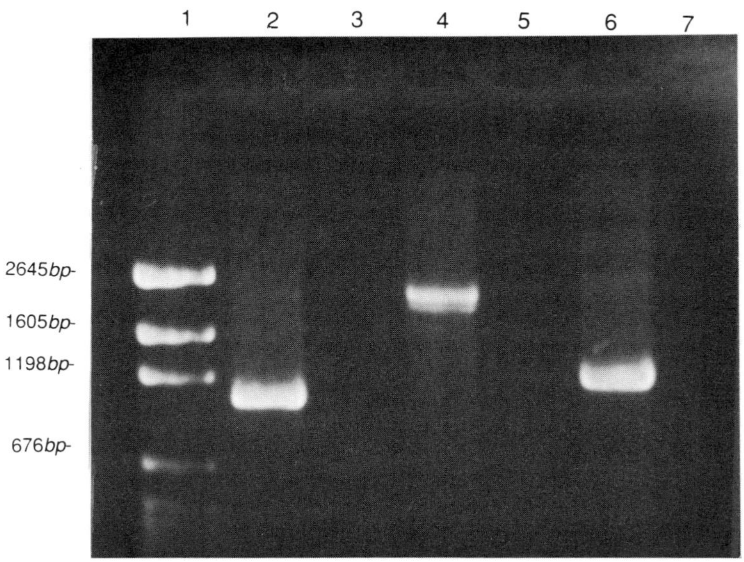

Figure 3. PCR Amplification of the C. difficile Toxin Genes. Lane 1 is the pGEM DNA marker (Promega, Madison, WI). Lanes 2, 4 and 6 represent the amplified 1.0kb toxin B fragment, the 1.2kb toxin A fragment, and the 2.3kb fragment containing the open reading frame from the toxic strain 10463. Lanes 3, 5 and 7 represent PCR reactions for toxin B, toxin A and the open reading frame using DNA from the non-toxic strain, 85.

THE OPEN READING FRAME

The function of the small open reading frame between toxin A and toxin B remains unknown. The entire region between toxin B and toxin A has been amplified using the primers BCT and AP4 (Figures 1, 3). This region included the 3′ end of toxin B, the ORF and the 5′ end of toxin A. Preliminary screening results indicate that 9 toxic strains amplify the ORF while 5 non-toxic strains do not (McMillin, unpublished). The results are consistent with the hypothesis that the toxic strains possess a 16.9kb (toxin B, ORF, toxin A, and linkers) stretch of DNA not present in non-toxic strains. These results also support the hypothesis that the toxin genes may be part of a single operon (Lyerly et al. 1985). Furthermore, since the ORF is found only in toxic strains, it is possible that this gene may play a role in toxin production or toxicity.

SIMULTANEOUS AMPLIFICATION OF BOTH TOXIN GENES

To date many diagnostic procedures have been published for C. difficile; however, there is still considerable interest in developing a rapid and sensitive diagnostic test for toxic strains. To insure that all toxic strains are detected, a test that monitors for both toxin determinants would be beneficial. Consequently a multiplex PCR procedure, which amplifies simultaneously the 1.0kb toxin B fragment and the 1.2kb toxin A fragment was developed in conjunction with a rapid DNA isolation procedure which could be conducted in diagnostic labs (McMillin et al. 1992). This multiplex procedure resulted in a distinct doublet band in 42 toxic strains which could not be detected in 10 non-toxic strains. The appearance of a distinctive doublet band would reduce the risk of misclassification resulting from spurious amplification near one of the toxin genes. The multiplex procedure would also be beneficial for studies looking for mutants of either the toxin A or toxin B gene.

PCR TYPING OF TOXIC STRAINS

A definitive and reliable method of typing strains of C. difficile is needed since nosocomial spread of PMC in hospitals and health care facilities occurs (Kim et al. 1983, Fekerty et al. 1981). Numerous methods have been used in an attempt to distinguish toxic strains. These methods include; SDS protein patterns (Tagaqchali et al. 1984), serogrouping (Delmee et al. 1985), restriction patterns (Wren and Tabaqchali 1987), and RFLP patterns (Bowman et al. 1991). Results from these typing procedures varied; however, none of the methods could clearly distinguish the toxic strains from the non-toxic strains. Arbitrary PCR primers were used in our laboratory to develop an amplification procedure which produced distinct DNA fingerprints for six toxic strains (McMillin and Muldrow 1992). As described earlier, toxic strains can be distinguished from non-toxic strains using PCR; therefore, the combination of amplification of the toxin genes, together with amplification using arbitrary primers could clearly distinguish individual toxic strains. This PCR fingerprinting could be used for epidemiological studies of toxic strains of C. difficile and ultimately assist in preventing the spread of PMC in health care facilities.

SUMMARY

C. difficile, the etiological agent of pseudomembranous colitis (PMC) in humans produces two toxins, toxin A and toxin B. A small open reading frame of unknown function separates the toxin A gene from the toxin B determinant in C. difficile's genome. These toxin genes and the ORF were amplified and characterized using the polymerase chain reaction (PCR). PCR primers were used to amplify the entire 7.1kb toxin B gene, as well as portions of the genes coding for toxin A, toxin B, and the open reading frame. Of the strains tested, all of the pathogenic strains contained the toxin A gene, the toxin B gene, and the open reading frame, but non-pathogenic strains usually did not contain these DNA sequences. One possible mutant non-toxic strain was found to amplify both toxin genes. Data on this mutant, along with the presence of the open reading frame in all of the toxin producing strains support the hypothesis that C. difficile toxins may be part of a single operon. Arbitrary PCR primers were also employed in this study to generate unique DNA fingerprints for six toxic strains. The PCR procedures used in this research could be adapted for any toxin producing bacteria in order to clone or characterize toxin determinants or to differentiate between different strains.

Acknowledgements

This work was supported by Public Health Service Grants R29AI26813, SO6RR08241, and 1RR03062.

REFERENCES

Barroso, L.A., Wang, S.Z., Phelps, C.J., Johnson, J.L., Wilkins, T.D. (1990). Nucleotide sequence of Clostridium difficile toxin B gene. Nucl. Acids Res., 18, 4004.

Borriello, S.P., Wren, B.W., Hyde, S., Seddon, S.V., Sibbons, P., Krishna, M.M., Tabaqchali, S., Manek, S., and Price, A.B. (1992) Molecular, Immunological, and biological characterization of a toxin A-negative, toxin B-positive, strain of Clostridium difficile. Infect.Immun., 60, 4192-4199.

Bowman, R.A., O'Neill, G.L. and Riley, T.V. (1991) Non-radioactive restriction fragment length polymorphism (RFLP) typing of Clostridium difficile. FEMS Microbiol. Lett., 79, 269-272.

Delmee, M., Homel, M. and Wauters, G. (1985) Serogrouping of Clostridium difficile strains by slide agglutination. J. Clin. Microbiol. 21, 323-327.

Dove, C.H., Wang, S.Z., Price, S.B., Phelps, C.J., Lyerly, D.M., Wilkins, T.D., Johnson, J.L. (1990). Molecular characterization of the Clostridium difficile toxin A gene. Infect. Immun., 58, 480-488.

Eichel-Streiber, C.V., Suckau, D., Wachter, M., and Hadding, U. (1989) Cloning and characterization of overlapping DNA fragments of the toxin A gene of Clostridium difficile. J. Gen. Microbiol., 135, 55-64.

Fekety, R., Kim, K., Brown, D., Batts, D.H., Cudmore, M., Silva, J. (1981). Epidemiology of antibiotic-associated colitis: Isolation of Clostridium difficile from the hospital environment. Am. J. Med., 70, 906-908.

Kato, N., Ou, C., Kato, H., Bartley, S.L., Brown, V.K., Dowell, V.R., Ueno, K. (1991). Identification of toxigenic Clostridium difficile by the polymerase chain reaction. J. Clin. Micro., 29, 33-37.

Kim, K., DuPont, H.L., Pickering, L.K. (1983). Outbreaks of diarrhea associated with Clostridium difficile and its toxin in day-care centers: evidence of person-to-person spread. J. Pediatr., 102, 377-382.

Lyerly, L.M., Saum, K.E., MacDonald, D.K., and Wilkins, T.D., (1985) Effects of Clostridium difficile toxins given intragastrically to animals. Infect. Immun., 47, 349-352.

Meador, J., and Tweeten, R.K. (1988) Purification and characterization of toxin B from Clostridium difficile. Infect. Immun., 56, 1707-1714.

McMillin, D.E., Muldrow, L.L., Leggette, S.J., Abdulahi, Y., Ekanemesang, U.M. (1991). Molecular screening of Clostridium difficile toxins A and B genetic determinants and identification of mutant strains. FEMS Microbiol. Lett., 78, 75-80.

McMillin, D.E., Muldrow, L.L., Leggette, S.J.. (1992) Simultaneous detection of toxin A and toxin B genetic determinants of Clostridium difficile using the multiplex polymerase chain reaction. Can. J. Microbiol., 38, 81-83.

McMillin, D.E., Muldrow, L.L. (1992). Typing of toxic strains of Clostridium difficile using DNA fingerprints generated with arbitrary polymerase chain reaction primers. FEMS Microbiol. Lett., 92, 5-10.

Muldrow, L.L., Ibeanu, G.C., Lee, N.I., Bose N.K. and Johnson, J. (1987) Molecular cloning of Clostridium difficile toxin A gene fragment in lambda gt11. FEBS Lett., 213, 249-253.

Phelps, C.J., Lyerly, D.L., Johnson, J.L., and Wilkins, T.D. (1991) Construction and expression of the complete Clostridium difficile toxin A gene in Escherichia coli. Infect. Immun., 59, 150-153.

Price, S.P., Phelps, C.J., Wilkins, T.D. and Johnson, J.L. (1987) Cloning of the carbohydrate-binding portion of the toxin A gene of Clostridium difficile. Curr. Microbiol., 16, 55-60.

Sauerborn, M., von Eichel-Streiber, C. (1990). Nucleotide sequence of Clostridium difficile toxin A. Nucl. Acids Res., 18 ,1629.

Tabaqchali, S., Holland, D., O'Farrell, S. and Silman, R. (1984) Typing scheme for Clostridium difficile: its application in clinical and epidemiological studies. Lancet 1, 935-938.

Wren, B.W., Clayton, C.L., Castledine, N.B., and Tabaqchali, S. (1990a) Identification of toxigenic Clostridium difficile strains by using a toxin A gene-specific probe. J. Clin. Microl., 28, 1808-1812.

Wren, B.W., Clayton, C.L., Mullany, P., and Tabaqchali, S. (1987) Molecular cloning and expression of Clostridium difficile toxin A in Escherichia coli K12. FEBS Lett., 225, 82-86.

Wren, B.W., Clayton, C.L., Tabaqchali, S. (1990b). Nucleotide sequence of Clostridium difficile toxin A gene fragment and detection of toxigenic strains by polymerase chain reaction. FEMS Microbiol. Lett., 90, 1-6.

Wren, B.W., Mullany, P., Clayton, C.L. and Tabaqchali, S. (1989) Nucleotide sequence of a chloramphenicol acetyl transferase gene from Clostridium difficile. Nucl. Acids Res., 17, 4877.

Wren, B.W. and Tabaqchali, S. (1987) Restriction endonuclease DNA analysis of Clostridium difficile. J. Clin. Microbiol., 25, 2402-2404.

CONSIDERATIONS IN THE DEVELOPMENT OF A MYCOLOGY LABORATORY

GERALD C. LLEWELLYN, Department of Forensic Science, The George Washington University, Washington, DC 20052, USA and The Delaware Department of Natural Resources and Environmental Control, New Castle, DE 19720, USA

CHARLES E. O'REAR, Department of Forensic Science, The George Washington University, Washington, DC 20052, USA

INTRODUCTION

The objective herein is to provide the conceptual basis as well as some specifics needed in the development of a mycology laboratory. It is acknowledged that physical constraints and fiscal limitations would have major effects on the development of an ideal mycology laboratory. However, it is possible, over several years, to plan and then gradually develop such a physical facility without the need for an immediate and substantial fiscal out-lay. It must be noted that substantial initial start-up funds for the development of such a laboratory would be needed if there is no physical facility currently in place or an building/area capable of being modified. At times, building modifications may be more expensive and have inherent restrictions which could make the additional cost for a new facility more palatable. Also, a new building offers numerous advantages in design and resultant quality of the laboratories housed therein. Certainly the mycology laboratory will be tied to the other laboratories and the facilities available, but creativity, planning, and flexibility can be exercised so that the mycology component is suitably functional and a quality laboratory.

A functional, self-contained mycology laboratory, whether part of a new building or a result of renovation could include three definitive areas: a clerical office-library-sample receiving area, the laboratory, and an isolation room. All such microbiological laboratories would need to provide for media preparation, autoclaving, glassware cleaning and disposal of materials. Also needed is clerical assistance, mail and telephone. Facsimile and photocopying would be important. Even though centrally-located, efficient, and effective support services are sometimes provided to laboratories, the laboratory described herein will be able, initially, to provide all of these services in one location, as well as the study of a variety of fungi. The space requirement for such a laboratory would be at least 2,500 square feet.

The development of this ideal mycology laboratory will be considered in the following way. First, the facility environmental components, important both in the immediate laboratory and in the surrounding area, will be presented. Secondly, the major components of the clerical office-library-sample receiving area and thirdly, the components and the design of the main mycology laboratory and isolation room will be noted. Additional supplies and support material needs will be noted. Many of the materials used in the main mycology laboratory should be duplicated and used in the isolation room also. Other topics addressed briefly include the need for a stock culture collection and maintenance, media and reagents, and preserved reference cultures.

FACILITY ENVIRONMENTAL COMPONENT

Environmental conditions in the facility such as lighting, temperature, and humidity must be under control. The main laboratory should always be under negative air pressure. The isolation room should be under even less barometric air pressure. Such conditions will prevent movement of airborne particles from the laboratory areas. Air flow is always into the laboratory and out the exhaust. The ventilation system should be designed so that air passes through the facility only once (Kent and Kubica, 1985). A "single pass" ventilation system moves air from outside the building, filters, heats, and humidifies as necessary. Usually 50% relative humidity is considered ideal. The air would flow into the laboratory and then into the isolation room. Air that enters the isolation room should be exhausted via a HEPA filter (Kent and Kubica, 1985). External air intakes should not be located near exhaust vents on the outside of the building. The consideration and planning for such a ventilation system is an important aspect for this type of a laboratory. Such a system certainly adds to the fiscal burden.

In addition, throughout the laboratory complex there is need for numerous and appropriate electrical service and outlets. It is generally recommended that during the initial construction and/or modification that such service be on the generous side. Also, the placement of telephone lines, hot and cold water runs, and drainage systems should be taken into account for the entire laboratory as a unit to provide for both efficiency and safety. Easy access to drains and traps will pay dividends.

CLERICAL OFFICE-LIBRARY-SAMPLE RECEIVING COMPONENT

The area designated as the office, usually about 300 square feet, would serve as a receiving area as well as an area for communications such as mail and telephone calls. Clerical assistants could use this area for associated needs, record keeping, inventory management, word processing, technical writing, and communication. A library of references, manuals, and journals would be appropriate in the office. In addition to the clerical personnel, technicians and mycologists may use this area for data analysis, technical writing, calculations, and planning. Each person working in the laboratory should have some designated work/storage space for files and personal belongings. Personal lockers and a refrigerator for food could add to the convenience and safety.

Office furnishings could include desks, work tables, chairs, telephones, facsimile equipment, file cabinets, book shelves, and storage cabinets. There should be one or more fully equipped desk top computer work stations. One computer station could be used by the clerical staff and others by the technicians and mycologists. There should be a laser jet type of printer available. A quality photocopier should be present as well as an electric typewriter. A small calculator with basic statistical capabilities would be a valuable addition. A coat rack as well as a separate storage area for laboratory would be recommended.

Library holdings could include standard procedural manuals and the development of a laboratory specific procedural manual for commonly run procedures within the laboratory. This in-house procedural manual may or may not duplicate methods and procedures found in reference manuals, depending upon modifications developed in the course of the studies. Reference materials should include a lexicon, a chemical dictionary, other procedure manuals, texts, catalogs, and journals. Since fungi are identified, their morphological structures are key factors in such processes, then appropriate references to support this effort are needed. It will be necessary to update the holdings of the library on a periodical basis.

THE MYCOLOGY LABORATORY COMPONENT

The laboratory will occupy approximately 1,800 square feet. The actual mycological work in the laboratory will be completed here. As much outside lighting as possible should be provided. Other important physical aspects include the traffic patterns, equipment placement, and storage. It is expected that most of the room would consist of work benches having suitable and readily cleanable tops and storage cabinets. The perimeter of the room will be discussed first. One wall would be expected to be a bank of incubators, refrigerator(s), and freezer(s). Bench top work areas could be increased by the use of work islands in the open, central room areas. The islands may also have overhead storage as well as lower storage areas. Most all storage areas should have glass doors for visibility of stored items. Some of the storage areas should lock. An area in this room should contain a small sterile hood.

Media components and glassware should be stored separately behind glass and a cataloging system should be put in place to provide an inventory of supplies. A computer program may function to this means and also maintain a record of purchases. The computer could also be useful in maintaining maintenance records and billing procedures. Commonly used spreadsheets would function well for these purposes.

Bench top work areas should be of high quality, being impervious, smooth, and resistant to disinfection agents. Duplex electrical outlets providing 115-125 volts and having appropriate ground-fault circuit interrupters should be spaced every three feet along above the counter areas. Permanent equipment with electrical motors may best derive their power from direct electric service runs having their own circuit breakers. Gas outlets need to be available at about six to 12 feet intervals. Care must be taken so that the gas outlet-placements do not encourage the use of an open flame directly underneath wall cabinets located above the work area. This can be avoided by placing the outlets out from or away from the wall and in the center of the bench top work areas. The laboratory should have at lease one or more 220 volt outlets. Hot water faucets should be color coded with their handle being red. Compressed air and vacuum lines would be a luxury but if funds permit, they should be included. Their lines and connectors should be color coded also and physically different is size to avoid the confusion of mixing the hookup when being used.

An area on the bench top should be designated for the following special needs: microscopy, analytical balances, and an area for notebook work. A laboratory desk and chair or a drop-bench region could serve this purpose as well as related needs for storage in this area. An area free of vibration will be necessary for best analytical balance functioning.

Large stainless steel sinks at opposite ends of the laboratory being deep and having foot controls for the water would be advisable. Both hot and cold water as well as deionized and distilled water should be available. Floor and wall materials should be capable of withstanding cleaning with disinfectants. Floors should be constructed of similar non-slip non-porous materials such as special ceramic or vinyl tiles, epoxy coated, or terrazzo-like coverings (Paul, 1975). A central floor drain, so located to function to carry away water via a gravity flow would be advantageous. Epoxy coatings for the walls and floors are recommended but non-slip floor applications are necessary.

Safety Issues

The presence of a safety shower in the laboratory or nearby is essential. Other laboratory safety equipment should include appropriate fire extinguisher, an eye wash station near a sink, first aid kits, spill retention supplies, a fire blanket, safety glasses and safety goggles, laboratory coats, paper face masks, rubber or latex gloves, disposal containers, safety containers, safety labels, and a respirator with an emergency air supply depending upon the chemicals and organisms used in the laboratory. A locking door having a small window should separate the rooms. For security reasons this window should be impregnated with wire. The door may also be protected with metal or plastic to protect it from the continual wear from contact with carts. Such a covering could serve well if decontamination were needed also. A telephone should be located in each room with emergency numbers posted. Fire or other emergency evacuation procedures should also be posted. OSHA work place information and the hazard communication standard or worker-right-to-know booklet/manual with material communication standard or worker-right-to-know booklet/manual with material safety data sheets, training records, the chemical inventory, safety procedures, and related required information should be at-hand. Utility service cut-off valves or switches for water, natural gas, and electricity should be clearly marked and accessible.

Large Laboratory Equipment

The laboratory would need a freezer capable of maintaining -20 degrees C, a standard refrigerator with 40 degree C refrigeration and a 4 degree C freezer, and incubators capable of being set and 37 degrees C and 40 degrees C. All temperature specific storage equipment should have alarm systems to indicate exceedences of desired ranges. Other equipment could include a pH meter, a shaker with temperature maintained via a water bath, an analytical balance with four to five place decimal readings for grams, and an electronic balance capable of reading at one half to one gram intervals and having a one to five kilogram maximum range.

Microscopes would include stereoscopic compound scopes with a built-in rheostatic-controlled light, zoom capability, and dual heads for simultaneous viewing and/or photography. Each type of microscope should be adaptable for photography via still and/or video cameras. A high quality single-lens reflex camera is needed for the laboratory. It should have both types of lens attachments; one for microscopic and a regular zoom lens for macroscopic documentation. Oftentimes laboratories have found that a camera capable of providing instant print development is useful. Microscope objectives available should be of high quality and present in most if not all of the following powers: 2.5x, 10x, 16x, 25x, 40x and 100x. Eye pieces should range above and below 10x. Microscope eye pieces can be equipped with calibrated micrometers and this would be advantageous. A fluorescent microscope with a variable light source, multiple filters and a camera attachment is needed. A colony counter should be provided in the microscope work area.

For cleaning and sterilization, a dishwasher, a pipette-washer, a bottle-washer, and large drainage racks over the sinks are needed, as well as a sterilizing oven, membrane filters and filtering apparatus. An autoclave is vital (Collins and Lyne, 1984).

Small Laboratory Equipment

Small equipment useful in the mycology laboratory could include the following: work carts with wheel locks, a bumper, side rails, and non-slip shelves; rubber work mats for the floor; metal kick pads and cart protectors on the doors; adhesive pads immediately outside the isolation room to catch dust from pedestrian traffic; plastic dish pans, probes; inoculating loops of various sizes; fingernail polish for sealing edges of the slide; plastic wash bottles; Bunsen burners; forceps; tripods; wire gauze (non-asbestos); spatulas; scissors; temperature sensitive tape; hot plates; and standard laboratory glassware (Dell'Orco, 1985). A few common tools such a set of pliers, vice grips, wire cutters, a screw driver set, a hammer, a set of open and box end wrenches, and an adjustable wrench would be helpful. Office supplies such as tape, paper clips, labels, and of course notebooks and writing equipment would be important.

THE ISOLATION ROOM COMPONENT

This component, the isolation room will require the remaining 300 square feet. It is in this area that the cultures will be prepared. This room should have a laminar flow hood with lights, a HEPA filter, an ultraviolet light source and appropriate utilities. A safety centrifuge operating at 2,000 x g will be needed.

Air flow through the hood makes the use of natural gas difficult so an electronic incinerator/heater will be useful inside the hood. This room will need a sink, storage cabinets, and an incubator. High quality bench tops such as those of stainless steel would be beneficial. One of the workbench storage cabinets should be a fireproof cabinet with a lock. Such a cabinet should be vent externally. It would serve as storage for chemicals. Electrical outlets on the benches should be well away from the work table surfaces so that these surfaces may be disinfected well.

The isolation room should be under an even lower level of atmospheric pressure than the main laboratory. Under such conditions, airborne particles can move only into the room and nothing can escape via the air from the room. The door should open inward and a glass window imbedded with wire should provide visibility and security. Natural and fluorescent lighting is needed. Temperature and humidity should be constant.

MYCOLOGY LABORATORY MAJOR SUPPLY AND SUPPORT COMPONENTS

An essential item to any microbiological laboratory undertaking mycological work are stock cultures. Stock isolates serve as a reference for identification and quality control. A well known and reputable source for these cultures is the American Type Culture Collection, 12301 Parklawn Drive, Rockville, MD 20852.

A mainstay of any microbiological or mycology laboratory are the media and reagents. These are necessary in the cultivation of the fungi. Media may be prepared or purchased commercially. Either way, it must be carefully controlled to provide correct performance. Pleomorphism and atypical microscopic structures are often caused by inappropriate and/or poor quality media.

SUMMARY

This report provides an overview of the major components needing consideration in the development of a mycological laboratory. It is assumed that such an endeavor is initiated from an initial and beginning point in time. Planning for such an undertaking is noted as being of ultimate importance. Acknowledged is the fact that the initial fiscal needs could be substantial but a gradual development of such facilities will reduce this hurdle. Not having to be involved as a portion of a larger facility under construction as a new laboratory building from the ground up provides some advantages in cost savings but renovation may contribute to expenses and infringe upon quality. Common problems encountered in renovation include such areas as ventilation system needs and service supplies for utilities such as electrical needs, gas, vacuum, compressed air, and negative pressure conditions. The prime components discussed herein include the desired environmental conditions needed in such a facility, the clerical office-library-sample receiving area, the mycology laboratory itself including the isolation room. Considered also are furnishings, small equipment, large equipment, safety concerns, prime support supplies, and suggested conveniences.

Acknowledgments

This article is a summary as well as a revision in part with updated additions to a comprehensive manuscript originally prepared for inclusion in Analytical Mycology of Foods edited by P. Mislivec and R. Bandler.

REFERENCES

Bacteriological Analytical Manual of the Division of Microbiology. Sixth Ed. (1984). Center for Food Safety and Applied Nutrition. U.S. Food and Drug Administration. Association of Official Analytical Chemists, Inc., Arlington, VA.

Collins, C.H. and Lyne, P.M. (1984). Microbiological Methods. Fifth Ed. Butterworth & Co. Ltd. London, pp. 10-28.

Dell'Orco, R.T. (1985). Layout and Equipment of a Cell Culture Laboratory. In: Setting Up and Maintenance of Tissue and Cell Cultures. Elsevier Scientific Publishers, Ireland Ltd. pp. 1-13.

Haley, L.D. and Callaway, C.S. (1978). Laboratory Methods in Medical Mycology. HEW Publication No. 78-8361 (CDC), Washington, DC.

International Equipment Company. (1980). Installation, Operation and Service Manual for General Purpose Refrigerated Centrifuge Model DPR 6000. Damon/IEC Division. Needham Heights, MA.

Kent, P.T. and Kubica, G.P. (1985). Public Health Mycobacteriology: A Guide for the Level III Laboratory. U.S. Department of Health and Human Services, Centers for Disease Control, Atlanta, GA 30333, pp. 5-20.

Llewellyn, G.C., Warren, N.G., and O'Dell, V.L. Requirements for a Properly Functioning Laboratory of Mycology, In: Analytical Mycology of Foods. Edited by P. B. Mislivec and R. Bandler. Marcel Dekker, NY. (Provided upon invite, 1989).

Melvin, D.M. and Brooke, M.M. (1982). Laboratory Procedures for the Diagnosis of Intestinal Parasites. U.S. Department of Health and Human Services, Centers for Disease Control, Atlanta, GA 30333, pp. 71-72.

Paul, J. (1975). Cell and Tissue Culture. Fifth Ed. Churchill Livingstone, 23, Ravelston Terrace, Edinburgh, UK. pp. 162-174.

DISTURBANCES IN TRYPTOPHAN METABOLISM FOLLOWING CHRONIC INGESTION OF A HIGH COPPER DIET BY MALE RATS

L. BRUCE WEEKLEY, College of Veterinary Medicine, Virginia Polytechnic and State University, Blacksburg, VA 24061, USA and Department of Veterinary Science, Colorado State University, Fort Collins, CO 80523, USA

T. D. KIMBROUGH, Department of Biology, Virginia Commonwealth University, Richmond, VA 23284 USA

CHARLES E. O'REAR, Department of Forensic Science, The George Washington University, Washington, DC 20052, USA

GERALD C. LLEWELLYN, Department of Forensic Science, The George Washington University, Washington, DC 20052, and Delaware Department of Natural Resources and Environmental Control, New Castle, DE 19720, USA

INTRODUCTION

Copper (Cu) is required in the diet as a trace mineral to prevent hypochromic microcytic anemia secondary to defective hemoglobin synthesis (Hart et al., 1928). This metal is also required for the normal function of a number of oxidative enzymes including catalase, peroxidase, and cytochrome oxidase. Following absorption from the gastrointestinal tract, copper is weakly bound to albumin and later more firmly bound to ceruloplasim, a specific Cu transport protein (Linder and Moor, 1977). Bile is the normal excretory route for copper and plays a primary role in copper homeostasis while liver and bone normally store excess copper.

Although copper homeostasis is normally maintained within relatively narrow limits, occasional outbreaks of copper toxicity occur for various reasons including deterioration of pipes and containers. Storage of acid drinks (e.g. Whiskey Sour) in copper containers has lead to several outbreaks of acute toxicity each year (C.D.C., 1975). Trifolium subteranean, subterrean clover, has a tendency to accumulate Cu from soils high in Cu and may produce a toxicosis when ingested by grazing animals (Kingsbury, 1975). Furthermore, serum copper has been shown to increase with age in man (Shroeder et al., 1966) and Harman (1965) has suggested that Cu may accelerate atherosclerosis. Copper has also been shown to enhance the tumorigenesis of N-hydroxy-2-acetylaminofluorene and topical benzanthracene (Fare, 1965; Stanton, 1967) and may possess some inherent carcinogenicity (Agnese et al., 1959; Gilman and Ruckenbauer, 1962). Indeed, Arnold and Sasse (1961) found elevated levels of copper, zinc, and iron in malignant tumors induced by organic copper complexes are produced by organic carcinogens while Station (1967) reported that organic copper complexes are produced in tumors of rat bone and lung. An increased incidence of stomach cancer has been reported in areas where the zinc/copper ratios exceeded certain limits (Stocks and Davies, 1960).

Altered Cu metabolism, with a resulting copper accumulation in tissues, occurs in Wilsons Disease. Cooper accumulates in liver, cornea and brain, and produces hepatitis, cirrhosis, intravascular hemolysis and necrosis of the corpus striatum and brain damage (Luckey and Venugopal, 1977).

Because of the potential hazard of ingestion of high Cu levels and the fact that such exposure may come from a number of sources, this study was undertaken to determine the effect of long term ingestion of high Cu levels on hematopoietic parameters and central and systemic biogenic amines.

MATERIALS AND METHODS

Weanling male Holtzman rats (Holtzman Co., Madison, WI) were divided into two groups, ten animals per group of approximately equal weights. Animals were housed individually in stainless steel cages and maintained in a well ventilated room at 27 $\pm$ 4°C on a 12:12, L:D photoperiod throughout the experiment. Animals were offered double distilled deionized water ad libitum and a diet of powdered Purina Laboratory Meal for Rats, Mice and Hamsters to which had been added copper acetate (0.5% w/w) in the experimental diet.

Body weight, feed and water consumption were determined on a weekly basis. At 21 weeks, the animals were sacrificed by decapitation and heparinized blood collected from the cervical stump. The liver, heart, spleen, kidneys, lungs, testes and a small section of proximal duodenum were separated from fascia, washed in Ringers Solution and weighed. Sections of each organ were then fixed in buffered (Na_2CO_3 5g/l) neutral formalin prior to hematoxylin and eosin staining for histopathological evaluation. The right medial lobe of the liver was assayed for tryptophan 2, 3 dioxygenase activity (Know, 1966).

An aliquot of blood was used for a blood cell count performed with a model S Coulter Counter. The remaining blood was centrifuged (4°C) and the plasma used for assay of albumin (Doumas et al., 1971), total cholesterol (Ness et al., 1964), Vitamin A and Vitamin E (Hansen and Warwick, 1969).

The calvarium was opened by an incision along the sagittal suture of the parietal bone. The pineal was removed, the brain lifted, and the adenohypophysis and neurohypophysis removed; any hypophyseal tissue not connected at the median eminence was removed from the sella turcica. The brain was dissected on a cold glass plate into cerebellum, cerebral hemispheres (cortex or telencephalon), diencephalon (including hypothalamus and midbrain), and brainstem (rhombencephalon). Tissue serotonin (5-HT), 5-hydroxyindole-3-acetic acid (5-HIAA), and norepinephrine (NE) were extracted and assayed fluorometrically (Maickel et al., 1968; Ansell and Beeson, 1968).

All data were analyzed using a t-test for two independent samples and data were considered significant if $P \leq 0.05$.

The incisions were made in the organs to assure penetration by the preservative. Hematoxylin and eosin-stained liver and other organ sections were prepared for light microscopy.

RESULTS

Analysis of total growth as mean weight gain during the 21-week test period revealed a 22.9% reduction in weight gain in the copper-supplemented animals. The Quetelet Index of Obesity (Billiwicz et al, 1962) indicated a significantly reduced ($P \leq 0.01$) growth rate in Cu-supplemented rats (2010 $\pm$ 90 in controls compared to 1630 $\pm$ 110 in Cu-supplemented rats). Furthermore, accumulative feed intake and accumulative water intake throughout the 21-week period paralleled the changes in body weight (Figure 1). Weighings of the heart and parenchymous internal organs at the end of the 21-week period indicated that ingestion of a high Cu diet significantly increased the weight of the testis and spleen when expressed as a percent of total body weight (Table 1). None of the erythrocyte or leukocyte parameters measured were altered by chronic ingestion of a high Cu diet (Table 2). Vitamins A and E, albumin, and cholesterol are reported in Table 3. Values for vitamin E and cholesterol were significantly elevated in the Cu supplemented animals. Urinary excretion of kynurenine and kynurenic acid were not altered by a high Cu diet after 21 weeks. Also, liver tryptophan 2, 3 dioxygenase activity was not altered by a high Cu diet (Table 4). Analysis of biogenic amines indicated that pineal norepinephrine is decreased in rats fed a high copper diet (Table 5), while brainstem serotonin was decreased in Cu fed animals (Table 6).

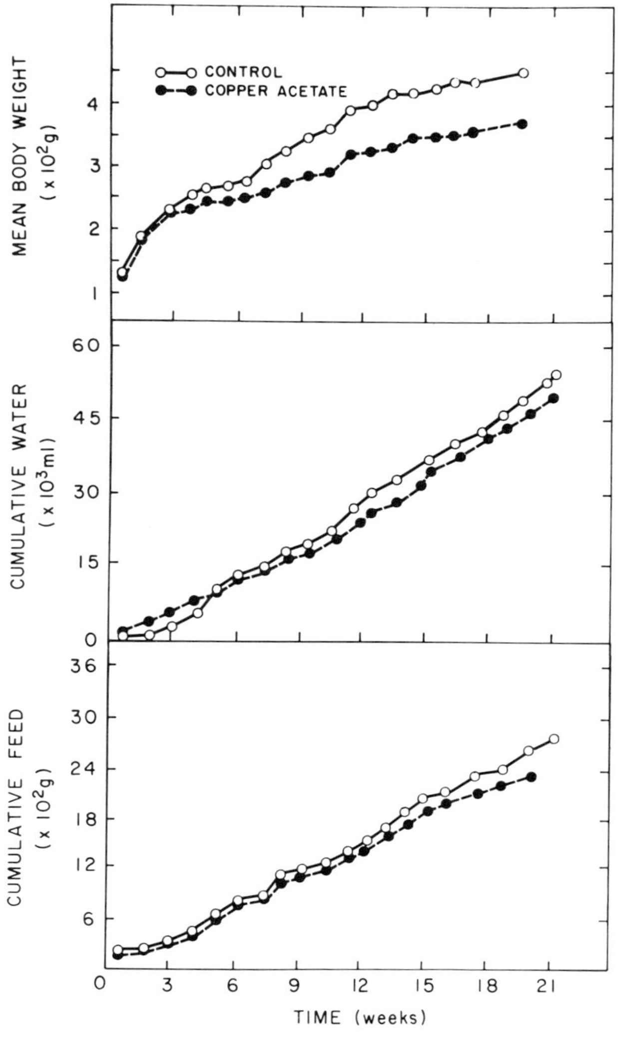

Figure 1. Mean Body Weight, Cumulative Feed Intake, and Cumulative Water Intake During the Course of the Study for Each Treatment Group.

Table 1. Organ Weights in Control and Copper Acetate Supplemented Rats Expressed as a Percent of Body Weight.

Organs	Control Diet Treatment[a]	Copper Acetate Diet Treatment[a]
Testis	0.42 $\pm$ 0.025%	0.51 $\pm$ 0.05[b]
Heart	0.31 $\pm$ 0.03 %	0.32 $\pm$ 0.04
Kidney	0.33 $\pm$ 0.04 %	0.37 $\pm$ 0.03
Spleen	0.14 $\pm$ 0.02 %	0.18 $\pm$ 0.02[b]
Lungs	0.06 $\pm$ 0.02 %	0.07 $\pm$ 0.01
Liver	3.10 $\pm$ 0.20 %	3.09 $\pm$ 0.23

[a]Each value is presented as the mean $\pm$ S.D. ($n = 10$).
[b]Value is significantly different from controls at $p \leq 0.01$.

Table 2. Erythrocytic and Leukocytic Parameters Measured in Control and Copper Acetate Supplemented Rats.

Erythocytic/Leukocytic Parameter-Measured	Control Diet Treatment[ab]	Copper Acetate Diet Treatment[ab]
Hemoglobin	13.26 $\pm$ 3.95 g/100 ml	13.61 $\pm$ 1.32
Hematocrit	43.36 $\pm$ 6.29 %	41.11 $\pm$ 1.32
Mean Cell Volume	58.36 $\pm$ 5.47 $\mu 3$	53.78 $\pm$ 2.35
Mean Cell Hemoglobin Concentration	34.4 $\pm$ 1.90 g/100 cc RBC	33.13 $\pm$ 0.43
Erythrocyte count (x106)	6.62 $\pm$ 1.97 cells/100 ml	7.66 $\pm$ 0.93
Leukocyte count (x103)	11.88 $\pm$ 7.47 cells/100ml	8.18 $\pm$ 4.37

[a]Each value is presented as the mean $\pm$ S.D. ($n = 10$).
[b]All blood values were determined with a Coulter counter Model S.

Table 3. Plasma Albumin, Total Cholesterol, Vitamin A and Vitamin E in Control and Copper Acetate Supplemented Rats.

Parameters Measured	Control Diet Treatment[a]	Copper Acetate (0.5%) Diet Treatment[a]
Albumin	31.2 $\pm$ 2.30 g/100 ml	32.3 $\pm$ 3.90
Cholesterol (total)	9.85 $\pm$ 0.12 mg/ml	1.19 $\pm$ 0.10[b]
Vitamin A	12.6 $\pm$ 0.50 μg/100 ml	13.2 $\pm$ 1.80
Vitamin E (total)	13.0 $\pm$ 0.80 μg/ml	14.1 $\pm$ 1.90[b]

[a]Each value is presented as the mean $\pm$ S.D. ($n = 10$).
[b]Value is significantly different from controls at $p \leq 0.05$.

Table 4. Liver Tryptophan 2, 3 Dioxygenase Activity, Urinary Excretion of Kynurenine, and Kynurenic Acid in Control and Copper Supplemented Rats.

Metabolite Evaluated	Control Diet Treatment[a]	Copper Acetate (0.05%) Diet Treatment[a]
Liver tryptophan 2, 3 dioxygenase[bc]	4.79 ± 1.50 μMkyn/gdw/hr	3.90 ± 0.63
Kynurenine (urine)	3.53 ± 1.21 μg/ml	2.74 ± 1.10
Kynurenic acid (urine)	1.93 ± 1.61.μg/ml	2.17 ± 1.77

[a]Each value is presented as the mean # S.D. (n = 10).
[b]Tryptophan 2, 3 dioxygenase was extracted from liver and assayed as described by Know (1966). Enzyme activity is expressed as uM kynurenine formed per gram dry weight of liver per hour of incubation at 37°C.
[c]These levels were determined by comparing urine extract fluorescence in UV light to standard solutions following separation by monodimensional paper chromatography in B:A:W (4:1:5). Fluid intake and urine volume were not significantly different between groups.

Table 5. Pineal and Pituitary Content of Biogrenic Amines in Control and Copper Acetate Supplemented Rats.

Tissue Metabolites	Control Diet Treatment[a]	Copper Acetate (0.5%) Diet Treatment[a]
Pineal 5-HT	2.36 ± 0.66 μg/pineal	1.76 ± 0.70
Pineal 5-HIAA	0.94 ± 0.47 μg/pineal	0.47 ± 0.19[b]
Pineal NE	0.95 ± 0.13 μg/pineal	0.76 ± 0.05
Pineal NE/5-HT Ratio	0.45 ± 0.18	0.50 ± 0.23
Pituitary 5-HT	1.38 ± 0.20 μg/pituitary	1.46 ± 0.27
Pituitary 5-HIAA	0.95 ± 0.10 μg/pituitary	0.44 ± 0.25[b]
Pituitary NE	0.85 ± 0.15 μg/pituitary	0.72 ± 0.07
Pituitary NE/5-HT Ratio	0.52 ± 0.03	0.50 ± 0.09

[a]Each value is presented as the mean ± S.D. (n = 10).
[b]Value is significantly different from controls at $p \leq 0.05$.

Table 6. Serotonin Levels in Various Brain Regions and Duodenum of Control and Copper Acetate Supplemented Rats.

Brain Region Evaluated	Control Diet Treatment[a]	Copper Acetate (0.5%) Diet Treatment
Cerebellum	3.41 ± 2.68 μg/g	2.92 ± 1.28
Diencephalon	3.23 ± 1.21 μg/g	3.85 ± 2.05
Cortex	0.91 ± 0.31 μg/g	0.96 ± 0.24
Brainstem	5.76 ± 0.31 μg/g	4.39 ± 1.19[d]
Whole Brain[b]	2.19 ± 0.44 μg/g	2.49 ± 0.86
	Intestional Region Evaluated	
Duodenum	8.73 ± 5.37 μg/g	5.43 ± 1.35[c]

[a]Each value is presented as the mean ± S.D. (n = 10).
[b]Whole brain 5-HT levels were calculated from the sum of individual regions.
[c]Value is significantly different from controls at $p \leq 0.05$.
[d]Value is significantly different from controls at $p \leq 0.01$.

Histopathological evaluation indicated lymphocytic infiltration into the liver of animals fed a 0.5% Cu acetate diet. The other organs of Cu-fed animals were normal. Gross pathology revealed no abnormalities in control or Cu-treated groups. The histopathology of control animals was normal (Table 7).

Table 7. Liver Histopathology Found in Control and Copper Acetate Supplemented Rats.

Treatment	Histopathology
Control Diet Animals (18 ppm Cu)	Liver tissue appeared normal grossly and microscopically Liver showed no congestion Liver showed no lobular demarcation other organs were normal.
Copper Acetate Animals	Liver tissue in 5 of 10 animals showed mild inflitration of inflammatory cells although the liver appeared normal grossly. Liver infiltration involved inflammatory cells in and around the interlobular spaces; interstitual nephritis occurred in 1 of 10 animals.

DISCUSSION

Both groups displayed significant gains in body weight during the 21-week experimental period; however, animals ingesting a 0.5% Cu acetate diet had a substantial reduced mean body weight as compared to controls from 9 weeks onward. Furthermore, the Quetelet Index of Obesity indicated a reduction of 17.7% in Cu acetate-treated animals at termination of the experiment.

Comparison of our data with that of Fare and Woodhouse (1963) suggests that the male Holtzman rat is more sensitive than the female albino rats. However, Petering et al., (1977) reported body weight gain reductions in rats were directly related to the level of dietary Cu acetate fed to the animals, accounting for the decrease in mean body weight independent of gender. Whether Cu acetate alters feed intake as a result of decreased palatability or secondary to some metabolic disturbance or reduced intestinal absorption of nutrients is not clear. Cumulative water intake was reduced by 8.16% at the end of six weeks and 11.1% at the end of 21 weeks. The significance of the increase in relative testis weights in Cu treated rats is not clear although the increase in relative spleen weight may be involved with the hematopoietic system since Cu is involved in hemoglobin synthesis. Indeed, the erythrocyte count is slightly increased in Cu-treated rats and mean cell volume is decreased while mean cell hemoglobin concentration is unaltered.

The increase in total serum cholesterol in Cu acetate-supplemented rats is probably due to increased acetate availability since it is a substrate in cholesterol synthesis (West et al., 1970). However, Petering et al. (1977) found an inverse relationship between serum Cu or ceruloplasmin and serum cholesterol in rats fed varying levels of Cu in the diet. The lack of any differences in urinary excretion of kynurenine and kynurenic acid suggests that tryptophan metabolism is not altered by chronic Cu ingestion. Although both Brady et al. (1972) and Gray and Daniel (1973) have reported that tryptophan 2, 3 dioxygenase requires Cu for normal activity, the slight decrease in tryptophan 2, 3 dioxygenase activity observed in these experiments may occur secondary to a metabolic disturbance induced by chronic Cu toxicity or increased turnover of the enzyme. Suttle and Mills (1966) reported that a 250 ppm Cu sulfate diet will cause a Cu toxicosis in pigs while elevated zinc or iron will alleviate the toxicity, probably by competing with Cu for absorption (Linder and Moor, 1977).

Cooper toxicity in rats generally results in increased Cu levels in the liver, small intestine and brain, all of which handle ionic Cu directly (Williams, 1965). Although serum ceruloplasmin is incorporated directly into cells to provide Cu for cytochrome oxidase (Linder and Moor, 1977), excesses of ceruloplasmin do not produce cellular damage as seen in Wilsons Disease unless Cu-chelating proteins are altered (e.g., sodium azide exposure) rendering them unable to bind exogenous metal (Palladini et al., 1977).

The experiments reported herein, provide a functional correlate to the neuropathology seen in chronic Cu toxicity in that pineal norepinephrine levels are reduced concomitant with reductions in brainstem serotonin with a trend toward reduced duodenal serotonin.

The biochemical site of action producing alterations in biogenic amines for this high Cu diet is not clear, although dopamine-B-hydroxylase, an intermediate enzyme in the synthesis of norepinephrine, requires copper for activity and monoamine oxidase which degrades biogenic amines also requires Cu.

In general, chronic Cu toxicity (0.5% or 5000 ppm) albeit from any source including deteriorations of pipes and storage containers, reduces the rate of weight gain concomitant with reductions in food and water intake. The relative weights of testis and spleen are increased while most blood parameters are unaltered. The similarity of serum albumin in control and Cu-supplemented rats indicates that hepatic synthesis is normal, major liver physiologic functions are probably normal and renal retention of protein (i.e., glomerular function) is normal. The latter indicating no development of a nephrotic syndrome at this stage of toxicity. However, changes in biogenic amine levels in pineal and brainstem indicates some direct and/or indirect neuropathologic changes in response to this level of Cu toxicity. The fact that all biochemical indices requiring Cu for activity were reduced is interesting and suggests that additional complex metabolic disturbances may have occurred following chronic ingestion of a 0.5% Cu acetate diet. Clearly, additional highly refined studies need to be undertaken to determine how Cu may be altering Cu requiring enzyme systems and central and systemic biogenic amine turnover.

SUMMARY

A diet high in copper as copper acetate was provided to male rats for 21 weeks. Copper-treated animals showed a substantial reduced mean body weight gain. Also, testis and spleen weights were increased in this group. Blood parameters that were changed included among others, for this treatment group, a slighted increased erythrocyte count and increased serum cholesterol. The urinary excretion of kynurenine and kynurenic acid were not altered in the copper-treated animals. Tryptophan 2, 3 dioxygenase activity was slightly decreased. Changes in biogenic amine levels in pineal and brainstem indicates that these are some neuropathologic changes associated with this level of copper treatment.

Acknowledgements

The assistance of Thomas Cheatle and Steven Gagliardio are acknowledged and appreciated.

REFERENCES

Agnese, T., Vreis, B., and Stanolini, B. (1959). Incidence of lung cancer in relation to occupation in Genoa. Ind. Hygiene Mod. 52:149-160.

Ansell, G., and Beeson, M. (1968). A rapid and sensitive procedure for the combined assay of norepinephrine, dopamine and serotonin in a single brain sample. Analyt. Biochem. 23:196-206.

Arnold, M., and Sasse, D. (1961). Quantitative and histochemical analysis of Cu, Zn and Fe in spontaneous and induced primary tumors of rats. Cancer Res. 21:761-766.

Billiwicz, W.Z., Kemsley, W.F., and Thomson, A.M. (1962). Indices of adiposity. Brit. J. Prev. Soc. Med. 16:183-188.

Brady, F.O., Monaco, M.E., and Forman, H.J. (1972). On the role of copper in activation of and catalysis by tryptophan 2, 3 dioxygenase. J. Biol. Chem. 247:7915-7922.

Centers for Disease Control (1975). Acute copper poisoning. Morb. Mort. Week. Rep. Mar. 15:99.

Chuttani, H.K., Gupti, P.S., and Gultati, S. (1965) Acute copper sulfate poinsoning. Amer. J. Med. 39:849-854.

Doumas, B.T., Waxton, W.A., and Biggs, H.C. (1971). Albumin standards and the measurement of serum albumin with bromcresol green. Clin. Chem. Acta. 31:87-96.

Fare, G. (1965). Copper acetate as an accelerator in mouse skin carcinogenesis by 9, 10 dimethyl-12-benzanthracene. Experientia. 21:415-416.

Fare, G., and Woodhouse, D.L. (1963). The distribution of copper in the tissues of the rat: The effects of age and of feeding with P-dimethylaminoazobenzene with and without copper acetate. Br. J. Cancer 17:775-786.

Gilman, J.P., and Ruckenbauer, G.M. (1962). Metal carcinogensis: Observations on the carcinogenicity of a refinery dust cobalt oxide and ollodial thorium oxide. Cancer Res. 22:152-157.

Gray, L.F., and Daniel, L.J. (1973). Effect of copper status on the urinary excretion of tryptophan metabolities and amino acids by the rat. J. Nutr. 103:1764-1771.

Hansen, L.G., and Warwick, W.J. (1969). A fluorometric micromethod for serum Vitamins A and E. Ameri. J. Clin. Natr. 51:538-541.

Harman, D. (1965). The free radical theory of aging. Effect of age on serum copper levels. J. Gerontol. 20:151-153.

Hart, E.B., Steenbock, H., Waddell, J., and Elvehjem, C.A. (1928). Iron in nutrition: Copper as a supplement to iron for hemoglobin building in the rat. J. Biol. Chem. 77:797-800.

Kingsbury, J.M. (1975). Toxicology the basic science of poisons. L. Casarett and J. Powell, ed. Macmillan Pub. New York, NY, pp. 591-603.

Knox, W.E. (1966). Hormonal control of tryptophan pyrrolase. Adv. Enz. Regl. 4:287-297.

Linder, M.C., and Moor, J.C. (1977). Plasma ceruloplasmin: Evidence for its presence in and uptake by heart and other organs of the rat. Biochm. Biophys. Acta. 499:329-366.

Luckey, T.D., and Venugopal, B. (1977). Metal toxicity in mammals-1: Physiologic and chemical basis for metal toxicity. Plenum Press, Neqw York, NY, pp. 166.

Maickel, R.P., Cox, R.H., Saillant, J., and Miller, F.P. (1968). A method for the determination of serotonin and norepinephrine in discrete areas of rat brain. Int. J. Neuropharmacol. 7:275-281.

Ness, A.T., Pastewka, J.V., and Peacock, A.C. (1964). Evaluation of a recently reported stable Liebermann-Burchard Reagent and its use for the direct determination of serum total cholesterol. Clin. Chem. Acta. 10:229-237.

Pallidini, G., Venturini, A., and Conforti, A. (1977). Copper and the nervous system: An experimental study. Path. Biol. 25:299-306.

Petering, H.G., Lalitha, M., and O'Flaherty, E. (1977). Influence of dietary zinc and copper on rat lipid metabolism. J. Agr. Food Chem. 25:1105-1109.

Schroeder, H.A., Nason, A., Tipton, A., and Balassa, J.J. (1966). Essential trace metals in man: Copper. J. Chronic Dis. 19:1007-1034.

Stanton, M.F. (1967). Primary tumors of bone and lungs in rats following local deposition of copper chelated N-hydroxy-2-acetylaminofluorene. Cancer Res. 28:1000.

Stocks, P., and Davies, R.I. (1960). Epidemiological evidence from chemical and spectrographic analysis that soil is concerned in the causation of cancer. Br. J. Cancer 14:8-22.

Suttle, N.F., and Mills, C.F. (1966). Studies on the toxicity of copper to pigs. Br. J. Nutr. 20:135-140.

West, S.W., Todd, W.F., and Mason, H.S. (1970). Textbook of Biochemistry. MacMillan Co. Toronto, Canada, pp. 1023.

Williams, A.O. (1965). Studies on copper, ceruloplamsin and cirrhosis in relation to Wilsons Disease. Brit. J. Exp. Pathol. 46:564-568.

AUTHOR INDEX

SUBJECT INDEX